Communications in Computer and Information Science 728

Commenced Publication in 2007
Founding and Former Series Editors:
Alfredo Cuzzocrea, Xiaoyong Du, Orhun Kara, Ting Liu, Dominik Ślęzak,
and Xiaokang Yang

More information about this series at http://www.springer.com/series/7899

Beiji Zou · Qilong Han
Guanglu Sun · Weipeng Jing
Xiaoning Peng · Zeguang Lu (Eds.)

Data Science

Third International Conference
of Pioneering Computer Scientists,
Engineers and Educators, ICPCSEE 2017
Changsha, China, September 22–24, 2017
Proceedings, Part II

 Springer

Editors

Beiji Zou
Central South University
Changsha
China

Qilong Han
Harbin Engineering University
Harbin
China

Guanglu Sun
Harbin University of Science
and Technology
Harbin
China

Weipeng Jing
Northeast Forestry University
Harbin
China

Xiaoning Peng
Huaihua University
Huaihua, Hunan
China

Zeguang Lu
Sciences of Country Tripod Institute
of Data Science
Harbin
China

ISSN 1865-0929   ISSN 1865-0937 (electronic)
Communications in Computer and Information Science
ISBN 978-981-10-6387-9   ISBN 978-981-10-6388-6 (eBook)
DOI 10.1007/978-981-10-6388-6

Library of Congress Control Number: 2017953389

Printed on acid-free paper

This Springer imprint is published by Springer Nature
The registered company is Springer Nature Singapore Pte Ltd.
The registered company address is: 152 Beach Road, #21-01/04 Gateway East, Singapore 189721, Singapore

Preface

As the general and program co-chairs of the Third International Conference of Pioneer Computer Scientists, Engineers, and Educators 2017 (ICPCSEE 2017, originally ICYCSEE), it is our great pleasure to welcome you to the conference, which was held in Changsha, China, 22–24 September 2017, hosted by Central South University and the Computer Education Committee of the Hunan Higher Education Federation. The goal of this conference is to provide a forum for computer scientists, engineers, and educators.

The call for papers of this year's conference attracted 420 paper submissions. After the hard work of the Program Committee, 112 papers were accepted to appear in the conference proceedings, with an acceptance rate of 26.7%. The main topic of this conference is data science. The accepted papers cover a wide range of areas related to basic theory and techniques for data science including mathematical issues in data science, computational theory for data science, big data management and applications, data quality and data preparation, evaluation and measurement in data science, data visualization, big data mining and knowledge management, infrastructure for data science, machine learning for data science, data security and privacy, applications of data science, case study of data science, multimedia data management and analysis, data-driven scientific research, data-driven bioinformatics, data-driven healthcare, data-driven management, data-driven eGovernment, data-driven smart city/planet, data marketing and economics, social media and recommendation systems, data-driven security, data-driven business model innovation, and social and/or organizational impacts of data science.

We would like to thank all the Program Committee members, 216 coming from 147 institutes, for their hard work in completing the review tasks. Their collective efforts made it possible to attain quality reviews for all the submissions within a few weeks. Their diverse expertise in each individual research area helped us to create an exciting program for the conference. Their comments and advice helped the authors to improve the quality of their papers and gain deeper insights.

Great thanks should also go to the authors and participants for their tremendous support in making the conference a success. We thank Lanlan Chang and Jian Li from Springer, whose professional assistance was invaluable in the production of the proceedings.

Besides the technical program, this year the ICPCSEE offered different experiences to the participants. We welcome you to the Central South China to enjoy the beautiful summer in Changsha. We hope you enjoy the conference proceedings.

July 2017

Min Li
Fangxiang Wu
Qilong Han
Ronghua Shi

Organization

The Third International Conference of Pioneering Computer Scientists, Engineers, and Educators (ICPCSEE 2017, originally ICYCSEE) –http://2017.icpcsee.org– was held in Changsha, China, September 22–24, 2017, and hosted by Central South University and the Computer Education Committee of the Hunan Higher Education Federation.

ICPCSEE 2017 Steering Committee

Yaoxue Zhang	Central South University, China
Jianer Chen	Central South University, China
Yi Pan	Central South University, China
Jianxin Wang	Central South University, China

General Chair

Beiji Zou	Central South University, China

Program Chairs

Min Li	Central South University, China
Fangxiang Wu	Central South University, China
Qilong Han	Harbin Engineering University, China
Ronghua Shi	Central South University, China

Organization Chairs

Kehua Guo	Central South University, China
Xiaoning Peng	Huaihua University, China
Junfeng Man	Hunan University of Technology, China
Zeguang Lu	Sciences of Country Tripod Institute of Data Science, China

Publication Chairs

Hongzhi Wang	Harbin Institute of Technology, China
Guanglu Sun	Harbin University of Science and Technology, China
Weipeng Jing	Northeast Forestry University, China

Publication Co-chairs

Xianhua Song	Harbin University of Science and Technology, China
Wei Xie	Harbin University of Science and Technology, China

| Yong Wang | Central South University, China |
| Liangwu Shi | Hunan University of Commerce, China |

Education Chairs

| Jiawei Huang | Central South University, China |
| Minsheng Tan | University of South China, China |

Industry Chair

| Yue Shen | Hunan Agricultural University, China |

Demo Chairs

| Jiazhi Xia | Central South University, China |
| Ying Xu | Hunan University, China |

Panel Chairs

| Jiawei Luo | Hunan University, China |
| Shaoliang Peng | National University of Defense Technology, China |

Registration/Financial Chairs

| Ya Huang | Central South University, China |
| Chengzhang Zhu | Central South University, China |

Post/Expo Chair

| Renren Liu | Xiangtan University, China |

ICYCSEE Steering Committee

Jiajun Bu	Zhejiang University, China
Jian Chen	PARATERA, China
Xuebin Chen	North China University of Science and Technology, China
Wanxiang Che	Harbin Institute of Technology, China
Tian Feng	Institute of Software, Chinese Academy of Sciences, China
Qilong Han	Harbin Engineering University, China
Yiliang Han	Engineering University of CAPF, China
Yinhe Han	Institute of Computing Technology, Chinese Academy of Sciences, China
Weipeng Jing	Northeast Forestry University, China
Hai Jin	Huazhong University of Science and Technology, China
Wei Li	Central Queensland University, Australia

Yingao Li	Neuedu, China
Junyu Lin	Institute of Information Engineering, Chinese Academy of Sciences, China
Zeguang Lu	Sciences of Country Tripod Institute of Data Science, China
Haiwei Pan	Harbin Engineering University, China
Shaoliang Peng	National University of Defense Technology, China
Haoliang Qi	Heilongjiang Institute of Technology, China
Pinle Qin	North University of China, China
Zhaowen Qiu	Northeast Forestry University, China
Yanjuan Sang	Beijing Gooagoo Technology Service Co., Ltd., China
Zheng Shan	The PLA Information Engineering University, China
Guanglu Sun	Harbin University of Science and Technology, China
Hongzhi Wang	Harbin Institute of Technology, China
Tao Wang	Peking University, China
Xiaohui Wei	Jilin University, China
Lifang Wen	Beijing Huazhang Graphics & Information Co., Ltd., China
Yu Yao	Northeastern University, China
Xiaoru Yuan	Peking University, China
Yingtao Zhang	Harbin Institute of Technology, China
Yunquan Zhang	Institute of Computing Technology, Chinese Academy of Sciences, China
Liehuang Zhu	Beijing Institute of Technology, China
Min Zhu	Sichuan University, China

Program Committee Members

Chunyu Ai	University of South Carolina Upstate, USA
Jiyao An	Hunan University, China
Ran Bi	Dalian University of Technology, China
Zhipeng Cai	Georgia State University, USA
Yi Cai	South China University of Technology, China
Zhao Cao	Beijing Institute of Technology, China
Richard Chbeir	LIUPPA Laboratory, France
Wanxiang Che	Harbin Institute of Technology, China
Wenliang Chen	Soochow University, China
Chunyi Chen	Changchun University of Science and Technology, China
Wei Chen	Beijing Jiaotong University, China
Zhumin Chen	Shandong University, China
Hao Chen	Hunan University, China
Shu Chen	Xiangtan University, China
Bolin Chen	Northwestern Polytechnical University, China
Hao Chen	Hunan University, China
Xuebin Chen	North China University of Science and Technology, China
Siyao Cheng	Harbin Institute of Technology, China
Byron Choi	Hong Kong Baptist University, Hong Kong, China
Xinyu Dai	Nanjing University, China

Lei Deng	Central South University, China
Vincenzo Deufemia	University of Salerno, Italy
Xiaofeng Ding	Huazhong University, China
Jianrui Ding	Harbin Institute of Technology, China
Qun Ding	Heilongjiang University, China
Xiaoju Dong	Shanghai Jiao Tong University, China
Hongbin Dong	Harbin Engineering University, China
Zhicheng Dou	Renmin University of China, China
Jianyong Duan	North China University of Technology, China
Xiping Duan	Harbin Normal University, China
Lei Duan	Sichuan University, China
Junbin Fang	Jinan University, China
Xiaolin Fang	Southeast University, China
Guangsheng Feng	Harbin Engineering University, China
Jianlin Feng	Sun Yat-Sen University, China
Weisen Feng	Sichuan University, China
Guohong Fu	Heilongjiang University, China
Jing Gao	Dalian University of Technology, China
Dianxuan Gong	North China University of Science and Technology, China
Yu Gu	Northeastern University, China
Yuhang Guo	Beijing Institute of Technology, China
Jiafeng Guo	Institute of Computing Technology, Chinese Academy of Sciences, China
Meng Han	Georgia State University, USA
Qi Han	Harbin Institute of Technology, China
Xianpei Han	Chinese Academy of Sciences, China
Zhongyuan Han	Harbin Institute of Technology, China
Tianyong Hao	Guangdong University of Foreign Studies, China
Shizhu He	Chinese Academy of Sciences, China
Jia He	Chengdu University of Information Technology, China
Qinglai He	Arizona State University, USA
Liang Hong	Wuhan University, China
Zhang Hu	Shanxi University, China
Chengquan Hu	Jilin University, China
Wei Hu	Nanjing University, China
Hao Huang	Wuhan University, China
Lan Huang	Jilin University, China
Shujian Huang	Nanjing University, China
Ruoyu Jia	Sichuan University, China
Bin Jiang	Hunan University, China
Jiming Jiang	King Abdullah University of Science & Technology, Kingdom of Saudi Arabia
Wenjun Jiang	Hunan University, China
Feng Jiang	Harbin Institute of Technology, China
Weipeng Jing	Northeast Forestry University, China
Shenggen Ju	Sichuan University, China

Hanjiang Lai	Sun Yat-Sen University, China
Wei Lan	Central South University, China
Yanyan Lan	Institute of Computing Technology, Chinese Academy of Sciences, China
Min Li	Central South University, China
Mingzhao Li	RMIT University, Australia
Zhixu Li	Soochow University, China
Rong-Hua Li	Shenzhen University, China
Zhixun Li	Nanchang University, China
Xiaoyong Li	Beijing University of Posts and Telecommunications, China
Jianjun Li	Huazhong University of Science and Technology, China
Peng Li	Shaanxi Normal University, China
Qiong Li	Harbin Institute of Technology, China
Zhenghua Li	Soochow University, China
Qingliang Li	Changchun University of Science and Technology, China
Chenliang Li	Wuhan University, China
Xuwei Li	Sichuan University, China
Moses Li	Jiangxi Normal University, China
Mohan Li	Jinan University, China
Hua Li	Changchun University, China
Hui Li	Xidian University, China
Zheng Li	Sichuan University, China
Guoqiang Li	Norwegian University of Science and Technology, Norway
Xiaofeng Li	Sichuan University, China
Yan Liu	Harbin Institute of Technology, China
Yong Liu	HeiLongJiang University, China
Guanfeng Liu	Soochow University, China
Hailong Liu	Northwestern Polytechnical University, China
Yang Liu	Tsinghua University, China
Bingqiang Liu	Shandong University, China
Yanli Liu	Sichuan University, China
Shengquan Liu	XinJiang University, China
Ming Liu	Harbin Institute of Technology, China
Wei Lu	Renmin University of China, China
Binbin Lu	Sichuan University, China
Junling Lu	Shaanxi Normal University, China
Zeguang Lu	Sciences of Country Tripod Institute of Data Science, China
Jizhou Luo	Harbin Institute of Technology, China
Zhunchen Luo	China Defense Science and Technology Information Center, China
Jiawei Luo	Hunan University, China
Jianlu Luo	Officers College of PAP, China
Huifang Ma	NorthWest Normal University, China
Yide Ma	Lanzhou University, China
Hua Mao	Sichuan University, China
Xian-Ling Mao	Beijing Institute of Technology, China

Jun Meng	Dalian University of Technology, China
Tiezheng Nie	Northeastern University, China
Haiwei Pan	Harbin Engineering University, China
Jialiang Peng	Norwegian University of Science and Technology, Norway
Wei Peng	Kunming University of Science and Technology, China
Xiaoqing Peng	Central South University, China
Fei Peng	Hunan University, China
Yuwei Peng	Wuhan University, China
Jianzhong Qi	University of Melbourne, Australia
Shaojie Qiao	Southwest Jiaotong University, China
Zhe Quan	Hunan University, China
Yingxia Shao	Peking University, China
Qiaomu Shen	The Hong Kong University of Science and Technology, Hong Kong, China
Hongwei Shi	Sichuan University, China
Hongtao Song	Harbin Engineering University, China
Wei Song	North China University of Technology, China
Xianhua Song	Harbin Institute of Technology, China
Yanan Sun	Sichuan University, China
Chengjie Sun	Harbin Institute of Technology, China
Guanglu Sun	Harbin University of Science and Technology, China
Minghui Sun	Jilin University, China
Xiao Sun	Hefei University of Technology, China
Guanghua Tan	Hunan University, China
Wenrong Tan	Southwest University for Nationalities, China
Jintao Tang	National University of Defense Technology, China
Dang Tang	Chengdu University of Information Technology, China
Binbin Tang	Works Applications, Japan
Xifeng Tong	Northeast Petroleum University, China
Yongxin Tong	Beihang University, China
Vicenc Torra	Högskolan i Skövde, Sweden
Leong Hou U	University of Macau, China
Chaokun Wang	Tsinghua University, China
Chunnan Wang	Harbin Institute of Technology, China
Dong Wang	Hunan University, China
Hongzhi Wang	Harbin Institute of Technology, China
Jinbao Wang	Harbin Institute of Technology, China
Xin Wang	Tianjin University, China
Yunfeng Wang	Sichuan University, China
Yingjie Wang	Yantai University, China
Yongheng Wang	Hunan University, China
Zhifang Wang	HeiLongJiang University, China
Zhewei Wei	Renming University, China
Zhongyu Wei	Fudan University, China
Yan Wu	Changchun University, China
Zhihong Wu	Sichuan University, China

Huayu Wu	Institute for Infocomm Research, China
Rui Xia	Nanjing University of Science and Technology, China
Min Xian	Utah State University, USA
Tong Xiao	Northeastern University, China
Yi Xiao	Hunan University, China
Degui Xiao	Hunan University, China
Sheng Xiao	Hunan University, China
Minzhu Xie	Hunan Normal University, China
Jing Xu	Changchun University of Science and Technology, China
Jianqiu Xu	Nanjing University of Aeronautics and Astronautics, China
Dan Xu	University of Trento, Italy
Ying Xu	Hunan University, China
Yaohong Xue	Changchun University of Science and Technology, China
Mingyuan Yan	University of North Georgia, USA
Bian Yang	Norwegian University of Science and Technology, Norway
Yajun Yang	Tianjin University, China
Gaobo Yang	Hunan University, China
Lei Yang	HeiLongJiang University, China
Ning Yang	Sichuan University, China
Bin Yao	Shanghai Jiao Tong University, China
Yuxin Ye	Jilin University, China
Minghao Yin	Northeast Normal University, China
Dan Yin	Harbin Engineering University, China
Zhou Yong	China University of Mining and Technology, China
Jinguo You	Kunming University of Science and Technology, China
Lei Yu	Georgia Institute of Technology, USA
Dong Yu	Beijing Language and Culture University, China
Ye Yuan	Northeastern University, China
Kun Yue	Yunnan University, China
Lichen Zhang	Shaanxi Normal University, China
Yongqing Zhang	Chengdu University of Information Technology, China
Meishan Zhang	Singapore University of Technology and Design, Singapore
Xiao Zhang	Renmin University of China, China
Huijie Zhang	Northeast Normal University, China
Kejia Zhang	Harbin Engineering University, China
Yonggang Zhang	Jilin University, China
Jiajun Zhang	Institute of Automation, Chinese Academy of Sciences, China
Yu Zhang	Harbin Institute of Technology, China
Haixian Zhang	Sichuan University, China
Yi Zhang	Sichuan University, China
Boyu Zhang	Utah State University, USA
Wenjie Zhang	The University of New South Wales, Australia
Xiaowang Zhang	Tianjin University, China
Tiejun Zhang	Harbin University of Science and Technology, China
Dongxiang Zhang	University of Electronic Science and Technology of China, China

Contents – Part II

Extracting Chinese Explanatory Expressions with Discrete
and Neural CRFs 1
   Da Pan, Mengqi Wang, Meishan Zhang, and Guohong Fu

Incremental Influence Maximization for Dynamic Social Networks 13
   Yake Wang, Jinghua Zhu, and Qian Ming

Method of Relevance Judgment for App Software's User Reviews 28
   Qixin Xiang, Ying Jiang, Meng Ran, and Jiaman Ding

Topic Model Based Text Similarity Measure for Chinese
Judgment Document 42
   Yue Wang, Jidong Ge, Yemao Zhou, Yi Feng, Chuanyi Li,
   Zhongjin Li, Xiaoyu Zhou, and Bin Luo

Utilizing Crowdsourcing for the Construction of Chinese-Mongolian
Speech Corpus with Evaluation Mechanism 55
   Rihai Su, Shumin Shi, Meng Zhao, and Heyan Huang

A Cluster Guided Topic Model for Social Query Expansion............. 66
   Wenyu Zhao and Dong Zhou

A Framework of Mobile Context-Aware Recommender System 78
   Caihong Liu and Chonghui Guo

Build Evidence Chain Relational Model Based on Chinese
Judgment Documents.. 94
   Siyuan Kong, Yemao Zhou, Jidong Ge, Zhongjin Li, Chuanyi Li,
   Yi Feng, Xiaoyu Zhou, and Bin Luo

Research and Development of Virtual Instruments System Based
on Depth Camera 108
   Xiao-li Xu, Ming-hui Sun, Xin-yue Sun, Wei-yu Zhao, and Xiaoying Sun

Text Understanding with a Hybrid Neural Network Based Learning 115
   Shen Gao, Huaping Zhang, and Kai Gao

Towards Realizing Mandarin-Tibetan Bi-lingual Emotional Speech
Synthesis with Mandarin Emotional Training Corpus................. 126
   Peiwen Wu, Hongwu Yang, and Zhenye Gan

Mining Initial Nodes with BSIS Model and BS-G Algorithm on Social
Networks for Influence Maximization   138
   Xiaoheng Deng, Dejuan Cao, Yan Pan, Hailan Shen, and Fang Long

Critical Value Aware Data Acquisition Strategy in Wireless
Sensor Networks.   148
   Ran Bi, Guozhen Tan, and Xiaolin Fang

An Energy Efficient Routing Protocol for In-Vehicle Wireless
Sensor Networks.   161
   Chundong Wang, Zhentang Zhao, Likun Zhu, and Honglei Yao

Energy-Conserving Transmission Network Model Based
on Service-Awareness   171
   *Huyin Zhang, Chenghao Li, Tianying Zhou, Long Qian,
and Jingcai Zhou*

A Multi-objective Optimization Data Scheduling Algorithm
for P2P Video Streaming   184
   Pingshan Liu, Xiaoyi Xiong, and Guimin Huang

A Novel Range-Free Jammer Localization Solution in Wireless Network
by Using PSO Algorithm.   198
   Liang Pang, Xiao Chen, Zhi Xue, and Rida Khatoun

An Algorithm for Hybrid Nodes Barrier Coverage Based on Voronoi
in Wireless Sensor Networks   212
   Xiaochao Dang, Rucang Ma, Zhanjun Hao, and Meixiu Ma

Measurement Analysis of an Indoor Positioning System Based on LTE   230
   Jiahui Qiu, Qi Liu, Wenhao Zhang, and Yi Chen

Urban Trace Utilizing Mobile Sequence.   241
   Yukun Ma, Bin Xu, and Qi Li

An Extension to ns-3 for Simulating Mobile Charging with Wireless
Energy Transfer   256
   *Ping Zhong, Yating Li, Weile Huang, Xiaoyan Kui, Yiming Zhang,
and Yingwen Chen*

Design and Implementation of Distributed Broadcast Algorithm Based
on Vehicle Density for VANET Safety-Related Messages   271
   Wei Wu, Zhijuan Li, Yunan Zhang, Jianli Guo, and Jing Zhao

Prediction of Cell Specific O-GalNAc Glycosylation in Human   286
   Yuanqiang Zou, Kenli Li, Taijiao Jiang, and Yousong Peng

Supervised Learning for Gene Regulatory Network Based on Flexible
Neural Tree Model   293
    Bin Yang and Wei Zhang

Predicting the Antigenic Variant of Human Influenza A(H3N2) Virus
with a Stacked Auto-Encoder Model   302
    Zhiying Tan, Beibei Xu, Kenli Li, Taijiao Jiang, and Yousong Peng

A Novel Statistical Power Model for Integrated GPU with Optimization   311
    Qiong Wang, Ning Li, Li Shen, and Zhiying Wang

Application of OFDM-CDMA in Multi-user Underwater Acoustic
Communication Based on Time Reversal Mirror   325
    Yonggang Wang, Jingwei Yin, Zhengrong Pan, and Pengyu Du

Hypergraph-Based Data Reduced Scheduling Policy for Data-Intensive
Workflow in Clouds   335
    *Zhigang Hu, Jia Li, Meiguang Zheng, Xinxin Zhang, Hui Kang,
    Yong Tao, and Jiao Yang*

Software System Rejuvenation Modeling Based on Sequential Inspection
Periods and State Multi-control Limits   350
    Weichao Dang and Jianchao Zeng

Research on Power Quality Disturbance Signal Classification Based
on Random Matrix Theory   365
    Keyan Liu, Dongli Jia, Kaiyuan He, Tingting Zhao, and Fengzhan Zhao

DCC: Distributed Cache Consistency   377
    Shenling Liu, Chunyuan Zhang, and Yujiao Chen

Harmonic Pollution Level Assessment in Distribution System
Using Extended Cloud Similarity Measurement Method   388
    Tianlei Zang, Yan Wang, Zhengyou He, and Qingquan Qian

Fusion of Multimodal Color Medical Images Using Quaternion Principal
Component Analysis   401
    Qamar Nawaz, Xiao Bin, Li Weisheng, and Isma Hamid

Research on Adaptive Mobile Collaborative Learning System   414
    Ling Luo, You Yang, and Yan Wei

Plagiarism Detection in Homework Based on Image Hashing   424
    *Ying Chen, Liping Gan, Shiqing Zhang, Wenping Guo,
    Yuelong Chuang, and Xiaoming Zhao*

A Multi-objective Genetic Algorithm Based on Individual
Density Distance.   433
    Lianshuan Shi and Huahui Wang

An Improved Binary Wolf Pack Algorithm Based on Adaptive Step
Length and Improved Update Strategy for 0-1 Knapsack Problems   442
    Liting Guo and Sanyang Liu

Reform of Teaching Mode in Universities Based on Big Data   453
    Bing Zhao and Li Fu

The Construction and Application of MOOCs University Computer
Foundation in Application-Oriented University   459
    Ying San, Hui Gao, Qilong Han, and Junyu Lin

Empirical Analysis of MOOCs Application in Sino-Foreign Cooperative
Design Major Teaching.   467
    Tiejun Zhu

Crossing-Scene Pedestrian Identification Method Based on Twice FAS   483
    Yun Chen, Xiaodong Cai, Yan Zeng, and Meng Wang

Vehicle Type Recognition Based on Deep Convolution Neural Network   492
    Lei Shi, Yamin Wang, Yangjie Cao, and Lin Wei

A Biomechanical Study of Young Women in High Heels with Fatigue
and External Interference   503
    Panchao Zhao and Zhongqiu Ji

Data Clustering Algorithm Based on Artificial Immune Network   516
    Zongkun Li and Dechang Pi

Multi-step Reinforcement Learning Algorithm of Mobile Robot Path
Planning Based on Virtual Potential Field   528
    Jun Liu, Wei Qi, and Xu Lu

A Novel Progressive Secret Image Sharing Method
with Better Robustness   539
    Lintao Liu, Yuliang Lu, Xuehu Yan, and Wanmeng Ding

The NCC: An Improved Anonymous Method for Location-Based Services
Based on Casper.   551
    Wenqi Liu, Mingyu Fan, Jie Feng, and Guangwei Wang

Baymax: A Mental-Analyzing Mobile App Based on Big Data.   568
    Fangyi Yuan, Hongzhi Wang, Shucun Tian, and Xin Tong

Ensemble Learning-Based Wind Turbine Fault Prediction Method
with Adaptive Feature Selection   572
   Shiyao Qin, Kaixuan Wang, Xiaojing Ma, Wenzhuo Wang, and Mei Li

Author Index   583

Contents – Part I

A Fine-Grained Emotion Analysis Method for Chinese Microblog 1
Rui Zhou, Hu-yin Zhang, and Gang Ye

Research of Detection Algorithm for Time Series Abnormal Subsequence . . . 12
Chunkai Zhang, Haodong Liu, and Ao Yin

An Improved SVM Based Wind Turbine Multi-fault Detection Method 27
Shiyao Qin, Kaixuan Wang, Xiaojing Ma, Wenzhuo Wang, and Mei Li

GPU Based Hash Segmentation Index for Fast T-overlap Query 39
*Lianyin Jia, Yongbin Zhang, Mengjuan Li, Jiaman Ding,
and Jinguo You*

A Collaborative Filtering Recommendation Algorithm Based
on the Difference and the Correlation of Users' Ratings. 52
Zhao-hui Cai, Jing-song Wang, Yong-kai Li, and Shu-bo Liu

Research on Pattern Matching Method of Multivariate Hydrological
Time Series 64
Zhen Gai, Yuansheng Lou, Feng Ye, and Ling Li

Further Analysis of Candlestick Patterns' Predictive Power 73
Tao Lv and Yongtao Hao

Partial Least Squares (PLS) Methods for Abnormal Detection
of Breast Cells 88
Yuchen Zhu, Shanxiong Chen, Chunrong Chen, and Lin Chen

Desktop Data Driven Approach to Personalize Query Recommendation 100
Xiao-yun Li and Ying Yu

Disease Prediction Based on Transfer Learning in Individual Healthcare 110
Yang Song, Tianbai Yue, Hongzhi Wang, Jianzhong Li, and Hong Gao

Research on Fuzzy Matching Query Algorithm Based on Spatial
Multi-keyword 123
Suzhi Zhang, Yanan Zhao, and Rui Yang

A New Approach to Dense Spectrum Analysis of Infrasonic Signals 134
Kaiyan Xing, Kaixue Hao, and Mei Li

Research on XDR Bill Compression Under Big Data Technology   144
  Bing Zhao, Sining Zhang, and Jun Zheng

The Scalability of Volunteer Computing for MapReduce
Big Data Applications   153
  Wei Li and William Guo

An Improved FP-Growth Algorithm Based on SOM Partition   166
  Kuikui Jia and Haibin Liu

A Novel Recommendation Service Method Based on Cloud Model
and User Personality   179
  Jing Yao, Zhigang Hu, Hua Ma, and Bingting Jiang

A Cooperative Abnormal Behavior Detection Framework Based
on Big Data Analytics   192
  Naila Marir and Huiqiang Wang

Composite Graph Publication Considering Important Data   207
  Yuqing Sun, Hongbin Zhao, Qilong Han, and Lijie Li

Hierarchical Access Control Scheme of Private Data Based
on Attribute Encryption   220
  Xi Lin and Yiliang Han

Secret Data-Driven Carrier-Free Secret Sharing Scheme Based
on Error Correction Blocks of QR Codes   231
  Song Wan, Yuliang Lu, Xuehu Yan, Hanlin Liu, and Longdan Tan

Template Protection Based on Chaotic Map for Face Recognition   242
  Jinjin Dong, Xiao Meng, Meng Chen, Zhifang Wang, and Linlin Tang

A Fast and Secure Transmission Method Based on Optocoupler
for Mobile Storage   251
  Lu Zou, Dejun Zhang, Fazhi He, and Zhuyang Xie

Android Malware Detection Using Local Binary Pattern
and Principal Component Analysis   262
  Qixin Wu, Zheng Qin, Jinxin Zhang, Hui Yin, Guangyi Yang,
  and Kuangsheng Hu

Elderly Health Care - Security and Privacy Issue   276
  Kaiyu Wan, Vangalur Alagar, and Peter Oyikanmi

Secure Multi-party Comparison Protocol and Application   292
  Jing Zhang, Shoushan Luo, and Yixian Yang

Security Analysis of Secret Image Sharing   305
   Xuehu Yan, Yuliang Lu, Lintao Liu, Song Wan, Wanmeng Ding,
   and Hanlin Liu

PRS: Predication-Based Replica Selection Algorithm
for Key-Value Stores   317
   Liyuan Fang, Xiangqian Zhou, Haiming Xie, and Wanchun Jiang

A General (k, n) Threshold Secret Image Sharing Construction Based
on Matrix Theory   331
   Wanmeng Ding, Kesheng Liu, Xuehu Yan, and Lintao Liu

A Real-Time Visualization Defense Framework for DDoS Attack   341
   Yiqiao Jin, Qidi Liang, Jian Zhang, and Ou Jin

A Research and Analysis Method of Open Source
Threat Intelligence Data   352
   Ruyue Liu, Ziping Zhao, Chengjun Sun, Xiaoyu Yang, Xiaoli Gong,
   and Jin Zhang

An Improved Data Packet Capture Method Based on Multicore Platform   364
   Xian Zhang, Xiaoning Peng, and Jia Liu

Research on Linux Kernel Version Diversity
for Precise Memory Analysis   373
   Shuhui Zhang, Xiangxu Meng, Lianhai Wang, and Guangqi Liu

Unsupervised Anomaly Detection for Network Flow Using Immune
Network Based K-means Clustering   386
   Yuanquan Shi, Xiaoning Peng, Renfa Li, and Yu Zhang

The Research on Cascading Failure of Farey Network   400
   Xiujuan Ma and Fuxiang Ma

Optimal Task Recommendation for Spatial Crowdsourcing
with Privacy Control   412
   Dan Lu, Qilong Han, Hongbin Zhao, and Kejia Zhang

An Information-Aware Privacy-Preserving Accelerometer Data Sharing   425
   Mingming Lu, Yihan Guo, Dan Meng, Cuncai Li, and Yin Zhao

A Range-Threshold Based Medical Image Classification Algorithm
for Crowdsourcing Platform   433
   Shengnan Zhao, Haiwei Pan, Xiaoqin Xie, Zhiqiang Zhang,
   and Xiaoning Feng

A New Method for Medical Image Retrieval Based on Markov
Random Field.   447
   Tiaodi Wang, Haiwei Pan, Xiaoqin Xie, Zhiqiang Zhang,
   and Xiaoning Feng

Three-Dimensional Reconstruction of Wood Carving Cultural
Relics Based on CT Tomography Data   462
   Guiling Zhao, Zongji Deng, Jun Shen, Zhaowen Qiu, and Jing Huang

Text Feature Extraction and Classification Based on Convolutional
Neural Network (CNN)   472
   Taohong Zhang, Cunfang Li, Nuan Cao, Rui Ma, ShaoHua Zhang,
   and Nan Ma

Predicting Big-Five Personality for Micro-blog Based on Robust
Multi-task Learning.   486
   Shuguang Huang, Jinghua Zheng, Di Xue, and Nan Zhao

Automatic Malware Detection Using Deep Learning Based
on Static Analysis.   500
   Liu Liu and Baosheng Wang

High-Level Multi-difference Cues for Image Saliency Detection   508
   Jianwei Sun, Junfeng Wu, Hong Yu, Meiling Zhang, Qiang Luo,
   and Juanjuan Sun

Nonlinear Dimensionality Reduction via Homeomorphic
Tangent Space and Compactness.   520
   Shaoqun Zhang and Wanyun Xie

Selective Image Matting with Scalable Variance and Model Rectification. . . .   534
   Xiao Chen, Fazhi He, Yiteng Pan, and Haojun Ai

Side-Channel Attacks Based on Collaborative Learning   549
   Biao Liu, Zhao Ding, Yang Pan, Jiali Li, and Huamin Feng

Research on Hydrological Time Series Prediction Based
on Combined Model   558
   Yi Cheng, Yuansheng Lou, Feng Ye, and Ling Li

A Cross-View Model for Tourism Demand Forecasting
with Artificial Intelligence Method   573
   Siming Han, Yanhui Guo, Han Cao, Qian Feng, and Yifei Li

Computational Intensity Prediction Model of Vector Data Overlay
with Random Forest Method   583
   Qian Wang, Han Cao, and Yan-Hui Guo

An Implementation and Improvement of Convolutional Neural Networks
on HSA Platform   594
  Zhenshan Bao, Qi Luo, and Wenbo Zhang

An Enhanced Transportation Mode Detection Method
Based on GPS Data.   605
  Jing Liang, Qiuhui Zhu, Min Zhu, Mingzhao Li, Xiaowei Li,
  Jianhua Wang, Silan You, and Yilan Zhang

The Triangle Collapse Algorithm Based on Angle Error Metrics.   621
  Xiaorong Yan, Yuansheng Lou, and Ling Li

Spatial-Temporal Event Detection Method with Multivariate
Water Quality Data.   633
  Yingchi Mao, Zhitao Li, Xiaoli Chen, and Longbao Wang

Context-Aware Technology of Disabled Health Service
for Intelligent Community   646
  Yao Tan and Wenbi Rao

DFDVis: A Visual Analytics System for Understanding
the Semantics of Data Flow Diagram.   660
  Hao Xiong, Haocheng Zhang, Xiaoju Dong, Lingxi Meng,
  and Wenyang Zhao

Design and Implementation of Medical Data Management System   674
  Jie Wang, Jianqiao Liu, Jian Li, Jian Zhang, and Qi Lei

Recognition of Natural Road Sign Based on the Improved
Curvature Feature   689
  Yanqing Wang, Hao Zheng, and Weiwei Chen

Evaluating Cities' Independent Innovation Capabilities Based
on Patent Using Data Analysis Methods   696
  Yan Zhang, Ping Yuan, and Bin Yu

Research on Target Extraction Technology of Fruit and Vegetable
Images in the Complex Environment.   708
  Yanqing Wang and Hao Zheng

Route Guidance for Visually Impaired Based on Haptic Technology
and Their Spatial Cognition   718
  Guansheng Wang, Jianghua Zheng, and Hong Fan

Prediction of Passenger Flow at Sanya Airport Based
on Combined Methods.   729
  Xia Liu, Xia Huang, Lei Chen, Zhao Qiu, and Ming-rui Chen

Research on the Copyright Protection Technology of Digital
Clothing Effect Diagram   741
    Yongqiang Chen and Lihua Peng

Visualization Analysis Framework for Large-Scale Software
Based on Software Network   751
    Shengbing Ren, Mengyu Jia, Fei Huang, and Yuan Liu

Author Index   765

Extracting Chinese Explanatory Expressions with Discrete and Neural CRFs

Da Pan, Mengqi Wang, Meishan Zhang, and Guohong Fu[✉]

School of Computer Science and Technology, Heilongjiang University,
Harbin 150080, China
pandacs@live.cn, mqwang1226@foxmail.com, mason.zms@gmail.com,
ghfu@hotmail.com

Abstract. Recent work on opinion mining typically focuses on subtasks such as aspect mining or polarity classification, ignoring the detailed explanatory evidences that account for one certain user opinion. In this paper, we study the extraction of explanatory expressions, by modeling the problem based on conditional random field (CRF). We compare the effectiveness of both discrete and neural features, and further integrate them. We evaluate the models on two datasets from two different domains which have been annotated with ground-truth explanatory expression. Results show that the neural CRF model performs better than the discrete CRF. After a combination of the discrete and neural features, our final CRF mode achieves the top-performing results.

Keywords: Conditional random field · Explanatory expression extraction · Neural network

1 Introduction

Opinion mining has received much attention in recent years [4,13,19–21]. A number of research works have been presented for this topic. Typically, given a piece of user-generated text, the main goal of opinion mining is to extract the users' opinions that the text expresses. One opinion can be structural, including its polarity, holder, target and etc. [12,17,31,32]. Yet few work has focused on the explanatory expressions of an opinion [9,15,16], which shows the evidence of one opinion. Table 1 shows several examples of the opinion explanatory expressions in the raw texts of phone domain.

Intuitively, these explanatory expressions can be potentially highly useful. On the one hand, these expressions can facilitate opinion mining itself. First, the explanatory expressions of a certain opinion can be helpful for inferring it's polarity. For example, given this instance "The screen is OK, but isn't very clear under the strong sunlight", if we only consider "The screen is OK", without following explanatory comment, we possibly identify it as a positive case when judging the polarity. But if we have the explanatory comment as well, we can easily identify this sentence is a negative case. Without these information, opinion mining rely heavily on external lexicon resources to help capturing the real

© Springer Nature Singapore Pte Ltd. 2017
B. Zou et al. (Eds.): ICPCSEE 2017, Part II, CCIS 728, pp. 1–12, 2017.
DOI: 10.1007/978-981-10-6388-6_1

Table 1. Examples of explanatory opinion expression

反应 不 够 快 ，   可能 是 全 触屏 的 原因 。
(The response is too slow, may be because it's full-touch-screen.)
电池 太 小 了 ，   才 **1000** 毫安 。
(The battery capacity is too small, only 1000mA.)
建议 别 开免提 ，   免 提 效果 非常 差 。
(Suggest to not use handfree mode, since it's effect is very poor.)
屏幕 还 可以 ，   **就是 在 强光 下 看 不清**。
(The screen is OK, but isn't very clear under the strong sunlight)

opinion polarities [24]. On the other hand, the explanatory information can be served as additional references for up-stream applications. For example, a manufacturer may improve its products according to the explanatory expressions, without which one should guess the reason behind of comments.

Explanatory expressions extraction is closely related expressions, opinion targets and opinion holders, which can be all modeled as sequence labeling problems. For each explanatory expression in a sentence, we denote the inside words by using a sequence of labels according to their boundaries. For example, 'B' denotes the beginning, 'M' denotes the middle word, etc. Thus we can have one label for each sentential word, and our task is formulated as a sequence labeling problem.

Conditional random field (CRF) is one representative framework for sequence labeling, and it has achieved state-of-the-art performances in number of tasks [2,12,18,22,26]. The real challenge is the features for explanatory expression extraction. Careful feature engineering is crucial for the performance at the task. Traditionally, we can exploit discrete by manually combining the basic sentential words and their pos tags.

Another line of work of feature representation is the neural network based frameworks. These frameworks are free of feature engineering, by exploiting embeddings and neural operations. Especially, the widely-used pre-trained word embeddings, which are learnt from large-scale corpus, have been shown their capabilities of capturing syntactic and semantic information in several NLP studies [20,35]. In addition, the recent advancing work of neural CRF is valuable for our task [7], which can be directly applied in out tasks with minor modifications.

In this work, we focus on explanatory expression extraction of opinions, studying the effectiveness of the discrete and neural CRF models. We build two different models for our task, which exploiting discrete features and the other exploiting neural features, respectively, Further, we compare the two models empirically and make a combination of the two models via simple feature integration according to the findings in the work of Turian et al. [27], who have shown that improved performance can be achieved by such a combination.

We conduct experiments on a manually annotated corpus, which consists of reviews two different products, namely phone and hotel, respectively. Experimental results show that the neural CRF model achieves better performance than the discrete model, and the combination model can obtain the best performances. We conduct feature ablation as well in order to examine the effects of various features in our models.

2 Related Work

Our work belongs to the opinion mining area, which aims to extract opinions from raw texts. Past work of opinion mining focuses on determining the opinion polarities, extracting opinion holders, opinion expressions and opinion targets [12,14,23,31,32], treating the problem as sequence labeling problem. We focus on the extraction of explanatory expressions, which is equally important for opinion mining, since they can offer the underlined evidences for certain opinion points.

Traditionally, the extraction of opinion expressions, holders and targets are modeled by using same CRF framework to output the final results. For examples, Jakob et al. [12] exploit the CRF model to identify opinion expressions. Similarly, Choi et al. [2] uses the CRF model to recognize opinion targets. Recently Yang et al. [32] improved the performance of opinion expression extraction by semi-Markov CRFs. While we use a similar CRF model for explanatory expressions extraction, with differences in concrete features.

Our study also falls into the line of work that exploits neural network structures for opinion mining. Because neural structures can avoid feature sparsity problem and are free of feature engineering, they have been studied for opinion mining as well. Irsoy and Cardie [11] exploits deep recurrent neural networks (RNN) for opinion expression extraction. Liu et al. [20] investigates LSTM-RNNs to jointly extracting opinion expressions and targets. Zhang et al. [35] incorporates feed-forward neural networks into the CRF framework for jointly recognizing opinion targets and expression. Our work combines LSTM-RNNs and CRF together, and apply the method for explanatory expression extraction.

3 Discrete CRF Model

3.1 Description

In this work, we formulate the task of explanatory expression extraction as a word-level sequence labeling problem. Concretely, we use a 'BMESO' tag set to label each sentential word, where 'B' denotes a explanatory expression's first word, 'M' for a word that occurs in the middle of a explanatory expression, 'E' denotes a explanatory expression's end word, 'S' for a word that occurs as a single-word explanatory expression, 'O' denotes a word that is not a part of explanatory expression. We exploit CRF to model the sequence labeling problem. Figure 1 shows the framework of our discrete model, where f_i is discrete features, and y_i is output label.

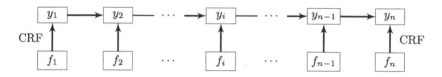

Fig. 1. The framework of our discrete model.

Table 2. Feature templates for the discrete model, where w_i denotes the word at each position i, similarly, p_i denotes the POS tagging.

Feature templates
w_0, $w_{-1}w_0$, w_0w_1, $w_{-2}w_{-1}$, w_1w_2, $w_{-1}w_0w_1$, p_0, $p_{-1}p_0$, p_0p_1, $p_{-2}p_{-1}$, p_1p_2, $p_{-1}p_0p_1$, w_0p_0

Given an input of a product review word sequence $X = x_1x_2\cdots x_n$, we extract discrete one-hot discrete features at each position i ($1 \leq i \leq n$) according to manually-designed feature templates which are listed in Table 2. As a whole, we use three kinds of feature sources, including the word forms w_i, the characters c_i in a word and the POS tags p_i of words.

Let $\boldsymbol{y} = y_1y_2\cdots y_n$ be a corresponding resulting sequence of explanatory expression extraction, we compute its score by the following equation.

$$Score(\boldsymbol{y}) = \Theta \cdot f_i(x, y_i) + T[y_{i-1}][y_i] \tag{1}$$

where Θ is one kind parameter vector denoting the weights of discrete features, T is a matrix, which is also a model parameter, denoting the weights of output label transitions. $f_i(x, y_i)$ be the features functions at position i. With the above score computation function, our goal of decoding is to find an output tag sequence with the highest score, which can be achieved by Viterbi algorithm.

3.2 Training

In order to train the model, the common method is using maximum conditional likelihood estimation with a set of labeled training examples [3,5]. It can be equated to maximization of the log-likelihood, which is more convenient to achieve. The objective function is to maximize $L(X, \Theta) = logP(\boldsymbol{y}_g)$, where the $P(\boldsymbol{y}_g) = e^{Score(\boldsymbol{y}_g)}/Z(X)$, and $Z(X)$ is an instance-specific normalization function that ensures $P(\boldsymbol{y}_g)$ is a probability distribution:

$$Z(X) = \sum_Y \prod_{i=1}^{n} \exp\left\{\Theta \cdot f_i(x, y_i) + T[y_{i-1}][y_i]\right\}. \tag{2}$$

Thus we define the log-likelihood as:

$$L(\boldsymbol{y}_g|X, \Theta) = \sum_{i=1}^{n}\left(\Theta \cdot f_i(x, y_i) + T[y_{i-1}][y_i]\right) - log(Z) \tag{3}$$

We use online learning to train model parameters Θ, updating parameters using the AdaGrad algorithm [6]. In particular, we use the same objective function and learning algorithm in our neural CRF models for all layers, back-propagating the derivatives of the loss function through chain rules.

4   Neural CRF Models

In this section, we introduce our proposed neural model for explanatory expression extraction. Actually, we can regard the neural models as an extension for the discrete model, by substituting the manually-crafted features of $f_i(x, y_i)$ with neural features h_i, yet we can also reserve the discrete features. By this way, we obtain an enhanced neural model with both neural and discrete features. Figure 2 shows the overall framework. Following, we first introduce how to compute the neural feature vector h_i, then we briefly show the combination of the neural and discrete features.

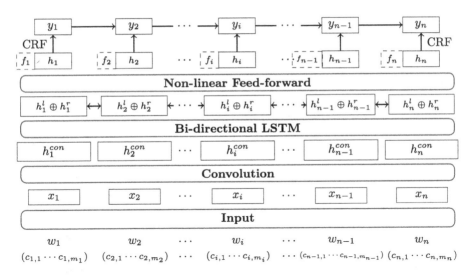

Fig. 2. The overall framework of our neural CRF models.

4.1   Basic Neural Structure

Overall, our basic neural structure consists of the follow five parts: (1) word representation enhanced with character embeddings; (2) input vector sequence obtained by input layer; (3) local context features captured by convolutional layer; (4) long-distance features extracted and by bi-directional LSTM-RNNs layer and (5) output hidden features combined by non-linear feed-forward layer. Below we describe the five steps in detail.

Word Representation. Character embedding have been proved their effectiveness in sentiment analysis [25]. We propose to enhance word representation with character embeddings. Given a word sequence of input $w = w_1 w_2 \cdots w_n$, our word representation r_{word} is composed by two parts, one $\mathbf{e}_{original}$ is obtained from the lookup table E_{word} directly, the other one $\mathbf{e}_{pooling}$ is consist of character embeddings.

Concretely, given a word $w_i = c_1 \cdots c_j \cdots c_m$, we first obtain the character embeddings of each character from lookup table E_{char}, and then we use three different pooling functions to projects the variable length hidden character vector into a fixed-dimensional vector, namely max, min and average, respectively. The pooling function can be formed by $\mathbf{e}_{pooling} = \sum_{j=1}^{m} \alpha_j \odot e(\mathbf{c}_j)$. Different α_j corresponding to different pooling function and we can obtain three different character-based word embedding $\mathbf{e}_{pooling}^{max}$, $\mathbf{e}_{pooling}^{min}$, $\mathbf{e}_{pooling}^{avg}$. We concatenate them with the dense vector $\mathbf{e}_{original}$ which is obtained from the lookup table E_{word} for the final word representation.

$$\mathbf{r}_{word} = \mathbf{e}_{pooling}^{max} \oplus \mathbf{e}_{pooling}^{min} \oplus \mathbf{e}_{pooling}^{avg} \oplus \mathbf{e}_{original}.$$

Input Layer. We build input layer to model the input vector sequence by integrating POS tag embedding. Formally, $\mathbf{x}_i = \mathbf{r}_i^{word} \oplus \mathbf{e}_i^{POS}$. Finally, we feed the input vector sequence $\mathbf{x}_1 \mathbf{x}_2 \cdots \mathbf{x}_n$ into convolution layer to capture context features.

Convolution Layer. Given the input sequence as an example, we first use a window size of five to capture the context features by concatenation such as ngram features. Then we use a non-linear layer to combine these context features to obtain the vector sequence $h_1^{con} h_2^{con} \cdots h_n^{con}$ for LSTM-RNNs layer as following:

$$\mathbf{h}_i^{con} = tanh(\mathsf{W}^c \cdot [\mathbf{x}_{i-2} \oplus \mathbf{x}_{i-1} \oplus \mathbf{x}_i \oplus \mathbf{x}_{i+1} \oplus \mathbf{x}_{i+2}])$$

where W^c denotes a parameter, $tanh$ is a standard element-wise nonlinearity.

Bi-directional LSTM-RNNs Layer. Our neural models are based on the LSTM-RNNs [10], a variant of recurrent neural network, which uses dense vectors to represent features. Unlike common recurrent neural network, the LSTM-RNNs can avoid the gradient vanishing and exploding problem during training, because of the gate techniques. Recently, the bi-directional LSTM-RNNs has been applied for a number of NLP tasks [8,11,28,30,36], which have shown its effectiveness.

There are two LSTM-RNNs in our neural architecture, one computing from left-to-right, the other one computing from right-to-left. Both of them can capture semantic and syntactic information from the input sequence automatically. Taking left-to-right one as example, the hidden vector sequence is computed as following:

$$\mathbf{i}_i^l = \sigma(\mathsf{W}_i^l\mathbf{h}_i^{con} + \mathsf{U}_i^l\mathbf{h}_{i-1}^l + \mathsf{V}_i^l\mathbf{c}_{i-1}^l + \mathbf{b}_i^l)$$

$$\mathbf{f}_i^l = \sigma(\mathsf{W}_f^l\mathbf{h}_i^{con} + \mathsf{U}_f^l\mathbf{h}_{i-1}^l + \mathsf{V}_f^l\mathbf{c}_{i-1}^l + \mathbf{b}_f^l)$$

$$\tilde{\mathbf{c}}_i^l = \tanh(\mathsf{W}_{\tilde{c}}^l\mathbf{h}_i^{con} + \mathsf{U}_{\tilde{c}}^l\mathbf{h}_{i-1}^l + \mathbf{b}_{\tilde{c}}^l)$$

$$\mathbf{c}_i^l = \mathbf{f}_i^l \odot \mathbf{c}_{i-1}^l + \mathbf{i}_i^l \odot \tilde{\mathbf{c}}_i^l$$

$$\mathbf{o}_i^l = \sigma(\mathsf{W}_o^l\mathbf{h}_i^{con} + \mathsf{U}_o^l\mathbf{h}_{i-1}^l + \mathsf{V}_o^l\mathbf{c}_i^l + \mathbf{b}_o^l)$$

$$\mathbf{h}_i^l = \mathbf{o}_i^l \odot \tanh(\mathbf{c}_i^l),$$

where $\mathbf{h}_i^{con}$ is the hidden vector of convolutional layer, $\mathsf{W}^l, \mathsf{U}^l, \mathsf{V}^l, \mathbf{b}^l$ are all the model parameters, and $\odot$ denotes the element-wise Hadamard product. The input $\mathbf{i}$ and output $\mathbf{o}$ gate scale how much of information should be let through from previous hidden vector, previous cell state and current input. Similarity, the forget gate $\mathbf{f}$ scales how much of information should be forgotten, $\tilde{\mathbf{c}}$ is a temporary variable for computing cell structure $\mathbf{c}$ which is used to deliver the long-term information, and it is controlled by input and forget gates. Finally, the $\mathbf{h}_i^l$ is compute by the previous hidden vector, current cell state and current input which is controlled by three gates. The idea of gate is a key point of LSTM-RNNs to solve long-term dependency problem.

Similarity, we can obtain the right-to-left hidden vector sequence $(\mathbf{h}_1^r\mathbf{h}_2^r\cdots\mathbf{h}_n^r)$ by the reverse order, and the corresponding model parameters are $\mathsf{W}^r, \mathsf{U}^r, \mathsf{V}^r, \mathbf{b}^r$.

Non-linear Feed-Forward. After we obtain two different direction features during walk through entire sequence, we concatenate them at each position. Then, we use a non-linear feed-forward layer to automatically combine this information, so that the features can be utilized more effectively. The non-linear hidden layer can be computed as:

$$\mathbf{h}_i = tanh(\mathsf{W}_n \cdot [\mathbf{h}_i^l \oplus \mathbf{h}_i^r])$$

where W_n is a kind of model parameter. Thus we obtain the final hidden vector sequence $\mathbf{h}_1\mathbf{h}_2\cdots\mathbf{h}_n$. Then, we feed the vector sequence into CRF layer, which is the same for discrete model.

4.2 Combination

We build two models for explanatory expression extraction above. Both of them have the common objective function and learning algorithm. Their difference is the feature representation method. The discrete model uses one-hot feature representation which is denoted by $f_1f_2\cdots f_n$ and the neural model uses the continuous feature representation which is denoted by $h_1h_2\cdots h_n$. We except the combination of two feature representations can improve performance.

There are several studies show that discrete and neural features can be mutually complementary with each other and bringing better performance for specified tasks [7,27,29,33,34]. In this work, we conduct the combination as well. Similar with the work of [27], we concatenate the neural and the discrete features

Table 3. Corpus statistics.

	Phone		Hotel	
	Sents	Segments	Sents	Segments
Train	3500	4355	3500	3749
Develop	500	638	500	542
Test	1000	1217	1000	1075
Total	5000	6220	5000	5366

Table 4. Hyper-parameter values.

Type	Hyper-parameters
Neural	$\mathbf{e}_{char} = 30$
	$\mathbf{e}_{original(hotel)} = 100$
	$\mathbf{e}_{original(phone),pos} = 50$
	$\mathbf{h}_{lstms} = 100$, $\mathbf{h} = 50$
Training	$\lambda = 10^{-8}$, $\alpha = 0.001$

at the penultimate layer as well. Then, we can obtain a new feature sequence $\tilde{h}_1 \tilde{h}_2 \cdots \tilde{h}_n$, where $\tilde{h}_i = f_i \oplus h_i$ and feed them together into the CRF layer. Just as shown by the dotted part of Fig. 2, this method is simple and easy to understand, and has been shown being effective in the past work.

5   Experiments

5.1   Experimental Settings

Data Collection. To date there is no dataset available for explanatory expression extraction. In order to evaluate our models, we collect two domains of product reviews from phone and hotel, respectively. Then we annotate the explanatory expression in raw texts by ourselves. The statistics of our dataset is shown in Table 3. Every sentence contains at least one explanatory expression. And in phone domain, the number of explanatory segments is much more than that in hotel. This indicates that in phone domain, some sentences have more explanatory segments than hotel. We use LTP [1] to segment and POS tag the sentences.

Evaluation Method. In our task, we adopt strict evaluation to evaluate our models. It means the predicted segment is correct only when it exactly match the gold standard result, which is judged by human.

Concretely, we report three main metrics, namely precision (P), recall (R) and F-score (F), where P is the percentage of explanatory segments which are correctly extracted, among the total number of explanatory segments we have predicted, R is the percentage of explanatory segments which are correctly extracted, out of the total number of explanatory segments in gold standard result, and $F = 2 * P * R / P + R$.

Hyper-parameters. There are several hyper-parameters in our discrete and neural models, which are tuned by the developmental performances in Sect. 5.2. The values of parameters are shown in Table 4. Both our discrete and neural models use the same regularization parameter λ and initial updating value α. The other parameters only in neural models, including the dimension size $\mathbf{e}_{char}$ of character embedding, dimension size of original word and POS tag embeddings $\mathbf{e}_{original, POS}$, dimension size $\mathbf{h}_{lstm}$ of bi-directional LSTM-RNNs layers and dimension size $\mathbf{h}$ of non-linear feed-forward layer are shown in Table 4 as well.

Embeddings. All the matrices and parameters are initialized randomly by uniform sampling in $(-0.01, 0.01)$, except the lookup table of word embedding. The values of word embedding $E_{original}$ in two domains is respectively pre-training from a large-scale segmented corpus, which is segmented by LTP [1] automatically. In this work, we use *word2vec*[1] to pre-train the word embeddings.

5.2 Development Results

In this section, we tune the hyper parameters and report several experiment performances based on development dataset. Thus we can understand our discrete and neural models better.

Neural Model. Figure 3(a) and (b) are the performance of our neural models in different dimensions and dropout value. We can see that our neural model has the highest F-score in 50 dimensions and 0.4 dropout. This indicates that when a dimension is too high or too low, the performance won't be well. The reason is that higher dimension may result in redundancy, and lower dimension contain only a few semantic information. And it also indicates that using dropout can prevent overfitting.

And the Table 5 shows the performances of our discrete and neural models. We can see that our neural model consistently performs the discrete model in both domains. This is consistent with our expectations. Compared to discrete model, the neural model can use pre-trained word embeddings which are learnt

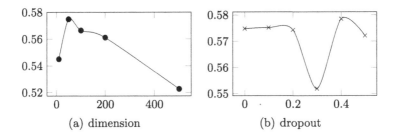

(a) dimension   (b) dropout

Fig. 3. The F-score of different dimensions and dropout in phone development results.

Table 5. Development results.

Model	Phone			Hotel		
	P	R	F	P	R	F
Discrete	51.25	45.48	48.19	77.19	70.58	73.74
Neural	60.48	52.14	56.00	84.44	82.87	83.65
Combine	60.89	54.04	**57.26**	86.39	81.56	**83.91**

---

[1] http://word2vec.googlecode.com/.

from large-scale unlabeled corpus. In addition, neural model don't need hand-crafted features which can be able of free of feature engineering, and it can capture syntactic and semantic information.

Combination Model. Based on the performance of neural model, we except the combination of discrete and neural models can achieve a better performance. As shown in Table 5, our combination model obtain the highest F-score for the explanatory expression extraction on both phone and hotel domains. In the combination model, we integrate one-hot and continuous feature representations at the penultimate layer, which can improve the final prediction performance.

5.3   Final Results

The final results on the test dataset is shown in Table 6. We present the performance of our discrete and neural CRF model and their combination. The results are consistent with development results. Our neural CRF model gives better performance compared to the discrete model by significantly increasing the recall, which shows that the bi-directional LSTM-RNNs succeed in modeling long-term dependency sequence. Further, our combination model achieves highest F-score by integrating discrete and neural features in both car and phone domains, which gaining an improvement of 12% compared to the discrete model. The performance of combine model is just a little better than neural model. It probably because that the sparse features are not enough. And we definitely have reason to believe that once we get more and richer sparse features, the performance will be much better.

We can also see that the performance of hotel is higher than phone by 23%. By analyzing the corpus, we think the reason is phone reviews are more complex, more flexible, people describe in many ways. While hotel corpus's constitution is unitary, words that people use are basically the same, which makes easier to extract.

Table 6. Final results.

Model	Phone			Hotel		
	P	R	F	P	R	F
Discrete	46.48	43.74	45.07	69.65	66.26	67.91
Neural	59.04	52.64	55.66	79.87	81.59	80.72
Combine	58.98	55.42	**57.14**	82.57	79.25	**80.88**

6   Conclusion

In this paper, we focus on explanatory expression extraction of phone and hotel reviews, which has been generally ignored in previous works. We formulate the task as a word-level sequence labeling problem and investigate the effectiveness of discrete and neural CRF models. And we add character and POS tagging embeddings in neural setting, which can improve the performances in both two domains. We further integrate discrete features into neural model and achieve the highest performance.

Acknowledgments. We thank the anonymous reviewers for their constructive comments, which helped to improve the paper. This work was partially funded by National Natural Science Foundation of China (Nos. 61672211, 61602160 and 61170148), Natural Science Foundation of Heilongjiang Province (No. F2016036), and the Returned Scholar Foundation of Heilongjiang Province.

References

1. Che, W., Li, Z., Liu, T.: LTP: a Chinese language technology platform. In: Coling 2010: Demonstrations, pp. 13–16 (2010)
2. Choi, Y., Cardie, C., Riloff, E., Patwardhan, S.: Identifying sources of opinions with conditional random fields and extraction patterns. In: Proceedings of the HLT-EMNLP, pp. 355–362. Association for Computational Linguistics (2005)
3. Collobert, R., Weston, J., Bottou, L., Karlen, M., Kavukcuoglu, K., Kuksa, P.: Natural language processing (almost) from scratch. JMLR **12**, 2493–2537 (2011)
4. Ding, X., Liu, B., Yu, P.S.: A holistic lexicon-based approach to opinion mining. In: Proceedings of the WSDM, pp. 231–240. ACM (2008)
5. Do, T., Artieres, T., et al.: Neural conditional random fields. In: International Conference on Artificial Intelligence and Statistics, pp. 177–184 (2010)
6. Duchi, J., Hazan, E., Singer, Y.: Adaptive subgradient methods for online learning and stochastic optimization. JMLR **12**, 2121–2159 (2011)
7. Durrett, G., Klein, D.: Neural CRF parsing. In: Proceedings of the 53rd ACL-IJCNLP (Volume 1: Long Papers), pp. 302–312, Beijing, China, July 2015
8. Graves, A., Schmidhuber, J.: Framewise phoneme classification with bidirectional LSTM and other neural network architectures. Neural Netw. **18**(5), 602–610 (2005)
9. He, Y., Pan, D., Fu, G.: Chinese explanatory segment recognition as sequence labeling. In: Wang, H., Qi, H., Che, W., Qiu, Z., Kong, L., Han, Z., Lin, J., Lu, Z. (eds.) ICYCSEE 2015. CCIS, vol. 503, pp. 159–168. Springer, Heidelberg (2015). doi:10.1007/978-3-662-46248-5_20
10. Hochreiter, S., Schmidhuber, J.: Long short-term memory. Neural Comput. **9**(8), 1735–1780 (1997)
11. Irsoy, O., Cardie, C.: Bidirectional recursive neural networks for token-level labeling with structure. CoRR abs/1312.0493 (2013)
12. Jakob, N., Gurevych, I.: Extracting opinion targets in a single-and cross-domain setting with conditional random fields. In: EMNLP, pp. 1035–1045 (2010)
13. Jin, W., Ho, H.H., Srihari, R.K.: A novel lexicalized HMM-based learning framework for web opinion mining. In: Proceedings of the 26th ICML, pp. 465–472 (2009)
14. Johansson, R., Moschitti, A.: Syntactic and semantic structure for opinion expression detection. In: CoNLL, pp. 67–76 (2010)
15. Kim, H.D., Castellanos, M., Hsu, M., Zhai, C., Dayal, U., Ghosh, R.: Compact explanatory opinion summarization. In: CIKM, pp. 1697–1702. ACM (2013)
16. Kim, H.D., Castellanos, M.G., Hsu, M., Zhai, C., Dayal, U., Ghosh, R.: Ranking explanatory sentences for opinion summarization. In: SIGIR, pp. 1069–1072 (2013)

17. Kobayashi, N., Inui, K., Matsumoto, Y.: Extracting aspect-evaluation and aspect-of relations in opinion mining. In: EMNLP-CoNLL, vol. 7, pp. 1065–1074 (2007)
18. Lafferty, J., McCallum, A., Pereira, F.: Conditional random fields: probabilistic models for segmenting and labeling sequence data. In: Proceedings of the Eighteenth ICML, vol. 1, pp. 282–289 (2001)
19. Liu, B.: Sentiment analysis and opinion mining. Synth. Lect. Hum. Lang. Technol. **5**(1), 1–167 (2012)
20. Liu, P., Joty, S., Meng, H.: Fine-grained opinion mining with recurrent neural networks and word embeddings. In: Proceedings of the EMNLP, pp. 1433–1443, Lisbon, Portugal, September 2015
21. Pang, B., Lee, L.: Opinion mining and sentiment analysis. Found. Trends Inf. Retr. **2**(1–2), 1–135 (2008)
22. Peng, F., Feng, F., McCallum, A.: Chinese segmentation and new word detection using conditional random fields. In: Proceedings of Coling 2004, pp. 562–568 (2004)
23. Popescu, A.M., Etzioni, O.: Extracting product features and opinions from reviews. In: Kao, A., Poteet, S.R. (eds.) Natural Language Processing and Text Mining, pp. 9–28. Springer, Heidelberg (2007). doi:10.1007/978-1-84628-754-1_2
24. Rao, D., Ravichandran, D.: Semi-supervised polarity lexicon induction. In: Proceedings of EACL, pp. 675–682. Association for Computational Linguistics (2009)
25. Ren, Y., Zhang, Y., Zhang, M., Ji, D.: Context-sensitive twitter sentiment classification using neural network. In: AAAI (2016)
26. Tseng, H., Chang, P., Andrew, G., Jurafsky, D., Manning, C.: A conditional random field word segmenter for sighan bakeoff 2005. In: Proceedings of the Fourth SIGHAN Workshop, pp. 168–171 (2005)
27. Turian, J., Ratinov, L.A., Bengio, Y.: Word representations: a simple and general method for semi-supervised learning. In: Proceedings of the ACL, pp. 384–394 (2010)
28. Wang, D., Nyberg, E.: A long short-term memory model for answer sentence selection in question answering. In: Proceedings of the 53rd ACL and the 7th IJCNLP (Volume 2: Short Papers), pp. 707–712, Beijing, China, July 2015
29. Wang, M., Manning, C.D.: Effect of non-linear deep architecture in sequence labeling. In: Proceedings of the Sixth International Joint Conference on Natural Language Processing, pp. 1285–1291 (2013)
30. Wang, P., Qian, Y., Soong, F.K., He, L., Zhao, H.: Learning distributed word representations for bidirectional LSTM recurrent neural network. In: Proceedings of ICASSP (2016)
31. Wiebe, J., Wilson, T., Cardie, C.: Annotating expressions of opinions and emotions in language. Lang. Resour. Eval. **39**(2–3), 165–210 (2005)
32. Yang, B., Cardie, C.: Joint inference for fine-grained opinion extraction. In: ACL (1), pp. 1640–1649 (2013)
33. Zhang, M., Zhang, Y.: Combining discrete and continuous features for deterministic transition-based dependency parsing. In: Proceedings of the 2015 Conference on Empirical Methods in Natural Language Processing, pp. 1316–1321 (2015)
34. Zhang, M., Zhang, Y., Fu, G.: Transition-based neural word segmentation. In: Proceedings of the 54th ACL (2016)
35. Zhang, M., Zhang, Y., Vo, D.T.: Neural networks for open domain targeted sentiment. In: Proceedings of the 2015 Conference on EMNLP, pp. 612–621 (2015)
36. Zhou, J., Xu, W.: End-to-end learning of semantic role labeling using recurrent neural networks. In: Proceedings of the 53rd ACL and the 7th IJCNLP (Volume 1: Long Papers), pp. 1127–1137, Beijing, China, July 2015

Incremental Influence Maximization for Dynamic Social Networks

Yake Wang[1], Jinghua Zhu[1,2(✉)], and Qian Ming[1,2]

[1] School of Computer Science and Technology,
Heilongjiang University, Harbin 150001, China
zhujinghua@hlju.edu.cn
[2] Key Laboratory of Database and Parallel Computing of Heilongjiang Province,
Harbin, China

Abstract. Influence maximization is one fundamental and important problem to identify a set of most influential individuals to develop effective viral marketing strategies in social network. Most existing studies mainly focus on designing efficient algorithms or heuristics to find Top-K influential individuals for static network. However, when the network is evolving over time, the static algorithms have to be re-executed which will incur tremendous execution time. In this paper, an incremental algorithm DIM is proposed which can efficiently identify the Top-K influential individuals in dynamic social network based on the previous information instead of calculating from scratch. DIM is designed for Linear Threshold Model and it consists of two phases: initial seeding and seeds updating. In order to further reduce the running time, two pruning strategies are designed for the seeds updating phase. We carried out extensive experiments on real dynamic social network and the experimental results demonstrate that our algorithms could achieve good performance in terms of influence spread and significantly outperform those traditional static algorithms with respect to running time.

Keywords: Influence maximization · Dynamic social network · Linear threshold model · Pruning strategy

1 Introduction

In recent years, large social networks have sprung up not only as an fundamental medium for people to exchange information, make friends, but also as an important business platform allowing businessmen to display and sell merchandise. In order to reach the largest scope of products advertisement, businessmen usually choose a small part of influential people in social networks and provide them with free products to make them recommend products to their friends. For example, some business men provide free commodities for some stars, because they usually have a lot of fans in the social network, their recommendations will enhance the product reputation among fans who may be interested in these products and buy them.

Although many existing researches have proposed a number of efficient algorithms for influence maximization, but most of them are based on static social networks. As a

© Springer Nature Singapore Pte Ltd. 2017
B. Zou et al. (Eds.): ICPCSEE 2017, Part II, CCIS 728, pp. 13–27, 2017.
DOI: 10.1007/978-981-10-6388-6_2

matter of fact, real-world social networks keep evolving over time, new accounts are created and some users will establish or lose contacts, the quantity of users and relationships are constantly changing. What's more, no measure can accurately predict the dynamic evolve of social networks because of two reasons: first, the relationships between users are randomly established; second, those relationships may change. Previous static algorithms can't capture and deal with these topology changes.

Although one could possibly run any of the static influence maximization algorithms to find the new Top-K influential nodes when the social network changes, the running time of the static algorithms on large scale social network will be extremely long and whenever the network topology changes, we need to recalculate the influence spreads for all nodes which will lead to quite high cost.

To address the challenges posed by the rapidly and unpredictable changing topology for dynamic social network, we proposed an efficient incremental influence maximization algorithm especially for dynamic social networks called Dynamic Influence Maximization (DIM) in this paper. DIM algorithm includes two stages: initial seeding (Init_Seeding) and incrementally seed updating (Inc_SeedUpdate). At time $t = 0$, we run Init_Seeding algorithm to get the initial seeds set for the static social network. Init_Seeding first gets simple paths by travelling network graph and then calculates the influence spread for each node, finally it outputs the initial seeds and derives the necessary conditions for the second phase. At the subsequent time t, we run Inc_SeedUpdate algorithm continuously to update the seeds. Inc_SeedUpdate calculates the influence spread change of nodes caused by the network change efficiently and quickly finds the new Top-K influential nodes in the evolution based on previously known information. In order to narrow the search space into nodes only experiencing major spread change, we put forward two pruning strategies: influence value increment pruning strategy and degree pruning strategy. The experimental results show that DIM algorithm can achieve as much as 30 speedup in execution time comparing to the static influence maximization algorithm while maintaining good performance in terms of influence spread.

To summary, the main contributions of this paper are as follows:

- We design an incremental influence maximization algorithm under Linear Threshold Model called DIM for dynamic social network.
- We propose two pruning strategies: influence value increment pruning strategy and degree pruning strategy to narrow the search space and improve the time efficiency of DIM algorithm.
- We conduct extensive experiments on NetHEPT, Facebook, Flixster and Flickr social network and the experimental results show that our algorithms can achieve 30 speedup in execution time while providing matching influence spread compared with the state-of-the-art static algorithms.

The rest of this paper is organized as follows. Section 2 reviews related work, Sect. 3 gives some preliminaries and the problem statement. We introduce the dynamic influence maximization algorithm DIM in Sect. 4. The experimental results as well as the analysis are given in Sect. 5. We make conclusions and outline future works in Sect. 6.

2 Related Work

Influence maximization on static networks has attracted a lot of attentions. Viral marking, first introduced to by Richardson and Domingos [1] is a significant marking strategy that promotes commodity by giving free or discount products to a small subset of influential users. Kempe [2] proved that the influence maximization problem is NP-Hard and proposed hill-climbing greedy algorithm with a provable approximation ratio $(1-1/e-\varepsilon)$. Leskovec [3] proposed CELF algorithm which is 700 times faster than the greedy algorithm. Chen [4] proposed heuristic algorithm MIA to enhance the scalability of the algorithm using maximum influence path of each pair of nodes. Kim et al. [5] proposed a scalable and parallelizable influence approximation algorithm based on independent paths. Yang et al. [6] extended the influence maximization problem and proposed a coordinate descent method to solve the transmission cost problem. All the above algorithms are designed for static networks, without considering the dynamic changes of the network.

There are a few influence maximization algorithms designed for dynamic networks. Chen [7] proposed a dynamic social network model which keeps involving during the influence propagation. Zhuang et al. [8] argued that the changes of the network can be obtained by traversing some probing nodes whose topological changes can approximately reflect the evolution of the whole network. You et al. [9] found that under certain incentives, constructing new relationships in social network are helpful for the influence diffusion process. Tong et al. [10] modeled the influence diffusion in social network as the probability event, both the activation of node and the probability on edges obey a certain distribution, influence spread is the expected number of the probability event. Liu et al. [11] proposed an incremental approach to identify the top-K influential individuals based on maximum influence path MIA.

All the above static or dynamic influence maximization algorithms are under the Independent Cascade Model. Goyal et al. [12] used the simple path between neighbor nodes to estimate the influence propagation spread. Lu et al. [13] presented an approximation algorithm to estimate the influence spread, an exact algorithm to compute the influence spread of node within four step.

3 Preliminaries and Problem Statement

3.1 Linear Thread Model

In this paper, a social network is represented as a weighted directed graph $G = (V, E, W)$, here V is node set, each node represents an individual in social network. $E \subseteq V \times V$ is edge set, each edge represents relationships between individuals. For example, a directed edge (v_i, v_j) will be established from node v_i to v_j if v_i is followed by v_j which represents that v_j may be influenced by v_i. W: E $\rightarrow$ [0, 1] represents the influence probability on edges, each edge $(e_i, e_j) \in E$ is associated with an influence probability $w(e_i, e_j)$, $\sum_{u \in V} w(u, v) \leq 1$. Nodes in network have two states: active or inactive. Once a node is activated, it will try to activate its inactive neighbors, if the activation succeeds, its neighbors will become active. Influence diffusion model is the operational model for the spread of an idea or innovation through a social network, which is the basis of influence maximization research. Currently, there are two diffusion models IC (Independent Cascade Model) and LT (Linear Threshold Model). In this paper, we study the influence maximization problem under LT model.

LT model is based on the use of node-specific threshold. Because the active neighbors co-decide whether other nodes will become active, the activation of nodes are independent and satisfy $\sum_{u \in pre(v)} w(u, v) \leq 1$, here $pre(v)$ represent precursor node set of node v, $w(u, v)$ is the probability that u successfully active v.

Each node v chooses a threshold θ_v, $\theta_v \in [0, 1]$, that intuitively represents the different latent tendency of nodes to adopt the innovation when their neighbors do. If θ_v is big then activation of node v is difficulty, on the contrary, activation is easy. At step $t = 0$, only nodes in $S \subseteq V$ is active. If x become active at step $t-1$, x may activate its inactive out-neighbor v. Vertex v is activated at step t only if the weighted number of its activated in-neighbor reaches its threshold, i.e. $\sum_x w(x, v) \geq \theta_v$, here x represents the active nodes in $pre(v)$. This process stops when no more node can be activated. We use $\sigma(S)$ to denote the influence spread of the initial seed set S, $\sigma(S)$ can be approximated by the expected number of active nodes in S.

Kempe et al. [2] proposed live-edge model: given an influence graph $G = (V, E, W)$, for every vertex v, select at most one of its incoming edges at random, such that edge (u, v) is selected with probability $w(u, v)$, and no edge is selected with probability $1 - \sum_u w(u, v)$. The selected edges is called live edge and all other edges are called blocked edge. G_L denotes the spanning subgraph which includes all vertices in V and all live edges selected. If vertex u can reach vertex v in G_L, there exists live path from u to v which consists of all live edges. Kempe et al. [3] proved that, given a seed set S, Linear Threshold Model and live-edge model can achieve the same influence spread.

$$\sigma(S) = \sum_{G_L} pro[G_L] \cdot \sigma_{G_L}(S) \qquad (1)$$

Here $pro[G_L]$ represents the probability of live-edge graph G_L appeared, $\sigma_{G_L}(S)$ denotes the expected number of active nodes starting from S in G_L.

3.2   Influence Maximization

Given an influence graph $G = (V, E, W)$ and integer k, influence maximization problem is to find a set of top-k influential nodes in social network so that their aggregated influence is maximized as shown in this formula:

$$S^* = \arg\max_{S \subseteq V, |S|=k} \sigma(S) \qquad (2)$$

Kempe et al. [2] proved the influence maximization problem is NP-Hard under Linear Threshold Model and proposed a greedy algorithm to solve it. The marginal influence (MI) of any node $v \in V$ given seed set S is defined as $MI(v|S) = \sigma(S \cup \{v\}) - \sigma(S)$. They used Monte-Carlo simulation to estimate the influence spread. The monotonicity and submodularity of $\sigma(S)$ guarantee that the greedy algorithm has approximation ratio $(1 - \frac{1}{e} - \varepsilon)$, that is, it returns a seed set S^g such that $\sigma(S^g) \geq (1 - \frac{1}{e} - \varepsilon)\sigma(S^*)$, for any small $\varepsilon > 0$, where ε accommodates the inaccuracy in Monte-Carlo estimation.

3.3   Influence Maximization for Dynamic Network

Real-world social networks are not static and keep evolving over time. Rapid changes in social networks have brought many new challenges to influence maximization problem, one of the most interesting and urgent problem is that how to quickly identify the most influential users in dynamic social network. So this paper focus on studying the influence maximization in dynamic network.

Dynamic Influence Maximization (DIM): We define an evolving network $G^S = \{G^1, G^2, G^3, \ldots, G^t\}$ as a sequence of network snapshots evolving over time, where $G^t = (V^t, E^t, W^t)$ is an influence graph snapshot at step t, $\Delta G^t = (\Delta V^t, \Delta E^t, \Delta W^t)$ represents the topological change of network graph G^t, obviously, $G^{t+1} = G^t \cup \Delta G^t$. Given an influence graph G^t at step t, the topological change ΔG^t of network graph G^t, the top-K influential nodes S^t in G^t, then we identify the influential nodes $S^{t+1} \subset V^{t+1}$ of size K in G^{t+1} at time $t + 1$.

When network changes, using traditional static influence maximization algorithm to find the seed set will lead to large computational overhead. This paper proposes an incremental algorithm DIM, which identifies the most influential nodes set for dynamic networks by incremental method.

4   Dynamic Influence Maximization Algorithm DIM

The DIM algorithm proposed in this paper is divided into two stages. The first stage is to obtain the initial seed set in static network and the second stage is to incrementally update the initial seed set after the network evolved. In order to further reduce the computational overhead, two optimization strategies are proposed in Sect. 4.3.

4.1   Init_Seeding Algorithm

Social networks keep evolving over time. But if we observe the changes of social network from discrete time step perspectives, in each time t, the social network topology is a static graph. At time $t = 0$, social network is an unchanged initial graph, according to formula (2), the influence spread of seed set S in influence graph G is $\sigma(S) = \sum_{G_L} pro[G_L] \cdot \sigma_{G_L}(S)$, here:

$$\sigma_{G_L}(S) = \sum_{v \in V} I_{G_L}(S, v) \tag{3}$$

$I_{G_L}(S, v)$ represents whether there exists live path from S to v in G_L. If live path is existed then $I_{G_L}(S, v) = 1$, otherwise 0. Therefore,

$$\sigma(S) = \sum_{v \in V} \sum_{G_L} pro[G_L] \cdot I_{G_L}(S, v) = \sum_{v \in V} \sigma(S, v) \tag{4}$$

$\sigma(S, v)$ is the activation probability of vertex v after S is selected as seed set, that is, the influence of set S on node v. Nodes in S may active node v through the path in live-edge model. Let $P = <v_u, v_1, \ldots v_m>$ denote a path, $(v_i, v_j) \in P$ indicates edge (v_i, v_j)

belongs to path P. The simple path is a path whose nodes are all different. The probability of a path P being live path is defined as follows:

$$pro[P] = \prod_{(v_i,v_j)\in P} w(v_i, v_j) \tag{5}$$

The influence of vertex u on v is:

$$\sigma(u, v) = \sum_{P\in Path_{u,v}} pro[P] \tag{6}$$

Here $Path_{u,v}$ is the path set from node u to node v. Therefore, the influence spread of node u can be represented as follows:

$$\sigma(u) = \sum_{v\in V} \sigma(u, v) \tag{7}$$

As an example, we consider the influence graph shown in Fig. 1. Assume that the probability of each edge is 0.2, then the influence of node1 on node 4 is: $\sigma(v_1, v_4) =$ pro[P = <1, 2, 3, 4>] + pro[P = <1, 2, 3, 5, 4>] = 0.008 + 0.0016 = 0.0096, and the influence spread of node1 is:

$$\sigma(v_1) = \sigma(v_1, v_1) + \sigma(v_1, v_2) + \sigma(v_1, v_3) + \sigma(v_1, v_4) + \sigma(v_1, v_5) + \sigma(v_1, v_6) + \sigma(v_1, v_7)$$
$$= 1 + 0.2 + 0.04 + 0.0096 + 0.008 + 0.0016 + 0.0016 = 1.2608.$$

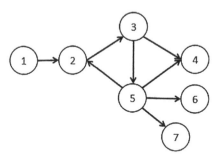

Fig. 1. Example of influence graph

The influence spread of set S is the sum of the spread of each node u in S.

$$\sigma(S) = \sum_{u\in S} \sigma^{V-S+u}(u) \tag{8}$$

The marginal influence $MI(v)$ of node v under seed set S, $MI(v|S) = \sigma(S\cup\{v\}) - \sigma(S)$. To compute $\sigma(S\cup\{v\})$, we need to compute $\sigma^{V-S-v+x}(S\cup\{v\})$ for each $x \in S\cup\{v\}$ on subgraph induced by $V - S - v + x$ according formula (4–6). Each node in S except node x, when adding new node to S, needs recalculation using formula

(4–6) even the basic subgraph has slight difference. Goyal et al. [18] proposed that the expected number of influence node will not be affected too much for slightly changed subgraph. The calculation formula of $\sigma(S \cup \{v\})$ is derived as follows:

$$\sigma(S \cup \{v\}) = \sum_{x \in S \cup \{v\}} \sigma^{V-S-v+x}(x) = \sigma^{V-S}(v) + \sum_{x \in S} \sigma^{V-S-v+x}(x)$$

$$= \sigma^{V-S}(v) + \sum_{x \in S} \sigma^{V-S-v+x}(x) = \sigma^{V-S}(v) + \sigma^{V-v}(S) \tag{9}$$

For example, as shown in Fig. 1, if seed set $S = \{v_5\}$, then $\sigma(S) = \sigma(v_5) = 1.848$. After node v_3 is incorporated into S, $\sigma(S \cup \{v_3\}) = \sigma^{V-v_5}(v_3) + \sigma^{V-v_3}(v_5) = 1.2 + 1.8 = 3.0$.

At time $t = 0$, the social network is an unchanged initial graph, we call algorithm GetAllPath to determine the simple path set and compute the influence spread for each node. The problem of enumerating all simple paths in graph is #P-Hard [5]. In social network, the probability of a path being live decreases rapidly as the length of the path increases. Thus, the majority of the influence can be captured by exploring the paths within a small neighborhood, where the size of the neighborhood can be controlled by the error tolerated. Parameter η represents a tradeoff between efficiency and accuracy. Given a probability threshold η, we can filter out those paths whose probabilities are lower than η. In this way, we can efficiently control the size of path set and reduce the computation cost.

GetAllPath algorithm is based on classical backtrack idea to enumerate all simple paths in graph. It starts from node u and traverses u's out-neighbors in depth-first fashion. If the probability of current path $P = <u, \ldots x, v>$ belows η, the algorithm will backtrack to node x, if node x has other out-neighbors, it will continue to traverse follow those out-neighbors, otherwise it will backtrack. In the process, the influence of node u can be estimated by accumulating those probabilities of paths. If it backtracks to starting node u, then the algorithm stops.

Since the probability of path decreases rapidly with increasing length, intuitively, for each change in the graph, the nodes affected by the topology change must be in a small neighborhood. In order to accurately and quickly determine which nodes are affected by topology change, we compute $IN[v]$ for each $v \in V$ which represents the predecessors of node v. Nodes in $IN[v]$ may influence v within the control threshold η. At the beginning, $IN[v]$ is null for each node, the idea of UpdateIN(v) algorithm that update IN set of node is described as follows:for each node $v \in V$, when the path set of v changed, each path in $Path[v]$ will be traversed. If the path contains node u, which means that v can reach u by the path, then add node v to $IN[u]$ set. If $IN[u]$ already contains node v, it is no need to add v. In such a way, we get all the nodes arriving at node u through the path. When the network is evolving, we can quickly confirm those influenced nodes that need to re-estimate influence spread using IN, we don't have to traverse the entire network and save a lot of running time.

Algorithm Init_Seeding calculates the influence spread of each node at the beginning by calling algorithm GetAllPath, and updates IN set according $Path$ set. In each iteration, according to the greedy idea, the node with the largest marginal influence are added to the seed set until a given size is reached.

The pseudo code of Init_Seeding is described as follows:

Init_Seeding (G, η, k)
Input: graph G, path length threshold η, size k
Output: seed set S

---

1) for each $u \in V$ do
2)   calculate $\sigma(u)$ by calling GetAllPath algorithm;
3)   UpdateIN(u);
4)   add u to CELF queue;
5) $S \leftarrow \varnothing$; $spd \leftarrow 0$;
6) While $|S| < k$
7) $u \leftarrow$ top node in CELF queue;
8)   for $s \in S$ //compute $\sigma^{V-u}(S)$
9)   for each $P \in Path[s]$ and $u \in P$
10) $\sigma^{V-u}(s) -= pro[P]$;
11) $spd += \sigma^{V-u}(s)$;
12)   for each $P \in Path[u]$ and $P \cap S \neq \varnothing$ //compute $\sigma^{V-S}(u)$
13) $\sigma^{V-S}(u) -= pro[P]$;
14)   compute $\sigma(S+u)$ by using formula(4-7);
15) $MI(u) = \sigma(S+u) - spd$;
16)   if $MI(u) > \sigma(u)$
17) $S \leftarrow S \cup u$; remove u from CELF queue;
18)   else
19)   re-insert u in CELF queue;
     return S

In line 1–4, we calculate the influence spread of each node in graph G by calling GetAllPath algorithm, update IN set for each node and add them into CELF queue. We initialize seed set S and influence spread spd in line 5 and get the top node in CELF queue in line 7. In line 8–11, as the path set of each node has been obtained, we calculate the influence spread of seed node in the subgraphs induced by $V - u$ equates to making the path contains u starting from the seed node invalid, then we calculate $\sigma^{V^{-u}}(s)$ by subtracting the influence probability of invalid path from the original influence, and compute spd according to the formula (4–6). We compute the influence of node u in subgraphs induced by $V - S$ in line 12–13 and estimate the marginal influence in line 14–15. In line 16–19, if $MI(u)$ of node u is larger than the top node in queue, we directly add this node to seed set, otherwise add it into CELF queue again.

4.2 Inc_SeedUpdate Algorithm

In Social network, new users might join while old users might withdraw, and people will make new friends with each other, thus the topology of network keep evolving. We define an evolving network as $G^S = \{G^1, G^2, G^3, ..., G^t\}$ from the perspective of discrete time, where $G^{t-1} = (V^{t-1}, E^{t-1}, w^{t-1})$ is the influence graph at step $t - 1$, $\Delta G^{t-1} = (\Delta V^{t-1}, \Delta E^{t-1}, \Delta w^{t-1})$ is the topological change of network graph G^{t-1}, thus $G^t = G^{t-1} \cup \Delta G^{t-1}$. So the influence graph in step t is new relating to the last time $t - 1$ for any $t > 0$. Liu et al. [16] propose six types of topology change as AddEdge,

DelEdge, AddVertex, DelVertex, IncWeight, DecWeight. AddEdge(u, v, w) is to establish a link with weight w from user u to v. DelEdge(u, v, w) is to delete a edge from user u to v whose weight is w. AddVertex and DelVertex correspond to add or delete nodes respectively. IncWeight(u, v, w) is to increase the weight of edge(u, v) to w, and DecWeight(u, v, w) is to decrease the weight to w. For those six changes of the social network, we only explain AddEdge and DelEdge operation in this paper because the other operations are similar.

The pseudo code of AddEdge(u, v, w) is as follows:

AddEdge (u, v, w)
Input: new edge (u, v, w)
Output: the influence spread change $\sigma^{\Delta}(x)$, $x \in C$

---

1) for each $x \in C$ do

2) $\sigma^{\Delta}(x)=0$;

3)   for $P_{x,u} \in Path[x]$ do

4)   if $pro[P_{x,u}] \cdot w(u, v) > \eta$ and $P_{x,v}$ is a simple path

5)   add $P_{x,v}$ into $temp$;

6)   if $P_{x,v} \cap S = \emptyset$

7) $\sigma^{\Delta}(x) += pro[P_{x,v}]$;

8)   if $temp \neq \emptyset$

9)   for $P_{x,v} \in temp$ do

10)   for $P_v \in Path[v]$ do

11)   if $pro[P_{x,v}] \cdot pro[P_v] > \eta$, $(u, v) \in P_x$

12)   add P_x into $temp$;

13)   if $P_x \cap S = \emptyset$

14) $\sigma^{\Delta}(u) += pro[P_x]$;

15)   copy $P \in temp$ into $Path[x]$;

16)   UpdateIN(x);

   return $\sigma^{\Delta}(x)$

In AddEdge algorithm, we use parameter w to represent the probability of the new edge (u, v), set C consists of nodes affected by network change, $temp$ is used to temporarily store the new path, $\sigma^{\Delta}(u)$ is the influence spread change of node u, $P_{x,u}$ represents the path from the node x to node u, and P_v represents the path whose source node is v. We compute influence spread change $\sigma^{\Delta}(u)$ for each $u \in C$ and update IN set according to the latest $Path$ collection. In line 3–7, we find one path from node x to node u and expand the path by adding new edge (u, v). If the probability of the new path from x to v is greater than threshold η and it is a simple path, then we add this path to $temp$. In line 8–14, if $temp$ is null, that means the probability of those paths are too small or there exits ring among those paths, then the computation will end. Otherwise, we try to connect the path $P_{x,v}$ with P_v, and perform the same operation. We store those new paths into $Path[x]$ set in line 15, and update $IN[x]$ set based on new $Path[x]$ set of node x in line 16.

Now we analyze the time complexity of AddEdge algorithm. Assume that the maximum number of paths in the path set is P_{max}, the number of temporary path is up to $O(P_{max})$ in *temp*, that is, original paths all can be extended. Then the time complexity of circulate in line 3–7 is $O(P_{max})$, the time complexity of dual-layer cycle in line 8–15 is $O(P_{max} \cdot P_{max})$, the time complexity of updating IN in line 16 is $O(P_{max})$, so the time complexity is $O(|C| \cdot P_{max}^2)$.

The pseudo code of DelEdge is as follows:

DelEdge (u, v, w)
Input: an edge (u, v)
Output: the influence spread change $\sigma^\Delta(x)$, $x \in C$

---
1) for $x \in C$ do
2) $\sigma^\Delta(x)=0$;
3)   for $P_x \in Path[x]$ and $(u,v) \in P_x$
4) $\sigma^\Delta(x) -= pro[P_x]$; delete P_x;
5)   UpdateIN(x);
      return $\sigma^\Delta(x)$

In DelEdge algorithm, we confirm all invalid paths of node x that is influenced by the change, and compute influence spread change $\sigma^\Delta(x)$ of node x and delete all invalid paths. We update IN set based on new $Path$ set of node x in line 5. The time complexity is $O(P_{max})$ for line 3–5. The time complexity of DelEdge is $O(2 \cdot |C| \cdot P_{max})$.

The pseudo code of seed set update algorithm Inc_SeedUpdate is as follows:

Inc_SeedUpdate $(S^{t-1}, G^{t-1}, G^t, k)$
Input: seed set S^{t-1},graph $G^{t-1}$and G^t in time t -$1$and t, size k
Output: seed set S^t in time t

---
1) $S^t \leftarrow \emptyset$;
2) While $|S^t|<k$
3)   for each change c from G^{t-1} to G^t do
4)   obtain influenced node set C
5)   for each $u \in C$ do
6)   compute $\sigma^\Delta(u)$;
7) $S^C \leftarrow S^C \cup u$;
8)   for each $u \in S^C$ do
9) $MI(u)=\sigma(S^t \cup \{v\})-\sigma(S^t)$;
10) $S^t=S^t \cup \text{argmax } MI(u)$;
      return S^t

In Inc_SeedUpdate algorithm, S^C is the candidate set, S^t is the seed set at step t. We select seed only from S^C, which can narrow the search space and reduce calculation cost. In line 3–7, according to the six types of topology change, we call the corresponding algorithms to compute influence spread change of the node, and filter candidate node. In line 8–10, we select the seed node with the largest *MI*.

4.3 Pruning Strategies

Real social networks like Facebook and Flickr usually have a huge number of users, however the size of seeds set is relatively small for financial reason. Each iteration to compute influence spread of all nodes might cause enormous cost. Although DIM algorithm only re-evaluate the influence spread of nodes affected by network change, those nodes who has small influence spread change still has little chance to be seed nodes. In order to further narrow the search space of influential nodes, we put forward two pruning strategies for DIM algorithm.

Least Influence Increment Pruning Strategy (LIIP): In the ith iteration, the influence spread change of seed set S^{t-1} in graph G^{t-1} is positive, if the influence spread change of node v is greater than any seed node in S^{t-1} set, node v is reserved as the candidate node, otherwise it will be filtered out. In most cases, the most influential nodes in graph G^{t-1} attract large number of nodes and create new connections. Thus the influence spread increase and the influence spread change are positive. If the influence spread and influence spread change of node v are lower than that of any seed node in S^{t-1} at time $t - 1$, then node v must be lower than any seed node in S^{t-1} at time t, that means node v will never become seed node and should be filtered out. Liu et al. [16] find that social network is based on the preferential attachment principle, so the least influence increment pruning strategy can filter out a large number of nodes.

Degree Pruning Strategy (DP): According to the preferential attachment principle, the new-coming edges prefer to attach to nodes with higher degree [16], thus nodes with large degree are more easily influenced by the topology change than nodes with small degree. If the influence spread change of S^{t-1} is negative, it means that the influence spread of S^{t-1} is decreased. In addition to the first optimization measure, the reservation node should also meet one of the following two conditions: First, Degree of nodes is among the top 5% of all nodes in G^t. Second, Degree increase ratio of nodes is among the top 5% of all nodes in G^t. The degree increase ratio of node v is defined as $degree^t(v)/degree^{t-1}(v)$, where $degree^t(v)$ represents the degree of node v in G^t.

We can extend the Inc_SeedUpdate algorithm by adding **LIIP** or **DP** pruning strategy before line 7. We only choose the valid candidate nodes to join S^C. The pruning strategies can further narrow the candidate nodes and improve the efficiency of the algorithm while guaranteeing relatively high accuracy. We call the extended DIM algorithm with pruning strategy Opt-DIM algorithm.

5   Experiments

5.1   Experimental Setup

In this section, we compare our algorithm with the static algorithms in terms of efficiency and effectiveness. We examine two metrics, influence spread and running time.

Influence spread is the final expected number of influenced nodes activated by seed nodes, representing the accuracy of different algorithm; running time is the time to identify the most influential k nodes, reflecting efficiency of algorithm.

We choose four datasets to test the performance of different algorithms, Tables 1 and 2. summarize the statistical information of the four datasets, the growth rate of nodes and edges can intuitively reflect that the real-world social network are rapidly change with time.

Table 1. Table of node information

Datasets	Nodes		
	Initial number	Final number	Growth
NetHEPT	15,634	18,557	18.7%
Facebook	59,736	83,983	40.3%
Flixster	99,825	147,328	48.5%
Flickr	771,738	1,037,995	34.5%

In this paper, we compare our algorithm with two static influence maximization algorithm LDAG and SIMPATH. LDAG algorithm estimate the influence spread of node based on local directed acyclic graph [17]. SIMPATH algorithm finds the Top-K influential nodes based on simple path [18].

Table 2. Table of edge information

Datasets	Edges		
	Initial number	Final number	Growth
NetHEPT	62,836	89,415	42.3%
Facebook	576,653	994,149	72.4%
Flixster	978,265	1,811,249	85.2%
Flickr	4,938,687	7,106,122	43.8%

The active threshold of node v is generated uniformly at random in range (0, 1). We set the weight of every incoming edge of v to be $\frac{1}{d_v}$.

5.2   Experimental Results and Analysis

The first group experiments evaluate the running time to identify 50 most influential nodes of different algorithms. DIM and Opt-DIM call Init_Seeding only on the first snapshot, and call Inc_SeedUpdate in the following seven network snapshots. As illustrated in Figs. 2 3, 4 and 5, the running time of DIM and Opt-DIM on the first snapshot are longer than on the other snapshots, and much shorter than the other two algorithms on the four datasets. Among the four compared algorithms, LDAG algorithm has the largest time cost. While the running time of DIM algorithm and Opt-DIM algorithm are relatively stable. Compared with the static algorithm LDAG and SIM-PATH, when the dynamic network changes, dynamic algorithm DIM and Opt-DIM are 6.3 times, 13.6 times, 17.4 times and 31 times faster on average than static algorithm on the four datasets.

Although the time costs of DIM and Opt-DIM algorithm are also increased in large scale networks, but the running time still have obvious advantages in large-scale dynamic network. And the perform of Opt-DIM algorithm is better than DIM algorithm on four datasets. On Facebook dataset, Opt-DIM algorithm is 1 time faster than DIM algorithm, while achieving double speedup on Flixster and Flickr datasets on average, that means the pruning strategies are effective.

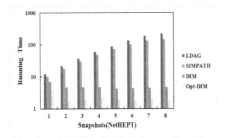

Fig. 2. Running time on NetHEPT dataset

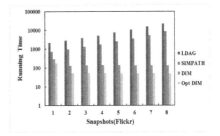

Fig. 3. Running time on Facebook dataset

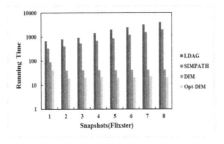

Fig. 4. Running time on Flixster dataset

Fig. 5. Running time on Flickr dataset

We use influence spread to measure the accuracy of different algorithms. The influence spread is the number of nodes influenced by the top-50 influential nodes. Figures 6 7, 8 and 9 show the results of the experiments. As shown in Figs. 6 and 7, DIM and Opt-DIM are basically consistent with the static algorithms on NetHEPT and Facebook datasets. However, in Figs. 8 and 9, the influence spread of static algorithm LDAG and SIMPATH have a slight advantage over dynamic algorithms on larger datasets, Flixster and Flickr. This is because LDAG algorithm and SIMPATH algorithm have to estimate the influence spread of all nodes in network in each iteration. This also explains that DIM and Opt-DIM can lift the execution speed ratio at the cost of a little accuracy decrease. The influence spread of SIMPATH algorithm outperforms that of LDAG algorithm on Flixster and Flickr datasets as shown in Figs. 8 and 9. On the Flixster and Flickr datasets, the advantage of DIM algorithm is more obvious because the pruning strategy based on influence spread change can narrow the search space of influential nodes.

The influence spread of DIM algorithm and Opt-DIM algorithm are smaller than algorithm SIMPATH 4.1%, 5.2% and 5.9% on the Facebook, Flixster and Flickr datasets, respectively. Although the accuracy of the dynamic influence maximization algorithm DIM and Opt-DIM is lower than the static algorithms, according to the operation time showed from Figs. 2, 3, 4 and 5, the dynamic DIM algorithm is more suitable for solving influence maximization problem in large social network.

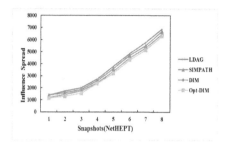

Fig. 6. Influence spread of NetHEPT

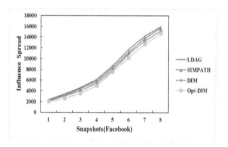

Fig. 7. Influence spread of Facebook

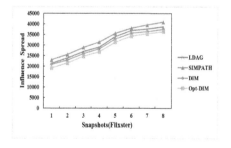

Fig. 8. Influence spread of Flixster dataset

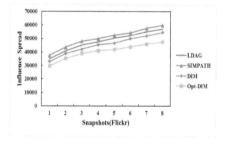

Fig. 9. Influence spread of Flickr dataset

6   Conclusions and Future Work

In this paper, we propose a dynamic influence maximization algorithm DIM, including two parts, Init_Seeding and Inc_SeedUpdate. Init_Seeding algorithm obtains the path set of all nodes by traversing the network graph, and computes the marginal influence of nodes according to the path set. Inc_SeedUpdate algorithm determines the influence spread change to incrementally update the seeds set when network changed. We also present two pruning strategies to further narrow the search space of influential nodes. The effectiveness and efficiency of DIM and Opt-DIM algorithm are verified by experiments.

In the future, we will study the probability distribution of edges and consider the influence maximization algorithm based on the probability distribution variation of edges.

Acknowledgment. This work was supported in part by the National Science Foundation of China (61632010, 61100048, 61370222), the Natural Science Foundation of Heilongjiang Province (F2016034), the Education Department of Heilongjiang Province (12531498).

References

1. Richardson, M., Domingos, P.: Mining knowledge-sharing sites for viral marketing. In: 7th ACM SIGKDD International Conference on Knowledge Discovery and Data Mining, pp. 61–70. ACM, Edmonton (2002)
2. Kempe, D., Kleinberg, J., Tardos, É.: Maximizing the spread of influence through a social network. In: 9th ACM SIGKDD International Conference on Knowledge Discovery and Data Mining, pp. 137–146. ACM, Washington D.C. (2003)
3. Leskovec, J., Krause, A., Guestrin, C.: Cost-effective outbreak detection in networks. In: 13th ACM SIGKDD International Conference on Knowledge Discovery and Data Mining, pp. 420–429. ACM, San Jose (2007)
4. Chen, W., Wang, C., Wang, Y.: Scalable influence maximization for prevalent viral marketing in large-scale social networks. In: 16th ACM SIGKDD International Conference on Knowledge Discovery and Data Mining, pp. 1029–1038. ACM, Washington D.C. (2010)
5. Kim, J., Kim, S.K., Yu, H.: Scalable and parallelizable processing of influence maximization for large-scale social networks? In: 29th International Conference on Data Engineering, pp. 266–277. IEEE Computer Society, Washington, D.C. (2013)
6. Yang, Y., Mao, X., Pei, J., et al.: Continuous influence maximization: what discounts should we offer to social network users? In: 26th International Conference on Management of Data, pp. 727–741. ACM, San Francisco (2016)
7. Chen, W., Wei, L., Zhang, N.: Time-critical influence maximization in social networks with time-delayed diffusion process. Chin. J. Eng. Des. **19**(5), 340–344 (2015)
8. Zhuang, H., Sun, Y., Tang, J., et al.: Influence maximization in dynamic social networks. In: 13th International Conference on Data Mining, pp. 1313–1318. IEEE, Austin (2013)
9. You, Q., Hu, W., Wu, O.: Influence maximization in human-intervened social networks. In: 24th International Conference on Social Influence Analysis, pp. 9–14. IJCAI, Buenos Aires (2015)
10. Tong, G., Wu, W., Tang, S., et al.: Adaptive influence maximization in dynamic social networks. IEEE/ACM Trans. Netw. **99**, 112–125 (2015)
11. Liu, X., Liao, X., Li, S., et al.: On the shoulders of giants: incremental influence maximization in evolving social networks. Comput. Sci. (2015)
12. Goyal, A., Wei, L., Lakshmanan, L.V.S.: SIMPATH: an efficient algorithm for influence maximization under the linear threshold model. In: 11th International Conference on Data Mining, pp. 211–220. IEEE Computer Society, Washington, D.C. (2011)
13. Lu, Z., Fan, L., Wu, W., et al.: Efficient influence spread estimation for influence maximization under the linear threshold model. Comput. Soc. Netw. **1**(1), 1–19 (2014)

Method of Relevance Judgment for App Software's User Reviews

Qixin Xiang[1,2], Ying Jiang[1,2(✉)], Meng Ran[1,2], and Jiaman Ding[1,2]

[1] Yunnan Key Lab of Computer Technology Application, Kunming 650500, China
jy_910@163.com
[2] Faculty of Information Engineering and Automation,
Kunming University of Science and Technology, Kunming 650500, China

Abstract. In order to judge whether the user reviews are relevant to App software, this paper proposed a method to judge the relevance of user reviews based on Naive Bayesian text classification and term frequency. Firstly, the keywords sets of App software's user reviews are extracted. Then, the keywords sets are optimized. Finally, the relevance score of the user reviews are calculated, and whether the user reviews are relevant is judged. Through the experiment, this method is proved that can judge the relevance of App software's user reviews effectively.

Keywords: App software · User reviews · Relevance judgment · Naive Bayesian text classification · Term frequency

1 Introduction

With the popularity of intelligent terminals, the number of mobile terminal users grows rapidly. Ovum forecasted that global smartphone sales would reach 1700 million in 2017, in which Android mobile phones account for half. Nowadays, App software has become the major application in the mobile terminals. Until June 2016, the number of App software on App Store was more than two million.

At present, many users will give the software's review after using App software. As time goes on, the number of user reviews for App software will increase rapidly. When the user is choosing the App software for himself, viewing the user reviews is one of the major methods. However, there are many irrelevant reviews, such as '你哪里人啊？' and etc. In addition, some users would advertise through reviews, such as '好消息招聘网络兼职，30–300日结，**QQ：2216484195**'. It is difficult for users to understand App software when they are viewing these irrelevant reviews. Many users are browsing relevant user reviews to evaluate the App software before choosing software. Furthermore, these reviews can also provide the references to developers. So it is necessary to judge the relevant reviews from the massive user reviews of App software.

B. Zou et al. (Eds.): ICPCSEE 2017, Part II, CCIS 728, pp. 28–41, 2017.
DOI: 10.1007/978-981-10-6388-6_3

2   Related Work

At present, many researches aimed at the analysis technology of products user reviews. Lin et al. [1] introduced three research fields of user reviews quality which are the quality evaluation of reviews, the summary of reviews and the detection of rubbish reviews. Then the major solutions to the above research fields were presented. Finally, the irrelevant reviews were classified through the solutions. Hu and Zheng [2] de-noised the user reviews by analyzing the frequency of the candidate feature words in product user reviews, and filtered the user reviews correlation to the product. Jiang and Zhang [3] proposed a method to analysis the relevance of reviews based on complex network and the credibility of reviews among the semantics relations. Li and Fu [4] established social graphs by studying user preference relation. In order to identify the fake user reviews, the social graphs were combining with a simple linear classifier based on text classification.

The above researches are mainly aimed at user reviews of goods and products. The researches about user reviews based on term frequency and semantic association analysis have been very mature. However, App software is different from the normal products. App software is a kind of new software. Users can not understand the quality information of this App software through advertisement and irrelevant user reviews, which has negative affection on the users' choosing of App software. Therefore, it is important for users to judge the relevance of App software's user reviews.

In the researches of App software's user reviews, most focus on the emotional tendency of users. Pagano and Maalej [5] analyzed the positive and negative attitude of users in the reviews which served as the basis for developers to improve App software requirements. Leopairote et al. [6] evaluated the quality of App software by mining the positive and negative emotions in user reviews. Harman et al. [7] proposed the method that matched the whole sentence or a part. Then the classification of reviews was elaborated. Finally, combining supervised and unsupervised techniques, the negative and positive attitudes of App software's user reviews were judged on the methods of entity mapping and polarity mining. AlQuwayfili et al. [8] divided the user reviews of App software into trusted category or untrusted category by analyzing the similarity among reviews. Gao and Xu [9] extracted the topic reviews from different user reviews of App software and sorted them with classification in order to judge the authenticity of user reviews. It is considered that the emotional tendency of users about App software in the reviews would influence the selection of App software for others in [5–7]. However, they did not consider whether the reviews are aimed at this App software, which would impact users' choice of App software. It is pointed out that the credibility of App software's user reviews influenced the user selection in [8,9]. For words that can express users' emotional tendency or some characteristics of App software, whether these words would influence the relevance judgment of App software's user reviews was not considered.

The analysis of its user reviews is still in the research stage. Currently, the researchers usually use the methods of traditional products to process App software, such as emotional tendency and topic mining. There are plenty of words describing the features of App software and users' emotion in the user reviews. Moreover, the high-frequency words will affect the relevance judgment of App software's user reviews. Therefore, the relevance judgment of App software's user reviews will be judged by analyzing the words describing the features and the high-frequency words in this paper.

3   Relevance Judgment of App Software's User Reviews

App software's user reviews are typical online reviews with greater freedom and arbitrariness. Compared with other products, the user reviews of App software are more brief and their emotional tendency are polarized. In addition, there are a large number of user reviews in App software that are irrelevant to the App software, which is inconvenient for users to browse the reviews. So, it will take difficulty to choose App software for users. In order to help users choosing App software effectively, it is very important to find out the relevant reviews about App software. In this paper, the relationship between the user reviews and the App software is defined as the relevance of App software's user reviews.

In order to judge the relevance of App software's user reviews, the words that can describe the users' emotional tendency and the features of App software are extracted. Then, the words above are optimized to improve efficiency of the relevance judgment. Finally, the optimized words are analyzed to judge the relevance of App software's user reviews.

3.1   The Keywords Set Extraction of User Reviews

At present, user reviews of App software are usually described in natural language. The relevance of user reviews cannot be judged directly. Therefore, the user reviews of App software need to be processed through structured or semi-structured method. Usually, some representative words or word-pairs of a user review can express the main meaning of user review.

Han et al. [10] found that nouns, verbs and adjectives in Chinese texts can express the main content of the text. There are many nouns, verbs and adjectives in App software's user reviews, which describe different characteristics of App software. So these words can serve as a basis for judging the relevance of App software's user reviews.

(1) The verbs and nouns in the user reviews of App software always describe the features of App software, such as '安装', '音效'.
(2) The adjectives in the user reviews can describe how the features are, such as '清晰' in '画面清晰', or express the user's emotion about the whole App software, such as '很好'.

In this paper, App software's user reviews are segmented into words and tagged into part-of-speech. And the nouns, verbs, and adjectives are extracted to serve as keywords, which is defined as follows.

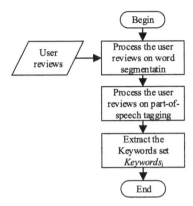

Fig. 1. The flow diagram of the keywords set extraction for App software's user reviews

Definition 1. Keywords Set $Keywords_i = \{w_0/f_0, w_1/f_1, \ldots, w_k/f_k\}$. $k = 0, 1, \ldots, K - 1, K$ is the segmented words amount of the ist user review, w_k is the kst segmented word of the reviews, f_k is the part-of-speech of w_k.

In this paper, the words that describe the features of App software or express user's emotion, which are called as feature words. The flow diagram of the keywords set extraction for App software's user reviews is shown in Fig. 1.

In order to extract the keywords sets of App software's user reviews, the user reviews are crawled from App software markets. Then, the user reviews are segmented into words and tagged into part-of-speech by using word segmentation tool. Finally, the keywords sets are extracted on the results of word segmentation by judging results of part-of-speech.

In order to extract the keywords set $Keywords_i$ of each user reviews, the part-of-speech of the segmented words need to be judged. If the segmented word of the part-of-speech is a noun, verb or adjective, this word and its part-of-speech becomes one part of the keywords set $Keywords_i$ of the ist user review, which is formed as 'w_k/f_k'.

3.2   The Keywords Set Optimization of User Reviews

Analyzing the keywords sets extracted from the App software's user reviews, we found out that the keywords set is a set of adjectives or a set of nouns, verbs and adjectives. Generally, the set of adjectives is emotional expression of App software while decorating the nouns and verbs. And nouns and verbs always describe one or more of the features about App software. In the paper, nouns, verbs and adjectives are called as n, v and a when describing the part-of-speech in the keywords sets. In order to improve the efficiency of judging relevance in user reviews, two categories of reviews keywords set are processed.

(1) Keywords set of all a

In the keywords set of a, the part-of-speech only has a. The users make the comprehensive reviews about the App software, which omitting the specific attributes or characteristics of the App software. The a is used to make emotional tendency about the App software. For example, {完美/a} is the keywords set of '十分完美'. '完美' is the only word and the only a in the keywords set. '完美' expresses the positive attitude of user about App software.

(2) Keywords set of not all a

In the keywords set, the part-of-speech has $n + a, v + a, n + v$ or $n + v + a$. n and v always talk about the specific attributes or features of App software that user cares. However, the a often decorates one of the attributes or features. At this point, only n and v extracted from reviews can judge the relevance of user reviews in App software. For example, {下载/v, 慢/a} is the keywords set of '下载慢'. '下载' is the v and '慢' is the a. '下载' describes the feature of App software and '慢' decorates '下载'.

Definition 2. The optimized keywords set $new\_Keywords_i = \{w_0, w_1, \ldots, w_j\}$. $j = 0, 1, \ldots, J - 1, J$ is the amount of optimized keywords in the ist user review, w_j is jst word in the optimized keywords set. The optimized part-of-speech set $new\_Features_i = \{f_0, f_1, \ldots, f_j\}$. f_j is the part-of-speech of w_j.

The flow chart of the keywords set optimization for App software's user reviews is shown in Fig. 2.

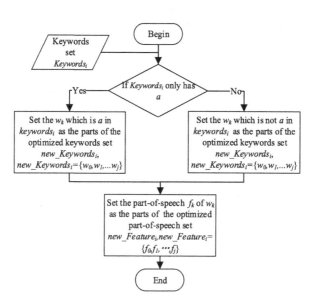

Fig. 2. The flow chart of the keywords set optimization for App software's user reviews.

In order to obtain the two $new\_Keywords_i$, the a is the judgment basis. In the first $new\_Keywords_i$, the a is the only data. However, there is no a in the second $new\_Keywords_i$. So, the second $new\_Keywords_i$ does not have the same a with the first one. Aiming at the keywords sets optimization of two categories, the methods are described as follows.

(1) Keywords set of all a

Aiming at this case, the a in the keywords set are served as the words in the optimized keywords set $new\_Keywords_i$. The optimized keywords set $new\_Keywords_i$ includes all the w_k in the keywords set $Keywords_i$. And the optimized part-of-speech set $new\_Feature_i$ includes the part-of-speech a of w_k.

(2) Keywords set of not all a

Aiming at this case, the words which are not a in the keywords set are served as the factors of the optimized keywords set. The optimized keywords set $new\_Keywords_i$ includes the w_k which are not a in the keywords set $Keywords_i$, i.e., all n and v in $Keywords_i$ are extracted to be parts of $new\_Keywords_i$ rather than a. And the optimized part-of-speech set $new\_Feature_i$ includes the part-of-speech f_k of w_k.

3.3 The Relevance Judgment of User Reviews

In order to determine whether the App software's user reviews are relevant or irrelevant, the relevance of App software's user reviews is judged based on the relevant degree between the user reviews and App software. In this paper, we use the method of text classification to judge the relevance of App software's user reviews. The common text classification algorithms include Naive Bayesian text classification, k-NN, Support Vector Machine (SVM). Zhang et al. [11] compared the classification results of Naive Bayesian text classification and SVM in App software's user reviews and found the former method is better when the review is short term. Wang [12] compared Naive Bayesian text classification and k-NN in Weibo text classification and found the former is faster when the two classification is accurate enough.

Based on the characteristics of App software's user reviews, we use Naive Bayesian text classification to judge the relevance of App software's user reviews. There are many repeated words in the online reviews used by many users, such as '很好', '垃圾'. However, Naive Bayesian text classification is a simple classification algorithm without considering the influence of a large number of repeated words from App software's user reviews. In this paper, repeated words are defined as high-frequency words.

The term frequency can be effectively to reflect how important the words is in text classification. So it will be used to judge the characteristic of App software's user reviews. In the judgment of relevance about App software's user reviews, term frequency will be one of the judgment factors.

In order to calculate the frequency of the features in the App software's user reviews, the corresponding feature lexicons are constructed according to the part-of-speech of the feature, including the verb feature lexicon, the noun feature lexicon and the adjective feature lexicon. Each feature lexicon contains features of App software and the frequency of each feature. For different App software, the feature lexicons are established.

As one of the judgment factors, term frequency of words needs to be gotten from the feature lexicons. The first step in judging the relevance of App software's user reviews is to get the term frequency amount of w_j in optimized keywords set $new\_Keywords_i$ in feature lexicon of f_j. In addition, Naive Bayesian text classification is an important part of the method that judging the relevance of App software's user reviews. The factors of Naive Bayesian text classification should be obtained from the feature lexicons and texts, such as the amount of texts. The amount of texts containing w_j and the total amount of texts need to be obtained, too. Then, the relevance score of App software's user reviews is calculated. Finally, the relevance of App software's user reviews is judged according to the score.

Based on Naive Bayesian text classification, Di and Duan [13] proposed that the prior probability has little effect on the classification result. So the formula calculation ignores the prior probability. In this paper, we consider the respective characteristics of Naive Bayesian text classification and term frequency statistics and combine both methods to judge the relevance of App software's user reviews, which the formula is as follows.

$$score = \prod_{j=1}^{J} \left(\frac{N_j + 1}{N + 1} + \frac{times_j + 1}{m_j + 1} \right) \tag{1}$$

N_j is the amount of text that contains w_j and N is the total amount of texts. $times_j$ means the term frequency amount of w_j. m_j denotes the average frequency of features in f_j feature lexicon of w_j. J is the optimized keywords amount of the ist user review.

According to the Eq. 1, whether the relevance of App software's user reviews score is greater than the threshold value α is judged. If the score is greater than α, the App software's user review is relevant comment; otherwise it is irrelevant review.

In this paper, the flow diagram of the relevance judgment for App software's user reviews is shown in Fig. 3.

If the user review is judged as relevant review, the feature lexicon should be updated in order to enlarge the amount of the feature words and improve the accuracy rate of the further relevance judgment. The feature lexicon will be updated every time we judge that the App software's user review is relevant in Fig. 3. The updated case is divided into two types:

(1) if the feature word exists in the corresponding feature lexicon, the frequency of the feature word will add at the times it exists in the optimized keywords set.
(2) if the feature word does not exist in the corresponding feature lexicon, it is added to the corresponding feature lexicon and its frequency is set to 1.

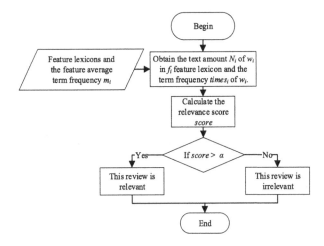

Fig. 3. The flow diagram of the relevance judgment for App software's user reviews.

4   Experimental Result' and Analysis

In order to verify the validity of this method, we crawled 17934 user reviews of 9 App software from Android Market, MyApp and Baidu Mobile Market. The information library of App software was established which shown in Table 1.

Based on the App software's user reviews in Table 1, the information library of App software's user reviews was established which shown in Table 2.

As can be seen from Table 2, the 4th user review is irrelevant and the 5th user review is an advertisement review. So the irrelevant user reviews about the App software are existed.

Table 1. Information library of App software (some examples)

Classifi-cation	App Name	Introduction	Version No.	Review Amount
Video	优酷	看视频，用优酷…	6.3.2	1632
	爱奇艺	【新用户必看！免费送VIP体验，一口…	8.2	1839
	土豆视频	土豆视频-每个人都是生活的导演!…	5.8	1328
Travel	谷歌地图	为Android 手机和平板电脑量身…	9.48.1	2351
	高德地图	中国专业的手机地图，超过7亿用户…	8.0.1	2134
	百度地图	专业手机地图，智能导航躲避拥堵…	9.7.4	1533
Shopping	手机淘宝	手机淘宝(Android 版)是阿里巴巴专…	6.5.0	2602
	手机京东	新人大礼：新人专享188 元大礼包…	5.7.6	1648
	天猫	天猫APP，现在下载，新用户…	5.31.0	2867

Table 2. Information library of App software's user reviews (some examples)

No.	App Name	User Review
1	微信	非常好用
2	优酷视频	太耗流量啦
3	水果忍者	几分钟就通关
4	百度地图	啊啊
5	支付宝	100日结，qq3735168476
6	高德地图	太卡了，太多垃圾消息通知
7	百度地图	两分钟的路因为它走了半个小时，整整的绕了一大圈
8	手机淘宝	不错，很好用
9	支天猫	闪退！！！

Table 3. Word segmentation and part-of-speech tagging of user reviews (some examples)

No.	User Review	Word Segmentation and Part-of-Speech Tagging
1	非常好用	非常/d 好/a 用/v
2	啊啊	啊/o 啊/o
3	100日结，qq3735168476	100/m 日/qt 结/v ，/wd qq3735168476/n
4	太卡了，太多垃圾消息通知	太/d 卡/v 了/y ，/wd 太/d 多/a 垃圾/n 消息/n 通知/n
5	不错，很好用	不错/a ，/wd 很/d 好/a 用/v
6	闪退！！！	闪/vi 退/v ！/wt ！/wt ！/wt

In this paper, ICTCLAS 2016 is used as the preprocessing tool for experimental data, which completing the word segmentation and part-of-speech tagging. Based on the user reviews in Table 2, parts of results are shown in Table 3.

For the App software's user reviews, the method in Sect. 3 is used to extract the keywords sets, optimize the keywords sets and judge the relevance of user reviews. The experimental results of keywords sets extraction and optimization are shown in Table 4.

In Table 4, the keywords set of 2nd user review is a typical keywords set of all a, which expressing the emotional tendency of user. Other keywords of user reviews are keywords sets of not all a, which describing the features of App software.

The experimental results of the relevance judgment for App software's user reviews are shown in Table 5.

The threshold value α is 1 when calculating the relevant score. In Table 5, the scores of 2nd, 5th, 10th, 11th, 12th, 15th, 17th and 18th user reviews do not

Table 4. Keywords sets and optimized keywords sets (some examples)

No.	User Review	Keywords Set	Optimized Keywords Set	Optimized Part-of-Speech Set
1	为什么下载了没有在桌面上出现？	下载/v 桌面/n 出现/v	下载 桌面 出现	v n v
2	非常不错	不错/a	不错	a
3	担心会不会有账户安全啊	担心/v 会/v 账户/n 安全/an	担心 会 账户	v v n
4	扫一扫付款很方便	中扫/v 付款/vi 方便/v	扫 付款 方便	v vi v
5	闪退！！！	闪/vi 退/v！/wt！/wt！/wt	闪 退	vi v

exceed α and the scores of other user reviews exceed α. So, the user reviews whose scores exceed α are judged as relevant user reviews and the other user reviews are judged as irrelevant user reviews. In addition, the 3rd and 4th user reviews belong to '爱奇艺'. The 3rd score is 1.9433 and the 4th is 32.8655. Because of the more feature words and the higher term frequencies in 4th user review, the score of 4th is high than that of 3rd, which showing that it can be more recognized on the method.

According to the calculation of the relevance judgment about App software's user reviews, we compare the results with manual marking and obtain the amounts of correct judgment and the accuracy rate of relevance judgment about each App software's user reviews. The amounts of correct judgment about each App software are shown in Fig. 4.

The Accuracy rates of relevance judgment about each App software are shown in Fig. 5.

In Figs. 4 and 5, there are the results of the relevance judgment for App software's user reviews on three classifications. In each classification, we found that the more amount of reviews, the higher accuracy rate of the relevance judgment.

In order to find the relation between the judgment accuracy of App software user reviews and the amount of user reviews, this paper took '微信' as an example and crawled 2116 user reviews from Android Market and MyApp to judge the relevance. The accuracy rate of the relevance judgment for '微信' user reviews is shown in Fig. 6.

It is found from Fig. 6 that the accuracy rate is decreasing when the count of reviews is less than 106 while it is increasing and gradually tends to be about 0.87 when the count is greater than 106. Through the analysis, we found that there exists greater randomness of the relevance judgment about App software's user reviews when the count of reviews (the sample number) is small. It will impact the accuracy of relevance judgment. When the count of user reviews is larger than a critical point, the accuracy rate of relevance judgment about App software's

Table 5. The relevance judgment of App software's user reviews (some examples)

App Name	No.	User Review	Feature Words	Score	Relevance Judgment	Manual Mark
优酷	1	很方便	方便	2.4324	relevant	relevant
	2	无意中。		0.9	irrelevant	irrelevant
爱奇艺	3	一个字，差!	差	1.94337	relevant	relevant
	4	网络整天连不进服务器，消息接受延迟	网络、服务器、消息、延迟	32.8655	relevant	relevant
土豆视频	5	彭江。		0.815	irrelevant	irrelevant
	6	今天安装好几次没装上真头疼	安装、装、头疼	7.8063	relevant	relevant
谷歌地图	7	非常好	好	10.0000	relevant	relevant
	8	这个怎么导航啊	导航	2.3533	relevant	relevant
高德地图	9	图片都是灰色的，看不到	图片、是、灰色、看不到	3.1503	relevant	relevant
	10	腌吖	腌、吖	0.4623	irrelevant	irrelevant
百度地图	11	卜一	卜	0.6371	irrelevant	irrelevant
	12	不…		1.0000	irrelevant	irrelevant
手机淘宝	13	怎么平板下载不下啊	平板、下载、下	12.3284	relevant	relevant
	14	马上就下完了，就下载失败	下、完了、下载、失败	15.9168	relevant	relevant
手机京东	15	哦，好哦好哦哈哈哈	好	0.8358	irrelevant	irrelevant
	16	不错，就是一个手机只能注册一个账号	不错、是、手机、只能、注册、账号	6.5962	relevant	relevant
天猫	17	艾米诺童装潮	艾、米、诺、童装、潮	0.4798	irrelevant	irrelevant
	18	lj		0.6682	irrelevant	irrelevant

user reviews will increase and tend to be stable on the increasing number of reviews. The experimental results are shown that the proposed method of App software's user reviews is effective.

The results in Figs. 4, 5 and 6 are shown that there are some inaccuracies in judgment relevance of App software's user reviews. Analyze the results and find the following reasons: (1) ICTCLAS 2016 words segmentation tool has the error, and user reviews has many online words because App software is a new software. ICTCLAS 2016 can not recognize the online words well leading that not

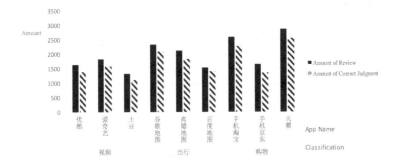

Fig. 4. The amounts of correct judgment about each App software.

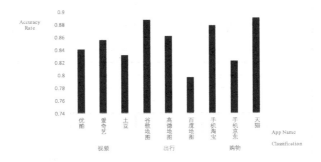

Fig. 5. The accuracy rate of relevance judgment about each App software.

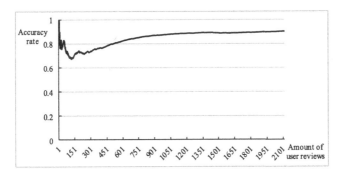

Fig. 6. The graph of accuracy rate about the relevance judgment of '微信' user reviews.

all words can be segmented into words and tagged into part-of-speech correctly. (2) Naive Bayesian text classification has a high accuracy comparing with other classification algorithm, but there are still some errors in itself. (3) Online word is the one of the characteristics about App software's user reviews. And this paper does not take into account the recognition of online words.

5 Conclusion

In this paper, we extracted the feature words of App software's user reviews and found the feature lexicon according to the part-of-speech of the feature words. Firstly, user reviews of different App software are crawled from App software markets. The user reviews are segmented into words and tagged into part-of-speech by word segmentation tool. And the keywords sets of user reviews are extracted by judging whether the part-of-speech of the words are n, v or a. Then, the keywords sets are optimized by analyzing whether keywords sets are keywords sets of all a, while the optimized part-of-speech sets are obtained. Finally, the relevant scores of App software's user reviews are calculated based on Naive Bayesian text classification and term frequency. In order to judge the relevance of App software's user reviews, the relevant score is compared with the threshold value α.

The experimental results are shown that the method can find out the relevant and irrelevant user reviews of App software. Compared with manual mark, the method average accuracy rate of the relevance judgment in this paper can reach to about 85%. In order to improve the results of the relevance judgment, the future work is the supplement of customized lexicons about online words. In order to improve the accuracy of relevance judgment about App software's user reviews, the customized lexicon of online words will be helpful to recognize more online words.

Acknowledgments. This research is sponsored by the National Science Foundation of China Nos. 61462049, 60703116, and 61063006.

References

1. Lin, Y., Wang, X., Zhu, T., Zhou, A.: Survey on quality evaluation and control of online reviews. J. Softw. **25**(3), 506–527 (2014)
2. Hu, Z., Zheng, X.: Product recommendation algorithm based on user's reviews mining. J. Zhejiang Univ. (Eng. Sci.) **47**(8), 1475–1485 (2013)
3. Jiang, W., Zhang, L.: Analyzing helpfulness of online reviews for user requirements elicitation. Chin. J. Comput. **36**(1), 119–131 (2013)
4. Li, Y., Fu, H.: Fake comments recognition based on social network graph model. J. Comput. Appl. **34**(S2), 151–153, 158 (2014)
5. Pagano, D., Maalej, W.: User feedback in the appstore: an empirical study. In: 2013 21st IEEE International on Requirements Engineering Conference (RE), pp. 125–134. IEEE (2015)
6. Leopairote, W., Surarerks, A., Prompoon, N.: Software quality in use characteristic mining from customer reviews. In: 2012 Second International Conference on Digital Information and Communication Technology and its Applications (DICTAP), pp. 434–439. IEEE (2012)
7. Harman, M., Jia, Y., Zhang, Y.: App store mining and analysis: MSR for app stores. In: Proceedings of the 9th IEEE Working Conference on Mining Software Repositories, pp. 108–111. IEEE (2012)

8. AlQuwayfili, N., AlRomi, N., AlZakari, N.: Towards classifying applications in mobile phone markets: the case of religious apps. In: 2013 International Conference on Current Trends in Information Technology (CTIT), pp. 177–180. IEEE (2013)

9. Gao, C., Xu, H.: AR-tracker: track the dynamics of mobile apps via user review mining. In: 2015 IEEE Symposium on Service-Oriented System Engineering (SOSE), pp. 284–290. IEEE (2015)

10. Han, P., Wang, D., Liu, Y.: Influence of part-of-speech on Chinese and English document clustering. J. Chin. Inf. Process. **27**(2), 65–73 (2013)

11. Zhang, L., Hua, K., Wang, H.: Sentiment analysis on reviews of mobile users. Procedia Comput. Sci. **34**, 458–465 (2014)

12. Wang, J.: Study of the application of text classification techniques on Weibo. Guangxi University (2015)

13. Di, P., Duan, L.: New Naive Bayes text classification algorithm. J. Data Acquis. Process. **29**(1), 71–75 (2014)

Topic Model Based Text Similarity Measure for Chinese Judgment Document

Yue Wang[1,2], Jidong Ge[1,2(✉)], Yemao Zhou[1,2], Yi Feng[1,2], Chuanyi Li[1,2], Zhongjin Li[1,2], Xiaoyu Zhou[1,2], and Bin Luo[1,2]

[1] State Key Laboratory for Novel Software Technology, Nanjing University, Nanjing 210093, China
gjdnju@163.com
[2] Software Institute, Nanjing University, Nanjing 210093, China

Abstract. In the recent informatization of Chinese courts, the huge amount of law cases and judgment documents, which were digital stored, has provided a good foundation for the research of judicial big data and machine learning. In this situation, some ideas about Chinese courts can reach automation or get better result through the research of machine learning, such as similar documents recommendation, workload evaluation based on similarity of judgement documents and prediction of possible relevant statutes. In trying to achieve all above mentioned, and also in face of the characteristics of Chinese judgement document, we propose a topic model based approach to measure the text similarity of Chinese judgement document, which is based on TF-IDF, Latent Dirichlet Allocation (LDA), Labeled Latent Dirichlet Allocation (LLDA) and other treatments. Combining with the characteristics of Chinese judgment document, we focus on the specific steps of approach, the preprocessing of corpus, the parameters choices of training and the evaluation of similarity measure result. Besides, implementing the approach for prediction of possible statutes and regarding the prediction accuracy as the evaluation metric, we designed experiments to demonstrate the reasonability of decisions in the process of design and the high performance of our approach on text similarity measure. The experiments also show the restriction of our approach which need to be focused in future work.

Keywords: Chinese judgment documents · Data science · Machine learning · Natural language processing · Text similarity · TF-IDF · Topic model · Latent Dirichlet Allocation · Labeled Latent Dirichlet Allocation

1 Introduction

In the recent informatization of Chinese courts, the huge amount of law cases and judgment documents, which were digital stored, has provided a good foundation for the research of judicial big data and machine learning. In law, a judgment is a decision of a court regarding the rights and liabilities of parties in a legal action or proceeding. Judgments also generally provide the court's explanation

© Springer Nature Singapore Pte Ltd. 2017
B. Zou et al. (Eds.): ICPCSEE 2017, Part II, CCIS 728, pp. 42–54, 2017.
DOI: 10.1007/978-981-10-6388-6_4

of why it has chosen to make a particular court order. Judgment document is the documented judgement with relevant content. In 2013, China Judgment Online System officially opened. Up to now, it has recorded more than 26 million electronic judgment documents and became the largest judgment document sharing website around the world. The achievements of the informatization of Chinese courts not only provides the benefit of digitization, but also is a great help to judges and relevant parties.

In this situation, some ideas about Chinese courts can reach automation or get better result through research of machine learning. For example, Judge can find the similar judgement documents by the basic situation of the case to contribute to the process of judgement; court can evaluate workload of a judge by the similarity of the judgement documents it handled; Even relevant parties can input the situation of case to view the relevant statutes. In trying to achieve all above mentioned, a text similarity measure for Chinese judgment documents is being called.

As an important category of natural language processing, text similarity has developed from String-based algorithms, Corpus-based algorithms, to Knowledge-based algorithms [1], including TF-IDF, topic model, distributed representation, etc. When the target of text similarity is changed to Chinese judgement document, there are some new challenges as follows:

1. It needs to focus on the semantic layer when measured the text similarity of judgement document.
2. Judicial specific words existing in various types of judgement documents may influence the text similarity.
3. Chinese judgement document is semi-structured, which means it includes not only expression with natural language, but also a relatively fixed standard. The standard may provide a chance to improve the result of text similarity measure.
4. In Chinese judgement document, besides the process of reasoning and judgement, the claims and evidence of pleadings also need be recorded. It's a critical factor to influence text similarity that how to judge the important of similarity measure for different part of judgement document.
5. Chinese legal system, which is embroidered on legislation and assisted by administration, is obviously different to other countries with adequate legal system [2]. Referencing civil law system, Chinese legal is grounded on statutory code instead of law precedent. It means Chinese judgement document depends on relatively fixed statutes, which may help the work of text similarity.

In this paper, we propose a topic model based approach to measure the text similarity of Chinese judgement document. For the challenges mentioned above, the approach is based on TF-IDF, Latent Dirichlet Allocation (LDA), Labeled Latent Dirichlet Allocation (LLDA) and other treatments. The approach can be used to develop corresponding applications, such as similar documents recommendation, workload evaluation based on similarity of judgement document, and prediction of possible relevant statutes.

The remainder of this paper is laid out as follows. Section 2 introduces related work. Section 3 introduces our approach in detail. Section 4 shows the implementation of our approach and the experiments and Sect. 5 makes conclusion and discusses the future work.

2   Related Work

Text similarity measure play an important role in natural language processing research and applications such as information retrieval, text summarization, text classification, topic detection and so on. Developing to this day, from String-based algorithms to Corpus-based algorithms to Knowledge-based algorithms [1], many text similarity measures have been proposed for using in different scenes.

Topic model, as a Corpus-Based algorithms, originated from Latent Semantic Indexing (LSI) [3]. Although LSI is not a probability model, Hofmann, based on the main idea of LSI, proposed Probabilistic Latent Semantic Indexing (pLSI) [4]. After that, Blei et al. proposed Latent Dirichlet Allocation [5], which introduces Dirichlet distribution to further improve topic model. Based on LDA, more improved topic model, such as Supervised LDA [6], Labeled LDA (LLDA) [7], Hierarchically Supervised LDA [8], etc., are proposed to solve the specific problems. Topic models have been applied successfully in documents modeling [9], image modeling [10, 11] and etc.

LDA, as one of the most important topic model, assume each document is modeled as probability distribution over an underlying set of topics, which in turn are modeled as probability distributions over words. LDA has been widely applied in various scenes, such as sentiment analysis [12], bug localization [13], image classification [14] and text segmentation [15]. In process of LDA training, finding the number of topics is a difficulty. For this problem, Arun et al. showed some observations [16] and some extensions of LDA, just like Hierarchical Dirichlet Processes, were designed [17].

LLDA is An extension of LDA with supervision [7]. This model allows corpus to be labeled by tags, and output a list of labeled topics. LLDA has been demonstrated the potentiality for fine grained topic modeling [18]. It also be applied to text classification [19] and social relation [20].

3   Approach

In this section, we describe the text similarity measure for Chinese judgment document in detail as follows. Subsection 3.1 presents an overview of the workflow of our approach. Subsection 3.2 describes the preprocessing of corpus. Subsection 3.3 introduces the choice of model, Subsect. 3.4 describes the process of topic model training. Subsection 3.5 introduces the meaning and method of corpus segmentation with different weight. Subsection 3.6 introduces the evaluation method for result.

3.1   Overview

Figure 1 presents an overview of workflow for Our approach. Because of the characteristics of Chinese natural language, the process is based on Chinese word segmentation. The steps of workflow is as follows:

1. Collect Chinese judgement documents, with structured information, including cause of action and category of cases, as the target corpus.
2. Divide the target corpus into two parts. The first part is used as the training corpus and the second part called test corpus, which needs to be labeled with practical similarity, is prepared for the evaluation of model.
3. Preprocess the judgment documents of the training corpus.
4. Choose the high reliable and important segments as the inputs.
5. Define and decide the parameters of training and model.
6. According to LDA or LLDA, Use the training corpus to complete training.
7. Use test corpus to evaluate the result of previous training model.

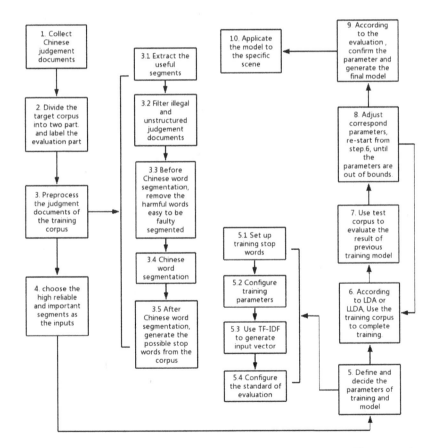

Fig. 1. Overview of the workflow for text similarity measure for Chinese judgment documents

8. Adjust correspond parameters, restart from step.6, until the parameters are out of bounds.
9. According to the evaluation of different parameters, confirm the parameter and generate the final model.
10. Applicate the model to the specific scene.

3.2 Preprocessing

The preprocessing step can be divided into four sub steps as follows: (1) Extract the useful paragraph; (2) Filter illegal and unstructured judgement documents; (3) Before Chinese word segmentation, remove the harmful words which are easy to be incorrectly segmented; (4) Chinese word segmentation; (5) After Chinese word segmentation, generate the possible stop words from the corpus. The purpose of this step is to further decompose the training corpus, minimize possible interference and prepare for training.

Chinese court has drew up a series of standards to define the structure of judgment document. It can help us to distinguish the corpus and identify low-quality judgement documents, as the content described in steps (1) and (2). With these steps, we can further filter the useless part of judgement document for semantic text similarity measure, just like the name of judge and the information of litigant participants, to focus on the case itself.

In Judgement document, a number of judicial specific words, just like prosecutor and defendant which occurs frequently, are not only meaningful, but also harmful for semantic similarity measure. Moreover, locale names and some ordinary names may also have influence on text similarity, especially the abbreviation of names for secrecy. Incorrect word segmentation of Chinese about the harmful words is another problem, which makes the target word split after word segmentation that can not be filtered by stop word list. The examples is shown in the following Table 1.

For these problems mentioned above, the steps (3) and (5) are necessary. In step (3), the main target is the name of litigant participants and some special judicial words. The formers can be extracted from judgement document by some rules and the specific words can be selected from the words list in step (5). In step (5), the most frequent terms are our candidates. Based on the segmented Chinese words, we can make statistics the frequency of terms and choose the stop words.

Table 1. Examples of incorrect word segmentation about the harmful words

Input	Correct Segmentation	Incorrect Segmentation	Main Affected Term
原告诉称	原告 | 诉称	原 | 告诉 | 称	告诉
李某某	李某某	李 | 某某	某某

3.3   Choice of Model

In step 6, we can choose LDA or LLDA to complete modeling. For Chinese judgement document, this choice is necessary. As a kind of text similarity measure, the approach generates different models for various similarity targets. The similarity targets based on statutes, which are explicit and finite tags of Chinese judgment documents, are an optional choice, which makes it possible to execute the process of evaluation automatically with little manual intervention. For this kind of similarity targets, such as statutes prediction, LLDA is more suitable than LDA, because of introduced supervision and fine grained topics.

With more manual intervention, such as the similarity targets based on manual classification or just the number of class, there will be different choices. In principle, if documents in corpus can be labeled explicitly, LLDA model will be recommended.

3.4   Topic Model Training

Each Chinese judgement document is associated with relatively fixed items, just like cause of action, category of case, codes and statutes. We can assume that Chinese judgement document is topic relevant, and based on this assumption, the topic model is an appropriate approach for semantic text similarity measure. We use LDA model to find the relation between number of topics and number of statutes. For verifying the conjecture and improving the accuracy of text similarity, we try to use supervised topic model named LLDA. This subsection including step 5 and 6 of our approach, which is hard to be separated described, is aimed to introduce the process of parameters choices and training.

There are four sub steps as follows in common: (1) Set up training stop words; (2) Configure training parameters, include the initial value and adjustment value, which is different for LDA and LLDA; (3) Use TF-IDF to generate input vector; (4) Configure the standard of evaluation. In the complete process, the training step and evaluation step should be execute repeatedly to confirm the parameters of model.

For LDA model, the most important parameter is topic number, which is also the difficulty for normal topic models. In this approach, we use a self-adaption method to choose the topic number. The brief steps is as follows: (1) Choose the initial topic number; (2) Start LDA training and evaluation the model; (3) Increase or decrease the topic number and return to step (2) until out of bound; (4) According the topic num and result of evaluation, choose the appropriate topic number. Besides the result about similarity, the perplexity [5] is also a important metrics.

For LLDA model, the additional work is to complete labelling. If the similarity target is based on statutes, the relevant statutes of Chinese judgement document will be the natural labels for supervised topic model. Because of the structuring of judgement document, the referenced statute can be extracted completely by fixed rules in most cases. Besides, the counted statutes should exclude the low frequency statutes with a threshold, because the target statutes, as the performance of topic in the approach, must have a certain number of occurrence in the corpus, or they will have no statistically significant.

3.5 Choice of Inputs

In the process of text similarity measure, because the importance and reliability of different parts of Chinese judgement document is various, which is attributed to the structure of judgement document, we should choose the appropriate segments as input.

For a Chinese judgement document, the core content is consist of evidence, fact, statute and judgement [21]. Corroborate evidence with each other; Deduct evidence or facts to facts; Relate facts to statutes; Generate judgement from statutes and facts. The structure of Chinese judgement document is as showed in Fig. 2. The judgement is result, the statute is explicit but the evidence and fact is full of uncertainty. For example, judgement document records a list of evidence provided by plaintiff and defendant, including not only the accepted evidence, but rejected part; The fact in fact finding segment is more credible than the fact recorded in judgement document from plaintiff.

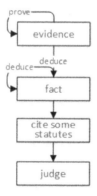

Fig. 2. Judgement document core structure

Ideally, the reliability of each evidence and fact can better reflect the content of Chinese judgement document. But considering the difficulty of this job and the redundancy of judgement document, we choose the segments as input which is confirm by judge, just like the fact finding segment and evidence segment of the base of case.

3.6 Result Evaluation

In step 2, as the concrete content presented in Fig. 3, some documents should be extracted from target corpus and labeled for evaluation. After the process of training, the model need been evaluated by designed evaluation method, which is described in detail in this subsection.

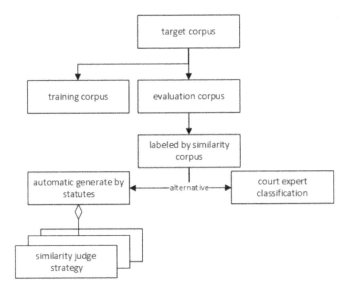

Fig. 3. Classification and labelling method for corpus

The process of similarity labelling is decided by two dimension. Labelling method as the first dimension, which means implementation methods of labelling, includes automatic labelling and artificial labelling. The former needs to formulate and implement correspond similarity strategy. The later demands the expert of court to finish the labelling. The other dimension includes digital labelling and non-digital labelling which is used to present if the label is represented by numbers called labelling granularity. For example, Comparing two judgement documents with the overlap of referenced statutes is an automatic digital labelling method, because the statutes can be extracted automatic and the result can be quantification; In another aspect, the classified judgement document handled by court expert is an artificial non-digital labelling method. The characteristics of the dimensions is showed in Table 2.

In this paper, we mainly focus on the labelling method with statutes, and the evaluation method is described below in general: The base of case in judgement document is the input of evaluation, and the result of evaluation depends on the accuracy of prediction on statutes. The reason for the choice is intuitive and universal for different model, includes TF-IDF, LDA and LLDA. The workflow of evaluation based on statute prediction is presented in Fig. 4.

Table 2. Advantage and disadvantage of labelling

	digital	non-digital	advantage	disadvantage
automatic	✔	✔	accurate, automatic	different standard for different target
artificial	✗	✔	currency	Lack of mathematical basis, Uncertain

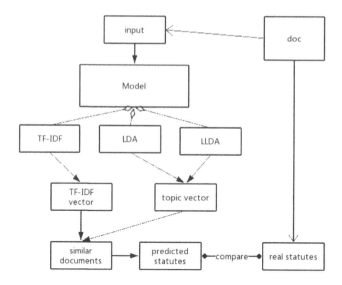

Fig. 4. Workflow of evaluation based on statute prediction

Based on various trained model, we can get the similar documents in training corpus from input of test document. For each test document, we sum the frequency of each unique statute by scanning all similar documents. And then sort the statutes in descending order. Without sufficient basis for truncating the sorted statutes, we choose top N to be the predicted statutes. The N is the number of real relevant statutes.

Perplexity, the exponentiation of the entropy, as a common metrics to evaluate the language model, is a reference metric for this approach. In Chinese judgment document, the statutes, which are associated with topic in our approach, can not be assumed independent. The correlation of statutes is common in judgment and the overlap is allowed. In this approach, we use perplexity to determine iteration number and to assist the evaluation of models.

4   Application and Evaluation

The purpose of this section is to implement topic model based text similarity measure for Chinese judgment documents, evaluate the result of experiments and provide support for our approach. The concerned points include: (1) the applicability of LLDA in this approach, (2) the performance of this approach in practical application, (3) The influence of specific preprocessing of corpus for Chinese judgment document on text similarity.

4.1   Preprocessing and Dataset

In lack of common corpus about Chinese judgement document, we collected documents in China Judgment Online System, the official website of Chinese court. To reduce the complexity of problem, In this experiment, we chosen the same type of documents and totally extracted 53000 first-instance civil judgment documents. 50000 of all is used to training corpus and last is used to evaluation.

In the programming tool aspect, we use jieba module to segment Chinese word, gensim library to implement TF-IDF and LDA, Stanford Topic Modeling Toolbox to implement LLDA.

The evaluation method is same as the Subsect. 3.6 described. The evaluation metrics is F-measure, For each category, F-measure is calculated by $2 *$ precision $*$ recall$/($precision $+$ recall$)$. F-measure represents the balance between precision and recall, the higher the F-measure of a category is, the better the performance of the classifier on this category is. In this situation, F-measure is same as precision and recall.

About the training parameters, $\alpha = \vec{l}/sl, \beta = 1/V$. The $\vec{l}$ is the vector of statutes frequency, the sl is the sum of statutes frequency and the V is number of words. The iteration number is depend on the perplexity of model.

4.2 Experiment and Result

For different threshold of statutes frequency, the number of counted statutes is showed in Fig. 5.

Based on the accounted statutes with different threshold of statutes frequency, we can generate corresponding LLDA models. In this scene, the topics of LLDA, as the output of model, are named after statutes in Chinese judgment documents. In another word, From LLDA, we obtain the probability distribution of statutes over the words in corpus, which can be used to predict statutes directly. Using the same idea as evaluation method, we can obtain the experiment results showed in Fig. 6.

The accuracy of direct statutes prediction based on LLDA is not ideal. For Chinese judgment document, one of the reasons, which is easily associated with,

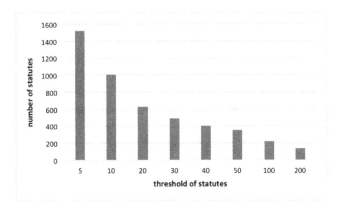

Fig. 5. Number of statutes with different threshold of statutes frequency

is the restriction of bag of words. Though being processed by TF-IDF, the model can not represent the logical relations between words, which is the source of statutes deduction. In another word, some kinds of statutes may not be predicted in current models, which is the emphasis of future work. As evidence, we manually culled some statutes which is hard to be predicted and obtained the better result.

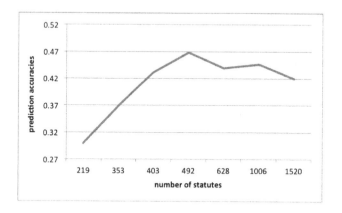

Fig. 6. Accuracies of direct statutes prediction based on LLDA

With the evaluation method mentioned in Subsect. 3.6, we experimented the statutes prediction on LLDA, LDA and independent TF-IDF. Figure 7 represents the comparison of LDA and LLDA. At least for statutes prediction in this scene, LLDA has better performance than LDA. Overall, the accuracy of statutes prediction based on LLDA has similar trend with the result of LDA model. Based on LLDA, the improvement of accuracies from direct statutes prediction to statutes prediction may shows that the statutes are not independent.

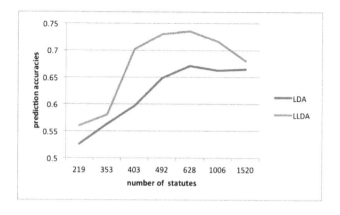

Fig. 7. Comparison of statutes prediction accuracy between LDA and LLDA

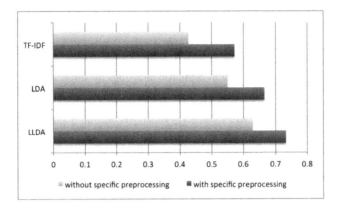

Fig. 8. Overall accuracies of statutes prediction with specific preprocessing and without specific preprocessing

As described in Sect. 3.2, besides the normal preprocessing, the approach asks more specific step for Chinese judgment document. In Fig. 8, it shows the overall accuracies of statutes prediction with specific preprocessing and without specific preprocessing. With specific preprocessing, the accuracies of statutes prediction improve obviously. Both LLDA and LDA has better performance than independent TF-IDF.

5   Conclusion and Future Work

In this paper, we propose a topic model based approach to measure the text similarity of Chinese judgement document, which is based on Latent Dirichlet Allocation (LDA) and Labeled Latent Dirichlet Allocation (LLDA), combining the characteristic of judgement document. In the experiments, we compared the result of statute prediction among TF-IDF based, LDA based and LLDA based approach. Both LDA and LLDA model have better performance than TF-IDF, and compared with LDA, LLDA improve a certain extent. The appearance also shows the word in Chinese judgement document has topic relevance on statutes.

However, the approach itself exits some defects as follows: (1) It is not a completely automatic approach. Manual intervention is required in the preprocess of corpus and the calculation of topic model parameters. (2) The whole model has some simplified assumption which need to be improve and perfected. (3) Some statutes generated from the logical relationship words, which can not be solved in word bag model or TF-IDF model, need the further research. These problems mentioned above left spaces for the future work.

Acknowledgement. This work was supported by the Key Program of Research and Development of China (2016YFC0800803).

References

1. Gomaa, W., Fahmy, A.: A survey of text similarity approaches. Int. J. Comput. Appl. **68**, 13–18 (2013)

2. Zhang, Z.: The construction of legal system in transitional China. China Legal Sci. **140**, 93 (2009)

3. Deerwester, S., Dumais, S., Furnas, G., et al.: Indexing by latent semantic analysis. J. Assoc. Inf. Sci. Technol. **41**, 391–407 (1990)

4. Hofmann, T.: Probabilistic latent semantic indexing. In: International ACM SIGIR Conference on Research and Development in Information Retrieval, pp. 50–57. ACM (1999)

5. Blei, D., Ng, A., Jordan, M.: Latent Dirichlet Allocation. J. Mach. Learn. Res. **3**, 993–1022 (2003)

6. Blei, D., Mcauliffe, J.: Supervised topic models. Adv. Neural Inf. Process. Syst. **3**, 327–332 (2010)

7. Ramage, D., Hall, D., Nallapati, R., et al.: Labeled LDA: a supervised topic model for credit attribution in multi-labeled corpora. In: Conference on Empirical Methods in Natural Language Processing, vol. 1, pp. 248–256. Association for Computational Linguistics (2009)

8. Perotte, A., Bartlett, N., Elhadad, N., et al.: Hierarchically supervised Latent Dirichlet Allocation. Adv. Neural Inf. Process. Syst. **24**, 2609–2617 (2011)

9. Li, F., Perona, P.: A Bayesian hierarchical model for learning natural scene categories. In: IEEE Computer Society Conference on Computer Vision and Pattern Recognition, pp. 524–531. IEEE Computer Society (2005)

10. Sivic, J., Russell, B., Efros, A., et al.: Discovering objects and their location in images. In: Tenth IEEE International Conference on Computer Vision, ICCV 2005, vol. 1, pp. 370–377. IEEE (2005)

11. Wang, C., Blei, D., Li, F.: Simultaneous image classification and annotation. In: IEEE Conference on Computer Vision and Pattern Recognition, CVPR 2009, pp. 1903–1910. IEEE (2009)

12. Lin, C., He, Y.: Joint sentiment/topic model for sentiment analysis. In: ACM Conference on Information and Knowledge Management, pp. 375–384. ACM (2009)

13. Lukins, S., Kraft, N., Etzkorn, H.: Bug localization using latent Dirichlet Allocation. Inf. Softw. Technol. **52**, 972–990 (2010)

14. Rasiwasia, N., Vasconcelos, N.: Latent Dirichlet Allocation models for image classification. IEEE Trans. Pattern Anal. Mach. Intell. **35**, 2665–2679 (2013)

15. Misra, H., Jose, J., Cappe, O.: Text segmentation via topic modeling: an analytical study. In: ACM Conference on Information and Knowledge Management, CIKM 2009, Hong Kong, China, pp. 1553–1556. DBLP, November 2009

16. Arun, R., Suresh, V., Veni Madhavan, C.E., Narasimha Murthy, M.N.: On finding the natural number of topics with Latent Dirichlet Allocation: some observations. In: Zaki, M.J., Yu, J.X., Ravindran, B., Pudi, V. (eds.) PAKDD 2010. LNCS, vol. 6118, pp. 391–402. Springer, Heidelberg (2010). doi:10.1007/978-3-642-13657-3_43

17. Teh, Y., Jordan, M., Beal, M., et al.: Hierarchical Dirichlet processes. J. Am. Stat. Assoc. **101**, 1566–1581 (2006)

18. Zirn, C., Stuckenschmidt, H.: Multidimensional topic analysis in political texts. Data Knowl. Eng. **90**, 38–53 (2014)

19. Li, W., Sun, L., Zhang, D.: Text classification based on labeled-LDA model. Chin. J. Comput. **31**, 620 (2008). Chinese Edition

20. Si, J., Mukherjee, A., Liu, B., et al.: Exploiting social relations and sentiment for stock prediction. EMNLP **14**, 1139–1145 (2014)

21. Zhou, F.: Reason, Jurisprudence, Sense and Writing. Shandong Justice (2007)

Utilizing Crowdsourcing for the Construction of Chinese-Mongolian Speech Corpus with Evaluation Mechanism

Rihai Su[1], Shumin Shi[1,2(✉)], Meng Zhao[1], and Heyan Huang[1,2]

[1] School of Computer Sciences and Technology, Beijing Institute of Technology, Beijing, China
bjssm@bit.edu.cn
[2] Beijing Engineering Research Centre of High Volume Language Information Processing and Cloud Computing Applications, Beijing, China

Abstract. Crowdsourcing has been used recently as an alternative to traditional costly annotation by many natural language processing groups. In this paper, we explore the use of Wechat Official Account Platform (WOAP) in order to build a speech corpus and to assess the feasibility of using WOAP followers (also known as contributors) to assemble speech corpus of Mongolian. A Mongolian language qualification test was used to filter out potential non-qualified participants. We gathered natural speech recordings in our daily life, and constructed a Chinese-Mongolian Speech Corpus (CMSC) of 31472 utterances from 296 native speakers who are fluent in Mongolian, totalling 30.8 h of speech. Then, an evaluation experiment was performed, in where the contributors were asked to choose a correct sentence from a multiple choice list to ensure the high-quality of corpus. The results obtained so far showed that crowdsourcing for constructing CMSC with an evaluation mechanism could be more effective than traditional experiments requiring expertise.

Keywords: Crowdsourcing · Chinese-Mongolian corpus · Speech corpus · WOAP · Evaluation mechanism

1 Introduction

Data acquring from the web for commercial and academic purposes has become more and more important, used for a wide variety of purposes in text and speech processing. However, to date, most of this data collection has been done for English and other High Resource Languages (HRLs). These languages are characterized by having extensive computational tools and large amounts of readily available web data and include languages such as French, Spanish, German and Japanese. Low Resource Languages (LRLs), although millions of people are much less likely and much more difficult to collect, due speak many largely to the smaller presence these languages have on the web. These include languages such as Igbo, Amharic, Pashto and what we are going to talk-Mongolian.

© Springer Nature Singapore Pte Ltd. 2017
B. Zou et al. (Eds.): ICPCSEE 2017, Part II, CCIS 728, pp. 55–65, 2017.
DOI: 10.1007/978-981-10-6388-6_5

The constructing process of traditional speech corpus is time-consuming and labor-intensive, which is recorded by speakers in professional studio [1–3]. Thus, the approach utilizing crowdsourcing technologies for corpus generation has gradually become a research hot topic [4].

In this paper, we present a new strategy that addresses the problem of acquiring large amounts of CMSC using WeChat Official Account Platform (WOAP), which can be served as a crowdsourcing system. Unlike current laboratory-generated corpus, WOAP provides a targeted collection pipeline for social networks and conversational style text. The purpose of this corpus collection is to augment the training data used by Automatic Speech Recognition (ASR) and for corpus resources for Mongolian NLP research. The more specific goal is to reduce the translation cost and improve the efficiency, and thus the corpus quality may improve sensibly. As we are aiming at spoken language, we try to collect audio recording from the real world environment. We performed an experiment of constructing CMSC via crowdsourcing technologies, all of them somewhat novel to the Mongolian NLP community but with potential for future research in other ethnic minority computational linguistics. Additionally, we discuss methods for validating corpus quality, finding the crowdsourced results are relatively better compared with controlled laboratory experiments.

The rest of the paper is organized as follows: Sect. 2 reviews the existing work; Sect. 3 describes the corpus constructing method in detail; Sect. 4 is the corpus evaluation part; we draw conclusions and discuss future work in Sect. 5.

2   Previous Research

The corpus research began in the 1970s. Along with the development of corpus linguistics, a large number of corpus such as Peking University Corpus, LDC Chinese Tree Bank, LOB and BROWN have been built up at home and abroad [5].

Although the construction of ethnic minority language corpus in our country started relatively late compared with the existing large-scale Chinese corpus, the ethnic minority speech corpus construction and speech recognition headed by Mongolian and Uygur have also been carried out fundamental research [6–10]. Yi et al. have described their work on the Mongolian multilingual speech corpus and its application in Japan [11]. Yu et al. have constructed "Mongolian speech corpus" towards Mongolian speech grammar research, which is made up of 10 million words recorded in various genres such as movies, jokes, stories, daily conversations and composition reading. Academy of Social Sciences of Inner Mongolia in china built the "Mongolian Speech Corpus". Recorders were selected from the eight provinces of Chinese 53 collection points, 10 points in Mongolia, the Russian Federation 7 points, recording the contents of free dialogue between native locals [12]. Yu et al. introduced the general situation of Mongolian standard speech-language dialogue corpus, and expounded some problems encountered in recording annotation and the way of solutions [13]. Yang et al. proposed a reasonable specification, a method of speech acquisition and annotation, and established a 300-h recording of speech Uighur language [14].

In the field of global NLP research, crowdsourcing has become a popular collaborative approach as well, which is utilized for the acquisition of annotated corpora and a wide range of other linguistic resources. The results of these studies have provided a good reference for the construction of CMSC [15–17]. It is possible to achieve ideal results by using crowdsourcing technology in corpus processing and annotation [18]. Posting on the platform AMT (Amazon's Mechanical Turk), a collection of parallel corpora between English and six languages from the Indian subcontinent: Bengali, Hindi, Malayalam, Tamil, Telugu, and Urdu have been built [19]. In the study of domestic crowdsourcing, a public opinion corpus of Uygur, Kazak and Kirgiz language based on crowdsourcing is proposed, which provides an essential resource support for the study of ethnic minority corpus [20].

3   Chinese-Mongolian Speech Corpus (CMSC)

Considering the Mongolian coding is not standardized, we turn to acquire spoken speech data. We utilize the crowdsourcing in the process of data acquiring and corpus evaluation. And we design and develop an interface-based WeChat Official Account Platform (WOAP) on which we can gather audio recordings sending from subscribers' mobile phone.

3.1   Basic Idea

WOAP is a convenient mobile Internet environment as an interactive platform, which can be used for interacting with WeChat subscribers to acquire data in a reasonable way. "Mother Tongue" is a WOAP oriented towards Mongolian users, on which we can interact with our followers unrestrictedly. And most of the subscribers are fluent in both Chinese and Mongolian, making it easy to meet the translation task requirements. Crowdsourcing enables real-world, large-scale corpus studies to be more affordable and convenient than expensive, time-consuming lab-based studies.

WOAP is flexible, relatively easy to use, and capable of collecting noisy, real-world audio data effectively. To sum up, this paper proposes a straightforward

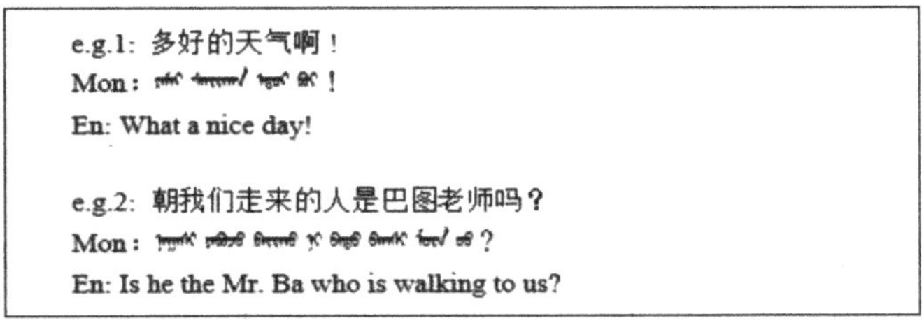

Fig. 1. Some examples of audio recording transcribed into text by Mongolian and English

method for gathering Chinese-Mongolian speech data, which can be utilized to form a large-scale, low-cost CMSC in a short time. Figure 1 contains two examples of audio recording transcribed into text by Mongolian and English.

3.2 Experiment Design

Contributor Selection. A significant part of selecting WeChat subscribers is to ensure them competently and conscientiously undertaking the tasks. As a quality-control measure, we release a Mongolian qualification test, consists of the basic subscriber attributes (WeChat nickname, user age, education level and location) and 15 test questions. The test comprises questions of short phrases, sentences in Chinese and answers in Mongolian. Candidates are asked to vote on one suitable translated answer among 4 choices, and candidates who pass the test can be considered as native speaker or proficient in Mongolian. The statistical results of the test show that 296 out of the 624 participants have passed the test, and these qualified subscribers are the main contributors in the process of corpus gathering.

Data Preparation. In this paper, our final corpus is aligned sentences of Chinese recordings and Mongolian recordings (partially transcribed). The main sources of data are Chinese conversations that embody the daily-life features in content. So the original data consists of following two parts: (1) Digitized forms of books series including Chinese-Mongolian daily conversations. (2) Texts downloaded from websites. According to the distribution of corpus and the frequency of use, we gather the original Chinese texts considering the categories and corpus proportion. Concerning about the robustness of ultimate corpus, it is necessary to ensure the collection of original data is extensive and diversiform. Table 1 provides a general quantitative description of the corpus.

Table 1. A quantitative description of the corpus.

Categories	#sens	#words
Education	6171	20918
Recreation	6682	24465
Tourism	7893	29789
Diet	6406	22431
Baidu Tieba	4320	17493
Total	31472	115096

Recording. After the preparation of original data, we start to perform the recording. On the basis of the management regulations, each WOAP like "Mother Tongue" can only push one article to their subscribers per day. After taking into account the data collection process and experimental cycle, we push an article contains 15 Chinese sentences by the Mother Tongue (WOAP) everyday.

Fig. 2. An article contains translating norms and daily sentences.

The article includes the translating specifications as follows: (1) each subscriber selects one or more of the 15 Chinese sentences in the article to translate, and response us with Mongolian audio recording. (2) Those who provide high-quality audio can receive monetary rewards. (3) Empty or very noisy responses wont be included in the final set. Figure 2, shows an article contains translating norms and daily sentences.

3.3 Incentive

There are many ways to motivate crowdsourcing participants, and we apply a traditional and common monetary incentive mechanism. In such crowdsourcing systems that base themselves on monetary payment, the incentive mechanism has to be effectively designed to be economical and fair. In general, each feedback audio recording of subscribers took on an average 20–40 s to complete and a subscriber was paid 5 cents per task including a bonus that was paid on completion of five tasks. It should be noted that subscribers who pass qualification review could receive the corresponding reward.

3.4 Quality Control

Acquiring high-quality corpus is one of the primary concerns in our crowdsourcing system. In this paper, not having any external mechanism can improve the quality of the corpus, we had to turn to the task design itself to perform a reasonable quality control.

Fig. 3. Test cases of the Mongolian speakers proficiency.

Firstly, different subscribers complete each Chinese sentence in the daily article. For every translation task, we ensure the audio recordings contributed by subscribers are more than one in this way, and select the most reasonable result from candidates with a feasible strategies detailed in Sect. 4. Secondly, the design of the Mongolian proficiency test above is another quality control mechanism. The difficulty of level test should be moderate; if the test is difficult, then the subscribers feel stressed and lost patience, resulting in the loss of the subscribers; if the test is too simple, then the filter-out process will become invalid and we cannot achieve high-quality results. Figure 3 shows the test cases of the Mongolian speaker's proficiency.

4   Evaluation

4.1   Crowdsourcing Based Evaluation Mechanism (CBEM)

Although a testing process is introduced before the translation task, there are inevitably some subscribers who misunderstand the task or intentionally provide some erroneous or random results. In order to achieve better results, we take full advantage of the crowdsourcing approach. Since speech corpus could be contributed by the public, the same can also be done to validate the results of the corpus. In the process of the evaluation, we use the public voting strategy to screen out further qualified results.

In practice, first given the correct Mongolian spoken sentences S and user-sent sentences S', then we need to judge if S' is the right translation of S. We detote it as a binary $\langle S, S' \rangle$. In this experiment, we randomly extracted

Table 2. Accuracy rate of manual annotation.

	P1	P2
100 binaries	57.4%	63.1%

100 sentence pairs from the corpus. First of all, the accuracy of the corpus was evaluated by manual annotation. Two definitions of the accuracy are defined as P1 and P2. In the definition of P1, only S' and S have the same meaning and the sentence is exactly the same, the binary then is judged to be positive. In the definition of P2, as long as S' and S have the same meaning, the binary is judged to be positive. The workers annotate 100 binaries according to the above criteria, the accuracy rate is shown in Table 2

From Table 2, we can see that the speech language contains a lot of noise, which is mainly caused by the participants' misunderstanding. In view of this phenomenon, we proposed a kind of corpus evaluation strategy based on crowd-sourcing assessment. For the results of the manual annotation obtained above, we used the crowdsourcing contributors to further carry out the voting activities to complete the sorting and selection of corpus. In practice, we presented an independent set of subscribers with an original Mongolian sentence and its four transcribed translations, and asked them to vote on which was best. Five independent subscribers voted on the translations of each source sentence. Similar to the online level test at the beginning of the experiment, the user only needs to complete a multiple-choice question. The more votes the results get, the more likely they are to be selected as the accurate results. Tallying the resulting votes, we found that roughly 65% of the sentences had five votes cast on just one or two of the translations, and about 95% of the sentences had all the votes cast on one, two, or three sentences. This suggests both (1) that there is a difference in the quality of the translations, and (2) the voters are able to discern these differences, and took their task seriously enough to report them. Figure 4 is an example of all the 5 votes cast on one translation.

In some cases, two translations obtained by the crowdsourcing process will get the same number of votes. We will not include these kinds of sentences into the final speech corpus, as these sentences are in small number, as shown in Fig. 5.

Fig. 4. An example of all the 5 votes cast on one translation; the parenthesized number indicates the number of votes. Translation: we can't go out, it's blowing sand outside.

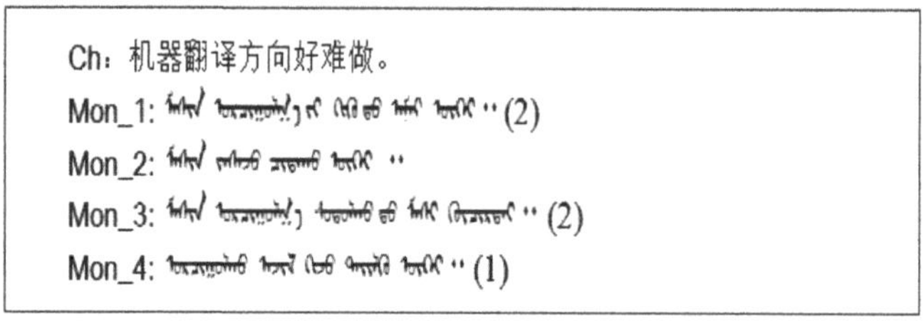

Fig. 5. An example of all the 5 votes cast on one translation; the parenthesized number indicates the number of votes. Translation: machine translation is difficult.

```
1 Initialize workers_ids
2 for worker1 ∈ workers_ids
3   do for worker2 ∈ workers_ids
4     do A[worker1, worker2]
        ←SIMILARITY(results[worker1], results[worker2])
5 Normalize x
6   while i < max_i
        do x_n ← NORMALIZE(A * x)
           dif ← x_n - x
7   for workerID ∈ workers_ids
        do scores[workerID] ← x[workerNum]
8   return scores
```

Fig. 6. An evaluation algorithm of crowdsourcing voting.

We use the algorithm similar to Google PageRank to evaluate the coherence of crowdsourcing workers, as shown in Fig. 6.

In total, the corpus contains 31472 sentences, in a recording time of about 30.8 h. The total cost of this work is 49.2$, the average cost of each recording is 0.01$. Our results are taken from 296 subscribers, ranging from age 14–65, with the majority of which belong to Inner Mongolia and Beijing. They represent a range of education levels, the majority had been to college: about 56% had master's degree, and 31% had bachelor's degree, proving that most of the contributors have a high degree of education. The compositon of the corpus is shown in Table 3.

Most subscribers responded an average of three audio recordings, with a few qualified chosen subscribers feedbacks plenty of recordings ranging from 20–50. It should be noted that few subscribers who contributed sufficient high-quality audio recordings are the candidates who achieved highest score in the proficiency test above. We did not strictly balance the speakers by their gender due to the difficulties in finding the qualified subscribers. Figure 7 shows the details.

Table 3. Detail composition of the speech corpus.

Categories	#sens	#hours	#Average voice length(s)
Education	6171	4.2	21
Recreation	6682	5.4	16
Tourism	7893	7.4	38
Diet	6406	4.6	25
Baidu Tieba	4320	3.4	10
Conversation	7345	5.8	31

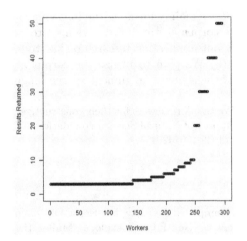

Fig. 7. About half of the subscribers returned more than 3 audio recordings and a few chosen subscribers accounted for most of our data.

4.2 Compared with Traditional Methods

In the research of Mongolian speech recognition, Inner Mongolia University constructed a corpus based on Mongolian standard pronunciation. The research intentionally select 20 students who major in Mongolian broadcasting, and most of them passed the Mongolian pronunciation examination. Then, the chosen speakers complete the recording-task at the experimental phonetics studio of Inner Mongolia University. Finally, a spoken corpus consisting of 30 h was constructed. Among which, 16 speakers read 3200 sentences in the training, and gathered a total of 24 h of recording. Then, 10 other speakers read the same 100 sentences per person for 6 h.

Based on the observation of the results above, we found that compared with the traditional professional-studio recording method, our WeChat-based method not only can collect large-scale voice data of real Mongolian speech environment, but also has the advantages of reducing construction costs and improving the quality of corpus.

5 Conclusion and Future Work

We can conclude that crowdsourcing platform such as WeChat can be significantly exploited for the resource construction in NLP area. We have present a method of constructing Chinese-Mongolian Speech Corpus (CMSC) based on WeChat Official Account Platform (WOAP), and collected a parallel speech corpus of 31472 utterances from almost 300 speakers and 30.8 h recordings. In order to ensure a high-quality speech corpus, a quality evaluation mechanism combining crowdsourcing contributors qualification and corpus evaluation is put forward. And our gathering of speech corpus in a real world environment shows a wider range of research and application value, compared with lab-generated speech corpus.

For the future work, we mainly focus on the following three aspects. (1) The scale and genres of the corpus is still limited, we need to expand it further. (2) We are trying to carry out application research on the tourism-oriented spoken language translation. (3) We plan to utilize the corpus as a training data to develop the research of the sentence-alignment speech language translation.

Acknowledgments. We thank reviewers for their constructive comments, and gratefully acknowledge the support of Natural Science Foundation of China (61671064) and BIT Basic Research Fund (20160742017).

References

1. Sigurbjörnsson, B., Kamps, J., De Rijke, M.: EuroGOV: engineering a multilingual web corpus. In: Peters, C., Gey, F.C., Gonzalo, J., Müller, H., Jones, G.J.F., Kluck, M., Magnini, B., de Rijke, M. (eds.) CLEF 2005. LNCS, vol. 4022, pp. 825–836. Springer, Heidelberg (2006). doi:10.1007/11878773_90
2. Crowdy, S.: Spoken corpus design. Literary Linguist. Comput. **8**(4), 259–265 (1993)
3. Adolphs, S., Knight, D.: Building a spoken corpus. In: The Routledge Handbook of Corpus Linguistics, pp. 38–52 (2010)
4. Howe, J.: The rise of crowdsourcing. Wired Mag. **14**(6), 1–4 (2006)
5. Kennedy, G.: An Introduction to Corpus Linguistics. Routledge, London (2014)
6. Bao, F.L., Gao, G.L., Bao, Y.L.: J. Inne. Mon. Sci. (NSE) **44**(3), 320–323 (2013)
7. Bao, F.L., Gao, G.L., Bao, Y.L.: J. Chin. Inf. Proc. **29**(1), 178–182 (2015)
8. Mu, R.G.W.: Research on Mongolian speech recognition. Dissertation (2013)
9. Zhao, J.D., Gao, G.L., Bao, F.L.: Research on hmm-based Mongolian speech synthesis. Comput. Sci. **1**, 014 (2014)
10. Tursun, R., Muhammat, I., Islam, W.: J. XJ Sci. (NSE) **30**(2), 199–203 (2013)
11. Yi, D., Zhang, Y., Zhang, S.: Processing of Mongolian by computer. J. Chin. Inf. Proc. **20**(4), 56–62 (2006)
12. Wu, J.X.: Lexical tagging of Mongolian corpus. S C Inne. Mon. 59–63 (2013)
13. Yu, R., et al.: Problems of recording tagging and solutions in "Mongolian Speech Corpus". PCC (2012)
14. Yang, Y.T., Dong, X.H., Wang, L.: Research on Uyghur speech language speech corpus of telephone chanel. Comput. Eng. Appl. **47**(23), 150–153 (2011)

15. Finin, T., Murnane, W., Karandikar, A., Keller, N., Martineau, J., Dredze, M.: Annotating named entities in Twitter data with crowdsourcing. In: Proceedings of the NAACL HLT 2010 Workshop on Creating Speech and Language Data with Amazon's Mechanical Turk, pp. 80–88. Association for Computational Linguistics (2010)
16. Sabou, M., Bontcheva, K., Derczynski, L., Scharl, A.: Corpus annotation through crowdsourcing: towards best practice guidelines. In: LREC, pp. 859–866 (2014)
17. Filatova, E.: Irony and sarcasm: corpus generation and analysis using crowdsourcing. In: LREC, pp. 392–398 (2012)
18. Munro, R., Bethard, S., Kuperman, V., Lai, V.T., Melnick, R., Potts, C., Schnoebelen, T., Tily, H.: Crowdsourcing and language studies: the new generation of linguistic data. In: Proceedings of the NAACL HLT 2010 Workshop on Creating Speech and Language Data with Amazon's Mechanical Turk, pp. 122–130. Association for Computational Linguistics (2010)
19. Post, M., Callison-Burch, C., Osborne, M.: Constructing parallel corpora for six Indian languages via crowdsourcing. In: Proceedings of the Seventh Workshop on Statistical Machine Translation, pp. 401–409. Association for Computational Linguistics (2012)
20. Chen, H.: Research on the Construction of Uygur, Kazak and Kirgiz Public Opinion Tagging Corpus Based on Crowdsourcing, MS thesis (2015)

A Cluster Guided Topic Model for Social Query Expansion

Wenyu Zhao and Dong Zhou[✉]

School of Computer Science and Engineering,
Hunan University of Science and Technology, Xiangtan 411201, Hunan, China
719727262@qq.com, dongzhou1979@hotmail.com

Abstract. As increasing amount of social data on today's social web systems, user-generated contents are not only getting richer, but also frequently inter-connected with users and other objects in various ways. Social data provides a perfect platform for personalized Web search. However, it is confronted with a great challenge named vocabulary mismatch problem. To overcome this problem, previous research has proposed many effective approaches utilizing social query expansion based on co-occurrence statistics, tag-tag relationships and semantic matching etc. Most of them focus on the statistical relationships between words/terms ignoring their truer semantics. In this paper, we propose a novel generative model which uses word embeddings to cluster words to enhance the latent topic model. Instead of just relying on the statistical rela-tionships of words, the approach tries to take into consideration of semantic information of context words and word clusters to construct user models for personalized query expansion. Experimental results on a large public social dataset show that the proposed method is more effective than other state-of-the-art baselines.

Keywords: Personalized search · Word embeddings · Word clusters · Query expansion

1 Introduction

Advances in new technologies have led to a rapid increase in the amount of data. With the advent of social web, users can optionally release any information worldwide. Social data has increased dramatically over the past few years. On the social data system such as delicious[1], users can label resources (or documents) by annotations (or tags). In the most of the cases, these marked resources together with annotations could not to be used to represent a user's real interests because the arbitrariness and ambi-guity exhibited in the tagging process. The net result is that searching through this collection will be less accurate. To deal with this problem, there are two methods can improve the accuracy of retrieved results. One is search results re-ranking [1–3], another is query expansion [4, 5]. Search results re-ranking needs the results retrieved are accurate at the first stage. Therefore, it is more attractive to expand original query

---

[1] http://www.delicious.com.

© Springer Nature Singapore Pte Ltd. 2017
B. Zou et al. (Eds.): ICPCSEE 2017, Part II, CCIS 728, pp. 66–77, 2017.
DOI: 10.1007/978-981-10-6388-6_6

from a practical perspective. How to effectively select relevant candidate words/terms to expand the original query is the key issue in personalized research utilizing social data.

From the query expansion perspective, previous studies pointed out that co-occurrence statistics between words in user model can be used for expanding query [6]. In these years, Zhou et al. proposed a query expansion framework to choose words from user model by the associations defined by tag-topic model [5]. Recently, Zhou et al. adopted a different approach to use an external corpus for query expansion using social data [7]. To deal with the cold start problem commonly exist in the social data, they proposed an enhanced user model by retrieving relevant documents from Wikipedia. Hence the expansion words are not only coming from a user's marked resources, but also from the information of external corpus. However, there exits some limitations in their work, the most important one is that their approaches are largely depended on the Latent Dirichlet Allocation (LDA) model [8]. As pointed out by other researchers [10], the LDA model tends to describe the statistical relationships of occurrence rather than real semantic information embedded in words. Instead, recent advances in neural language models like word embeddings [11] are playing an increasingly vital role in capturing the semantics of context words.

In this paper, we propose a novel generative model to construct user models for selecting related words to expand queries. Unlike other work by injecting word embeddings into the LDA model or vise verse [11–13], our model use word embeddings to cluster words to enhance the LDA model for user model generation. Subsequently, we can fetch more relevant words to expand queries based on the topics learned. To illustrate our methodology, we apply the approach to a real social dataset. Experimental results show that our proposed method is consistent and promising compared with the state-of-the-art techniques.

The rest of this paper is organized as follows. In Sect. 2, we briefly present related work on personalized search. We summarize the problem definition in Sect. 3. Then, Sect. 4 demonstrates our approach. Next, in Sect. 5, we describe the details about experiments and results. Finally, Sect. 6 conclude our paper and discuss the future works.

2 Related Work

To improve the accuracy and reduce the amount of irrelevant documents, many studies have been proposed in personalized search, and most of these studies are based on the social data [14]. Generally speaking, these studies can be broadly classified into two categories. One is personalized search results re-ranking, which reorder the documents in the first retrieval according to user's particular information and other social information. For example, researchers used information about user's interests from their tag models and bookmarks, then to compute the similarities between users' interests and topics of documents for re-ranking the results [1]. In [15] the authors considered gathering information from other online social system to enrich user's models for adapting the search results. Bouadjenek et al. proposed to use social data and user's relationships to enhance document representation for re-ranking purpose [2]. Carmel et al. re-ranked results based on the user's social relationships [16]. However, Search results re-ranking needs the results retrieved are accurate at the first stage.

Another group of work is query reformulation such as query expansion, which expand the original queries by extracting words similar to the original queries from user model. [9] selected words related to the original queries. Subsequently, [4] defined a query expansion that exploited relationships between users, documents tags. Chirita et al. pointed out that co-occurrence statistic can be taken into consideration for expanding query according to user's activities with social systems [6]. Bender et al. utilized tag-tag relationships for query expansion by adding similar tags from user's models to the query keywords [17]. Kuzi et al. used word embeddings to reweight the query words for expanding queries [18]. Zhou et al. proposed a query expansion framework to choose words from user model by the associations defined by tag-topic model [5]. Recently, Zhou et al. adopted a different approach to use an external corpus for query expansion using social data [7]. In their work, to deal with the cold start problem commonly exist in the social data, they proposed an enhanced user model by retrieving relevant documents from Wikipedia. Hence the expansion words are not only coming from a user's marked resources, but also from the information of external corpus.

However, most of these researches are largely depended on the LDA model, which tends to describe the statistical relationships of occurrence rather than real semantic information embedded in words. In this paper, we propose a novel generative model to construct user models for selecting related words to expand queries. Unlike other work by injecting word embeddings into the LDA model or vise verse, our model use word embeddings to cluster words, and then use these word clusters results to enhance the LDA model for user model generation. Subsequently, we can fetch more relevant terms to expand queries based on the topics learned.

3 Problem Definition

Formally, social data can be represented by a tuple $\mathcal{P} := (\mathcal{U}, \mathcal{D}, \mathcal{T}, \mathcal{A})$. $\mathcal{A} \subseteq \mathcal{U} \times \mathcal{D} \times \mathcal{T}$ is a ternary relation, whose elements are called tags or annotations. The set of annotations of a user can be defined as $\mathcal{A}_u := \{(t,d)|u,d,t \in \mathcal{A}\}$. The tag vocabulary of a user is given as: $\mathcal{T}_u := \{t|(d,t) \in \mathcal{A}_u\}$. A set of documents maked by a user u is defined as $\mathcal{D}_u := \{d|(d,t) \in \mathcal{A}_u\}$. We defined the terms extracted from a user's set of documents as $docterm^{\mathcal{D}_u} := \{w|w \in \mathcal{D}_u\}$, where w denotes a word or a term in documents annotated by a user. Similarly, we also can define terms extracted from a user's set of external documents as $ext\_term^{\mathcal{D}_{ext}^u} := \{w|w \in \mathcal{D}_{ext}^u\}$, where $\mathcal{D}_{ext}^u$ denotes a user's set of external documents from an external corpus $\mathcal{D}_{ext}$. In Table 1, we list all the definition of basic notations used in our paper.

In personalized search scenario, the general problem we address can be formulated as follows:

Input: Given an original query q, the set of annotations of a user $\mathcal{A}_u$, the tag vocabulary of a user $\mathcal{T}_u$, a set of documents marked by a user $\mathcal{D}_u$, a set of words in the user model consists of $docterm^{\mathcal{D}_u}$, $exte\_term^{\mathcal{D}_{exter}^u}$ and $\mathcal{T}_u$.

Table 1. Basic notations used in the paper

Name	Description	Name	Description
$\mathcal{U}$	Finite sets of users	q'	An expanded query
$\mathcal{D}$	Finite sets of documents	K	Number of topics
$\mathcal{T}$	Finite sets of tags	C	A word cluster
$\mathcal{A}$	A ternary relation, elements are tags	θ	Multinomial distribution of topics
u	A user	φ	Multinomial distribution of words
d	A document	α	The prior parameter of model
t	A tag	β	The prior parameter of model
w	A word/term	λ	The prior parameter of model
$\mathcal{D}_{ext}$	An external corpus	W_C	The set of word clusters
$\mathcal{D}^u_{ext}$	A user's set of external documents	d_r	The set of relevant documents
$\mathcal{A}_u$	The set of annotations of a user	Z_i	Topic associated with the i-th word
$\mathcal{T}_u$	The tag vocabulary of a user	Ψ	The function of influence of $\delta_{i,j}$
$\mathcal{D}_u$	A user's set of documents	w'	Query terms of a original query q
doc$term^{\mathcal{D}_u}$	Terms extracted from documents annotated by a user	$\delta_{i,j}$	The influence of words in C_i exerting on current topic sampling of word w_i
ext_$term^{\mathcal{D}^u_{ext}}$	Set of terms extracted from a user's set of external documents	$n^{(w_i)}_{k,\neg i}$	The number of times w_i generated by topic k
q	A original query	$n^{(\cdot)}_{k,\neg i}$	The number of times words generated by topic k in vocabulary dictionary
n	Number of words in the user model	$v^{(d)}_{k,\neg i}$	The number of times words generated by topic k in the document d
γ	Top terms from a user's documents with high tf-idf scores	ς	Top documents retrieved from external corpus

Output: A ranked list of candidate words to be added to the original query for the second round retrieval of results.

4   Social Query Expansion

In this section, we propose a novel generative model to construct user model for capturing accurate information between words, then describe the adopted query expansion approach.

4.1   User Model Generation

Same as in Zhou et al.'s work [7], our personalized approach consists of two main steps: external documents retrieval and user model generation. In step one, we retrieve documents from an external corpus. The procedure works as follows: at first, we concatenate all tags t in $\mathcal{T}_u$ into a query $q^{\mathcal{T}_u}$. Then, for each document d in $\mathcal{D}_u$, we extract top γ terms from a user's set of documents with the highest inverted document frequency scores for forming queries $Q^{\mathcal{D}_u}$. Next, we use query in $Q_{ext}(q^{\mathcal{T}_u} \cup Q^{\mathcal{D}_u})$ to retrieve an external corpus $\mathcal{D}_{ext}$ to fetch top ς number of documents to form $\mathcal{D}_{ext}^u$.

In step two, we construct a user model with a novel generative model. Firstly, we regard the tag vocabulary in $\mathcal{T}_u$ as a document, and take documents of $\mathcal{D}_{ext}^u \cup \mathcal{D}_u \cup \mathcal{T}_u$ as inputs of Word2Vec to elicit word vectors of all terms $\{w_1, w_2 \cdots w_n\} \in docterm^{\mathcal{D}_u} \cup ext\_term^{\mathcal{D}_{ext}^u} \cup \mathcal{T}_u$. Then we take all word vectors of words in $docterm^{\mathcal{D}_u} \cup ext\_term^{\mathcal{D}_{ext}^u} \cup \mathcal{T}_u$ and the value K as inputs of the K-Means++ algorithm [19], where K is the number of topic trained by topic model LDA in the remaining of paper. After implemented the words clustering process based on the K-Means++ algorithm, we can obtain the words clustering results denoted as $W_C = \{C_1, C_2 \cdots C_K\}$. Thus, we learn words based on word embeddings and words clusters about all words consisting of $docterm^{\mathcal{D}_u}$, $ext\_term^{\mathcal{D}_{ext}^u}$ and $\mathcal{T}_u$. This is the procedure of clustering word by using word embeddings. In the next, we will introduce the user model generation in more details.

As we known, LDA model and word embeddings can also capture the semantic of words in the corpus. But there exits difference between LDA model and word embeddings. LDA model bases on the statistic relationships of words in the corpus, while word embeddings take advantage of the semantics of context words. In this paper, we make the most use of the merits of each other. Thus, we integrate $\mathcal{D}_{ext}^u, \mathcal{D}_u, \mathcal{T}_u$ into a novel generative probabilistic topic model named A Cluster Guided Topic Model (CG-LDA), which uses the word clustering results and word embeddings as auxiliary information to improve the effectiveness of LDA model.

The plate notation of CG-LDA model is shown in Fig. 1. Denote the number of topics by K, the number of words of $\mathcal{D}_{ext}^u, \mathcal{D}_u, \mathcal{T}_u$ by n. θ is a length K vector indicating the proportions over all topics for document d of $\mathcal{D}_{ext}^u, \mathcal{D}_u, \mathcal{T}_u$. φ is a length n vector denoting the distribution over all words. C_i represents the words cluster sampled word w_i belongs to and $|C_i|$ is word size of cluster C_i.

To jointly model words and word embeddings produced by Word2Vec tools, CG-LDA model can learn latent topic to generate words in documents of $\mathcal{D}_{ext}^u, \mathcal{D}_u, \mathcal{T}_u$ and corresponding words clustering results. The model takes documents of $\mathcal{D}_{ext}^u, \mathcal{D}_u, \mathcal{T}_u$ and word clustering results as input. We use the Skip-Gram model [20] to learn the word embeddings before words clustering. Skip-gram model is usually used to predict context words of a target word in a sliding window. Each target word w will be associated with a vector $\vec{w} \in \mathbb{R}^{dim}$, where dim is the dimensionality of word embeddings. We regard the target word vector as a feature to predict the context words. Then we can use words vectors as inputs of the K-Means++ algorithm for clustering words. Thus we take words clustering results as auxiliary information about words to train LDA model in order to enhance the effectiveness of the LDA model.

With word embeddings trained by the Skip-Gram model and clustered words, the generation process of CG-LDA model can be summarized as follows (see Algorithm 1). Firstly, the model draws a distribution over words of documents. Then for each document of $\mathcal{D}_{ext}^u, \mathcal{D}_u, \mathcal{T}_u$, the topic distribution is chosen from Dirichlet distribution. Next, for each word w_i in each document, a particular topic Z_i can be sampled from document distributions, a word w_i is drawn from the topic distribution. If a word w_i in words cluster C_i, for each word w_j is not word w_i but in words cluster C_i, we sample $\delta_{i,j}$ from the function Ψ which represents the influence of words in words cluster C_i exerting on current topic sampling of word w_i. The process is repeated for all words in the each document of $\mathcal{D}_{ext}^u, \mathcal{D}_u, \mathcal{T}_u$.

Algorithm 1 Generative process for CG-LDA model

Require: tags of a user $\mathcal{T}_u$
Require: documents of a user $\mathcal{D}_u$
Require: a user's set of external documents $\mathcal{D}_{ext}^u$
Require: word embeddings calculated by Skip-Gram for all words in $\mathcal{D}_{ext}^u \cup \mathcal{D}_u \cup \mathcal{T}_u$
Require: words clusters information W_C
1. **for** each topic $k \in [1, K]$ **do**
2.   sample a distribution over words $\varphi \sim Dirichlet(\beta)$
3. **end for**
4. **for** each document $d \in (\mathcal{D}_{ext}^u, \mathcal{D}_u, \mathcal{T}_u)$ **do**
5.   draw a vector of topic proportions for this document $\theta | \alpha \sim Dirichlet(\alpha)$
6.   **for** each word w_i in document d **do**
7.   conditional on θ choose a topic $Z_i | \theta \sim Mult(\theta)$
8.   conditional on Z_i choose a word $w_i | Z_i \sim Mult(Z_i)$
9.   locate the words cluster C_i ,where $w_i \in C_i$
10.   for each word $w_{j,\neg i} (j \neq i)$ in C_i ,draw $\delta_{i,j} \sim \Psi(Z_{i,j})\alpha$,where $Z_{i,j}$ is the topic probability of word $w_{j,\neg i}$ at topic Z_i.
11. **end for**
12. **end for**

In the Algorithm 2, α, β and λ are the prior parameters of CG-LDA mode. θ and φ are the posterior distributions of latent variables. In this model, the posterior distributions of latent variable and a particular topic Z_i can be approximated by Gibbs sampling method [21]. In the sampling procedure, for each word the topic is sampled by utilizing the latent topic information of word clusters, the update rule is as follow:

$$P(z_i = k | W_C, \lambda) \propto \frac{n_{k,\neg i}^{(w_i)} + \beta}{n_{k,\neg i}^{(\cdot)} + N\beta} \times \frac{v_{k,\neg i}^{(d)} + \alpha}{v_{\cdot,\neg i}^{(d)} + K\alpha} \times \prod_{w_{j,\neg i} \in C_i} exp\left(\frac{\lambda}{|C_i|} \times Z_k^{(w_{j,\neg i})}\right) \quad (1)$$

In this formula, for each sampled document $d \in (\mathcal{D}_{ext}^u \cup \mathcal{D}_u \cup \mathcal{T}_u)$, $n_{k,\neg i}^{(w_i)}$ denotes the number of times word w_i has been observed with topic k, $n_{k,\neg i}^{(\cdot)}$ denotes the number of times words in the vocabulary dictionary are assigned to topic k, $v_{k,\neg i}^{(d)}$ represents the number of times words in the document d have been assigned to topic k, $v_{\cdot,\neg i}^{(d)}$ is the number of all words in the document d. After above sampling process, we can obtain the document-level latent topics by formula 2:

$$P(z_i = k) \propto \frac{v_k^{(d)} + \alpha}{v_{\cdot}^{(d)} + K\alpha} \quad (2)$$

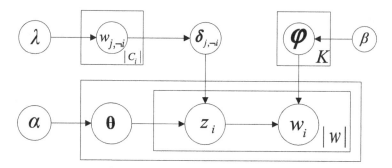

Fig. 1. Plate notion of CG-LDA model

where $P(z_i = k)$ indicates the topic probability of topic k in the topic vector of document d. Due to the high-quality word vectors learned by the Skip-Gram model, the words clusters information is very useful to elicit better distributed representations of documents based on the LDA model. After that we can calculate the posterior estimate of θ and φ.

4.2 Personalized Query Expansion

Our query expansion approach is also adopted from Zhou et al.'s work [7]. We assume that the query q consists of n independent query terms $\{w_1', w_2' \cdots w_n'\}$, the probability of a word w generated by the query q can be defined as:

$$P(w|q) = P(w|w_1', w_2' \cdots w_n') \propto \prod_{e=1}^{n} P(w|w_e') \tag{3}$$

We also can assume that there exit a set of relevant documents $\{d_r\}_{r=1}^{N}$ related to the query and the word, so we can use the set of documents converted the formula 3 into the equivalent form, we have:

$$P\left(w|w_e'\right) = \sum_{r=1}^{N} P(w|d_r)P(d_r|w_e') \propto \frac{1}{N}\sum_{r=1}^{N} P(w|d_r)P(w_e'|d_r) \tag{4}$$

Based on the step two, we can get a latent topic trained by our new CG-LDA model related to each document and each word in user model, so we can use the documents in user model as a set of relevant documents. Especially, because there is no direct dependency of w, w_e' on relevant document and the query, we can simplify formula 4 to get the probability of a word generated by the query q. We have:

$$P\left(w|w_e'\right) \propto \frac{1}{N}\sum_{r=1}^{N}\left(\sum_{k=1}^{K} P(w|Z_k)P(Z_k|d_r)\right) \times \left(\sum_{k=1}^{K} P\left(w_e'|Z_k\right)P(Z_k|d_r)\right) \tag{5}$$

All the terms $\{w_1, w_2 \cdots w_n\}$ in user model are ranked by their probability of being generated by the given query, and the top δ terms are chosen to expand the given query.

5 Evaluation

In this section, we present the experimental settings in details, then describe the compared state-of-the-art techniques and evaluation methodology. At last, we present and analyze the results.

5.1 Experimental Setup

We select two public available social datasets from social web system del.icio.us: socialbm0311 and deliciousT140, which are public, described and analyzed in [22, 23]. The deliciousT140 dataset contains 144,574 unique URLs and social tags, and another socialbm0311 dataset is made up complete bookmarking activities of almost 2 million users. We matched the documents in deliciousT140 dataset with bookmarking activities in socialbm0311 dataset because of no the actual web pages in the deliciousT140 dataset. After have done this, we can obtain a total of 5,153,720 bookmark activities, 259,511 users, 131,283 web pages, 137,870 tags. For preparation, we got the textual content by parsing all web pages with a public parser. All documents parsed from web pages and tags are pre-processed using English analyzer, Porter's stemmer and a stopword list. In addition, the corpus constructed by socialbm0311 and deliciousT140 dataset is indexed by using Terrier[2] toolkit.

We also constructed an external knowledge base from Wikipedia. A Wiki snapshot was obtained on the 14/08/2014, which contained a collection of 4,634,369 articles. To evaluate the effectiveness of our approach, we randomly selected one group of 100 users. For each user, 75% of a user's tags with annotated web pages were used to create the user model and the other 25% were used as a test collection.

We choose the following evaluation metrics to measure our approach and other baselines: normalized discounted cumulative gain (NDCG@1), mean reciprocal rank (MRR) and mean average precision (MAP). The average performance over all users is calculated. Statistically significant differences were determined using a paired t-test at a confidence level of 95%.

5.2 Experiment Runs

We compared our approach with several state-of-the-art methods on using social data for personalized query expansion.

LM: We use the quite popular language model as the retrieval model which has previously demonstrated good results [24].
CoOCS: This method expands the original queries based on co-occurrence statistics between query terms and terms in user model [6].

---

[2] http://www.terrier.org.

TagTM: Zhou et al. built the user model according to Tag-Topic model for all model terms and proposed a query expansion framework based on the semantic word association enhanced by using terms extracted from top-ranked documents [5].

SEUP: This is the method used in Zhou et al.'s paper. They proposed a enhanced user model based on LDA model [7]. We chose a model with the highest performance in their work, i.e. SEUP model.

The experiments described in this paper, the number of documents from an external corpus retrieved by each query is set to $\varsigma = 5$ empirically. The number of terms from a user's set of documents with the highest inverted document frequency scores is set to $\gamma = 10$. For parameters, α, β and λ are the prior parameters of CG-LDA mode are set to $K/50$, 0.01 and 2 respectively. In the query expansion process, the number of candidate terms δ are set to 50. The number of latent topics and word dimensions used in our experiments are set to 5 and 50 for all test collections. In other baseline, the parameters are set the same as the original paper or according to those obtaining best performance.

5.3   Results

Table 2 shows the performance of baseline and our proposed method. Statistically significant differences between our method and LM, CoOCS, TagTM, SEUP are indicated by *, c, t, m, s respectively. The experimental results show that our proposed method can achieve a better performance compared with different baselines. From Table 2, we observe that the LM model is the lowest performance for all metrics in this experiment. As expected, CoOCS method outperforms the LM model by considering the co-occurrence statistic relationships between words in user model. SEUP works consistently better than TagTM for the evaluation metrics of MAP, NDCG@1 and MRR. This demonstrates the power of enriched user model for query expansion. As we can see, our approach can get significantly better performance than the baselines which considered the statistic relationships in the corpus. Moreover, it can outperform the state-of-the-art TagTM and SEUP with the highest improvement reaching 51.8% and 23.4% in the terms of MAP. For NDCG@1, our proposed method achieves 16.15%, which are about 8.34% higher than TagTM and 2.4% higher than SEUP. In addition, our method beats the TagTM by 48.5% and SEUP by 21.5% for MRR.

Table 2. Evaluation results of various methods. Statistically significant differences between our method and LM, CoOCS, TagTM, MEUP, SEUP are indicated by *, c, t, m, s respectively.

	User50		
	MAP	NDCG@1	MRR
LM	0.0262	0.0104	0.0267
CoOCS	0.0682*	0.0469*	0.0732*
TagTM	0.0907*	0.0781*	0.1031*
SEUP	0.1441*, c, t	0.1377*, c, t	0.157*, c, t
CG-LDA	0.1882*, c, t, m, s	0.1615*, c, t, m, s	0.2*, c, t, m, s

The improvement of our method works consistently better than state-of-the-art baselines. The results illustrate that we cluster words by using K-Means++ algorithm based on word embeddings for learning more information about words, which makes the best use of the semantics of context word. Word clusters can help us to improve the performance of the LDA model, it benefits for selecting candidate terms to expand query.

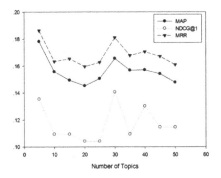

Fig. 2. Performance with different number of topics

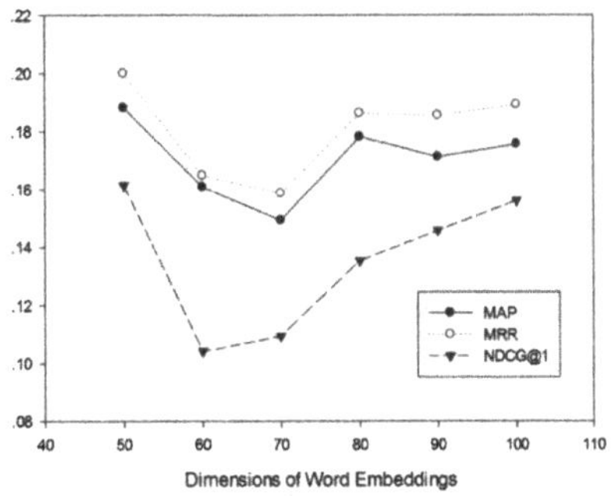

Fig. 3. Performance with different dimensions of word embeddings

5.3.1 Performance with Different Number of Topics
In this sub-section, we describe the impact of number of topic in our model on query expansion performance. We adjust the number of topics from 5 to 50. We evaluate the performance by the metrics MAP, NDCG@1, MRR. The results are shown in Fig. 2. As illustrated by the results, when the number of topics sets to 5, the performance of our approach achieves good results in terms of MAP and MRR.

5.3.2 Performance with Different Number of Dimensions of Word Embeddings
This sub-section examines the performance of our model on different dimensions of word embeddings. We set the dimensions of word embeddings from 50 to 100. We also use MAP, NDCG@1, MRR to evaluate the performance with different size of word embeddings. The results are shown in Fig. 3. As can be seen from the figure, the highest performance of dimensions of word embeddings is 50 in terms of MAP, NDCG@1 and MRR.

6 Conclusion

In this paper, we propose a novel generative model to construct user model for capturing its truer semantics of words in user model. The model can integrate neural language model with latent semantic models for improving better distributed representations. Because we cluster words by using K-Means++ algorithm based on word embeddings, which makes the best use of the semantics of context word. Word clusters can help us to improve the performance of the LDA model, it benefits for selecting candidate terms to expand query. The proposed model performed well on a real-world social dataset. Results show that our models significantly outperformed several state-of-the-art techniques. In the future, we consider exploring other complementary resources to automatically expand query for enhancing the performance of personalized social search.

Acknowledgement. The work described in this paper was supported by National Natural Science Foundation of China under Project No. 61300129, Scientific Research Fund of Hunan Provincial Education Department of China under Project No. 16K030, Hunan Provincial Natural Science Foundation of China under Project No. 2017JJ2101, Hunan Provincial Innovation Foundation For Postgraduate under Project No. CX2016B575.

References

1. Xu, S., Bao, S., Fei, B., Su, Z., Yu, Y.: Exploring folksonomy for personalized search. In: Proceedings of the 31st Annual International ACM SIGIR Conference on Research and Development in Information Retrieval, pp. 155–162 (2008)
2. Bouadjenek, M.R., Hacid, H., Bouzeghoub, M.: Sopra: a new social personalized ranking function for improving web search. In: Proceedings of the 36th International ACM SIGIR Conference on Research and Development in Information Retrieval, Dublin, Ireland, pp. 861–864. ACM (2013)
3. Xie, H., Li, X., Wang, T., Chen, L., Li, K., Wang, F.L., Cai, Y., Li, Q., Min, H.: Personalized search for social media via dominating verbal context. Neurocomputing **172**, 27–37 (2016)
4. Bouadjenek, M.R., Hacid, H., Bouzeghoub, M., Daigremont, J.: Personalized social query expansion using social bookmarking systems. In: Proceedings of the 34th International ACM SIGIR Conference on Research and Development in Information Retrieval, Beijing, China, pp. 1113–1114. ACM (2011)
5. Zhou, D., Lawless, S., Wade, V.: Improving search via personalized query expansion using social media. Inf. Retr. **15**(3–4), 218–242 (2012)
6. Chirita, P.-A., Firan, C.S., Nejdl, W.: Personalized query expansion for the web. In: Proceedings of the 30th Annual International ACM SIGIR Conference on Research and Development in Information Retrieval, Amsterdam, The Netherlands, pp. 7–14 (2007)
7. Zhou, D., Lawless, S., Wu, X., et al.: Enhanced personalized search using social data. In: Proceedings of the 2016 Conference on Empirical Methods in Natural Language Processing, pp. 700–710 (2016)
8. Blei, D.M., Ng, A.Y., Jordan, M.I.: Latent dirichlet allocation. J. Mach. Learn. Res. **3**, 993–1022 (2003)
9. Biancalana, C., Gasparetti, F., Micarelli, A., Sansonetti, G.: Social semantic query expansion. ACM Trans. Intell. Syst. Technol. **4**(4), 1–43 (2013)
10. Vulić, I., Moens, M.-F.: Monolingual and cross-lingual information retrieval models based on (Bilingual) word embeddings. In: Proceedings of the 38th International ACM SIGIR Conference on Research and Development in Information Retrieval, Santiago, Chile, pp. 363–372 (2015)
11. Mikolov, T., Sutskever, I., Chen, K., Corrado, G.S., Dean, J.: Distributed representations of words and phrases and their compositionality. In: Advances in Neural Information Processing Systems, pp. 3111–3119 (2013)

12. Das, R., Zaheer, M., Dyer, C.: Gaussian LDA for topic models with word embeddings. In: Proceedings of the 53rd Annual Meeting of the Association for Computational Linguistics, ACL 2015, pp. 795–804 (2015)

13. Liu, Y., Liu, Z., Chua, T.-S., Sun, M.: Topical word embeddings. In: Proceedings of the Twenty-Ninth AAAI Conference on Artificial Intelligence, AAAI 2015, Austin, Texas, USA, pp. 2418–2424 (2015)

14. Ghorab, M.R., Zhou, D., O'Connor, A., Wade, V.: Personalised information retrieval: survey and classification. User Model. User-Adap. Inter. **23**, 381–443 (2013)

15. Wang, Q., Jin, H.: Exploring online social activities for adaptive search personalization. In: Proceedings of the 19th ACM International Conference on Information and Knowledge Management, pp. 999–1008. ACM (2010)

16. Carmel, D., Zwerdling, N., Guy, I.: Personalized social search based on the user's social network. In: Proceedings of the 18th ACM Conference on Information and Knowledge Management, pp. 1227–1236. ACM (2009)

17. Bender, M., Crecelius, T., Kacimi, M., Michel, S., Neumann, T., Parreira, J.X., Schenkel, R., Weikum, G.: Exploiting social relations for query expansion and result ranking. In: IEEE 24th International Conference on Data Engineering Workshop, ICDEW 2008, pp. 501–506 (2008)

18. Kuzi, S., Shtok, A., Kurland, O.: Query expansion using word embeddings. In: Proceedings of the 25th ACM International on Conference on Information and Knowledge Management, pp. 1929–1932. ACM (2016)

19. Arthur, D., Vassilvitskii, S.: k-means++: the advantages of careful seeding. In: Proceedings of the Eighteenth Annual ACM-SIAM Symposium on Discrete algorithms. Society for Industrial and Applied Mathematics, pp. 1027–1035 (2007)

20. Mikolov, T., Chen, K., Corrado, G.S.: Efficient estimation of word representations in vector space. arXiv preprint arXiv:1301.3781. Cornell University Library (2013, 2016)

21. Heinrich, G.: Parameter estimation for text analysis. University of Leipzig, Technical report (2008)

22. Zubiaga, A., Garcia-Plaza, A.P., Fresno, V., Martinez, R.: Content-based clustering for tag cloud visualization. In: Proceedings of the IEEE International Conference on Advances in Social Network Analysis and Mining, ASONAM 2009, pp. 316–319 (2009)

23. Zubiaga, A., Fresno, V., Martinez, R., Garcia-Plaza, A.P.: Harnessing folksonomies to produce a social classification of resources. IEEE Trans. Knowl. Data Eng. **25**, 1801–1813 (2013)

24. Zhai, C., Lafferty, J.: Model-based feedback in the language modeling approach to information retrieval. In: Proceedings of the Tenth International Conference on Information and Knowledge Management, pp. 403–410. ACM (2001)

A Framework of Mobile Context-Aware Recommender System

Caihong Liu[1,2(✉)] and Chonghui Guo[1,3]

[1] Institute of Systems Engineering, Dalian University of Technology,
Dalian 116024, Liaoning, China
[2] College of Software, Dalian University of Foreign Languages,
Dalian 116041, Liaoning, China
lch@dlufl.edu.cn
[3] State Key Laboratory of Software Architecture (Neusoft Corporation),
Shenyang 110179, Liaoning, China

Abstract. Mobile users can be recommended services or goods precisely according to their actual needs even in different contexts. Therefore, it is necessary to construct a framework integrating following functions: context identification, context reasoning, services or product recommendations and other tasks for the mobile terminal. In this paper, we firstly introduce mobile context awareness theory, and describe the composition of context-aware mobile systems. Secondly, we construct a framework of mobile context-aware recommendation system in line with the characteristics of mobile terminal devices and mobile context-aware data. Then, we build a nested key-value storage model and an up-to-date algorithm for mining mobile context-aware sequential pattern, in order to find both the user's long-term behavior pattern and the new trend of his recent behavior, to predict user's next behavior. Lastly, we discuss the difficulties and future development trend of mobile context-aware recommendation system.

Keywords: Mobile context-aware · Long-term behavior pattern · Short-term behavior pattern · Recommendation system

1 Introduction

Currently, smartphone has become a kind of common portable intelligent terminal for people. Due to the fact that smartphone owns a variety of highly integrated sensors with mature technology, it has become the most suitable mobile platform for the development of context-aware applications. The combination of intelligent communication technology and sensor technology has driven the demand for the development of context-aware adaptive services on the mobile terminal, which collects the user's contexts and their changes, initiatively provides the user personalized information and services, so as to satisfy his needs. For example, the smartphone plays the user's favorite songs automatically when he takes the subway.

Applications of Context-aware system involve e-learning [1, 2], tourism [3, 4], transportation [5], health care [6, 7], entertainment [8] and so on. Recently, many system

© Springer Nature Singapore Pte Ltd. 2017
B. Zou et al. (Eds.): ICPCSEE 2017, Part II, CCIS 728, pp. 78–93, 2017.
DOI: 10.1007/978-981-10-6388-6_7

frameworks have been proposed. Context toolkit [9] which provides support for acquiring the context data from the heterogeneous sensors, but does not support fusion that is essential for context-aware system. Context-aware middleware [10, 11], serves as a software layer to abstract the heterogeneity of the lower layer and ease the complexity of developing a higher layer, but it ignores preservation of historical context data for further use or queries. Context agent architecture [12, 13], which usually requires multiple agents as well as a mediator to negotiate with them. Cloud-based context-aware infrastructure [14–16], relaxes the memory by making the cloud responsible for the storage and recognition of context, but it may consume more resources than is expected to save. As for existing frameworks of mobile context-aware application [17–19], left a lot of context management and computational burden on smart phones, such as, neither taking into account of the limited computing resources, storage space nor considering smart phone as the main context information provider.

Hence, this paper aims to propose a new mobile context-aware framework for providing the user of smart phone with personalized services, which is based on the user's behavior pattern and the current context. The contributions of this paper are listed as follows:

(1) We propose the definition of mobile context-aware computing, which identifies the difference among raw data, low level context and high level context, and then we divide the mobile context into four categories, it specifies our work.

(2) We present a general framework of mobile context-aware recommendation system, which can make the intelligent mobile terminal not only collect all kinds of sensor data, but also reason and apply contexts under complex context-aware conditions.

(3) We construct a non-relational data description model based on nested key-value, which is applicable to mobile intelligent terminal where multi-source and heterogeneous data grows rapidly while the storage capacity is limited.

(4) We design a sequential pattern mining algorithm, which incorporates the new concepts, classification and non-relational model to discover user's consistent behavior pattern and their recent changes from the multidimensional context and behavior information, it offers efficient and flexible instance of framework.

The rest of paper is organized as follows: Sect. 2 introduces the theory of context awareness and discusses the context-aware definition, classification and context-aware computing in the mobile terminal; Sect. 3 puts forward our new framework and key technologies of a mobile context-aware system, such as model and algorithm; In Sect. 4, two case studies are demonstrated so as to illustrate the usefulness and applicability of the proposed system framework, the effectiveness and the efficiency of the improved algorithm and mode; Sect. 5 concludes the research and discusses future research direction.

2 Definition and Classification of Mobile Context-Awareness

Context-aware, firstly proposed by Schilit et al. [20] in 1994, originates from pervasive computing. Many definitions of context have been proposed in various research fields according to the corresponding application [21]. The common definition, proposed by

Dey [22], describes the state of an entity's information. An entity can be a person, a place, or an object that is related to the interaction between the user and the application. Context-aware computing [23] is defined as automatic adjusting or adapting to the system application according to the user's contexts; context-aware system [24] is a system that utilizes context information to provide users with information or services.

This paper argues that mobile context-aware computing actively apperceives information of entity and changes of the information through built-in sensors of the mobile terminal, and then provides user services relate to the current context based on information management and processing. For instance, the smart phone detects ambient noise levels and automatically adjusts the volume of the ring when the user takes the subway.

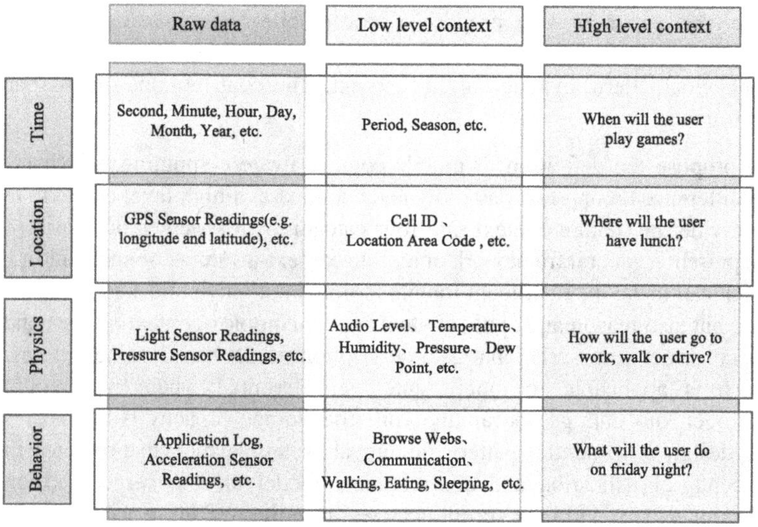

	Raw data	Low level context	High level context
Time	Second, Minute, Hour, Day, Month, Year, etc.	Period, Season, etc.	When will the user play games?
Location	GPS Sensor Readings(e.g. longitude and latitude), etc.	Cell ID, Location Area Code, etc.	Where will the user have lunch?
Physics	Light Sensor Readings, Pressure Sensor Readings, etc.	Audio Level, Temperature, Humidity, Pressure, Dew Point, etc.	How will the user go to work, walk or drive?
Behavior	Application Log, Acceleration Sensor Readings, etc.	Browse Webs, Communication, Walking, Eating, Sleeping, etc.	What will the user do on friday night?

Fig. 1. Classification of mobile context.

Researchers had proposed different context classification based on different perspectives [25, 26]. We suggest that data obtained directly from the sensor is raw data and information obtained through processing is low level context, such as location, time and behavior, the information found in the low level context is high level context. For example, readings of GPS sensor are regarded as the raw data, and the location expressed from readings is low level context, and the user's behavior reasoned by the geographical position is high level context.

In this paper, the mobile context is divided into four categories: time, location, physics and behavior, each of them can be further classified according to its specific characteristics, as shown in Fig. 1.

A detailed description is listed as follows:

(1) Time, description about the temporal information;
(2) Location, description about the user's location;
(3) Physics, description about the natural environment characteristics of the user;
(4) Behavior, description about the user's complete information of the service state or the activity concerning the current task.

3 Framework of Mobile Context-Aware Recommendation System

In this section, we analyze the basic concepts and requirements of context-aware framework, and then design a general framework for context-aware system that conforms to the mobile environment.

The main challenges of mobile context-aware recommendation system contain three points: firstly, original sensor signal without preprocessing and interpretation is usually useless, therefore, it is necessary to carry out cleaning, fusion and interpretation for sensor data; secondly, the data description model needs a strong expression ability considering multi-source heterogeneity and the spatiotemporal characteristics of mobile context-aware data; thirdly, mobile context is rich in attributes and the corresponding reasoning process is very complex.

The mobile context-aware recommendation system which we construct here is a human centered computing system, it automatically senses the user's environment information and its variation through the sensor technology and mobile technology, infers the high level context information from the low level context, predicts the user's intention by matching the user's current context, it initiatively provides services related with the current context and improves the quality of user's experience. The framework of mobile context-aware recommendation system is shown in Fig. 2.

The mobile context-aware recommendation system mainly includes four modules: context acquisition, context modeling, context reasoning and context application. The functions of each module are listed as follows:

(1) Context acquisition module: It collects original data through mobile intelligent devices of various embedded sensors, including physical sensors and virtual sensors. Data processing includes cleaning, feature fusion, aggregation and so on.
(2) Context modeling module: It expresses the context information in an unified and proper format which can be identified by the computer, and describes semantic meaning of the data.
(3) Context reasoning module: It learns modeled context in order to obtain the implicit semantics hidden behind the relevant features of the context, including user's services and behavior prediction, eventually, we can obtain high level context information from the low.
(4) Context application module: It adopts adaptive action by a combination of high level context information and the user's current context.

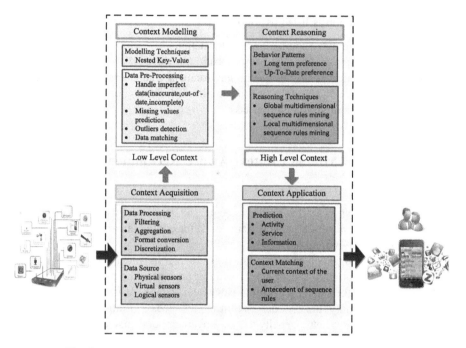

Fig. 2. Framework of mobile context-aware recommendation system.

In addition, the sensing system should also have the functions of transmission control, access control and security privacy protection. However, these are not related to our research, therefore we will not pay attention to those.

3.1   Context Acquisition

The way of context acquisition can be divided into three types: explicit, implicit and reasoning.

(1) The explicit acquisition: It obtains the user's data through direct inquiry, special system setting and web survey. This method is simple and direct, it can exactly reflect the information and relatively obtain accurate and reliable result; but this method needs users take time to express their contexts which are heterogeneous, so it is invasive and inflexible.

(2) The implicit acquisition: It automatically tracks the physical or virtual sensing devices, perceives the user's related data, and then obtains the user's information. This method does not require users to participate in the interaction process. However, the data may not correctly reflect the true context, because the user may be disgusted with automatic tracking, sensing and reject to use current mobile context-aware recommendation systems.

(3) The inference acquisition: It obtains information commonly by data mining and statistical techniques and conducts data analysis, acquires information which is based on constructing prediction model, training dataset and others. For example, the user does not want to be disturbed by the news notification between 10 p.m. and 6 a.m. because of sleeping.

For the context-aware system on a mobile terminal, the information acquisition adopts an implicit way to record journey between home and workplace. In this process, the sensors record the acceleration, atmospheric pressure, temperature, humidity, the sound of the surroundings and others. GSM network cell identification code (Cell ID) and location area code (LAC) records the user's location, all data is collected once a second, the low level context information is obtained by processing the raw sensor data as shown in Fig. 3.

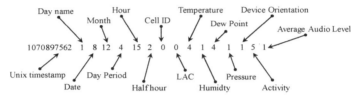

Fig. 3. The low level context information.

The advantage of using the built-in sensors to obtain the information is that it does not require the user to spend time in answering the question, however, the disadvantage is that the data may not be accurate, and there are missing values. Therefore, the collection of context information needs data cleaning, interpretation, description and other data processing techniques to carry out storage, reasoning and other operations.

3.2   Context Model

Mobile intelligent devices generate and accumulate a large amount of dynamic and heterogeneous data, which has obvious spatiotemporal characteristics. How to establish an unified abstract logic model, in order to deal with and store the context information of heterogeneous source, is one of the hotspot in the study of context-aware computing. Existing models of context information mainly contain key-value, markup scheme, graphical, object based, logic based, ontology based and so on.

Key-value model is the most basic data description model of computer software design, simple but very practical. In the key-value model, a global and unique identifier is assigned to all entities. The key is the information in the current environment of these entities, while the values are the entity and the information.

The context data of mobile context-aware recommendation system is continuous and real-time, various in attribute and abundant in temporal properties; but the behavior is occasional and very sparse by comparison, results in the data amount of context data and behavior is not balanced. In addition, with the development of context-aware applications, the database server of the context-aware system not only operates and manages database, but also responds to the high concurrent requests of the client.

Traditional database system mainly uses the relational model to store formatted data table, assigns all fields for each tuple, such a structure is easy to operate with multiple tables, but cannot support high concurrent access, which limits the whole system scalability. It will cause the system slow and instable when the amount of data increases exponentially. The structure of key-value model, as the representative of non-relational database, is not fixed. There is no need to associate tables with a relational database as multi-table query, just take the corresponding value out of the key even if getting the user's different information. In the key-value model, all entities and information of those entities in the current environment is assigned a globally unique identifier, known as key, entities and information such as the data itself is known as value, the key and the value must be one-to-one. Generally, type of the key is unified, it may be an integer or a character, and types of the value can be diverse according to the actual application, even empty.

RecordTable

RecordNum	LocationNum	DayNum	PeriodNum	TimeStamp
1	1	1	2	1001
1	1	1	2	1002
1	2	1	3	1003
2	3	2	2	1004
2	2	2	2	1005
......	...		...	...

LocationTable

LocationNum	Location
1	SubWay
2	BusStation
3	CoffeeBar

DayTable

DayNum	Day
1	Monday
2	Tuesday
3	Wednesday
4	Thursday
5	Friday
6	Saturday
7	Sunday

PeriodTable

PeriodNum	Period
1	Night
2	Morning
3	Afternoon
4	Evening

Nested Key-Value Model

Multidimensional Item	Timestamp
SubWay,Monday,Morning	{1:[1001,1002],......}
BusStation,Monday,Afternoon	{1:[1003],......}
CoffeeBar,Tuesday, Morning	{2:[1004],......}
BusStation, Tuesday, Morning	{2:[1005],......}
......	......

(a) Relational table storage   (b) Nested key-value storage

Fig. 4. Comparision between relational table and nested key-value storage.

We propose an improved data storage model based on nested key-value in the context modeling module, it collects the context-aware data and user's behavior as the key from smart phones, the corresponding value is another key-value pairs whose key is the sequence number and the value is the timestamp of context data in that sequence. A relational table such as Fig. 4(a) shows is converted to nested key-value storage model as shown in the Fig. 4(b). Multidimensional item described by location, day and period is the parent key, the value corresponding to the key is another pair of key-value, the sub key is the sequence number, and the sub value is the time information. The key-value stored in the serialization forms in the file can quickly query the key and its value.

Temporal features, which are obvious in the mobile context-aware system, can be stored by combining with the timestamp in key-value. In addition, key-value can be stored based on the column to compress the input volume due to a lot of data (such as air pressure, humidity, etc.) stays stable or changes a little over a period of time, it just needs to record one value on continuous time point, thus effectively reducing the space overhead.

3.3   Context Reasoning

Context reasoning is a method of deriving high level new knowledge from the existing low level context. Existing reasoning techniques can be divided into six categories: supervised learning, unsupervised learning, rule based reasoning, fuzzy logic, ontology based and probability logic. Due to the limits of mobile intelligent terminal in memory, power and other fields, a concise and powerful reasoning technology is needed to find user's consistent behavior habits and their recent trends. The Rule based method has become the most popular technology for its straightforward and understandability in knowledge reasoning, it is very suitable for the generation of high level information from the low level [27]. However, the traditional mining methods only attach importance to the order without considering the location, time, weather and other attributes, so it is not suitable for the mobile context-aware recommendation system to mine user's behavior pattern. In reasoning module, we construct a multidimensional sequential pattern mining algorithm UTDMSP (Up-to-Date multidimensional sequential pattern) to learn context information and user's behavior sequential pattern, it can not only discover the user's long-term behavior habits, but also capture the new trends of user's behavior or habits. In this algorithm, the maximal time interval between two adjacent items of sequence is termed as gap and the maximal time interval of the first item and the last item is termed as the width.

The algorithm UTDMSP is described as follows:

Step 1: We map the sequence database into nested key-value storage model after scanning it once, the key of the parent model is the sequence described by multidimensional context data, the value corresponding to the key is another key-value pair whose sub-key is the serial numbers of sequence and sub-value is the temporal information corresponding to the sequence.

Step 2: We join sequence Seq_1 with Seq_2 only if the subsequence obtained by dropping the first item of Seq_1 is the same as the subsequence obtained by dropping the last item of Seq_2. The candidate sequence is generated by extending the sequence Seq_1 with the last item of Seq_2, its serial number set is the intersection of Seq_1's and Seq_2's.

Step 3: We remove the candidate sequence which has a non-frequent subsequence, because it could not be frequent according to downward closure property of Apriori algorithm. We compute global support of each candidate sequence under both gap and width constraints in the sequence database $SeqDB = \{Seq_1, Seq_2,..., Seq_N\}$, if it is greater than or equal to the minimum support threshold, join it in frequent itemsets; otherwise, validate whether the support is frequent in a recent period, if true, we also join it in the frequent itemsets and record the serial numbers of sequences which it belongs.

Step 4: Repeat step 2–3 until the candidate item is not produced.

Step 5: We calculate the confidence of frequent sequences, if it meets the minimum confidence threshold and the last item of the sequence is one of user's behavior, then generate a rule and put it into the rule set R.

After that, we store the user's behavior sequence rules that are mined by the multidimensional sequential pattern mining algorithm UTDMSP from the knowledge base, wait for the context application module to trigger behavior prediction by matching user's current context with his behavior sequence rules.

3.4   Context Application

Häkkilä and Mäntyjärvi [28] divided the context application into two classes: the first is information retrieval by utilizing acquired data, and the second is device adjustment so as to respond to user' needs. Dey et al. [25, 29] assorted context application for three groups: active perception, passive perception, and personalization. Gu et al. [30] distributed context application into two kinds: the first focuses on physical perception, such as the location-based services, and the second concentrates on more abstract context application based on user-oriented behavior, and social, organizational environment, such as information retrieval.

In the context-aware system framework of this paper, the context application module constantly detects the user's context information, and compares the acquired information with user's behavior sequence rules of context reasoning module. When user's current context information matches the antecedent of the behavior rules on the condition of time interval and duration thresholds, we make the subsequent of the rules as the predication of user's next behavior, then the mobile context-aware recommendation system can initiatively provide services to user. In the process of matching the rules, the rule matching is failed whether the gap or the width exceeds a threshold.

4   Application of Mobile Context-Aware Recommendation System

In this section, we will illustrate how to use the framework of mobile context-aware recommendation system and related methods to recommend services in some fields such as smart phone service and user's behavior recognition.

Mobile context can be represented on different granularity according to the number of attributes contained in the mobile context, such as location, noise level, time and so on. Generally speaking, the more attributes in context, the more specific the context information. In this paper, we use four dimensional data with location, day, period and behavior, as shown in Table 1.

The sequence rules obtained using the nested key-value model and the multidimensional context-aware sequential pattern mining algorithm are shown in Table 2. Rule 1 represents a user attends academic discussion activities in the teaching building on Tuesday morning; Rule 3 represents a user often browses a website after visiting canteen and laboratory on Friday morning.

Table 1. Data description

Dimensions	Value domains/format
Location	'Area Code. Cell ID'/'[0–14] Location ID'
Day	[1–7] Monday = 1, Tuesday = 2, etc.
Period	[1–4] Night, Morning, Afternoon, Evening
Behavior	'a. [0–31] Application'
	'b. [0–13] Web service'
	'c. [0–4] Communication'
	'd. [0–12] Activity'

Table 2. Multidimensional sequence rules

No.	Multidimensional sequence rules	Data sets
1	('6', 3, 4) → '5'	Activity_log
2	('2', 4, 1) → '4'	Activity_log
3	('2.11', 5, 2), ('2.14', 5, 2) → 'b.2'	Nokia Context Data
4	('2.49', 2, 2), ('3.48', 2, 2), ('3.46', 2, 2) → 'a.12'	Nokia Context Data

The results are evaluated by precision, recall and F1. The precision illustrates the probability that the predicted interactive behavior happened actually, the recall presents the probability that the interactive behavior can be predicted, and the F1 reflects the overall index of two previous methods. All experiments were set minimum support MinSupp = 0.09, minimum confidence MinConf = 0.7.

4.1 Forecast of User's Activity

Activity Log Dataset [31], which is collected by the pervasive computing group of Northwestern Polytechnical University, records user's daily social activities by mobile phone and online blog. This dataset includes: type of activity (such as meeting, running, shopping, etc.), activity location (such as conference rooms, stadiums, shops, etc.), and activity time. We look for frequent subsequences from mobile context data and user's daily social activities based on the framework and algorithm UTDMSP proposed in this paper, then we can discovery useful behavior pattern of the user. For example, the user usually has a discussion with his teacher and schoolfellows on Monday's morning and takes a walk in the playground after dinner almost every day.

In the course of experiment, we maintain minimum support and minimum confidence unchanged, then we change gap and width thresholds: gap = 7200 s, width = 18000 s; gap = 10800 s, width = 36000 s; gap = 18000 s, width = 36000 s; gap = 36000 s, width = 72000 s, the result is presented in Fig. 5. The social activity prediction accuracy rate decreased, which lead to the decrease of the F1 value. The reason for the decline of accuracy was that the system reasoning module found more rules since we gradually relaxed two kinds of time thresholds, but some predictions of social activities did not happen.

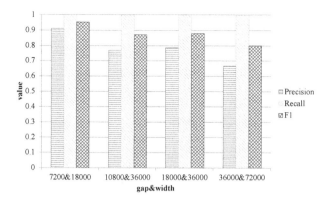

Fig. 5. Forecast of user's activities.

4.2 Forecast of Smart Phone Service

Nokia Context Data [32], which is collected by NOKIA Research Center, records the user's environment and smart phone services. For example, application program, communication service, and web browsing. The interaction behavior between the user and the mobile phone, such as web browsing, music listening, etc., can be captured and collected by the smart phone. The user's interaction information can be used to learn the user's behavior pattern, understand the user's preferences and change the passive waiting into the active recommendation.

In the course of experiment, we maintain minimum support and minimum confidence unchanged and set gap = 180 s and width = 300 s during the rule generation phase, but modify gap = 180 s and width = 300 s, gap = 300 s and width = 600, gap = 900 s and width = 1200 s, gap = 1800 s and width = 3600 s during the matching phase, the result is presented in Fig. 6. The accuracy, recall and F1 value increased gradually with the increase of the gap and width. Multidimensional sensing data can be used to accurately obtain the user's behavior habits because of the use of a

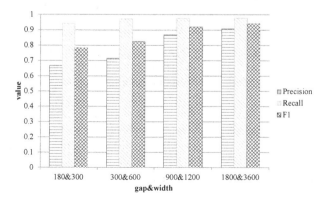

Fig. 6. Forecast of smart phone services.

variety of sources of context data. However, the more specific of the users' behavior pattern, the lower probability of the reemergence. Therefore, the rule matching will be restricted. In other words, some of the sequence rules will be temporarily shelved because there is no corresponding matching, and it is hard to prove the validity of the rule. So if we relax time constraints, we can achieve better evaluation results.

4.3   Performance Comparison

In order to examine the ability of the proposed framework, we apply mining algorithm with the existing work in literature, namely MSP [33], All the experiments are implemented on Activity Log Dataset, in the case of varying gap and width thresholds. The experimental results show that the UTDMSP algorithm always finds more multidimensional sequence rules than the MSP algorithm, as shown in Fig. 7. The same is true with respect to precision, as shown in Fig. 8.

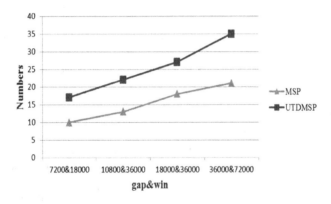

Fig. 7. Numbers of sequential rules.

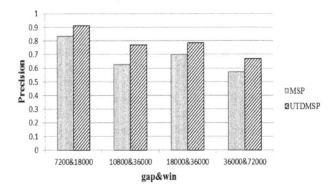

Fig. 8. Precision under different thresholds.

5 Summary and Challenges

Firstly, our research analyzed the status of acquisition, modeling and reasoning of the context-aware computing, and then put forward a new framework of mobile context-aware recommendation system; secondly, our research carried on a thorough discussion about each function module of the framework, and then proposed a nested key-value storage model so as to construct a multidimensional sequential pattern mining algorithm UTDMSP; finally, our research mined user's behavior pattern in mobile environment according to our proposed framework and improved technology, triggered services and pushed information combining in the light of the user's current context. The framework of mobile context-aware recommendation system we proposed is efficient in terms of scalability, reliability, flexibility and repeatability.

Mobile context-aware services, which integrate universality, flexibility and personalization, have received an increasing number of attentions from industry and research institutions. Especially in recent years, mobile context-aware services have pioneered tremendous business opportunities, and have been successfully applied to the fields of navigation, tourism, e-commerce, healthcare and others; context-aware services will also become a new and important growth point of profit in mobile commerce industry.

As a new research field, there are many research directions which can be further studied, which mainly includes:

(1) Data accuracy

Intelligent terminal collects the original data and the user's interactive information through the built-in sensors. Though the user's behavior is sparse, the collected context data is continuous. Sometimes the user appears in a scene just in a short time, such as the position a car passed by, but the smartphone's sensors have no time to collect corresponding information [34]. Besides, there may be some other conditions, temperature difference between the mobile phone and the external environment. For example, temperature is higher when the phone in the bag. Therefore, the accuracy of the information under complex conditions is the basic problem of context-aware computing.

(2) Service startup

Mobile terminal context-aware system emphasizes the timeliness of service. In case of less collected data, it is bound to be invalid if we carry out data mining and behavior's prediction, which will interfere with the user's normal behavior; If we start after collecting a huge amount of data, the overhead produced by computing will greatly affect the efficiency of the system; Besides, waiting too long will affect the user's experience feelings. Then, what is the right time to carry out the user behavior's prediction is a very practical problem in the application of mobile context-aware recommendation system.

(3) Attribute selection

There are different degrees of impact on user's preference for mobile environment. In the absence of explicit rules, we should use as little properties as possible to describe as much information as possible in order to reduce the computational complexity [35, 36]. The accuracy of the inference result is not necessarily improved with much more context. How to choose the most representative attribute is also need to study.

(4) Group recommendation

Services of the mobile context-aware recommendation system constructed in our research are for individuals. If you want to recommend a restaurant for a group of friends, you need to consider a group of people's preferences for recommendation, rather than just considering one personal preference [37, 38]. However, how to deal with the conflicts among the group members' preferences so as to obtain the accurate group preference and to complete the services recommendation for a group of users, is a very meaningful research direction.

(5) Negative sequential pattern

Negative sequential pattern (NSP) (such as missing medical treatments) are critical and sometimes much more informative than positive sequential pattern (PSP) (e.g. using a medical service) in many intelligent systems and applications such as intelligent transport systems, healthcare and risk management, as they often involve non-occurring but interesting behavior [39, 40]. However, discovering NSP is much more difficult than PSP due to the significant problem complexity caused by non-occurring elements, it is thus important to develop efficient framework and approaches.

Acknowledgments. This work was supported by the Science Research Program of the Education Department of Liaoning Province, China (Grant No. 2016JYT01), National Social Science Foundation of China (Grant No. 15BYY028) and the Open Program of State Key Laboratory of Software Architecture (SKLSAOP1703).

References

1. Verbert, K., Manouselis, N., Ochoa, X., Wolpers, M., Drachsler, H., Bosnic, I., Duval, E.: Context-aware recommender systems for learning: a survey and future challenges. IEEE Trans. Learn. Technol. **5**(4), 318–335 (2012)
2. Hsu, T.Y., Chiou, C.K., Tseng, J.C.R., Hwang, G.J.: Development and evaluation of an active learning support system for context-aware ubiquitous learning. IEEE Trans. Learn. Technol. **9**(1), 37–45 (2016)
3. Meehan, K., Lunney, T., Curran, K., McCaughey, A.: Context-aware intelligent recommendation system for tourism. In: 2013 IEEE International Conference on Pervasive Computing and Communications Workshops, vol. 18, pp. 328–331 (2013)
4. Colomo-Palacios, R., García-Peñalvo, F.J., Stantchev, V., Misra, S.: Towards a social and context-aware mobile recommendation system for tourism. Pervasive Mob. Comput. (2016)
5. Vahdat-Nejad, H., Ramazani, A., Mohammadi, T., Mansoor, W.: A survey on context-aware vehicular network applications. Veh. Commun. **3**, 43–57 (2016)
6. Sriram, R., Geetha, S., Madhusudanan, J., Iyappan, P., Venkatesan, V.P., Ganesan, M.: A study on context-aware computing framework in pervasive healthcare. In: Proceedings of the 2015 International Conference on Advanced Research in Computer Science Engineering & Technology, vol. 39 (2015)

7. Forkan, A.R.M., Khalil, I., Tari, Z., Bouras, A.: A context-aware approach for long-term behavioural change detection and abnormality prediction in ambient assisted living. Pattern Recogn. **48**(3), 628–641 (2015)

8. Colombo-Mendoza, L.O., Valencia-García, R., Rodríguez-González, A., Alor-Hernández, G., Samper-Zapater, J.J.: RecomMetz: a context-aware knowledge-based mobile recommender system for movie showtimes. Expert Syst. Appl. **42**(3), 1202–1222 (2015)

9. Dey, A.K., Abowd, G.D.: The context toolkit: aiding the development of context-aware applications. In: Workshop on Software Engineering for Wearable and Pervasive Computing, pp. 431–441 (2000)

10. Liang, G., Cao, J.: Social context-aware middleware: a survey. Pervasive Mob. Comput. **17**, 207–219 (2015)

11. Li, X., Eckert, M., Martinez, J.F., Rubio, G.: Context aware middleware architectures: survey and challenges. Sensors **15**(8), 20570–20607 (2015)

12. Chen, H., Finin, T., Joshi, A.: An ontology for context-aware pervasive computing environments. Knowl. Eng. Rev. **18**(03), 197–207 (2003)

13. Ranganathan, A., Campbell, R.H.: A middleware for context-aware agents in ubiquitous computing environments. In: ACM/IFIPUSENIX International Middleware Conference, pp. 143–161 (2003)

14. Wan, J., Zhang, D., Zhao, S., Yang, L., Lloret, J.: Context-aware vehicular cyber-physical systems with cloud support: architecture, challenges, and solutions. IEEE Commun. Mag. **52** (8), 106–113 (2014)

15. Naqvi, N.Z., Moens, K., Ramakrishnan, A., Preuveneers, D., Hughes, D., Berbers, Y.: To cloud or not to cloud: a context-aware deployment perspective of augmented reality mobile applications. In: Proceedings of the 30th Annual ACM Symposium on Applied Computing, pp. 555–562 (2015)

16. Forkan, A., Khalil, I., Tari, Z.: CoCaMAAL: a cloud-oriented context-aware middleware in ambient assisted living. Future Gener. Comput. Syst. **35**, 114–127 (2014)

17. Coppola, P., Mea, V.D., Gaspero, L.D., Mizzaro, S., Scagnetto, I., Selva, A.: Context-aware mobile applications on mobile devices for mobile users. In: Proceedings of the International Workshop on Exploiting Context Histories in Smart Environments (2005)

18. Korpipää, P., Mntyjrvi, J., Kela, J., Keranen, H., Malm, E.J.: Managing context information in mobile devices. IEEE Pervasive Comput. **2**, 42–51 (2003)

19. Hofer, T., Schwinger, W., Pichler, M.M., Leonhartsberger, G., Altmann, J., Retschitzegger, W.: Context-awareness on mobile devices-the hydrogen approach. In: Proceedings of the 36th Annual Hawaii International Conference on System Sciences, pp. 292–302 (2002)

20. Schilit, B., Adams, N., Want, R.: Context-aware computing applications. In: Workshop on Mobile Mobile Computing Systems and Applications, pp. 85–90 (1994)

21. Gu, J.Z.: Context-aware computing. J. East China Normal Univ. (Nat. Sci. Ed.) **5**, 1–20 (2009)

22. Dey, A.K.: Providing architectural support for building context-aware applications. Georgia Institute of Technology, vol. 25, pp. 106–111 (2000)

23. Ryan, N., Pascoe, J., Morse, D.: Enhanced reality fieldwork: the context-aware archaeological assistant. Comput. Appl. Archaeol. **750**, 269–274 (1999)

24. Schilit, B., Theimer, M.: Disseminating active map information to mobile hosts. IEEE Netw. **8**, 22–32 (1994)

25. Abowd, G.D., Dey, A.K., Brown, P.J., Davies, N., Smith, M., Steggles, P.: Towards a better understanding of context and context-awareness. In: Gellersen, Hans-W. (ed.) HUC 1999. LNCS, vol. 1707, pp. 304–307. Springer, Heidelberg (1999). doi:10.1007/3-540-48157-5_29

26. Perera, C., Zaslavsky, A., Christen, P., et al.: Context-aware computing for the internet of things: a survey. IEEE Commun. Surv. Tutor. **16**, 414–454 (2014)

27. Lin, T.N., Lin, P.C.: Performance comparison of indoor positioning techniques based on location fingerprinting in wireless networks. In: IEEE International Conference on Wireless Networks, Communications and Mobile Computing, vol. 2, pp. 1569–1574 (2005)

28. Häkkilä, J., Mäntyjärvi, J.: Developing design guidelines for context-aware mobile applications. In: Proceedings of the 3rd International Conference on Mobile Technology, Applications & Systems, p. 24. ACM (2006)

29. Dey, A.K.: Understanding and using context. Pers. Ubiquit. Comput. 5, 4–7 (2001)

30. Gu, T., Pung, H.K., Zhang, D.Q.: A service-oriented middleware for building context-aware services. J. Netw. Comput. Appl. 28(1), 1–18 (2005)

31. Flanagan, A.: Nokia Context Data, 13 December 2010. http://www.pervasive.jku.at/Research/Context_Database/index.php. Accessed 2004

32. Guan, D., Yuan, W., Lee, S., Lee, Y.K.: Context selection and reasoning in ubiquitous computing. In: IEEE International Conference on Intelligent Pervasive Computing, pp. 184–187 (2007)

33. Tang, H., Liao, S.S., Sun, S.X.: A prediction framework based on contextual data to support mobile personalized marketing. Decis. Support Syst. 56, 234–246 (2013)

34. Jenkins, M.P., Gross, G.A., Bisantz, A.M., Nagi, R.: Towards context aware data fusion: modeling and integration of situationally qualified human observations to manage uncertainty in a hard+ soft fusion process. Inf. Fusion 21, 130–144 (2015)

35. Pitarch, Y., Ienco, D., Vintrou, E., Bégué, A., Laurent, A., Poncelet, P., Teisseire, M.: Spatio-temporal data classification through multidimensional sequential patterns: application to crop mapping in complex landscape. Eng. Appl. Artif. Intell. 37, 91–102 (2015)

36. Zheng, Y., Mobasher, B., Burke, R.: Integrating context similarity with sparse linear recommendation model. In: Ricci, F., Bontcheva, K., Conlan, O., Lawless, S. (eds.) UMAP 2015. LNCS, vol. 9146, pp. 370–376. Springer, Cham (2015). doi:10.1007/978-3-319-20267-9_33

37. Zapata, A., Ndez, V., Ctor, H., Prieto, M.E., Romero, C.: Evaluation and selection of group recommendation strategies for collaborative searching of learning objects. Int. J. Hum Comput Stud. 76, 22–39 (2015)

38. Salehi-Abari, A., Boutilier, C.: Preference-oriented social networks: group recommendation and inference. In: Proceedings of the 9th ACM Conference on Recommender Systems, pp. 35–42 (2015)

39. Cao, L., Yu, P.S., Kumar, V.: Nonoccurring behavior analytics: a new area. IEEE Intell. Syst. 30(6), 4–11 (2015)

40. Cao, L., Dong, X., Zheng, Z.: e-NSP: efficient negative sequential pattern mining. Artif. Intell. 235, 156–182 (2016)

Build Evidence Chain Relational Model Based on Chinese Judgment Documents

Siyuan Kong[1,2], Yemao Zhou[1,2], Jidong Ge[1,2(✉)], Zhongjin Li[1,2],
Chuanyi Li[1,2], Yi Feng[1,2], Xiaoyu Zhou[1,2], and Bin Luo[1,2]

[1] State Key Laboratory for Novel Software Technology, Nanjing University,
Nanjing 210093, China
gjdnju@163.com
[2] Software Institute, Nanjing University, Nanjing 210093, China

Abstract. The reasoning of judgment documents is the touchstone of justice. Attaching importance to the reasoning of judgment documents is essentially the embodiment of judiciary civilization. In order to promote the reform of judgment documents reasoning and improve the level of it, the technology of automated judgment documents reasoning evaluation has to be studied on. How to build evidence chain relational model is the basis and key to this technology. An approach is proposed to build evidence chain relational model based on Chinese judgment documents. Using automated text preprocessing for Chinese judgment documents creates semi-structured XML documents and extracts evidence set and fact set. The method of key elements extraction is used to obtain the keywords of evidence and facts. Calculating the degree of association can work out the connection points of evidence chain relational model. Tabular display and graphical display of evidence chain relational model can be realized.

Keywords: Chinese judgment documents · Judicial evidence · Fait juridique · Reasoning of judgment documents · Evidence chain relational model

1 Introduction

Although the reasoning of judgment documents cannot solve the actual problems of judicature, it is the touchstone of justice [1]. The reasoning of judgment documents is not purely a legal or technical issue, but it is a political and institutional issue. Attaching importance to the reasoning of judgment documents is essentially the embodiment of judiciary civilization, especially the embodiment of judicial democracy and judicial rationality [2]. In the latest reform outline and project of information development released by the Chinese People's Court, promoting the reform of judgment documents reasoning and improving the level of judgment documents reasoning are both mentioned. In order to achieve the goals, we have to study on the technology of automated judgment documents reasoning evaluation.

The reasoning of judgment documents contains four aspects, which are fact of case, law of case, emotion of document and style of writing. The aspect of fact focuses on the description of evidence, fact and the relationship between them [3]. For example, how can judicial evidence become the basis of judgement? Judicial evidence must be

B. Zou et al. (Eds.): ICPCSEE 2017, Part II, CCIS 728, pp. 94–107, 2017.
DOI: 10.1007/978-981-10-6388-6_8

objective, relevant and legal. These three characteristics of judicial evidence are not only related to the theoretical problem, but also should be reflected in judgment documents [4]. The collection of evidence, the relationship between evidence and fact and the relationship between different evidence is called evidence chain. How to build evidence chain relational model is the basis and key to automated judgment documents reasoning evaluation.

The Study of judgment documents reasoning evaluation is mainly based on the extraction and analysis of evidence chain. Our research uses the legal language (language with the characteristics of legal terminology) and Chinese natural language processing technology, extracts key paragraphs of reasoning, builds evidence chain relational model and uses computer visualization technology to display the model. To highly efficiently handle historical big data, we need to use some big data processing technologies such as Hadoop or Spark.

In this paper, we propose an approach to build evidence chain relational model based on Chinese judgment documents. Our work includes: (1) propose an automated method to do text preprocessing based on Chinese judgment documents, convert unstructured judgment documents into semi-structured XML format files, extract evidence set and fact set, (2) propose a strategy of extracting key elements, obtain the keywords of evidence and facts, (3) propose an approach to create the connection points of evidence chain relational model, including finding the relationship between evidence and fact and finding the relationship between different evidence, (4) realize tabular display and graphical display of the evidence chain relational model.

To evaluate the performance of our approach, we use civil documents, criminal documents and administrative documents to test the system. We will prove the contribution of this paper by answering the research questions as follows:

(1) How to do the automated text preprocessing for Chinese civil documents, criminal documents and administrative documents? Which information should be included in the preprocessed semi-structured files?
(2) Which kinds of key elements should be selected? How to extract those key elements?
(3) How to find the relationship between evidence and fact? How to find the relationship between different evidence?
(4) How to display evidence chain relational models clearly?

The remainder of this paper is laid out as follows. Section 2 introduces related work of this paper. Section 3 introduces our approach in detail. Section 4 concludes with a discussion of future work.

2   Related Work

For a long time, the study of judgment documents had been limited to the relevant legal experts, until the development of computer automation technology. *Some Speculation about Artificial Intelligence and Legal Reasoning* published by Buchanan in 1970 marks the birth of a branch research on artificial intelligence and legal reasoning [5]. In recent years, as the artificial intelligence algorithm and natural language processing technology become a boom once again, a large number of domestic and international researches on judgment documents and semantic information spring up. The modeling of legal reasoning about evidence within general theories of feasible reasoning and argumentation was studied in 2003 [6]. An approach to formalize argumentative story-based analysis of evidence was proposed in 2007 [7]. In 2009, researchers explored the potential of using extended belief change operators for modeling the evolution of legal evidence [8]. Recent research has shown that argumentation can inform the construction of Bayesian networks. A position paper presents an investigation into the similarities, differences and synergies between Bayesian networks and argumentation diagrams and shows a first version of an algorithm to extract argumentation diagrams from Bayesian networks [9]. Because of the particularity of Chinese and Chinese legal system, many foreign research methods and achievements cannot be directly applied to the analysis of the Chinese judgment documents. In recent years, the domestic research on automated Chinese judgment documents processing gradually increases, research contents include the informationalized storage and management of archives, recognition and classification of judgment documents, detection and protection of privacy content in judgment documents and so on.

Evidence chain is a long-standing question in the field of law. In *Rationale of Judicial Evidence*, the author makes a thorough analysis of the rationality of judicial proof based on various kinds of judicial evidence, and puts forward the guidance of constructing reasonable evidence chain [10]. *The Study on Criminal Evidence Chain* is the representative work of the research on evidence chain in China. The author defines the basic conception of chain unit, main part of unit (chain unit body), key of unit (chain unit head) and connection point in evidence chain, and proposes different kinds of link, for example simple link, multiple link, net link, etc. [11]. There are also some researches on the combination of computer technology and evidence chain. A few researchers focus on the analysis of computer forensics, evidence from micro computer technology to refine the process chain of evidence, summarizing the main phase of formation of the acquisition and curing, identification, adjudication and archiving in the evidence chain [12]. An electronic judicial identification model based on three-dimensional trusted electronic evidence acquisition model is proposed. The model implements linear process control and evidence supervision. The model ensures evidence's own safety and legal effect on evidence supervision chain [13].

A very important step of our approach to build evidence chain relational model is extracting key elements. We refer to some existing research achievements during our work. The authors of *Extracting 5W1H Event Semantic Elements from Chinese Online News* propose a verb-driven approach to extract 5W1H (Who, What, Whom, When, Where and How) event semantic information from Chinese online news. This approach extracts event facts (i.e. 5W1H) by applying a rule-based method (verb-driven) and a supervised machine-learning method (SVM) [14]. The researchers who published *Chinese News Event 5W1H Semantic Elements Extraction for Event Ontology Population* propose a novel approach of 5W1H (Who, What, Whom, When, Where, How) event semantic elements extraction for Chinese news event knowledge base construction. The approach comprises a key event identification step, an event semantic elements extraction step and an event ontology population step [15]. Although the application scenarios of these papers are not as the same as this one, they are all based on Chinese text.

3 Approach

In this section, we present our approach to build evidence chain relational model based on Chinese judgment documents in detail as follows. Section 3.1 introduces evidence chain relational model and presents an overview of the flow diagram for Chinese judgment documents analysis. Section 3.2 introduces text preprocessing for Chinese judgment documents. Section 3.3 introduces the strategy of extracting key elements from evidence and facts. Section 3.4 introduces an approach to create the connection points. Section 3.5 introduces methods of model display.

3.1 Overview

Figure 1 presents the sample graph of evidence chain relational model. An evidence chain relational model has at least two different chain units. Each chain unit contains chain unit body and head. One chain unit body corresponds to certain evidence, and head is the key element of this evidence. Connecting several chain unit heads forms connection point which presents the relationship between different evidence. Evidence set is also connected to fact by connection point.

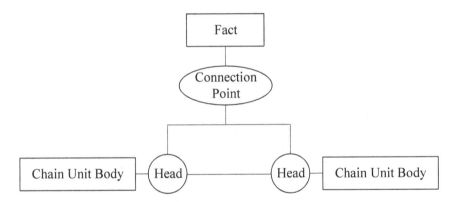

Fig. 1. Sample graph of evidence chain relational model

Figure 2 presents the flow diagram of Chinese judgment document analysis. Because the original judgment document is pure text format, in order to complete building evidence chain relational model, we have to do text preprocessing firstly. In Sect. 3.2.1 we will introduce the method we use to convert unstructured judgment documents into semi-structured XML format files and the evaluation results of this method. In the process of making court verdicts, the parties, agent ad litem and third parties can debate authenticity, legality, relevance of evidence they submit. The judge has to show the result of debate in his judgment document. Inadmissible evidence apparently can't be included in evidence chain relational model. We will introduce the approach to label admissibility of evidence in Sect. 3.2.2, and then we will introduce how to extract evidence set and fact set in Sect. 3.2.3.

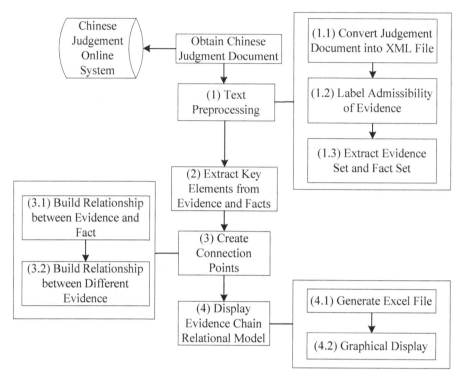

Fig. 2. Flow diagram for chinese judgment documents analysis

Connection points in evidence chain relational model are formed by connecting chain unit heads that mean key elements of evidence. So the most direct way to obtain correlation of different text is calculating contact ratio of key elements, therefore the strategy of extracting key elements is very important. Key elements have to embody characteristics of evidence and fact. We select 4W1H (What, Where, When, Who and How Much) as five key elements. In Sect. 3.3 we will introduce how to extract WHAT, WHERE, WHEN, WHO and HOW MUCH in turn and show the evaluation results of these methods.

As mentioned, how to organize statement while writing reasoning has no standard, so the judge may not express the relationship between evidence and fact it supports in his judgment document. We can obtain some clear relationships while we do text preprocessing, but unclear relationships need to be calculated. In Sect. 3.4.1 we will introduce the calculating method. We define the chain unit head as the key element of evidence, in other words, a chain unit head means a proof point. Each chain unit is allowed to have many heads and only one body. For example, a knife with fingerprints on it at the scene of the crime has two proof points. One is that the weapon of this crime is a knife, and another is criminal suspect's fingerprints. Different chain units may have same heads that embody the relationship between them. For example, an expert conclusion provided by an accreditation agency also proves fingerprints of criminal suspect. Section 3.4.2 will introduce the method to build the relationship between different evidence.

The final goal of building evidence chain relational model is to evaluate judgment document reasoning. For this purpose, the model should be easy to read, examine and evaluate. We propose two ways to display evidence chain relational model. The way of using Excel sheets is introduced in Sect. 3.5.1, and the way of using graphical interfaces is introduced in Sect. 3.5.2.

3.2 Text Preprocessing

In this section, we mainly focus on introducing the methods used to preprocess Chinese judgment documents.

3.2.1 Convert Judgment Document into XML File

According to case nature, judgment documents can be classified into three types: Civil, criminal and administrative. Moreover, judgment documents can be classified into certain types by trail procedure. For example, first instance, second instance and so on. For this reason, we set up six kinds of XML templates for all types of judgment documents.

First step of converting is paragraph splitting. According to the position of a paragraph and regex match results of keywords from first sentence; all paragraphs can be classified into seven parent sections: headline, party, litigation record, basics of case, trail process, verdict and end. Deeper analysis of each parent section is used for splitting child sections. For building evidence chain relational model, the most important parent section is basics of case. In first instance over civil judgment documents, this section includes seven child sections at most, such as plaintiff's appeal, defendant's defense, evidence section, fact section, etc. Take evidence section as an example, we are going to introduce how to identify whether a paragraph is evidence section in detail. If a paragraph contains "提出如下证据:", "证据如下:" or other similar phrases, it must be an evidence section. If a paragraph contains "证据", "证明", "证言" or other similar keywords, it may be an evidence section. In this case, other keywords need to be verified. If keyword "查明" is also contained in this paragraph, then this paragraph belongs to fact section rather than evidence section. In other cases, characteristics of context need to be considered. Furthermore, when we extract evidence sections and fact sections from basics of case, the relationship between evidence and fact mentioned in the judgment document will be recorded. For instance, if an evidence section starts with "认定上述事实的证据有", the fact in the section before it will be assigned to the same group as the evidence in it. We call the group "evidence-fact group", which will be mentioned again in Sect. 3.2.3.

Second step of converting is information item extraction. According to the requirement of judgment document analysis and the writing standard of judgment document, we set up about 630 information items that need to be extracted. Because judges have many undefined "hidden" habits while writing judgment documents, we have summarized these habits by reading a large number of judgment documents. In our work, information items are extracted by regular expression in light of writing standards and habits. Table 1 presents some samples of regular expressions we have studied and summarized for extracting information items.

Table 1. Some samples of regular expressions for extracting information items

Information item	Corresponding regular expression
Reference number of case	[(\\((〔]\\d{4}[])\\)〕)].+?[^鉴]字第?\\d+-?\\d+号
ID number	\\d{18}|\\d{17}(\\d|X|x)
Accusation	犯([\u4e00-\u9fa5]+?[,]*[\u4e00-\u9fa5]+?)*罪
Institution	[经被由].{1,13}?分局|[经被由].{1,18}?派出所|[经被由].{1,15}?局|[经被由].{1,18}?院|本院|被公安机关|[经被由].{1,15}?公安处|被指定居所

Table 2. Accuracies of splitting seven parent sections

Type of case	Headline	Party	Litigation record	Basics of case	Trail process	End
First instance over civil case	99%	93%	88%	94%	89%	91%
Second instance over civil case	99%	92%	90%	90%	91%	97%
First instance over administrative case	99%	94%	88%	92%	91%	98%
Second instance over administrative case	99%	97%	96%	96%	95%	99%
First instance over criminal case	99%	97%	96%	97%	98%	98%
Second instance over criminal case	99%	98%	96%	97%	98%	99%

To evaluate the accuracy of our approach, we use two ways: both existing tools and human labeling, to build test dataset that contains more than 5000 judgment documents. By comparing each section and each information item, the evaluation result shows that overall accuracy of our approach is 93.5%. As Table 2 presents, accuracies of splitting basics of case, which is the most important parent section for building evidence chain relational model are all higher than 90%, even up to 97%.

As Table 3 presents, accuracies of evidence section extraction and fact section extraction are all over 70%, even up to 98%.

Our approach can reach the level of mature tools in the industry. Furthermore, we can meet the requirements formulated for the evaluation of judgment document reasoning. Our approach can do more in-depth analysis of reasoning sections and realize some functions that other tools don't have.

Table 3. Accuracies of evidence section extraction and fact section extraction

Type of case	Evidence section of this trial	Fact section of this trial	Evidence section of previous trial	Fact section of previous trial
First instance over civil case	80%	81%	–	–
Second instance over civil case	72%	72%	83%	88%
First instance over administrative case	80%	75%	–	–
Second instance over administrative case	98%	73%	81%	80%
First instance over criminal case	88%	88%	–	–
Second instance over criminal case	90%	90%	94%	94%

3.2.2 Label Admissibility of Evidence

In face of much evidence, three characteristics are supposed to be considered as filter criteria. They are authenticity, legality and relevance. Evidence without these characteristics cannot be brought into evidence chain relational model [16].

The admissibility of evidence is always written at the end of evidence section or after each evidence. First of all, we have to extract the description of admissibility. The method to do this work is similar to the approach to extract child section from parent section.

Due to the improvement and development of Chinese criminal justice, the judicial interpretation is mainly concentrated on evidence rules of criminal proceeding [17]. It's harder to label the admissibility of evidence extracted from civil or administrative cases, because the situation that part of evidence can be adopted meanwhile the other part can't always occurs. In this case, processing requires following three steps:

(1) Divide the description of admissibility into single sentences and each sentence only embodies the admissibility of one or a set of evidence. Judges always use Arabic numerals or Chinese character numbers as identifiers. A set of evidence means identifiers of those evidence are connected by punctuation connectors such as ",", "~", "–" or Chinese connectors such as "和", "至", "到". If there are no obvious identifiers, word segmentation is needed. Segmented words are used to build a dictionary which is the basement of matching in the next step.

(2) If there are identifiers, use them in each sentence to relate the sentence to corresponding evidence. Otherwise, match words in each sentence based on the dictionary and find out corresponding evidence.

(3) Use keywords as characteristics to classify the result of admissibility. For example, "不采信", "不采纳", "不予采信" and "不予采纳" mean not adopted, on the contrary, those keywords without negative expression mean adopted.

3.2.3   Extract Evidence Set and Fact Set

Building evidence chain relational model needs every single fact and evidence, so extracting them from the section they belong to is required. Concrete steps are as follows:

(1) Read every evidence-fact group. If there is a fact section in it, we put the fact into fact set, and turn to step (2). Otherwise, turn to step (3).
(2) Put all evidence in this evidence-fact group into an evidence set related to the fact set we build in step (1).
(3) Put all evidence in this evidence-fact group into a specific evidence set called "unrelated evidence set".

3.3   Extract Key Elements

Because 4W1H have different characteristics, each kind of element needs respective method. In this section, we mainly focus on introducing the methods used to extract key elements. Same word may repeat in one text, so de-duplication must be considered. Make sure that the set of each kind of element doesn't contain repeated words.

3.3.1   Extract What

WHAT refers to entitative things mentioned in evidence and facts. For example, WHAT usually refers to drug name in drug cases, and refers to type and brand of vehicle in traffic cases. Words of WHAT have two characteristics, one is all of them are nouns; the other is most of them belong to subject or object.

   We use two methods to extract WHAT:

(1) RegExp (Regular Expression). For things have regular structure, such as file name between " 《" and "》 ", use corresponding regular expression to extract.
(2) Semantic Relation. We use constraint-based Chinese dependency parsing method to analyze sentence structure automatically based on maximum entropy model and maximum spanning tree. Select word itself, part of speech, distance between two words and dependency of two words as features to train the model.

   As mentioned above, subject and object are likely to be elements we look for. After statement structure analysis, obtain all of subject and object, filter out sites, names and words that are not noun. Chinese attribute V-N compounds which can be interpreted as nominals with V as the modifier and nominal phrases with a verbal modifier both need to be considered to select an appropriate completion for rest words afterwards [18].

3.3.2   Extract Where

WHERE refers to sites mentioned in evidence and facts. Words of WHERE have two characteristics, one is all of them are nouns or location words; the other is most of them appear after prepositions.

   We use two methods to extract WHERE:

(1) POS (Part of Speech). Unlike English, Chinese text requires word segmentation. We use Chinese tokenizer based on n-Gram, CRF (Conditional Random Field) and HMM (Hidden Markov Model) to segment words. After that, nature of every word is marked. If a word is labeled as "S" which means site, it is what we look for.

(2) Semantic Relation. In consideration of error rate of word segmentation, semantic relation is also required. We build a list of common prepositions to extract phrases after them.

3.3.3 Extract When

WHEN refers to time mentioned in evidence and facts. Words of WHEN have two characteristics, one is containing Arabic numerals or Chinese character numbers; the other is they have standard structure.

We use RegExp (Regular Expression) to extract WHEN. Regular expressions covered from the most precise time "X年X月X日X时X分" to the simplest time "X年" can match time phrases with different precision.

3.3.4 Extract Who

WHO refers to parties mentioned in evidence and facts. Parties are not only people but also companies and administrative organs.

We use POS (Part of Speech) to extract WHO. Names of people and organizations are both extracted. Furthermore, in consideration of the problem of excessive word segmentation, we use a similar method that is similar to the method mentioned above when we introduce how to extract WHAT to select an appropriate completion.

3.3.5 Extract How Much

HOW MUCH refers to quantitative phrases mentioned in evidence and facts. Words of HOW MUCH have two characteristics, one is containing Arabic numerals or Chinese character numbers; the other is they always appear before quantifiers.

We use RegExp (Regular Expression) to extract HOW MUCH. We build a list of common quantifiers to extract phrases before them.

3.3.6 Evaluation

To evaluate our methods, we conduct an experiment. The experiment is conducted on a self-constructed dataset which comprises of 1692 evidence and 337 facts from Chinese judgment documents. The evaluation results are shown in Table 4.

As the table presents, recall and precision are almost more than 70%. The evaluation results prove that our methods to extract key elements can lay a solid foundation for finding out connection points and calculating chain unit heads in later steps.

Table 4. Evaluation results of key elements extraction

Key elements	Recall	Precision	F1
What	58.5%	72.9%	64.9%
Where	81.2%	68.4%	74.3%
When	92.7%	100%	96.2%
Who	83.7%	90.6%	87.0%
How much	95.2%	98.7%	96.9%

However, what we should not ignore is that results of extracting key elements based on semantic relation and part of speech are not very satisfactory. Especially, how to extract WHAT can be further optimized.

One way to solve this problem is to change features for maximum entropy model training. Those features we select now are too focused on words themselves. Another way is changing a better corpus. Furthermore, we would like to try to use CRF (Conditional Random Field) or SVM (Support Vector Machine) as our main algorithm.

3.4   Create Connection Points

In this section, we mainly focus on introducing the methods used to create connection points in evidence chain relational model. Even for legal experts, it's not easy to create logical links when building evidence chain relational model. Our approach to create connection points cannot possibly obtain completely correct results. What we do is providing the most probable result.

3.4.1   Build Relationship Between Evidence and Fact

Evidence chain relational model we define contains all facts of a case, and each fact connects to related evidence chain by connection points, therefore, we have to build relationship between all of evidence and facts before building evidence chains. Because one of characteristics evidence chain has is uniqueness which means every chain proves only one fact [16]. For each of evidence in "unrelated evidence set", we find out the most probable relevant fact.

A method to obtain correlation of different Chinese texts is calculating similarity coefficient based on characteristics of literal words, such as statistical character and semantic feature [19]. We build relationship between evidence and fact according to the number of same words they have. More same words mean more possibilities, so we choose the fact that has most same words with the evidence and build relationship between them. In this method, two points need to be paid attention to:

(1) Semantic similarity of words. Two words to be compared must be the same kind of key elements.
(2) Weight factor of words. Different weights can be set for different kinds of key elements according to the cause of action. For example, in drug cases, criminal facts are always that a group of drug dealers sells a drug at different times. The separation of facts in this circumstances is WHEN instead of WHAT, so WHEN requires high weights.

3.4.2   Build Relationship Between Different Evidence

Relationship between different evidence means the mutual corroboration based on chain unit heads. In other words, the same key elements different evidence has been able to be chain unit heads. The algorithm we use to find out same key elements is Hash algorithm. We use key elements as keys of a HashMap, if evidence contains a certain key element, the identifier of this evidence will be saved into the value-list of relevant key. After completing mapping, the value-list of each key is read in turn to find out candidate set of chain unit heads.

There are many jargon terms in Chinese judgment documents. These words appear in lots of evidence, but as we all know they are not chain unit heads. Although many published sets of stop words can be used, such stop words cannot meet our requirements. For example, "证据" (evidence), "事实" (fact), "原告" (plaintiff), "被告" (defendant), etc. occur in almost every judgment documents, but they are not common stop words. Besides that, some words that are not jargon terms also cannot be chain unit heads. For example, "行为" (behavior), "内容" (content), "过程" (process), etc. These words don't have practical significance to be proof points.

To construct a stop words list that meets requirements talked above, we use 13571 judgment documents as our corpus and calculate IDF of each unique word. IDF (Inverse Document Frequency) refers to the inverse fraction of documents that contain a specific word. Words with low IDF can be regarded as potential stop words. After scanning all judgment documents and filtering common stop words, we calculate IDF of rest words and sort words in ascending order. Top N words have been chosen as stop words. The number N is chosen manually after we scan the result of sorting. Stop words we choose are added into the set of common stop words to construct the stop words list we use to filter candidate set of chain unit heads.

3.5   Display Evidence Chain Relational Model

In this section, we mainly focus on introducing the methods used to display evidence chain relational model. The Excel sheets can clearly present the facts and the contents of each evidence in detail and the graph can show the relationship in the model directly and comprehensively.

3.5.1   Generate Excel File

Evidence chain relational model should be saved as a format which is easy to store, read, understand and exchange, so we choose Excel. An evidence chain relational model built based on a judgment document is saved as an Excel file with two sheets:

(1) Evidence sheet only focuses on evidence in the judgment document. Identifier, name, content, type, submitter, admissibility result, reason of admissibility result, chain unit heads and relevant key text of each evidence are set out in the sheet in turn. The relationship between chain unit heads and key text is one to one. Each chain unit head is a word which contains few information. If readers want to know the real meaning of a chain unit head, they have to find its location in the text of evidence and read the context. To provide convenience for readers, we extract the clause with a chain unit head in it as the relevant key text of this chain unit head.

(2) Fact sheet focuses on not only facts but also relationship between evidence and fact. Identifier, name, content of each fact are set out in the sheet in turn. At the same time, relevant evidence and their chain unit heads are displayed after every fact.

3.5.2   Graphical Display

Actually, the research on evidence chain relational model has two directions. This paper introduces the part of reverse direction research. Forward direction research focuses on how to implement a system for judges to build evidence chain relational

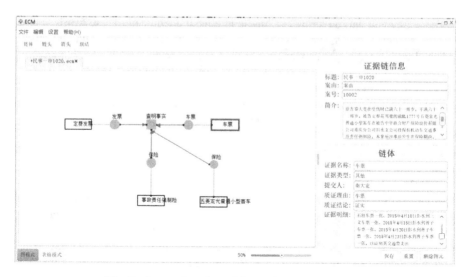

Fig. 3. System interface of forward direction research

models by themselves conveniently and how to translate graphical models into reasoning sentences automatically.

As Fig. 3 presents, we can use the system with evidence chain relational model in Excel format as input. The system analyzes the Excel file and extracts the model. After automatic typesetting, the evidence chain relational model we built can be displayed on the interface clearly.

4   Conclusion

In this paper, we propose an approach to build evidence chain relational model based on Chinese judgment documents. We can extend our study on reasoning modeling in two ways. First, combine our evidence chain relational model with evidential Bayesian network. Bayesian network provides a means to identify and evaluate the hypotheses that may have produced the available evidence in a case and to assess their plausibility [20]. Evidential Bayesian provides a means to estimate how much more strongly evidence supports one hypothesis over another, and how strongly the addition or removal of evidence would affect the relative level of support for alternative hypothesis [21]. So we can do the numerical evaluation of argumentation and the credibility of reasoning. Second, combine the analysis of main section of verdict to construct the reasoning logic diagram. The diagram contains every step of reasoning. Each step has relevant judicial evidence, fait juridique, law and so on. The diagram shows the relationship between evidence and inference, shows each layer of reasoning, so as to achieve the visualization of reasoning in judgment documents.

When we complete the study that we talk about above, we will make a solid foundation for the evaluation of judgment documents reasoning, we can also achieve the goal of evaluating 'fact of case' which is one of the aspects of reasoning. If there are other methods to evaluate 'laws of case' and 'style of writing', the whole system can do a comprehensive, accurate and efficient evaluation of judgment documents reasoning, to provide technical support for improving the level of judgment documents reasoning to the Chinese People's Court.

Acknowledgements. This work was supported by the Key Program of Research and Development of China (2016YFC0800803).

References

1. Hou, X.: On holistic concept of judgment document reasoning. J. Party Sch. Guiyang Comm. C.P.C **1**, 52–55 (2009)
2. Hu, Y.: On reason giving of judgements. J. Law Appl. **3**, 48–52 (2009)
3. Zhou, G.: Fact of case, law of case, emotion of document and style of writing—the way of judgment document reasoning. J. Shandong Judges Train. Inst. (Shandong Judgem.) **23**(5), 41–46 (2007)
4. Wang, Z.: On reasoning in judgement documents. J. Law Appl. **12**, 3–7 (2000)
5. Buchanan, B.G., Headrick, T.E.: Some speculation about artificial intelligence and legal reasoning. Stanf. Law Rev. **23**(1), 40–62 (1970)
6. Prakken, H., Reed, C., Walton, D.: Argumentation schemes and generalisations in reasoning about evidence. In: International Conference on Artificial Intelligence and Law, pp. 32–41. ACM (2003)
7. Bex, F.J., Prakken, H., Verheij, B.: Formalising argumentation story-based analysis of evidence. In: International Conference on Artificial Intelligence and Law, pp. 1–10. ACM (2007)
8. Williams, M.A.: Evidence transmutations: gathering admissible evidence using belief revision. In: International Conference on Artificial Intelligence and Law, pp. 216–217. ACM (2009)
9. Keppens, J.: On extracting arguments from Bayesian network representations of evidential reasoning (2011)
10. Bentham, J.: Rationale of judicial evidence (1827)
11. Chen, W.: The study on the evidence chain. J. National Prosec. Coll. **15**(4), 128–136 (2007)
12. Ma, G., Wang, Z., Wang, K.: On electronic judicial identification model based on evidence chain. J. Hebei Univ. (Nat. Sci. Ed.) **33**(3), 317–323 (2013)
13. Jia, Z., Bai, J.: Computer forensics and formation of evidence chain. Autom. Instrum. **8**, 124–126 (2014)
14. Wang, W., Zhao, D., Zou, L., et al.: Extracting 5W1H event semantic elements from Chinese online news. In: DBLP Proceedings of the Web-Age Information Management, International Conference, WAIM 2010, Jiuzhaigou, China, 15–17 July 2010, pp. 644–655 (2010)
15. Wang, W.: Chinese news event 5W1H semantic elements extraction for event ontology population, pp. 197–202 (2012)
16. Cai, Z.: Standard of evidence chain integrity and approach to examine. Lawyer World **3**, 11–13 (2003)
17. Chen J.: On fact-finding in the court and evidence. Southwest University of Political Science and Law (2009)
18. Shi, D.: Chinese attributive V-N compounds. Chin. Lang. **6**, 483–495 (2003)
19. Li, R., Wang, Z., Li, S., et al.: Chinese sentence similarity computing based on frame semantic parsing. J. Comput. Res. Dev. **50**(8), 1728–1736 (2013)
20. Bex, F.J., Koppen, P.J.V., Prakken, H., et al.: A hybrid formal theory of arguments, stories and criminal evidence. Artif. Intell. Law **18**(2), 123–152 (2010)
21. Keppens, J., Shen, Q., Price, C.: Compositional Bayesian modelling for computation of evidence collection strategies. Appl. Intell. **35**(1), 134–161 (2011)

Research and Development of Virtual Instruments System Based on Depth Camera

Xiao-li Xu[1] ⓘ, Ming-hui Sun[1,2(✉)], Xin-yue Sun[1], Wei-yu Zhao[1], and Xiaoying Sun[3]

[1] College of Computer Science and Technology, Jilin University, Changchun 130012, China
smh@jlu.edu.cn
[2] Key Laboratory of Symbol Computation and Knowledge Engineering of the Ministry of Education, Jilin University, Changchun 130012, China
[3] College of Communication Engineering, Jilin University, Changchun 130022, China

Abstract. This paper takes advantage of the depth camera of somatosensory kinect and sensors to implement gesture recognition and design a virtual instrument system. As long as the user waves his arm without the help of other equipments, our system can automatically recognize the hand gesture and make suitable sound. In order to achieve depth camera's detection of hands movement, this paper introduce the depth imaging technology *Light Coding* and bone tracking technology to obtain the actual position information and hand movement information of the human body. Feet movement detection uses sensor technology, different stampede strength outputs different digital number after AD conversion so that the intensity can be controlled. A series of experiments show that the system has good fluency and practicality and increased the fun of playing instruments.

Keywords: Virtual instrument · Kinect · Sensors · Depth data processing · Bone tracking

1 Introduction

1.1 Background

Musical instruments are the spiritual possessions that humans have owned long long ago and are constantly enriched with the evolution of mankind. It can be referred from archaeological excavations and murals, cliff paintings that hunting, signal transmission, sacrifice prayer, fighting cheering or celebration dance are tightly connected with pronunciation tools in which the percussion instruments are the most ancient one [1]. Some percussion instruments can not only produce rhythms, but also make melody and chorus effect [2].

With the increase in music lovers, the demand for musical instruments has gradually increased, but some large-scale instruments are in a high price and difficult to move. Many people had to give up their favorite large instruments in the choice of

© Springer Nature Singapore Pte Ltd. 2017
B. Zou et al. (Eds.): ICPCSEE 2017, Part II, CCIS 728, pp. 108–114, 2017.
DOI: 10.1007/978-981-10-6388-6_9

musical instruments [3], based on this, this paper discuss the feasibility of virtual instruments and take advantage of depth image processing technology to take the drum as an example, realizing the research and implementation of virtual drum.

1.2 The Present Research of Virtual Performance

Large musical instruments are still the majority, with high prices and poor mobility. Electronic enthusiasts abroad have taken advantage of accelerator to make a virtual electronic drum, but after all, the accelerator has been limited to achieve the effect [4].

Kinect is the hot spot in human-computer interaction and its gesture recognition technology has a wide range of applications, mainly in the blind sign language recognition, virtual environment interaction, multi-channel user interface [5], etc. Using Kinect depth data processing and skeletal tracking techniques, coding each space by the light source to determine the location of the object. Using the computer graphics visual technology to distinguish the human body from the background environment and identify the human body joints to depicting a skeleton system. Besides, using Poisson equation and other algorithms for noise filtering, smoothing algorithm to achieve automatic repair in order to present a smooth virtual portrait and track hands movement [6]. At the same time based on the piezoelectric effect of the piezoelectric ceramic, achieving the goal of converting the foot pressure into the sound of Bass Drum, Side Stick and Pedal Hi-Hat while controlling the intensity of the stampede.

Connect the Kinect and sensor technology to restore the performance more accurately. The users have no need to hold the real instrument, that is, facing the depth camera, two hands simulating percussion action, percussing the air in different locations so that the depth camera can recognize hands movements and pc can restore the drum sounds [7]. The system enables large instruments to be used in a wider range to achieve the perfect human-computer interaction.

2 System Architecture

The virtual drum system is composed of Kinect equipment, computer, Arduino development board and a number of buzzer. As shown in Fig. 1.

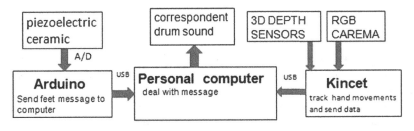

Fig. 1. Overall structure

2.1 Kinect

Kinect can provide three types of raw data information, including depth data stream, color video stream, the original audio data, corresponding to bone tracking, identification, voice recognition [8]. This system only use the technology of skeletal tracking to quickly build the player's trunk, limbs, head even fingers according to the bone joints, which is the basis of Kinect somatosensory operation.

2.2 Arduino

Arduino is a high portability, convenient and quick embedded development platform of open source electronics, a microcontroller that interacts with the physical world by connecting with the sensors to receive the information of sounds, positions, vibrations [9] and control the computer to have corresponding reaction, such as making sounds.

2.3 Piezoelectric Ceramic Buzzer

The buzzer operates on the principle of piezoelectric effect, which produces mechanical vibration when an alternating voltage is applied to it, on the other hand, it generates a weak voltage signal when a mechanical force is applied to it [10]. Therefore, the piezoelectric ceramic buzzer can be flexibly used as a vibration sensor.

3 Action Detection

3.1 Depth Image Imaging Principle

Depth data is the essence of Kinect. In this system, Kinect get the depth map by emitting near-infrared light, as long as there is an object shaped like Chinese character '大', Kinect will go to track it. Kinect has a similar process of firing, capturing, computing visual reproduction. Its "depth of the eye" is a combination of infrared projectors and infrared cameras and projection and reception overlap each other.

Kinect depth image in this system uses the Light Coding technology. It is a structure light technology which uses light source to code the space [11], but the calculation method of depth is not the same. Light Coding light source is known as the "laser speckle", which is the diffraction formed after laser illuminate to rough objects or through the ground glass.

These speckles have a high degree of randomness, and will change with the distance of the pattern. That is, any two places in the speckle pattern are different. As long as the space marked with such a structured light, the entire space has been marked. Putting an object into this space, as long as the object's speckle pattern is determined, you can know where the object is. Before this, the whole space speckle patterns should be recorded. So first, have a light source calibration. The calibration is to take a reference plane at regular intervals and record the speckle pattern on the reference plane [12]. As shown in Fig. 2.

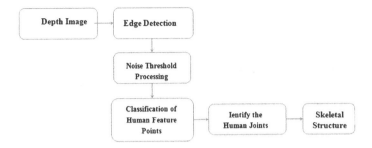

Fig. 2. Speckle pattern

3.2 The Formation of Skeletal Pattern

Kinect can identify the moving object that may be a human body in the field of view, that is, to generate a skeletal map from a depth image [13]. The first step in recognizing the human body is to separate the human body from the background environment in the depth image, which is a process of extracting useful information from the noise. The Kinect system will firstly analyze the region closer to Kinect. Then, the pixels of the depth image of these regions will be scanned point by point in order to identify every part of the body. The human body part is quickly classified by eigenvalues which is a computer graphics visual technology, including edge detection, noise threshold processing, classification of human feature points, and ultimately distinguish the human body from the background environment [14]. The next step is to further identify the human joints, the system generate a skeletal structure based on 20 joints, in this way Kinect can accurately assess the actual location of the human body. In this system, the application only needs to capture the upper body movement, so take the 'Seated Mode', only to capture the upper 10 bone points.

3.3 Arduino and Piezoelectric Ceramic Buzzer

Piezoelectric ceramic can detect the vibration of sound waves, output analog signals, like a percussion monitor, the greater the vibration, the greater the voltage [15]. When the feet tap on them, the computer will make different sounds. The Arduino detects the voltage signal generated by the vibrating piezoelectric ceramic. After the on-chip AD conversion, the pc play the sound of Pedal Hi-Hat, Bass drum 1 and Side Stick through software Virtual MIDI piano Keyboard [15] MIDI messages (as shown in Fig. 3) contain three parts, pitch, velocity and channel, different velocities correspond to different sounds. 36 (0×24) corresponds to Bass Drum, 44 ($0 \times 2C$) corresponds to Pedal Hi-Hat.

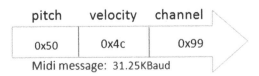

Fig. 3. MIDI messages

4 Test

4.1 Hand Movement Test

The hands simulate the percussion action, percuss in different positions in the air in the face of Kinect and Computer interface (that corresponds to the different drums on computer screen, as shown in Fig. 4), and it will make different drum sounds [16]. Movements test is showed in Fig. 5.

Fig. 4. Movements test

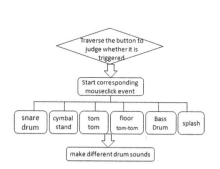

Fig. 5. Motion detection process

4.2 Feet Action

When piezoelectric ceramics is beat, it will not only output a substantial waveform once, it will be followed by a aftermath [17], as shown in Fig. 6. So the choice of threshold becomes particularly significant, neither too low nor too high.

Experiment results: When beating buzzer connected to A_0 pin of the analog port, the led lamp will be on and off, indicating that the analog value of the percussion signal is read by the development board [15]. The range value for the serial monitor is [50–500].

1	Knock: 130	Knock: 141	Knock: 148	Knock: 191
2	Knock: 175	Knock: 218	Knock: 415	Knock: 197
3	Knock: 177	Knock: 286	Knock: 240	Knock: 165
4	Knock: 214	Knock: 387	knock: 217	Knock: 116

As a result, this system sets threshold as 200.

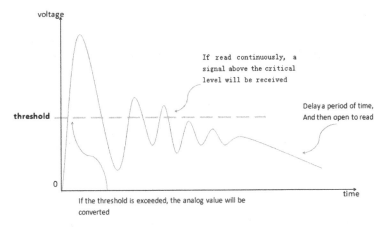

Fig. 6. Voltage fluctuation

5   Conclusion

Based on Kinect and Sensors, we adopted multi-sensor fusion technology, breaking through the traditional method of playing drums, putting forward a new method to solve the high price and large size of large instruments.

This paper introduce the depth image technology Light Coding [18] and bone tracking technology to obtain the actual position information and hands movement information of the human body to restore the performing more accurately. Practice shows it is feasible that the combination of Kinect and sensor technology can better restore the performance of the drummer.

At present, the expansion of the entire system is not completed, only to support the use of virtual drums. This skeletal tracking technology can also be applied to a wider range, such as other virtual percussion instruments, virtual dressing room and so on. List the table of Contents [19], the operator can choose their favorite application then enter a different system.

6   Work in the Future

There are still many shortcomings in this virtual drum system based on the depth camera.

6.1   Time Delay

When the hands waved, hitting the air in different locations, the computer shows a slight delay in the sound, resulting in hand movement can not be too fast, which is particularly evident when the two hands waving at the same time. The system should also be improved in the stability. In the process of testing, there have been a flash. There is still a space for improvement.

6.2   Lack of Fusion

The fusion of image recognition and sensor proposed in this paper is only a simple fusion based on interactive conditions, and there is no fusion of specific data, motion characteristics and other higher levels' data [20].

Acknowledgments. This study has been partially supported by National Key Research and Development Program of China (No. 2016YFB1001300), Postdoctoral Science Foundation of China (2014M561294) and the Science, Technology Development Program Funded Projects of Jilin Province (20150520065JH) and Fundamental Research Funds for the Central Universities.

References

1. Peng, J.: There should be an instrument in life. Cross-Century. Time **11**, 66 (2011)
2. Wu, Y.: A study on the identity, dissimilarity and harmony of western percussion and national percussion. The Youth Writers **5**, 125 (2013)
3. Song, T.: The choice of musical instruments and the age of beginners. Parents **10** (1986)
4. Gao, P.: Virtual musicians world guide. Trading Up **9**, 156–157 (2008)
5. Li, Q., Shen, M.: Gesture recognition technology and its application in human - computer interaction. Chin. J. Ergon. **8**(1), 27–29, 33 (2002)
6. Yan, L., Li, Y.: Extraction of human skeleton based on Kinect depth data. Electron. Measure. Technol. **38**(3), 39–42 (2015)
7. Luo, T.: Design and realization of percussion virtual performance and entertainment system based on human pose. Diss. Zhejiang University (2016)
8. Wu, G., Li, B., Yan, J.: KINECT Human Computer Interaction Development Practice. Post&Telecom Press (2013)
9. D'Ausilio, A.: Arduino: a low-cost multipurpose lab equipment. Behav. Res. Methods **44**(2), 305–313 (2012)
10. Zhao, Y.: Perfect graphic Arduino interactive design introduction. Science Press (2014)
11. Shi, M.: Research on the technology and working principle of Kinect. Nat. Sci. J. Harbin Normal Univ. **29**(3) (2013)
12. Fan, J., Zhou, G.: The research of gesture recognition based on Kinect skeleton tracking technology. J. Anhui Agri. Sci. **42**(11), 3444–3446 (2014)
13. Li, J.: Research on Human Body Motion Recognition Based on Kinect Depth Image. Beijing University of Posts and Telecommunications (2015)
14. Ye, C.: Research on the important role of computer image technology in visual communication design. Comput. CD Softw. Appl. **8**, 183–184 (2014)
15. Margolis, M.: Arduino Cookbook. Post&Telecom Press (2015)
16. Wang, S.: Introduction to Kinect somatosensory programming. Science Press (2015)
17. Li, Y.: The design and implementation of a wireless hand piano system based on Arduino and Arduino control. Yunnan University (2015)
18. Bai, Z.: Research and Implementation of Sports Teaching System Based on Kinect. Tianjin University (2012)
19. Zhang, S.: Study on gesture recognition and national stringed instruments virtual. Playing Hangzhou Normal University (2015)
20. Lu, S.: Research and development of virtual instruments system based on depth camera. Central China Normal University (2016)

Text Understanding with a Hybrid Neural Network Based Learning

Shen Gao[1], Huaping Zhang[1(✉)], and Kai Gao[2]

[1] College of Computer Science,
Beijing Institute of Technology, Beijing 100080, China
63388@qq.com, kevinzhang@bit.edu.cn
[2] School of Information Science and Engineering,
Hebei University of Science and Technology,
Shijiazhuang 050000, Hebei, China
gaokai@hebust.edu.cn

Abstract. Teaching machine to understand needs to design an algorithm for the machine to comprehend documents. As some traditional methods cannot learn the inherent characters effectively, this paper presents a new hybrid neural network model to extract sentence-level summarization from single document, and it allows us to develop an attention based deep neural network that can learn to understand documents with minimal prior knowledge. The proposed model composed of multiple processing layers can learn the representations of features. Word embedding is used to learn continuous word representations for constructing sentence as input to convolutional neural network. The recurrent neural network is also used to label the sentences from the original document, and the proposed BAM-GRU model is more efficient. Experimental results show the feasibility of the approach. Some problems and further works are also present in the end.

Keywords: Deep learning · Convolutional neural network · Recurrent neural network · Word embedding · Gated recurrent unit

1 Introduction

Textual compress representation or understanding usually needs to generate a headline or summarization which can capture the main ideas of the original article, and it usually needs to train the statistical models flexible enough to learn to exploit document content. As for text understanding, traditional approaches have been based on either hand engineered grammars [1], or information extraction methods of detecting predicate argument triples that can later be queried as a relational database [2]. And still others usually represent each document entirely with one bag-of-n-gram vector, using a language model. There usually exists an incorrect but necessary Markov and independent homogeneity assumption. As some traditional ways are handcrafted parsing program or statistically learnt model, many of them require labeled training data, but

© Springer Nature Singapore Pte Ltd. 2017
B. Zou et al. (Eds.): ICPCSEE 2017, Part II, CCIS 728, pp. 115–125, 2017.
DOI: 10.1007/978-981-10-6388-6_10

almost all original articles are unlabeled. Furthermore, as training over high dimension usually yields very sparse features vectors (i.e., one-hot representation), the traditional approaches are incredibly fragile. Recently, the neural network based approaches have been broadly used, and some methods based on neural networks have shown promising results for text understanding. For example, the sequence to sequence based model is a generating based method, but the main disadvantage is that the generated results may not conform to the grammar rules.

This paper proposes a hybrid neural network learning model for sentence-level extracting based summarization. First, a multi-layer based convolutional neural network (CNN) architecture has been proposed to produce a representation of a sentence. Second, a document can be transformed into a vector list by means of the recurrent neural network (RNN), and this approach can be viewed as a sequence labeling problem. That is to say, the label is the decision of whether the sentence should be used as a part of the extracted summarization. Hence, the main contributions of the paper are: (1) a pre-trained word embedding model has been used to initialize the word embedding which has been trained by the GloVe model; (2) the convolutional layer is used to learn a vector representation of a sentence; (3) the bidirectional GRU network is used to make a decision on whether the sentence should be extracted as a part of the summarization. In order to evaluate the performance, this paper compares different approaches, and the experiment results shows the feasible of the BAM-GRU based model, which can achieve the state-of-the-art results. Some problems and further works are also present in the end.

2 Related Work

There are two main different methods on summarization. One is the extracting based method which can extract some words or sentences from the original article. The other is the generating based method which can generate some new sentences, that is to say, as for summarization, words in the generated results may not appear in the original article. Some other methods [3, 4] have viewed the summarization as a sequence to sequence problem. TextRank [5] is a graph-based ranking model for summarization and extracting keywords. Deletion based method [6] uses a simple long short-term memory (LSTM) to label the word sequence into a single sentence for textual compression, and the label means the decision of whether the word should be removed from the sentence.

On the other hand, the deep learning based method has achieved success in many sequence to sequence learning tasks such as reading comprehension [7], and machine translation [8–10]. And the convolutional network has also been used in many fields such as semantic classification and sentence classification [11, 12]. A gated recurrent unit (GRU) is proposed in [13], and the experiment in [14] shows that the performance of these recurrent units is better than that of the traditional RNN units. Reference [15] proposes a bidirectional recurrent neural network. Reference [16] proposes a bidirectional LSTM model which is faster and more accurate than both the standard RNN and multilayer perceptron. To avoid overfitting, dropout [17] is also added to RNN layer. A few resent studies [18] have proposed a model which uses a pre-trained sentence embedding model to extract summary. As for embedding, the GloVe is an unsupervised learning algorithm for obtaining vector representations [19]. Training is performed on aggregated global word-word co-occurrence statistics from corpus, and the resulting representations present interesting linear substructures of the word vector space. Reference [20] uses convolutional layers to learn a vector representation of a sentence, and it also presents a RNN model to make a decision on whether the sentence should be extracted as the part of summary.

Differ from the above related works, this paper presents a hybrid model. A multi-layer convolutional network is used to extract more high-level features from the original sentences. And a bidirectional RNN network with attention has been used to make the decision of each sentence whether it should be extracted.

3   A Hybrid Neural Network Based Model

3.1   Convolutional Based Sentence Encoder

Words in one-hot representation need to be embedded into a lower dimensional space. A vector representation is associated to each character while words are represented as the sum of these representations. It needs to count all words in the whole corpus and build a dictionary V (the size is denoted as |V|). It also needs to retrieve each word w_i within the embedding matrix and then concatenate the word vectors to build the input matrix $S_j \in R^{D_s \times D_E}$, where the i-th column represents the word vector W_i of i-th word of the input sentence, and the parameter D_s and D_E denotes the length of the sentence and the dimensions of word embedding matrix, respectively. A sentence S_j can be denoted as: $S_j = [W_1, W_2, \ldots, W_i, \ldots, W_{D_s}]$. So the corresponding original article can be represented as $A_l = [S_1, S_2, \ldots, S_j, \ldots, S_{D_A}]$, where D_A denotes the number of sentences in the article. The reasons why the convolutional layer has been chosen is that this layer can be trained effectively without the long-term dependencies. Then the convolutional layer which has K kernels will be applied to each sentence in A_l, The formalized method is shown in Eqs. (1) and (2), and the symbol $\oplus$ means the concatenation operation, θ denotes the activation function, and $\otimes$ denotes the convolutional operation. After the convolutional layer, the output C will go through a max-pooling layer in order to extract some more useful features. Then the sentence S_j can be transferred to a real value vector representation.

$$S_j = (W_1 \oplus W_2 \oplus \ldots \oplus W_{D_s}) \in R^{D_s \times D_E} \tag{1}$$

$$C = \theta(K \otimes A_l) \tag{2}$$

The proposed model is shown in Fig. 1.

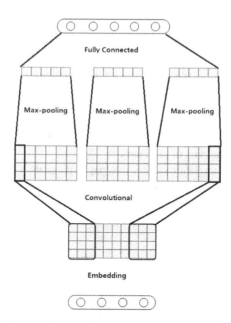

Fig. 1. Convolutional based sentence encoder

As the result matrix $C \in R^{n \times (|D_s| - K + 1)}$ obtained from the convolutional layer is usually too large for the final decision, this paper proposes the max-pooling layer to reduce the result from the convolutional layer so as to obtain some useful features. The aim is to capture the most relevant features and to return the maximum of the set of values, selecting the max value in the pooling window and then using this max value to represent the current whole values of the pooling window. In detail, along with the window sliding to the end, if the length of the remainder is shorter than the window, the valid-padding will be used to discard the remainder. Furthermore, a dropout is used to avoid over-fitting.

3.2   Recurrent Neural Network Based Sentence Labelling

As the extracted results have been viewed as a sequence labeling problem, the RNN model is used to determine which sentence should be extracted from the original document. Similar with the LSTM, the GRU has gating units which can modulate the flow of information inside the GRU unit. The GRU is faster than LSTM because it doesn't have a separate memory cells. As GRU works faster in terms of both the number of updates and actual CPU time, the recurrent neural network using GRU will be trained faster than the traditional LSTM based network. The input of the RNN is the sentence vector list which represents the corresponding original article, and the output will be sent to the fully connected layer, which will produce a one-hot vector which represents whether the corresponding sentence should be extracted or not.

The GRU based model will transfer the sentence to another vector, which will be used to decide whether the corresponding sentence should be extracted or not. Then the output of the RNN model will go through the fully connected layer so as to produce the final decision. In order to determine which sentence should be extracted, the three different RNN models are proposed.

Multi-layer GRU (M-GRU): This model is a simple RNN model which contains some stacked GRU layers. The embedded sentences are used as the input of M-GRU. Then the output of the stacked GRU layers is sent to a fully connected layer and softmax is used as the final activate function to make the final decision.

Attention based Multi-layer GRU (AM-GRU): Like the M-GRU model, this model adds an attention weight through the GRU. An attention weight vector is applied when the GRU generates the decision vector. It gives the GRU layer the ability to focus on the input information.

Bidirectional Attention based Multi-layer GRU (BAM-GRU): This is a bidirectional GRU network. Differ from the above AM-GRU, this model doubles the GRU layers, and this can make the model to use both the history and future information. The architecture has shown in Fig. 2.

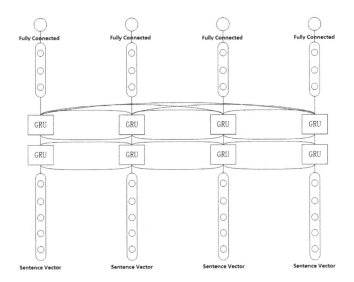

Fig. 2. BAM-GRU based sentence labeling

4 Experiment Results and Analysis

4.1 Dataset Overview and Preprocessing

In order to evaluate the proposed approach's performance on real dataset, we use the DailyMail dataset has been built by [20]. The dataset is split into the training and test set. The test set is not included in either the training set. This segmentation ensures that no nearly duplicate news is in both the training and validate set. Furthermore, the training data is randomly shuffled. News containing less than 10 sentences will be discarded. The original news articles contain different numbers of sentences. As the convolutional layer requires same length of input vectors, so the original sentence which has less than 30 words will be added some special token "UNK". The original documents which has less than 30 sentences will also be added some special sentences.

In detail, this dataset which has 193986 news articles for training while 10350 news articles for testing. This dataset contains news title, full text and highlight, which can be seen as the extracted summary of the original news. As for the final training, there are 180000 news articles for the model training. Pre-trained word embedding has been trained by GloVe model by using Wikipedia 2014 and Gigaword 5 corpus. It has 400 K words in vocabulary and 100 dimensions of each word vector.

A dictionary is needed, and the words in the dictionary are the same as the words in the pre-trained word embedding. Then we will replace the word in news article by the word index in the dictionary.

The hyper parameters are proposed in the Table 1, where the hyper parameter "Document length" denotes how many sentences in the documents.

Table 1. Hyper parameters

Hyper parameter name	Value
Vocabulary size	60000
Embedding size	50
Batch size	50
Document length	30
Word count of sentence	20
Convolutional kernels	4
GRU layer size	3

4.2   Evaluating Metrics

ROUGE. ROUGE is the abbreviation of Recall-Oriented Understudy for Gisting Evaluation, and it is a set of metrics used for evaluating automatic summarization and machine translation. The metric compares an automatically produced summary or translation against a reference or a set of references (human-produced) summary or translation [21]. This paper uses ROUGE-1, ROUGE-2, ROUGE-L to evaluate the performance on learning summary of documents. ROUGE-1 and ROUGE-2 are the ROUGE score for 1-gram and 2-gram, respectively.

BLEU. Ranging from 0 to 1, BLEU [22] metric reflects a fraction of the n-gram length difference between the original summary and the newly generated summary. Meanwhile, BLEU also considers the difference of the number of words between the extracted summary and the original summary. In this paper, BLEU implemented by NLTK[1] is used for performance evaluation.

---

[1] http://www.nltk.org/.

Loss. This paper uses the sampled softmax as the loss function of sentence labeling model. Since the number of target labels in the sentence labeling model is as the same size as that of the dictionary's, the target classification size is set to the same size of the dictionary. If the traditional softmax is used, the training will be very slow. So the sampled softmax will be used as the loss function.

Accuracy. This metrics shows how many sentences are correctly classified. The function $\mathbb{I}(x, y)$ means if x equals to y it returns 1, otherwise, it returns 0.

$$accuracy = \frac{\sum \mathbb{I}(true\ labels, predict\ labels)}{sentence\ count\ of\ document}$$

4.3 Experiment Result Analysis

Each of the proposed model has been trained 20000 steps. It is clear from the Fig. 3 that as the number of training steps increases, the accuracy of the model is also rising. But the BAM-GRU model has been at the highest accuracy rate. After 5000 steps, every model's accuracy will not show significant changes. As for the loss on training steps, it is also clear from Fig. 4 that the convergence rate of the BAM-GRU model is the fastest.

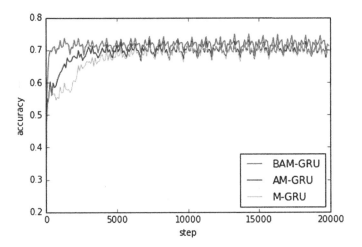

Fig. 3. Accuracy on training steps

From Tables 2, 3 and 4, we can see clearly that the BAM-GRU model is more efficient than the other two models. For example, the BAM-GRU model is two percentages higher than the other model in accuracy, see Table 2.

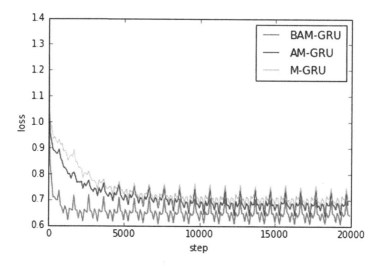

Fig. 4. Loss on training steps

Table 2. Accuracy and loss of each model

Model	Accuracy	Loss
M-GRU	0.7098	0.6353
AM-GRU	0.7142	0.6095
BAM-GRU	**0.7345**	**0.5771**

From Table 3, the ROUGE-L score indicates that means the fluency of the generated summary. The result shows that the summary produce by BAM-GRU model is more fluency than the other two models'. Generally, the BAM-GRU model is optimal for both accuracy and f-score, but the AM-GRU model is superior to other models in recall.

Table 3. ROUGE score of each model

Model	ROUGE-1			ROUGE-2			ROUGE-L		
	Precision	Recall	F-score	Precision	Recall	F-score	Precision	Recall	F-score
M-GRU	0.7468	0.8350	0.7657	0.6907	0.7803	0.7114	0.7034	0.7940	0.7244
AM-GRU	0.7404	**0.8520**	0.7698	0.6899	**0.8012**	0.7202	0.7018	**0.8146**	0.7326
BAM-GRU	**0.7916**	0.8135	**0.7902**	**0.7282**	0.7529	**0.7295**	**0.7409**	0.7658	**0.7420**

Table 4. BLEU score of each model

Experiment name	BLEU	
	Sentence	Modified
M-GRU	0.6215	0.7483
AM-GRU	0.6278	0.7416
BAM-GRU	**0.6682**	**0.7934**

4.4   Sample Results

Our model achieves the state-of-the-art results on three different models. In the Table 5, the sentences with underlined lines indicate that the model considers this sentences to be the result that need to be extracted. The sentences with shadows represent these

Table 5. Sample summary of BAM-GRU model

Original Document

@entity0 is considering his position as manager of @entity3 after the club suspended his assistant @entity5 on thursday @entity3 took two months to appoint @entity5 but he was told in a letter from director of football @entity7 yesterday morning that he was be suspended until the end of the season effectively ending his time at @entity13 after 105 days @entity5 is understood to have been told that he will be suspended until his contract expires at the end of the season he has also been informed that an option to renew his deal will not be taken up by the club @entity3 assistant manager @entity5 (left) has been suspended by the club @entity5 's suspension has left manager @entity0 considering his future at @entity13 a club statement read : ' the director of football @entity7 , has today (april 2) suspended the assistant coach @entity5 from his duties at the club this is an internal matter and the club will make no further comment on this internal issue ' @entity0 , the club 's third coach of the season , took training before attending a delayed press conference and said : ' it's become difficult for me it's not ideal timing i 've got to have a good think about my future now we have put things together me and @entity5 before it was a no brainer for me to stay on , not so much now ' i love this club , i was brought up on @entity3 but this situation is a difficult one for me ' @entity47 was informed this morning through a letter he's been suspended - for what , i don't know you 'd have to find that out from the club i do n't understand why he has been suspended it's a difficult one for me to take , i can't think for one minute why someone who has been part of this success has been sacked ' me and @entity5 were the ideal partnership , he's been great for me and i 'm bitterly disappointed that he's not here ' @entity0 's contract is also up at the end of the season and club president @entity62 is suspended until may 3 @entity64 heads home in a recent win for the @entity67 at @entity69 's @entity68 @entity3 sit 13th in the @entity71 table but were unbeaten in march and take on @entity72 on saturday it is also understood that @entity0 is under pressure not to select top scorer @entity75 who will trigger a bonus clause in his contract should he score two more goals @entity5 is taking advice from the @entity80 while @entity7 's own position is thought to be open to review at the end of the season .

Extract Summary

@entity0 is considering his position as manager of @entity3 after the club suspended his assistant @entity5 on thursday @entity3 took two months to appoint @entity5 but he was told in a letter from director of football @entity7 yesterday morning that he was be suspended until the end of the season effectively ending his time at @entity13 after 105 days @entity5 is understood to have been told that he will be suspended until his contract expires at the end of the season he has also been informed that an option to renew his deal will not be taken up by the club @entity3 assistant manager @entity5 (left) has been suspended by the club @entity5 's suspension has left manager @entity0 considering his future at @entity13 a club statement read : ' the director of football @entity7 , has today (april 2) suspended the assistant coach @entity5 from his duties at the club ' @entity0 , the club 's third coach of the season , took training before attending a delayed press conference and said : ' it's become difficult for me ' @entity0 's contract is also up at the end of the season and club president @entity62 is suspended until may 3 @entity64 heads home in a recent win for the @entity67 at @entity69 's @entity68 @entity3 sit 13th in the @entity71 table but were unbeaten in march and take on @entity72 on saturday @entity5 is taking advice from the @entity80 while @entity7 's own position is thought to be open to review at the end of the season .

Entity Mapping

@entity3=Leeds, @entity0=Neil Redfearn, @entity13=Elland Road, @entity5=Steve Thompson, @entity62=Massimo Cellino, @entity7=Salerno, @entity80=LMA, @entity47=Thommo, @entity68=Craven Cottage, @entity69=Fulham, @entity75=Mirco Antenucci, @entity67=Whites, @entity64=Sol Bamba, @entity71=Championship, @entity72=Blackburn Rovers

sentences will be contained in the final extracted summarization. The sentences with crossed lines represent these have been wrongly extracted. From the result of the BAM-GRU model, it is clear that this summary which is produced by the model is basically consistent with the summary of the human annotations.

5 Conclusion and Future Works

Deep neural networks are powerful models that have achieved excellent performance on learning tasks. Differ from the traditional methods, this paper presents a hybrid neural network method to extract summary from document. It is clearly that the proposed BAM-GRU model has achieved the state-of-the-arts. However, there also exists some problems. If the depth of the network is not enough, perhaps it tends to lead to the lack of learning ability. And the accuracy is not high enough. For the further works, we plan to increase the amount of data to train the model. And we also plan to use more convolutional layers and more fully connected layers to give the model more ability to learn.

Acknowledgment. This work is sponsored by National Natural Science Foundation of China (Grant No.: 61272362) and National Basic Research Program of China (973 Program, Grant No.: 2013CB329606). This work is also sponsored by National Science Foundation of Hebei Province (No. F2017208012) and Key Research Project for University of Hebei Province (No. ZD2014029).

References

1. Riloff, E., Thelen, M.: A rule-based question answering system for reading comprehension tests. In: Proceedings of the Workshop on Reading Comprehension, NAACL/ANLP-2000 (2000)
2. Poon, H., Christensen, J., Domingos, P., Etzioni, O., Hoffmann, R., Kiddon, C., Lin, T., Ling, X., Mausam, Ritter, A.: Machine reading at the University of Washington. In: NAACL HLT 2010 First International Workshop on Formalisms and Methodology for Learning by Reading, pp. 87–95 (2010)
3. Lopyrev, K.: Generating news headlines with recurrent neural networks, pp. 1–9 (2015). arXiv:1512.01712
4. Rush, A.M., Chopra, S., Weston, J.: A neural attention model for abstractive sentence summarization. In: EMNLP 2015, pp. 379–389 (2015)
5. Mihalcea, R., Tarau, P.: TextRank: bringing order into texts (2004)
6. Filippova, K., Alfonseca, E., Colmenares, C.A., Kaiser, L., Vinyals, O.: Sentence Compression by Deletion with LSTMs. In: EMNLP, pp. 360–368 (2015)
7. Moritz, K., Tom, H., Kay, W., Hermann, K., Kocisky, T., Grefenstette, E.: Teaching machines to read and comprehend. In: Advances in Neural Information, pp. 1–9 (2015)
8. Sutskever, I., Vinyals, O., Le, Q.V.: Sequence to sequence learning with neural networks. In: Advances in Neural Information Processing Systems, pp. 3104–3112 (2014)
9. Bahdanau, D., Cho, K., Bengio, Y.: Neural machine translation by jointly learning to align and translate. In: ICLR 2015, pp. 1–15 (2015)

10. Wu, Y., Schuster, M., Chen, Z., Le, Q.V., Norouzi, M., Macherey, W., Krikun, M., Cao, Y., Gao, Q., Macherey, K., Klingner, J., Shah, A., Johnson, M., Liu, X., Kaiser, Ł., Gouws, S., Kato, Y., Kudo, T., Kazawa, H., Stevens, K., Kurian, G., Patil, N., Wang, W., Young, C., Smith, J., Riesa, J., Rudnick, A., Vinyals, O., Corrado, G., Hughes, M., Dean, J.: Google's Neural Machine Translation System: Bridging the Gap between Human and Machine Translation, pp. 1–23 (2016). arXiv:1609.08144v2
11. Kim, Y.: Convolutional neural networks for sentence classification. In: Proceedings of the 2014 Conference on Empirical Methods in National Language Processing (EMNLP 2014), pp. 1746–1751 (2014)
12. Zhang, X., Zhao, J., LeCun, Y.: Character-level convolutional networks for text classification. In: Proceedings of the 28th International Conference on Neural Information Processing Systems, pp. 649–657. MIT Press, Cambridge (2015)
13. Rumelhart, D.E., Hinton, G.E., Williams, R.J.: Learning representations by back-propagating errors. Nature **323**, 533–536 (1986)
14. Chung, J., Gulcehre, C., Cho, K., Bengio, Y.: Empirical Evaluation of Gated Recurrent Neural Networks on Sequence Modeling, pp. 1–9 (2014). arXiv:1412.3555v1
15. Schuster, M., Paliwal, K.K.: Bidirectional recurrent neural networks. IEEE Trans. Signal Process. **45**, 2673–2681 (1997)
16. Graves, A., Schmidhuber, J.: Framewise phoneme classification with bidirectional LSTM networks. In: Proceedings of the International Joint Conference on Neural Networks, vol. 4, pp. 2047–2052 (2005)
17. Srivastava, N., Hinton, G.E., Krizhevsky, A., Sutskever, I., Salakhutdinov, R.: Dropout: a simple way to prevent neural networks from overfitting. J. Mach. Learn. Res. **15**, 1929–1958 (2014)
18. Kobayashi, H., Yatsuka, M., Taichi, N.: Summarization based on embedding distributions. In: Proceedings of the 2015 Conference on Empirical Methods in Natural Language Processing, pp. 1984–1989 (2015)
19. Pennington, J., Socher, R., Manning, C.D.: GloVe: global vectors for word representation. In: Proceedings of the 2014 Conference on Empirical Methods in Natural Language Processing, pp. 1532–1543 (2014)
20. Cheng, J., Lapata, M.: Neural Summarization by Extracting Sentences and Words, pp. 484–494 (2016). arXiv:1603.07252
21. Lin, C.Y.: Rouge: a package for automatic evaluation of summaries. In: Proceedings of the Workshop on Text Summarization branches out (WAS 2004), pp. 25–26 (2004)
22. Papineni, K., Roukos, S., Ward, T., Zhu, W.-J.: BLEU: A method for automatic evaluation of machine translation. In: Proceedings of the 40th Annual Meeting. Association for Computational Linguistics - ACL 2002, pp. 311–318 (2002)

Towards Realizing Mandarin-Tibetan Bi-lingual Emotional Speech Synthesis with Mandarin Emotional Training Corpus

Peiwen Wu, Hongwu Yang[✉], and Zhenye Gan

College of Physics and Electronic Engineering, Northwest Normal University,
Lanzhou 730070, China
yanghw@nwnu.edu.cn

Abstract. This paper presents a method of hidden Markov model (HMM)-based Mandarin-Tibetan bi-lingual emotional speech synthesis by speaker adaptive training with a Mandarin emotional speech corpus. A one-speaker Tibetan neutral speech corpus, a multi-speaker Mandarin neutral speech corpus and a multi-speaker Mandarin emotional speech corpus are firstly employed to train a set of mixed language average acoustic models of target emotion by using speaker adaptive training. Then a one-speaker Mandarin neutral speech corpus or a one-speaker Tibetan neutral speech corpus is adopted to obtain a set of speaker dependent acoustic models of target emotion by using the speaker adaptation transformation. The Mandarin emotional speech or the Tibetan emotional speech is finally synthesized from Mandarin speaker dependent acoustic models of target emotion or Tibetan speaker dependent acoustic models of target emotion. Subjective tests show that the average emotional mean opinion score is 4.14 for Tibetan and 4.26 for Mandarin. The average mean opinion score is 4.16 for Tibetan and 4.28 for Mandarin. The average degradation opinion score is 4.28 for Tibetan and 4.24 for Mandarin. Therefore, the proposed method can synthesize both Tibetan speech and Mandarin speech with high naturalness and emotional expression by using only Mandarin emotional training speech corpus.

Keywords: Mandarin-Tibetan cross-lingual emotional speech synthesis · hidden Markov model (HMM) · Speaker adaptive training · Mandarin-Tibetan cross-lingual speech synthesis · Emotional speech synthesis

1 Introduction

Speech is one of the most natural and ideal ways of human-computer interaction. Current speech synthesis technologies can synthesize natural and intelligible neutral speech by unit selection method or hidden Markov model (HMM)-based statistical parametric speech synthesis technology [1]. However, neutral speech is

© Springer Nature Singapore Pte Ltd. 2017
B. Zou et al. (Eds.): ICPCSEE 2017, Part II, CCIS 728, pp. 126–137, 2017.
DOI: 10.1007/978-981-10-6388-6_11

not suitable for face-to-face communication or spoken dialog system. Since emotional speech synthesis has strong potential in enhancing effective communication between human and computers in spoken dialog systems [2], it has been a hot topic of research in recent years [3]. Emotional speech synthesis mainly includes waveform unit selection method [4], prosodic feature modification method [5] and statistical parametric speech synthesis method [6]. Each method has its advantages and disadvantages. In the waveform unit selection method, a large emotional speech database needs to be built for the emotional speech synthesis system. However, it is not easy to establish a speech corpus with different emotions of multiple speakers. Furthermore, this method cannot be implanted or generalized to synthesize emotions for different languages [7]. The prosodic feature modification method integrates the prosody or phonetic strategy into the unit selection to build a small or mixed emotional corpus for modifying the contours of the target f0 and the duration. However, this method does not generate the emotional speech of any speaker. Some of the necessary sub-word units may not be included in the established small or mixed emotional corpus will result in reduced quality of synthesized speech [8]. Although the above two methods can synthesize emotional speech, they are relatively backward in terms of scalability. The HMM-based speech synthesis can be successfully applied to scalability tasks due to the nature of its parameters by speaker adaptation techniques and has been shown to significantly improve the perceived quality of synthesized speech [9]. Since all acoustic parameters are modeled in a single framework, the HMM-based speech synthesis can use interpolation [10], emotion vector multiple regression [11] and adaptive techniques [12,13] to easily transform or modify the speaker's style or emotion. The HMM-based statistical parametric speech synthesis is also adopted to train the emotional acoustic models by using emotional corpus [9]. Masuko [14] and Tamura [15] used the maximum likelihood linear regression (MLLR) or the maximum a posterior (MAP) criterion with a small amount of emotional corpus to modify the neutral average acoustic models [16] for obtaining the target emotional acoustics models. Jaime [2] proposed a HMM-based emotion transplantation method using an adaptive algorithm based on constrained structural maximum a posterior linear regression (CSMAPLR) to modify the parameters of the acoustic model [16]. Because HMM-based speech synthesis can synthesize different emotional speech of different languages in the case of limited speech data [17,18], this method has become a hot spot in cross-lingual speech synthesis and emotional speech synthesis.

One of the biggest problems for emotional speech synthesis is data acquisition especially for minority languages such as Tibetan. Because Tibetan and Mandarin are in Sino-Tibetan family, the phonetics and linguistics have many similarities on these two languages. Yang et al. [18] previously showed that a Mandarin-Tibetan cross-lingual speech synthesis can be realized by using a large Mandarin corpus and a small Tibetan corpus. Because Tibetan and Mandarin also have similarities in emotional expression [19], our research is focusing on extending the word in [18] to realize a Mandarin-Tibetan emotional speech synthesis. Because it is very difficult to record Tibetan emotional speech corpus, this paper only use a Mandarin emotional training speech corpus to train acoustic models not only for Mandarin but also for Tibetan. We firstly use a Mandarin emotional speech corpus recorded by several Mandarin speakers and a Tibetan one-speaker neutral speech corpus to obtain a set of mixed-language (Mandarin-Tibetan) average acoustic models of target emotion. The Mandarin or Tibetan one-speaker neutral speech corpus is then used to perform the speaker adaptation transformation to obtain a set of Mandarin or Tibetan speaker dependent acoustic models of target emotion for synthesizing the Mandarin or Tibetan speech of target emotion.

2   Framework of Mandarin-Tibetan Bi-lingual Emotional Speech Synthesis

The framework of the Mandarin-Tibetan bi-lingual emotional speech synthesis is shown in Fig. 1. We firstly use all Mandarin speaker's speech corpus of target emotion, a multi-speaker Mandarin neutral speech corpus and a one-speaker Tibetan neutral speech corpus to train a set of mixed-language (Mandarin-Tibetan) average acoustic models of target emotion by using the speaker adaptive training. The one-speaker Mandarin or Tibetan speech neutral corpus is then used to perform the speaker adaptation transformation to obtain a set of speaker dependent Mandarin or Tibetan acoustic models of target emotion. The Tibetan emotional speech or Mandarin emotional speech is finally synthesized from speaker dependent Tibetan or Mandarin acoustic models of target emotion.

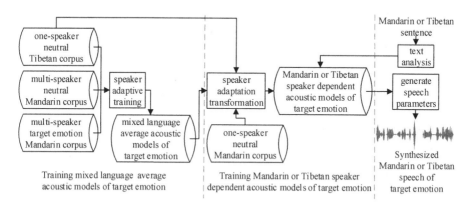

Fig. 1. Framework of the Mandarin-Tibetan cross-lingual emotional speech synthesis.

During the training of mixed-language average acoustic models of target emotion, the semi-hidden Markov model (HSMM)-based SAT algorithm [12] is used to improve the quality of the synthesized speech and reduce the influence of the differences between speakers and languages. The linear regression equations of the duration distribution and state outputs are shown in the Eqs. 1 and 2,

$$\widehat{d_i}^{\,s}(t) = \alpha^s d_i(t) + \beta_s = X_i^{\,s}(t)\,\xi_i(t)\,, \xi = [d_i, 1] \tag{1}$$

$$\widehat{o_i}^{\,s}(t) = A^s o_i(t) + b^s = W_i^{\,s}(t)\,\xi_i(t)\,, \xi = [o_i, 1] \tag{2}$$

where, $\widehat{d_i}^{\,s}(t)$ is the speaker s's mean vector of duration distribution, $\widehat{o_i}^{\,s}(t)$ is the speaker's mean vector of state output. $X = [\alpha, \beta]$, $W = [A, b]$ is the duration distribution and state outputs transformation matrices of model. d_i is average duration vector, o_i is average observation vector.

In the paper, we use the constrained maximum likelihood linear regression (CMLLR) [12] to train a set of context-dependent multi-space-distribution hidden semi-Markov models(MSD-HSMM) as the bi-lingual average acoustics models of target emotion.

After training the target emotional Tibetan and Mandarin average acoustic models, the MSD-HSMM-based CMLLR adaptive algorithm is applied to the specific speaker's training speech of target-language (Mandarin or Tibetan) to obtain a set of target emotional speaker dependent models of target language for realizing the emotion transformation from Mandarin target emotional speech to Tibetan or Mandarin neutral speech. The transformation equations of the state d and the feature vector o under the state are shown in Eqs. 3 and 4,

$$p_i(d) = N(d; \alpha m_i - \beta, \alpha \sigma_i^2 \alpha) = |\alpha^{-1}| N(\alpha \psi; m_i, \sigma_i^2) \tag{3}$$

$$b_i(o) = N(o; Au_i - b, A\textstyle\sum_i A^T) = |A^{-1}| N(W\xi; u_i, \textstyle\sum_i) \tag{4}$$

where, $\psi = [d, 1]^T$, $\xi = [o^T, 1]$, m_i is mean duration distribution, u_i is mean state outputs, $\sum_i$ is diagonal covariance matrix, $X = [\alpha^{-1}, \beta^{-1}]$ is transformation matrix of the duration distribution probability density, $W = [A^{-1}, b^{-1}]$ is linear transformation matrix of the state outputs probability density.

The fundamental frequency, spectrum and duration parameters of speech data can be transformed and normalized by MSD-HSMM-based adaptive transformation algorithm. For adaptive data O of length T, the maximum likelihood estimation of $\Lambda = (W, X)$ can be transformed as shown in Eq. 5,

$$\tilde{\Lambda} = (\tilde{W}, \tilde{X}) = \arg\max_{\Lambda} P(O\,|\lambda, \Lambda) \tag{5}$$

where, λ is the parameter set of HSMM.

Finally, the target-emotional speaker dependent models of the target-language speech are updated and modified by using the maximum a-posteriori (MAP) algorithm [11]. For a given model λ, if the forward probability and the backward probability are: $\alpha_t(i)$ and $\beta_t(i)$, under the state i, the probability $k_t^{\,d}(i)$ of its continuous observation sequence $o_{t-d+1}...o_t$ is shown in Eq. 6,

$$k_t^d(i) = \frac{1}{P(O\,|\lambda)} \sum_{\substack{j=1 \\ j\neq i}}^{N} \alpha_{t-d}(j) p(d) \prod_{s=t-d+1}^{t} b_i(o_s)\beta_t(i) \tag{6}$$

MAP estimation is shown in Eqs. 7 and 8,

$$\widehat{m}_i = \frac{\tau \bar{m}_i + \sum_{t=1}^{T} \sum_{d=1}^{t} K_t^d(i)d}{\tau + \sum_{t=1}^{T} \sum_{d=1}^{t} K_t^d(i)d} \tag{7}$$

$$\widehat{u}_i = \frac{\omega \bar{u}_i + \sum_{t=1}^{T} \sum_{d=1}^{t} K_t^d(i) \sum_{s=t-d+1}^{t} o_s}{\omega + \sum_{t=1}^{T} \sum_{d=1}^{t} K_t^d(i)d} \tag{8}$$

where, $\bar{m}_i$ and $\bar{u}_i$ is mean vector after linear regression, τ and ω is the MAP estimate parameter of the duration distribution and state outputs, $\widehat{m}_i$ and $\widehat{u}_i$ is the weighted average MAP estimate value of adaptive vector $\bar{m}_i$ and $\bar{u}_i$.

3   Context-Dependent Labels

We adopt a full context-dependent label format [18] of Mandarin to label Mandarin sentences and Tibetan sentences. A set of speech assessment methods phonetic alphabet (SAMPA) is designed for labeling initials and finals of Mandarin and Tibetan. The shared initials or finals by tow languages are labeled with the same SAMPAs. All initials and finals of Mandarin and Tibetan, including silence and pause, are used as the synthesis unit of the context-dependent MSD-HSMMs. A six level context-dependent label format is designed by taking into account the following contextual features.

- **unit level:** the {pre-preceding, preceding, current, succeeding, suc-succeeding} unit identity, position of the current unit in the current syllable.
- **syllable level:** the {initial, final, tone type, number of units} of the {preceding, current, succeeding} syllable, position of the current syllable in the current {word, prosodic word, phrase}.
- **word level:** the {POS, number of syllable} of the {preceding, current, succeeding} word, position of the current word in the current { prosodic word, phrase }.
- **prosodic word level:** the number of {syllable, word} in the {preceding, current, succeeding} prosodic word, position of the current prosodic word in current phrase.
- **phrase level:** the intonation type of the current phrase, the number of the {syllable, word, prosodic word} in the {preceding, current, succeeding} phrase.
- **utterance level:** whether the utterance has question intonation or not, the number of {syllable, word, prosodic word, phrase} in this utterance.

We extend a question set designed for the HMM-based Mandarin speech synthesis by adding the language-specific questions. Tibetan-specific units and Mandarin-specific units are asked in the question set. We also design the questions to reflect the special pronunciation of Tibetan. Finally we get more than 3000 questions. These questions cover all features of the full context-dependent labels.

4 Experiments

4.1 Experimental Conditions

In our work, We use 7 female speaker's EMIME Mandarin speech database [20] as the neutral Mandarin speech corpus in which each speaker records 169 training sentences, together with a female Tibetan neutral speech corpus and a multi-speaker Mandarin emotional speech corpus as the training data. We select all 9 female speaker's recordings as the Mandarin emotional training data in which 100 neutral utterances of one female are randomly selected as the testing set. A native female Tibetan Lhasa dialect speaker is invited to record the Tibetan neutral speech corpus in a sound proof studio. 800 Tibetan sentences are selected from recent year's Tibetan newspapers, in which 100 sentences are randomly selected as the testing set. We also use psychology method to stimulate the emotional speech by inner stimulated situation. 9 female Mandarin speakers who are not a professional actress are selected to record the emotional speech in a sound proof studio. There are 5 kinds of emotions including sadness, relax, joy, disgust and neutral. Each speaker records 100 Mandarin sentences of one emotion. The neutral speech is recorded firstly, and then the emotional speech. When recording neutral speech, the speaker is required to use an unchanged tone and speaking rate to read sentences. When recording other 4 kinds of emotional speech, the speaker is asked to read sentences after stimulating the speaker's relative emotion with a specific scene. We select 3, 6 and 9 speaker's speech(300, 600 and 900 utterances) from each kind of emotion as the emotional training corpus. The synthesized Mandarin and Tibetan emotional utterances are accordingly marked as M300, M600, M900 as well as T300, T600, T900. All recordings are saved in the Microsoft Windows WAV format as sound files (mono-channel, signed 16 bits, sampled at 16 kHz). We use 5-state left-to-right context-dependent multi-stream MSD-HSMMs. The TTS feature vectors are comprised of 138-dimensions: 39-dimensional STRAIGHT [21] mel-cepstral coefficients, log F0, 5 band-filtered aperiodicity measures, and their first-order and second-order difference.

4.2 Experimental Results

We synthesize 3 categories of Tibetan (T300, T600, T900) emotional speech and 3 categories of Mandarin (M300, M600, M900) emotional speech. Each category has 5 emotions. In each category, 100 utterances are synthesized for each emotion. A total of 3000 utterances (100 utterances * 5 emotions * 3 categories * 2 languages) are synthesized. 10 utterances of one emotion were randomly selected from each category to consist of an evaluation set that includes a total of 300 utterances (10 utterances * 5 emotions * 3 categories * 2 languages). We invite 9 Tibetan speakers and 9 Mandarin speakers as the subjects to evaluate the synthesized Tibetan and Mandarin emotional speech.

The Evaluation of Lingual Similarity. We use the degradation mean opinion score (DMOS) test to evaluate the lingual similarity of synthesized emotional speech. In the DMOS evaluation, all synthesized Tibetan and Mandarin emotional utterances and their original neutral utterances, which are a total of 600 utterances (300 original utterances + 10 utterances * 5 emotions * 3 categories * 2 languages), are used to participate in evaluation. A testing group is built with the synthesized Tibetan and Mandarin emotional utterances for each category from each kind of emotion and their original neutral utterances. We randomly play 30 sets of test files to the subjects in which the original neutral utterance is played firstly, and then the synthesized emotional utterance. Subjects need to carefully compare the lingual similarity of the two speech files to score the speech files according to a 5-point scale that uses the DMOS scoring standard in [22].

DMOS scores of synthesized Tibetan and Mandarin speech of 6 categories for all emotions are compared in Fig. 2. It can be seen from Fig. 2 that the DMOS score of synthesized emotional speech is high than other emotions for smooth colored emotion (such as relax, neutral), and the DMOS scores become more stable with the increasing of the Mandarin emotional training corpus.

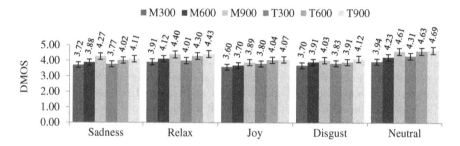

Fig. 2. DMOS score (and 95% confidence interval) of synthesized Tibetan and Mandarin emotional speech. (Color figure online)

The Evaluation of Speech Quality. We use the mean opinion score (MOS) test to evaluate the speech quality of synthesized emotional speech. In the MOS evaluation, we randomly play the synthesized emotional utterances of any 15 test groups (5 emotion * 3 categories) to the subjects. Subjects scores the naturalness of each synthesized emotional utterance according to [22].

MOS scores of each emotion for synthesized Tibetan and Mandarin emotional speech from 6 categories are shown in Fig. 3. It can be seen from the Fig. 3 that the MOS score of smooth colored emotion (such as relax, joy and neutral) achieves a higher MOS score than other emotions. The synthesized emotional speech can achieve high and stable MOS scores with the increasing of the Mandarin emotional training corpus.

The Evaluation of Emotion Similarity. We use the emotional mean opinion score (EMOS) test to evaluate the emotional expression of synthesized

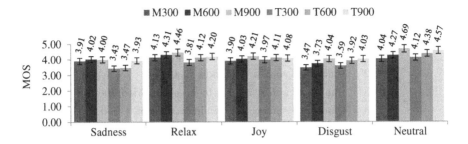

Fig. 3. MOS score (and 95% confidence interval) of the synthesized Tibetan and Mandarin emotional speech. (Color figure online)

emotional utterances. In the EMOS evaluation, we play each synthesized emotional utterances of Tibetan and Mandarin for 9 subjects of target language (Tibetan or Mandarin) to score the emotional expression of speech according to a 5-point scale standard shown in Table 1. The EMOS evaluation results are shown in Fig. 4. We can see from Fig. 4 that the EMOS scores of smooth colored emotion (such as relax and neutral) did not change significantly with the increasing of the original mandarin emotional training corpus while other emotions changed greatly.

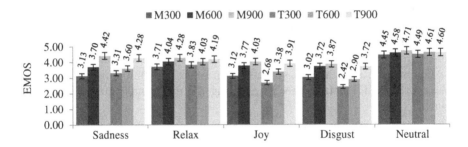

Fig. 4. EMOS score(and 95% confidence interval) of the synthesized Tibetan and Mandarin emotional speech. (Color figure online)

From the above evaluation we can see that the synthesized emotional speech can achieve high DMOS, MOS and EMOS when we increase the Mandarin emotional training corpus. This suggests that we can improve the naturalness and emotional expression of synthesized Tibetan or Mandarin speech by increasing Mandarin emotional training corpus. When 900 Mandarin emotional training utterance is used to training the acoustic models, we can achieve the highest MOS, DMOS and EMOS score. Therefore, proposed method can synthesize emotional Tibetan speech and Emotional Mandarin speech by using only Mandarin emotional training corpus.

Table 1. The evaluation standard of EMOS

Score	Evaluation criteria
0–1	Emotional similarity is unknown, Bad
1–2	Emotional similarity is indistinct, Poor
2–3	Emotional similarity can be accepted, Medium
3–4	Emotional similarity is willing to accept, Good
4–5	Emotional similarity is very well, Excellent

Objective Evaluation. Because we only have the emotional Mandarin corpus, we only analyze the root mean square error (RMSE) of duration and fundamental frequency of the synthesized Mandarin emotional speech as show in Table 2. From the Table 2, we can see that the RMSE of duration and fundamental frequency for the category M300, category M600 and category M900 are gradually reduced. This indicates that the synthesized emotional Mandarin speech is getting closer to the original emotional Mandarin speech with the increasing of Mandarin emotional training corpus. From the average RMSE of duration and fundamental frequency we can see that the proposed Mandarin-Tibetan cross-lingual emotional speech synthesis can also synthesize better emotional Mandarin speech by mixing the emotional Mandarin training corpus. For Tibetan, because there is no original Tibetan emotional corpus, we compare the pitch contour of the synthesized emotional Tibetan speech in the category T900 as show in Fig. 5 that come from a same Tibetan sentence. From the Fig. 5, we can see that the pitch contour of the synthesized emotional Tibetan speech is all different from the neutral. This indicates that the proposed Mandarin-Tibetan cross-lingual emotional speech synthesis can affect the fundamental frequency of the synthesized Tibetan speech by mixing the emotional Mandarin training corpus.

Table 2. The RMSE of duration (marked as durRMSE) and the RMSE of fundamental frequency (marked as f0RMSE) for the synthesized emotional Mandarin speech.

Synthesized speech	durRMSE (s)			f0RMSE (Hz)		
	M300	M600	M900	M300	M600	M900
Sadness	0.72	0.46	0.26	19.32	18.25	16.94
Relax	0.47	0.33	0.22	34.37	18.72	17.88
Joy	0.68	0.52	0.37	39.77	31.23	28.57
Disgust	1.08	0.76	0.62	28.32	24.93	23.48
Neutral	0.35	0.27	0.21	17.54	16.98	16.62
Average	0.66	0.47	0.34	27.86	22.02	20.69

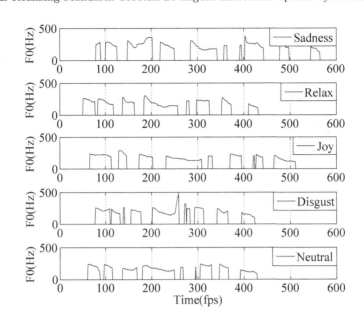

Fig. 5. The pitch contour of the synthesized emotional Tibetan speech.

5 Conclusions

This paper presents a method of HMM-based Mandarin-Tibetan bilingual emotional speech synthesis that realizes emotion transplantation from Mandarin emotional training corpus to Tibetan neutral utterance and Mandarin neutral utterance. Firstly, we have realized a Tibetan-Mandarin bi-lingual speech synthesis framework by using SAT algorithm. Then, we use Mandarin neutral and emotional training corpus as well as Tibetan neutral training corpus to train a set of mixed-language (Mandarin-Tibetan) average acoustic models of target emotion. After that, we obtain a set of Mandarin or Tibetan speaker dependent acoustic models of target emotion from the mixed-language average acoustic models with Tibetan neutral training corpus or Mandarin neutral training corpus to synthesize target Tibetan emotional speech or Mandarin emotional speech. Future work will focuses on improving the speech quality of the synthesized emotional speech.

Acknowledgments. The research leading to these results was partly funded by the National Natural Science Foundation of China (Grant No. 11664036, 61263036) and Natural Science Foundation of Gansu (Grant No. 1506RJYA126).

References

1. Barra-Chicote, R., Yamagishi, J., King, S., et al.: Analysis of statistical parametric and unit selection speech synthesis systems applied to emotional speech. Speech Commun. **52**, 394–404 (2010)

2. Lorenzo-Trueba, J., Barra-Chicote, R., San-Segundo, R., et al.: Emotion transplantation through adaptation in HMM-based speech synthesis. Comput. Speech Lang. **34**, 292–307 (2015)
3. Schröder M.: Emotional speech synthesis: a review. In: Interspeech, pp. 561–564 (2001)
4. Adell, J., Escudero, D., Bonafonte, A.: Production of filled pauses in concatenative speech synthesis based on the underlying fluent sentence. Speech Commun. **54**, 459–476 (2012)
5. Hamza, W., Eide, E., Bakis, R., et al.: The IBM expressive speech synthesis system. In: Interspeech (2004)
6. Zen, H., Tokuda, K., Black, A.W.: Statistical parametric speech synthesis. Speech Commun. **51**, 1039–1064 (2009)
7. Pitrelli, J.F., Bakis, R., Eide, E.M., et al.: The IBM expressive text-to-speech synthesis system for American English. IEEE Trans. Audio Speech Lang. Process. **14**, 1099–1108 (2006)
8. Strom, V., King, S.: Investigating Festival's target cost function using perceptual experiments (2008)
9. Yamagishi, J., Onishi, K., Masuko, T., et al.: Acoustic modeling of speaking styles and emotional expressions in HMM-based speech synthesis. IEICE Trans. Inf. Syst. **88**, 502–509 (2005)
10. Tachibana, M., Yamagishi, J., Masuko, T., et al.: Speech synthesis with various emotional expressions and speaking styles by style interpolation and morphing. IEICE Trans. Inf. Syst. **88**, 2484–2491 (2005)
11. Takashi, N., Yamagishi, J., Masuko, T., et al.: A style control technique for HMM-based expressive speech synthesis. IEICE Trans. Inf. Syst. **90**, 1406–1413 (2007)
12. Yamagishi, J., Kobayashi, T., Nakano, Y., et al.: Analysis of speaker adaptation algorithms for HMM-based speech synthesis and a constrained SMAPLR adaptation algorithm. IEEE Trans. Audio Speech Lang. Process. **17**, 66–83 (2009)
13. Tokuda, K., Nankaku, Y., Toda, T., et al.: Speech synthesis based on hidden Markov models. Proc. IEEE **101**, 1234–1252 (2013)
14. Masuko, T., Tokuda, K., Kobayashi, T., et al.: Voice characteristics conversion for HMM-based speech synthesis system. In: 1997 IEEE International Conference on Acoustics, Speech, and Signal Processing, ICASSP 1997, vol. 3, pp. 1611–1614. IEEE (1997)
15. Tamura, M., Masuko, T., Tokuda, K., et al.: Adaptation of pitch and spectrum for HMM-based speech synthesis using MLLR. In: 2001 IEEE International Conference on Acoustics, Speech, and Signal Processing, Proceedings, (ICASSP 2001), vol. 2, pp. 805–808. IEEE (2001)
16. Lorenzo-Trueba, J., Barra-Chicote, R., Yamagishi, J., Montero, J.M.: Towards cross-lingual emotion transplantation. In: Navarro Mesa, J.L., Ortega, A., Teixeira, A., Hernández Pérez, E., Quintana Morales, P., Ravelo García, A., Guerra Moreno, I., Toledano, D.T. (eds.) IberSPEECH 2014. LNCS, vol. 8854, pp. 199–208. Springer, Cham (2014). doi:10.1007/978-3-319-13623-3_21
17. Yamagishi, J., Kobayashi, T.: Average-voice-based speech synthesis using HSMM-based speaker adaptation and adaptive training. IEICE Trans. Inf. Syst. **90**, 533–543 (2007)
18. Yang, H., Oura, K., Wang, H., et al.: Using speaker adaptive training to realize Mandarin-Tibetan cross-lingual speech synthesis. Multimedia Tools Appl. **74**, 9927–9942 (2015)
19. Russell, J.A.: Pancultural aspects of the human conceptual organization of emotions. J. Pers. Soc. Psychol. **45**, 1281 (1983)

20. Wester, M.: The emime bilingual database. University of Edinburgh (2010)
21. Kawahara, H., Masuda-Katsuse, I., De Cheveigne, A.: Restructuring speech representations using a pitch-adaptive time-frequency smoothing and an instantaneous-frequency-based F0 extraction: Possible role of a repetitive structure in sounds. Speech Commun. **27**, 187–207 (1999)
22. Loizou, P.C.: Speech quality assessment. In: Lin, W., Tao, D., Kacprzyk, J., Li, Z., Izquierdo, E., Wang, H. (eds.) Multimedia Analysis, Processing and Communications. SCI. Springer, Heidelberg (2011). doi:10.1007/978-3-642-19551-8_23

Mining Initial Nodes with BSIS Model and BS-G Algorithm on Social Networks for Influence Maximization

Xiaoheng Deng[✉], Dejuan Cao, Yan Pan, Hailan Shen, and Fang Long

School of Computer Science and Engineering, Central South University,
Changsha 410083, China
{dxh,caodejuan,panyan,hailansh,164611128}@csu.edu.cn

Abstract. Influence maximization is the problem to identify and find a set of the most influential nodes, whose aggregated influence in the network is maximized. This research is of great application value for advertising, viral marketing and public opinion monitoring. However, we always ignore the tendency of nodes' behaviors and sentiment in the researches of influence maximization. On general, users' sentiment determines users behaviors, and users' behaviors reflect the influence between users in social network. In this paper, we design a training model of sentimental words to expand the existing sentimental dictionary with the marked-comment data set, and propose an influence spread model considering both the tendency of users' behaviors and sentiment named as BSIS (Behavior and Sentiment Influence Spread) to depict and compute the influence between nodes. We also propose an algorithm for influence maximization named as BS-G (BSIS with Greedy Algorithm) to select the initial node. In the experiments, we use two real social network data sets on the Hadoop and Spark distributed cluster platform for experiments, and the experiment results show that BSIS model and BS-G algorithm on big data platform have better influence spread effects and higher quality of the selection of seed node comparing with the approaches with traditional IC, LT and CDNF models.

Keywords: Social networks · Influence maximization · Behavior tendency · Sentiment tendency · Greedy algorithm

1 Introduction

With the development of online social networks, such as Twitter, Facebook and Microblog, the social media platforms are becoming important channels for information dissemination. As information is disseminated in social networks through nodes, influence is spread by word-of-mouth or viral marketing in the cascade mode. One of the tough problems is how to make the best use of the features of information dissemination and discover the personal influence of every node in online social networks, so the research of influence maximization is of great

© Springer Nature Singapore Pte Ltd. 2017
B. Zou et al. (Eds.): ICPCSEE 2017, Part II, CCIS 728, pp. 138–147, 2017.
DOI: 10.1007/978-981-10-6388-6_12

significance. With the rapid development of data science, big data technology provide effective supports for the research of influence maximization on social networks.

Currently, most of the works is making improvements on the heuristic strategy based on IC and LT model, or improving the performance of the algorithm based on the structure of social network. Considering both the tendency of users' behaviors and sentiment is of great significance for influence maximization on social networks. However, the existing sentiment dictionary for sentiment analysis is not fit to analyze the comment data set of social networks, we build a training model of sentimental words to expand the existing dictionary. So based on the behavior tendency and the sentiment analysis of users in social network, we propose an influence spread model named as BSIS (Behavior and Sentiment Influence Spread Model) to describe and compute the influence between users. Combining with greedy algorithm and BSIS model, we design an algorithm named as BS-G (BSIS with Greedy Algorithm) to detect the most influential seed nodes. In the experiments, using the marked-comment data set from Joybuy and two real social network data sets from Flickr and Microblog, we implement BSIS and BS-G on the Hadoop and Spark platform.

2   Related Work

In [1,2], Domingos and Richardson first modeled the Influence Maximization (IM) as an optimization algorithmic problem. Based on a heuristic strategy, Kempe et al. in [3] aimed to formulate the IM problem as an optimization problem to improve the operation efficiency. Moreover, they presented two cascade models, i.e., IC and LT models, and showed that the greedy approximation algorithm whose influence spread is within $(1 - 1/e)$ of the optimal result. Based on the properties of monotonicity of the influence functions, Leskovec et al. in [5] proposed CELF (Cost-effective Lazy Forward) optimization, and Amit et al. further improved the CELF schema and named it as CELF++ optimization to improve the efficiency significantly [9]. Combining with the calculation on the optimization model based on the local tree structures, PMIA algorithm was proposed in [4,7] which overcome the inefficiency of operation successfully together with heuristic algorithms such as LDAG in [6], but comparing with greedy algorithm, these heuristic algorithms sacrificed the accuracy to improve the efficiency. In the real scene, it is reasonable that users' influence needs to be evaluated by many characters, Yan Pan et al. in [15] evaluated users' influence by user activity, user sensitivity, user affinity, and proposed CD-NF model and GNF algorithm. Besides, combining with the limit constraints and time constraints in the process of influence spread, they used credit distribution and users' behavior records to model the process of influence spread and proposed CDTC model in [16]. [8,10,12] combined users' individual preferences of feature information and proposed different models as an extension if traditional IC model. The research of IM has many important applications. Yang DN in [11] proposed an algorithm to make users to contact with each other based on recommending uses to each other. Li Hui in [14] proposed a parallel computing model which divided the network into many misalignment subgraphs and evaluated the influence probability with influence and recognition. Liu Qi in [13] used matrix transformation to compute the influence between nodes and proposed two algorithms to select the most influential initial nodes.

3   Influence Spread Model

3.1   Influence Maximization

Influence is spread by word-of-mouth, so with the spread of influence in social network, different members will be affected on different levels. On general, the social network can be modeled as a graph of $G(V, E)$, where V represents nodes set, and E represents edges of nodes. The aim of the research of IM is to find the amount of seed nodes, which makes the greatest influence in total. Therefore, the two keys of IM are how to detect and find the most influential seed nodes and how to evaluate the influence the seed nodes produce in the process of influence spread, which are confirmed as an NP-hard problem.

3.2   User Influence of Behavior

In social networks, users perform with execution behaviors which can be concluded as {post, thumb up, repost, comment}, and users interact with interaction behaviors which can also be concluded as {thumb up, repost, comment}. In general, the interaction behaviors records reflect the influence between users visually. While the frequency of interaction behaviors is an important factor to evaluate the intimacy between nodes, and the delay time factor nodes interact also reflects the sensitivity between users, so we can utilize these two key factors to evaluate the influence between nodes.

In social networks, given two adjacent user nodes u and v, assuming that when u posts a message, v may repost, make some comments or thumb up after some delay-time, then we can use the frequency of the interaction behaviors between u and v to infer the motivate probability of interaction behaviors u make on v named as $IBMP_{u,v}$.

$$IBMP_{u,v} = \frac{|B(u) \cap B(v)|}{|B(u)|} \tag{1}$$

Meanwhile, we can calculate the occupancy of interaction behaviors between u and v named as $OIB_{u,v}$.

$$OIB_{u,v} = \frac{|B(u) \cap B(v)|}{|B(v)|} \tag{2}$$

where $|B(u)|$ represents the number of execute behaviors of u, $|B(v)|$ represents the number of execute behaviors of v, and $B(u) \cap B(v)|$ represents the number of interaction behaviors of u and v.

Hence, the possibility of v interacts u is directly proportional to $IBMP_{u,v}$ and $OIB_{u,v}$, which means the influence u makes on v is in positive relationship with $IBMP_{u,v}$ and $OIB_{u,v}$. However, if u and v are not active nodes, there is a possibility that $IBMP_{u,v}$ and $OIB_{u,v}$ is high and make mistakes. Besides, the delay-time of the interaction behaviors of u and v is also determining the influence between them. We can infer that the influence between u and v is in inverse correlation with the average delay-time of them. So the user influence of behavior with the time factor of u and v named as $BTInf_{u,v}$ can be computed as below.

$$BTInf_{u,v} = \{\lambda \cdot IBMP_{u,v} + (1 - \lambda) \cdot OIB_{u,v}\} \cdot e^{-\frac{t_{v,u} - T_v}{T_v}} \tag{3}$$

where $t_{v,u}$ represents v's average delay-time towards u, T_v represents v's average delay-time towards all the neighbors of v. It can be seen that if $t_{v,u}$ is much smaller than T_v, the exponent is close to 1, and the influence becomes larger, on the contrary, if $t_{v,u}$ is much larger than T_v, the exponent is close to negative infinity, and the influence becomes smaller.

3.3   User Influence of Sentiment

Nowadays, people prefer to express their personal thoughts on some people, products, social events or else on social network rather in real world. Therefor, we can use the sentiment word dictionary to make sentiment analysis on the comment between nodes to build the user influence of sentiment. People used to use some oral words or new words to comment on others on social networks, and we can not find these words in the existing sentiment dictionary. Based on the existing sentiment dictionary like NTUSD and HowNet, we design a training model of sentimental words to train and obtain more sentimental words adding to the existing dictionary. We fetch the marked-comment data set with star-tags from the Joybuy with web crawlers as the training and testing data set. So the training model of sentimental words can be described as below:

Step 1: Use the ANSJ as the word segment tool, and count up the occurrence number of the segmented word in the positive and negative comments respectively.

Step 2: Count up the total number of positive and negative comments as N, set the threshold value as r, and set the max sentiment weight as max.

Step 3: Given a segmented word of $word$ with its occurrence number on positive comments as N_P and its occurrence number on negative comments as N_N, if $\frac{N_P + N_N}{N} > r$, execute step $4, 5, 6$.

Step 4: Get the positive probability of $word$ as:

$$P_P = \frac{N_P}{N_P + N_N}. \tag{4}$$

Step 5: Compute the sentimental weight of $word$: w_1 and w_2

$$w_1 = 2P_P - 1 = \frac{N_P - N_N}{N_P + N_N}. \tag{5}$$

$$w_2 = \begin{cases} log_2 \dfrac{2 \cdot N_P}{N_P + N_N} & (N_P \geq N_N) \\[3mm] -log_2 \dfrac{2 \cdot N_N}{N_P + N_N} & (N_P < N_N) \end{cases} \tag{6}$$

Step 6: Compute the final sentimental weight of *word* based on w_1 and w_2:

$$w = \begin{cases} 0.5 \cdot \{ \dfrac{N_P - N_N}{N_P + N_N} + log_2 \dfrac{2 \cdot N_P}{N_P + N_N} \} \cdot max & (N_P \geq N_N) \\[3mm] 0.5 \cdot \{ \dfrac{N_P - N_N}{N_P + N_N} - log_2 \dfrac{2 \cdot N_N}{N_P + N_N} \} \cdot max & (N_P < N_N) \end{cases} \tag{7}$$

Under normal conditions, persons' behaviors are dependent on persons' positive or negative emotions, and the sentiment between users determines the degree a user is influenced by another user. On social networks, given u, v, and $comment_j$ which is a comment v makes on u, make word segment on $comment_j$ as $\{word_1, word_2, ..., word_n\}$, look up its weight in the extended sentimental dictionary as $\{w_1, w_2, ..., w_n\}$, then the sentiment score of $comment_j$ can be computed as:

$$SentiSocre_{u,v,comment_j} = \sum_{i=1}^{n} w_i. \tag{8}$$

It is worth that there are some adverbs of degree just as "not" in comments on social networks, which can turn the sentiment tendency of the word they modify to be opposite. It is necessary to find the context of these adverbs of degree, get the set of sentimental words they are adjacent to and its sentimental weights in the expanded sentiment dictionary as $\{r\_w_1, r\_w_2, ...r\_w_m\}$, and amend the sentiment score of $comment_j$ as:

$$ReSentiSocre_{u,v,comment_j} = \sum_{i=1}^{n} w_i - 2 \cdot \sum_{k=1}^{m} w_k. \tag{9}$$

Thus, in order to consider all the comments synthetically, we get the average user influence of sentiment by:

$$SInf_{u,v} = \frac{1}{n} \cdot \sum_{j=1}^{n} \frac{1}{1 + e^{-o \cdot | \sum\limits_{w_i \in comment_j} w_i - 2 \cdot \sum\limits_{r\_w_k \in comment_j} r\_w_k |}}. \tag{10}$$

where the value of o is between 0 and 1.

3.4   User Comprehensive Influence

Considering both the tendency of users' behaviors with time-delay factor and users' sentiment, the user comprehensive influence u makes on v them can be denoted as:

$$Comp\_Inf_{u,v} = SInf_{u,v} + BTInf_{u,v}. \tag{11}$$

4   BS-G Algorithm

4.1   Node Margin Gain

As described above, we can compute the influence between nodes in social network by BSIS model. Note that we use $I(u)$ to denote the nodes set which have influence on u. Therefor, for u and v, the influence u makes on v can be denoted as:

$$\phi_{u,v}(u) = \sum_{w \in I(v)} \phi_{u,w}(u) \cdot Comp\_Inf(w,v) \tag{12}$$

We compute the use comprehensive influence by BSIS model, and use $\sigma_{Comp\_Inf}(S)$ to denote the influence spread function, the value of which is the total influence the seeds set S make on the social networks. Therefore, the margin gain of u which is the total influence u makes on the nodes out of S is given by:

$$\sigma_{BSIS}(S+u) - \sigma_{BSIS}(S) = (1 - \sum_{s \in S} \phi_{S,u}(S)) \cdot \sum_{v \in A-S} \phi_{u,v}(u) \tag{13}$$

4.2   BS-G Algorithm

As greedy algorithm can guarantee the accuracy of seed node for influence maximization. Hence, combining the BSIS model and greed algorithm, we design an algorithm for influence maximization named as BS-G to find the seed nodes.

5   Experimental Evaluation

We conduct our experiments on real online social networks to evaluate the BSIS model and the BS-G algorithm we proposed. More specifically, we focus on validating the influence spread consequence of seed set identified by our approach compared with other solutions, such as IC, LT and CDNF models. All models and algorithms are conducted on the distributed platform on Hadoop2.5.2 and Spark2.0.2.

5.1   Dataset Describe

We conduct our experiments using the comment data set from Joybuy to extend the existing sentiment word dictionary, which contains 17052 pieces of positive comments and 14937 pieces of negative comments. Meanwhile, we conduct our experiment with two real social network data set from Flickr and Microblog, while the data set from Flickr contains 40808 user nodes, and 75269 users' behavior records, and the data set from Microblog contains 412952 user nodes and 500977 users' behavior records.

Algorithm 1. BS-G Algorithm

Require: G, k

Ensure: seed set S.

1: set $S \leftarrow \varnothing, Q \leftarrow \varnothing$

2: **for** each $u \in V$ **do**

3:    compute: $Comp\_Inf(u, v)$;

4: **end for**

5: **for** each $u \in V$ **do**

6:    compute the margin gain: $\sigma_{BSIS}(S + u) - \sigma_{BSIS}(S)$;

7:    $u.flag \longleftarrow 0$, push u into queue Q in decreasing order;

8: **end for**

9: **for** $i = 1$ to k **do**

10:    $S \leftarrow pop(Q)$;

11:    **if** $u.flag = |S|$ **then**

12:        $S = S \bigcup \{u\}$;

13:        update influence distribution of $V - S$, resort Q;

14:    **else**

15:        re-compute marginal gain of u: $(\sigma_{BSIS}(S + u) - \sigma_{BSIS}(S))$;

16:        $u.flag \leftarrow |S|$, insert u into Q, resort Q;

17:    **end if**

18: **end for**

19: return S;

5.2   Experimental Results and Analysis

We perform our BS-G algorithm on the BSIS model. Comparing with three different models (CDNF [15], IC, LT) and heuristic algorithms on the same dataset, we want to explore the performances of our approach.

Influence Spread Consequence. In order to compare the influence spread effects of the seed nodes picked by different methods, we conduct our experiments using the data set from Flickr under BSIS, CDNF, IC and LT models. Figure 1(a) shows the single influence of top 50 seed node picked by different models, while the experiments illustrates the single influence of seed node picked by BSIS model is almost high than other models. Figure 1(b) shows the total influence of seed nodes selected by different algorithm. Figure 2(a) and (b) shows the single influence and total influence of seed nodes selected by BSIS with the data set from weibo. As BSIS model take both the users' behavior tendency and sentiment tendency into consideration, the experiments results proves that the proposed BSIS model and BS-G algorithm have better influence spread effects.

Real Influence Spread Consequence. In real social networks, we usually measure the influence of nodes by its clicked number, commented number reposted number or else. We use the sum of these three indexes named as action_number as the reference point of the users' real influence spread consequence in social network, and make analysis on the seed nodes about the

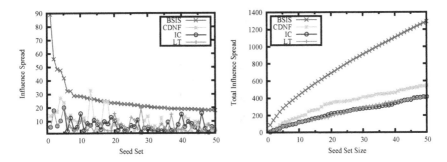

Fig. 1. Influence spread consequences achieved by different models on Flickr

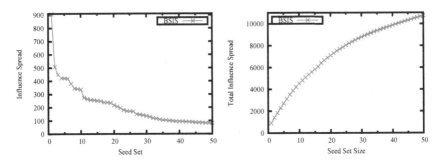

Fig. 2. Influence spread consequences on Microblog

action_number. Figure 3(a) and (b) shows the single action_number and the total action_number of seed node using the data set from Flickr selected by BSIS, CDNF, IC and LT, while it indicates that the real influence effects of seed nodes selected by BSIS is much higher than others. Figure 4(a) and (b) shows the single action_number and total action_number of seed nodes by BSIS using the data set from Microblog. Obviously, the experiment proves that BSIS can ensure the high quality of seed nodes.

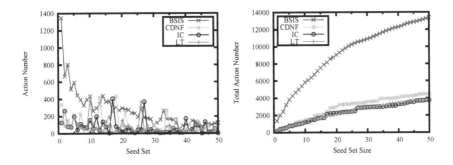

Fig. 3. Action_number achieved by different models on Flickr

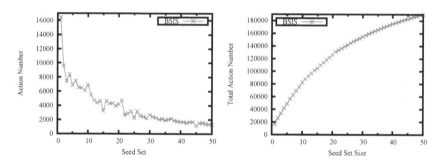

Fig. 4. Action_number on Microblog

6   Conclusion

In this work, we mining the user influence by analyzing the users' records and leveraging the temporal nature of influence spread for influence maximization. Specifically, we describe user influence from various aspects which cover user' behaviors and sentiment analysis of users. Furthermore, we present the BSIS model and first adopt the BS-G algorithm to identify the seed nodes which would be activated in prime. After calculating and simulating the influence spread in the whole network, we identify the nodes which have maximum marginal gain to form the seed set S. It can be proved that the BS-G algorithm implemented in BSIS model has better performance in seed set quality.

Acknowledgment. This work is supported by the National Natural Science Foundation of China (Grant No. 61379058).

References

1. Domingos, P., Richardson, M.: Mining the network value of customers. In: ACM SIGKDD International Conference on Knowledge Discovery and Data Mining, pp. 57–66. ACM (2001)
2. Richardson, M., Domingos, P.: Mining knowledge-sharing sites for viral marketing. In: Eighth ACM SIGKDD International Conference on Knowledge Discovery and Data Mining, pp. 61–70. ACM (2002)
3. Kempe, D., Kleinberg, J., Tardos, É.: Maximizing the spread of influence through a social network. In: ACM SIGKDD International Conference on Knowledge Discovery and Data Mining, pp. 137–146. ACM (2003)
4. Chen, W., Wang, Y., Yang, S.: Efficient influence maximization in social networks. In: ACM SIGKDD International Conference on Knowledge Discovery and Data Mining, Paris, France, 28 June – 1 July 2009, pp. 199–208. DBLP (2009)
5. Leskovec, J., Krause, A., Guestrin, C., et al.: Cost-effective outbreak detection in networks. In: ACM SIGKDD International Conference on Knowledge Discovery and Data Mining, pp. 420–429. ACM (2007)

6. Chen, W., Yuan, Y., Zhang, L., et al.: Scalable influence maximization in social networks under the linear threshold model. In: 2010 IEEE 10th International Conference on Data Mining (ICDM), Los Alamitos, CA, United states, pp. 88–96 (2010)
7. Chen, W., Collins, A., Cummings, R., Ke, T., Liu, Z., Rincon, D., Sun, X., Wang, Y., Wei, W., Yuan, Y.: Influence maximization in social networks when negative opinions may emerge and propagate In: SDM, Mesa, AZ, United states, vol. 11, pp. 379–390 (2011)
8. Goyal, A., Bonchi, F., Lakshmanan, L.V.S.: A data-based approach to social influence maximization. Proc. VLDB Endow. 5(1), 73–84 (2011)
9. Goyal, A., Lu, W., Lakshmanan, L.V.S.: CELF++: optimizing the greedy algorithm for influence maximization in social networks. In: International Conference Companion on World Wide Web, pp. 47–48. ACM (2011)
10. Zhang, H., Dinh, T.N., Thai, M.T.: Maximizing the spread of positive influence in online social networks. In: IEEE, International Conference on Distributed Computing Systems, pp. 317–326. IEEE (2013)
11. Yang, D.N., Hung, H.J., Lee, W.C., et al.: Maximizing acceptance probability for active friending in online social networks. In: Proceedings of the 19th ACM SIGKDD International Conference on Knowledge Discovery and Data Mining, pp. 713–721. ACM (2013)
12. Goyal, A., Bonchi, F., Lakshmanan, L.V.S.: Learning influence probabilities in social networks. In: International Conference on Web Search and Web Data Mining, WSDM 2010, New York, NY, USA, pp. 241–250. DBLP, February 2010
13. Liu, Q., Xiang, B., Chen, E., et al.: Influence maximization over large-scale social networks: a bounded linear approach. In: ACM International Conference on Conference on Information and Knowledge Management, pp. 171–180. ACM (2014)
14. Li, H., Bhowmick, S.S., Sun, A.: Conformity-aware greedy algorithm for influence maximization in online social networks. In: International Conference on Extending Database Technology, pp. 323–334 (2013)
15. Deng, X., Pan, Y., Wu, Y., et al.: Credit distribution and influence maximization in online social networks using node features. In: International Conference on Fuzzy Systems and Knowledge Discovery, pp. 2093–2100. IEEE (2016)
16. Pan, Y., Deng, X., Shen, H.: Credit distribution for influence maximization in online social networks with time constraint. In: IEEE International Conference on Smart City/SocialCom/SustainCom, pp. 255–260. IEEE (2016)

Critical Value Aware Data Acquisition Strategy in Wireless Sensor Networks

Ran Bi[1(✉)], Guozhen Tan[1], and Xiaolin Fang[2]

[1] School of Computer Science and Technology, Dalian University of Technology, Dalian 116024, China
{biran,gztan}@dlut.edu.cn
[2] School of Computer Science and Technology, Southeast University, Nanjing 211189, China
xiaolin@seu.edu.cn

Abstract. To monitor the physical world, Equi-Frequency Sampling (EFS) methods are widely applied for data acquisition in sensor networks. Due to the noise and inherent uncertainty of the environment, EFS based data acquisition may result in misconception to the physical world, and high frequency scheme produces massive sensed data, which consumes substantial cost for transmission. This paper proposes a novel sensed data model. Based on maximum likelihood estimation, the model can minimize measurement error. It is proved that the proposed model is asymptotic unbiased. Furthermore, this paper proposes Model based Adaptive Data Collection (MADC) Algorithm and designs a distributed lightweight computation algorithm named Distributed Adaptive Data Collection Algorithm (DADC). Based on the error of prediction, both algorithms can adaptively adjust the cycle of data collection. Performance evaluation verifies that the proposed algorithms have high performance in terms of accuracy and effectiveness.

Keywords: Data model · Adaptive collection · Sensor networks

1 Introduction

Wireless sensor networks (WSNs) bridge the physical world and the computing systems, which open up the potential to monitor the large geographical areas inexpensively. In most practical applications, there exist two stages for observing the physical world. In the first stage, each sensor node samples the data from the physical world based on equi-frequency sampling (EFS) technique. In the second stage, the raw data collected by the sensor nodes are forwarded to the base station. EFS based data acquisition technique is well known and widely applied in the research domain of sensor networks. To reduce the communication cost, most works focus on the technique of data processing, such as approximate computing algorithm [1–3], data transmission [4,5].

Due to the noise and inherent uncertainty of the environment, EFS based data acquisition may result in misconception to the physical world. As shown

© Springer Nature Singapore Pte Ltd. 2017
B. Zou et al. (Eds.): ICPCSEE 2017, Part II, CCIS 728, pp. 148–160, 2017.
DOI: 10.1007/978-981-10-6388-6_13

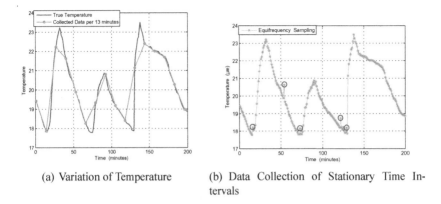

(a) Variation of Temperature

(b) Data Collection of Stationary Time Intervals

Fig. 1. Performance of EFS based data acquisition method (Color figure online)

in Fig. 1(a) and (b), we give examples to see the problem. In Fig. 1(a), blue curve describes the variation of true temperature at Massive Data Computing Lab located in Harbin. Red star points represent the sensed data sampled per 13 min. Figure 1(a) shows that some crucial values are missed, such as maximal and minimal points. Data acquisition of the low sampling frequency may miss the crucial points of the physical world. And it is desirable to sample data more frequently around crucial value points. Temperature changed slowly at the interval of 30–70 min, so the data sampled at lower frequency can reflect the tendency of temperature variation. Then low sampling rate is satisfactory. In Fig. 1(b), the red star points represent the temperature values which were collected per minute. The red star in blue circle presents the large deviation from the true temperature, which is caused by measurement error and noise.

In general, physical world contiguously changes and it presents strong spatial-temporal correlations. Prediction algorithm is on the basis of the observation that the correlations among sensed data create the possibility of training and using predictors in a distributed method [7]. Schemes that exploit temporal and spatial correlations in wireless sensor networks have been widely discussed [6,9]. There exist many models and prediction based algorithms. These algorithms imply that sensor nodes do not need to transmit sensed data unless the deviation between predicted value and sensed value is larger than a predefined threshold [8].

Currently, there are only several works that studied adaptive sampling for sensed data acquisition [6]. According to statistical modeling, literature [10] proposed adaptive sampling strategy. But the method is also based on the EFS approach. [6] provided an adaptive sampling method for reconstructing the physical world. The approach depends on the estimation of the first and second derivatives of the physical process.

In many applications, such as environmental monitoring and fire detection, users always pay attention to the extreme values [6], which implies unusual event in the physical world. High-frequency strategy can collect extreme value points as much as possible. However, this scheme incurs substantial cost for transmission.

Sensed data shows highly spatial-temporal correlation. Then it indicates that model based prediction opens out prospect for adaptive data collection. The proposed strategy aims to forward extreme value points and reduce the transmission of the sensed data whose effect on comprehending physical process is negligible. Model based approaches can efficiently take advantage of the spatio-temporal correlation among sensed data. Motivated by this, model based adaptive data collection algorithms are proposed. Contribution of this paper can be summarized as follows:

- Sensed data model is proposed in this paper. The novelty of the model is that it integrates measurement error and correlations of sensed data among different sensor nodes. The proposed model efficiently coordinates the factors that affect the sensed data of node.
- Maximum likelihood estimation is applied to achieve the estimations of parameters in sensed data model. Through this method, the model can minimize measurement error and the estimator with guarantee of asymptotic approximation is proved.
- On the basis of prediction error, Model based Adaptive Data Collection algorithm is proposed. MADC algorithm adaptively adjusts the cycle of data collection. Moreover, a distributed lightweight computation algorithm named Distributed Adaptive Data Collection algorithm (DADC) is designed. DADC algorithm makes the tradeoff between energy cost and accuracy of prediction.
- Performance evaluation verifies that the proposed algorithms have high performance in terms of accuracy and effectiveness.

The rest of this paper is structured as following. Section 2 introduces Sensed Data Model. Section 3 describes the mathematical foundations of adaptive data collection algorithm. MADC and DAFC algorithm are given in Sect. 4. Experimental results are illustrated in Sect. 5, and Sect. 6 concludes this paper.

2   Sensed Data Model

2.1   Problem Definition

It is assumed that there are m sensor nodes in a local monitoring area and the ids of sensor nodes are $1, 2, ..., m$. At a period of time, the sensed data of node i are $d_{i1}, d_{i1}, ..., d_{in}$, where d_{ij} represents the sensed data of node i at time t_{ij}. Then sensed the data model is defined as following

$$d(t_{ij}) = \alpha(t_{ij}) + v_i(t_{ij}) + \varepsilon_i(t_{ij})$$

where, $i = 1, 2, ..., m; j = 1, 2, ..., n_i$. $\alpha(t_{ij})$ models the mean sensed value of node i at time t_{ij} and $v_i(t_{ij})$ models the individual deviation of the sensed value from the mean value $\alpha(t_{ij})$. $\varepsilon_i(t_{ij})$ is the measurement error of node i at time t_{ij}, which is incurred by noise. In practical application of sensor networks, there exists spatial-temporal correlation among sensed data. Hence $v_i(t)$ can be regarded as

Gaussian process with mean μ_i and covariance $Cov_i(t)$. $\varepsilon_i(t)$ can be seen as an uncorrelated Gaussian process with mean 0 and variance function $\sigma^2(t)$.

Denote $\bar{d} = \frac{\sum_{i=1}^{m}\sum_{j=1}^{n_i} d_{ij}}{\sum_{i=1}^{m} n_i}$, $\bar{d}_i = \frac{\sum_{j=1}^{n_i} d_{ij}}{n_i}$, and let $\Delta_i = \bar{d}_i - \bar{d}$, then $\mu_i \approx \Delta_i$. Then we have the following.

$$
\begin{aligned}
s_i(t_{ij}) &= d_i(t_{ij}) - \Delta_i \\
&= \alpha(t_{ij}) + (v_i(t_{ij}) - \Delta_i) + \varepsilon_i(t_{ij}) \\
&= \alpha(t_{ij}) + \varphi_i(t_{ij}) + \varepsilon_i(t_{ij})
\end{aligned}
\tag{1}
$$

where $j = 1, ..., n; i = 1, ..., m$. It is known that $\varphi_i(t)$ and $\varepsilon_i(t)$ are independent, where $\varphi_i(t)$ is a Gaussian process with mean 0. In this paper, the proposed sensed data model is as follows.

$$
s_i(t_{ij}) = \alpha(t_{ij}) + \varphi_i(t_{ij}) + \varepsilon_i(t_{ij})
\tag{2}
$$

Based on Taylor Expansion about t_{ij}, $\alpha(t_{ij})$ and $\varphi_i(t_{ij})$ can be approximated by k-th order polynomials within a neighborhood of t.

$$
\begin{aligned}
\alpha(t_{ij}) &= \alpha(t) + \alpha'(t)(t_{ij} - t) + \cdots + \frac{\alpha^{(k)}(t)}{k!}(t_{ij} - t)^k + R_k(t_{ij} - t) \\
\varphi_i(t_{ij}) &= \varphi_i(t) + \varphi_i'(t)(t_{ij} - t) + \cdots + \frac{\varphi_i^{(k)}(t)}{k!}(t_{ij} - t)^k + R_k(t_{ij} - t), i = 1, ..., m
\end{aligned}
\tag{3}
$$

$R_k(t_{ij} - t)$ is Peano Remainder term, where $R_k(t_{ij} - t) = o\left((t_{ij} - t)^k\right)$. Therefore, we can achieve the following approximations of $\alpha(t_{ij})$ and $\varphi_i(t_{ij})$.

$$
\begin{aligned}
\alpha(t_{ij}) &\approx \alpha(t) + \alpha'(t)(t_{ij} - t) + \cdots + \frac{\alpha^{(k)}(t)}{k!}(t_{ij} - t)^k \\
\varphi_i(t_{ij}) &\approx \varphi_i(t) + \varphi_i'(t)(t_{ij} - t) + \cdots + \frac{\varphi_i^{(k)}(t)}{k!}(t_{ij} - t)^k, i = 1, ..., m
\end{aligned}
\tag{4}
$$

Denote $\tau_{ij} = (1, t_{ij} - t, ..., (t_{ij} - t)^k)^T$, $\gamma = (\alpha(t), \alpha'(t), ..., \frac{\alpha^{(k)}(t)}{k!})^T$, $\psi_i = (\varphi_i(t), \varphi_i'(t), ..., \frac{\varphi_i^{(k)}(t)}{k!})^T$, we have the following.

$$
\alpha(t_{ij}) \approx \tau_{ij}^T \gamma, \varphi_i(t_{ij}) \approx \tau_{ij}^T \psi_i
$$

$\alpha(t_{ij})$ and $\varphi_i(t_{ij})$ are approximated by the local k-th order polynomial within a neighborhood of t, then we can obtain the following sensed data model.

$$
s_i(t_{ij}) = \tau_{ij}^T(\gamma + \psi_i) + \varepsilon_i(t_{ij}), j = 1, ..., n_i; i = 1, ..., m
\tag{5}
$$

Let $S_i = (s_i(t_{i1}), s_i(t_{i2}), ..., s_i(t_{in_i}))^T$, $\lambda_i = (\tau_{i1}, \tau_{i2}, ..., \tau_{in_i})^T$, $\varepsilon_i = (\varepsilon_i(t_{i1}), \varepsilon_i(t_{i2}), ..., \varepsilon_i(t_{in_i}))^T$, we can obtain the following equation.

$$
S_i = \lambda_i(\gamma + \psi_i) + \varepsilon_i, i = 1, ..., m
\tag{6}
$$

where ψ_i follows the distribution of $N(\mathbf{0}, D_i)$ and $D_i = E(\psi_i\psi_i^T)$. Thus we can get to know.

$$\begin{pmatrix} S_1 \\ \vdots \\ S_m \end{pmatrix} = \begin{pmatrix} \lambda_1(\gamma + \psi_1) \\ \vdots \\ \lambda_m(\gamma + \psi_m) \end{pmatrix} + \begin{pmatrix} \varepsilon_1 \\ \vdots \\ \varepsilon_m \end{pmatrix} \tag{7}$$

For convenience, let $\mathbf{S} = (S_1^T, ..., S_m^T)^T$, $\boldsymbol{\varepsilon} = (\varepsilon_1^T, ..., \varepsilon_m^T)^T$. And similar notations are introduced for λ and ψ. That is $\boldsymbol{\lambda} = (\lambda_1^T, ..., \lambda_m^T)^T$ and $\boldsymbol{\psi} = (\psi_1^T, ..., \psi_m^T)^T$. It is obviously that $\lambda_i = \lambda_j, i, j \in \{1, 2, ..., m\}$. Let $\boldsymbol{\Gamma} = diag(\lambda_1, ..., \lambda_m)$, we can obtain

$$\mathbf{S} = \boldsymbol{\lambda}\gamma + \boldsymbol{\Gamma}\boldsymbol{\psi} + \boldsymbol{\varepsilon} \tag{8}$$

Based on the analysis of sensed data model, we know that ψ_i follows normal distribution of $N(\mathbf{0}, D_i)$, in which $D_i = E(\psi_i\psi_i^T)$. Then ψ follows normal distribution $N(\mathbf{0}, \mathbf{D})$, where $\mathbf{D} = diag(D_1, D_2, ..., D_m)$. Assuming the distribution of measurement error incurred by noise is uniform, then $\boldsymbol{\varepsilon}$ follows normal distribution $N(\mathbf{0}, \mathbf{R})$, in which $\mathbf{R} = diag(\sigma_1^2 I_{n_1}, .., \sigma_m^2 I_{n_m})$ I_n and is an identity matrix of n dimension. Based on the above discussion, we can obtain the joint distribution of ψ and $\boldsymbol{\varepsilon}$, then we have

$$\begin{pmatrix} \psi \\ \varepsilon \end{pmatrix} \sim N\left(\begin{pmatrix} \mathbf{0} \\ \mathbf{0} \end{pmatrix}, \begin{pmatrix} \mathbf{D} & \mathbf{0} \\ \mathbf{0} & \mathbf{R} \end{pmatrix}\right) \tag{9}$$

3   Mathematical Foundation

3.1   Parameter Estimation Based on Maximum Likelihood

Theorem 1. *Based on maximum likelihood estimation, the estimations for γ and ψ are equivalent to the solutions of the following linear equations.*

$$\begin{bmatrix} \boldsymbol{\lambda}^T\Phi_h\boldsymbol{\lambda} & \boldsymbol{\lambda}^T\Phi_h\boldsymbol{\Gamma} \\ \boldsymbol{\Gamma}^T\Phi_h\boldsymbol{\lambda} & \boldsymbol{\Gamma}^T\Phi_h\boldsymbol{\Gamma} + \mathbf{D}^{-1} \end{bmatrix} \begin{bmatrix} \hat{\gamma} \\ \hat{\psi} \end{bmatrix} = \begin{bmatrix} \boldsymbol{\lambda}^T\Phi_h\mathbf{S} \\ \boldsymbol{\Gamma}^T\Phi_h2\mathbf{S} \end{bmatrix} \tag{10}$$

where $\boldsymbol{\lambda} = (\lambda_1^T, ..., \lambda_m^T)^T$, $\mathbf{S} = (\mathbf{S_1^T}, \mathbf{S_2^T}, ..., \mathbf{S_m^T})^T$, $\Phi_h = diag[\Phi_1^h, ..., \Phi_m^h]$, $\Phi_i^h = K_h^{1/2} R_i^{-1} K_h^{1/2}$, $R_i = E(\varepsilon_i\varepsilon_i^T) = \sigma_i^2 \mathbf{I_n}$. K_h *is a kernel function, h is the bandwidth of kernel function.*

Proof. According to formula 8, we know the log-likelihood function of $\mathbf{S}$ given ψ excepting constant terms.

$$L(\mathbf{S}|\boldsymbol{\psi}) = -\frac{1}{2}\log|\Phi^{-1}| - \frac{1}{2}[\mathbf{S} - (\boldsymbol{\lambda}\gamma + \boldsymbol{\Gamma}\boldsymbol{\psi})]^T\Phi[\mathbf{S} - (\boldsymbol{\lambda}\gamma + \boldsymbol{\Gamma}\boldsymbol{\psi})] \tag{11}$$

In which $\Phi = diag[R_1^{-1}, ..., R_m^{-1}]$, and $R_i = E(\varepsilon_i\varepsilon_i^T) = \sigma_i^2\mathbf{I_n}$. Since the sensed data is finite, we apply the maximum penalized likelihood estimation on γ and ψ. We derive the following.

$$L(\gamma, \mathbf{S}, \psi) = L(\mathbf{S}|\psi) - \frac{1}{2}\psi^T \mathbf{D}^{-1}\psi$$

$$= -\frac{1}{2}\log|\Phi^{-1}| - \frac{1}{2}[\mathbf{S} - (\lambda\gamma + \Gamma\psi)]^T \Phi [\mathbf{S} - (\lambda\gamma + \Gamma\psi)] \quad (12)$$

$$- \frac{1}{2}\psi^T \mathbf{D}^{-1}\psi$$

Actually, maximum penalized likelihood estimation maximizes formula (12) and it is equivalent to minimizing the following.

$$\sum_{i=1}^{m} [S_i - \lambda_i(\gamma + \psi_i)]^T R_i^{-1}[S_i - \lambda_i(\gamma + \psi_i)] + \psi_i^T D_i^{-1}\psi_i$$

$$+ \sum_{i=1}^{m} \log|D_i| + \log|R_i| \quad (13)$$

We notice that the first term in formula (13) is the weighted sum of squares, therefore maximum penalized likelihood based estimation can minimize the measurement error. In the regression analysis, kernel function is a prevalent weighting function. Based on the propinquity of variable neighborhood, kernel function assigns the weights of the contribution for estimated value. For effective utilization of neighborhood information, we take kernel function into consideration. Let $K_h = diag(K_h(t_1 - t), ..., K_h(t_n - t))$, h is the bandwidth of kernel function K_h. We can derive the following.

$$\sum_{i=1}^{m} [S_i - \lambda_i(\gamma + \psi_i)]^T K_h^{1/2} R_i^{-1} K_h^{1/2}[S_i - \lambda_i(\gamma + \psi_i)]$$

$$+ \sum_{i=1}^{m} \psi_i^T D_i^{-1}\psi_i + \log|D_i| + \log|R_i| \quad (14)$$

Differentiation on γ and ψ, we can achieve the following linear equations.

$$\begin{bmatrix} \lambda^T \Phi_h \lambda & \lambda^T \Phi_h \Gamma \\ \Gamma^T \Phi_h \lambda & \Gamma^T \Phi_h \Gamma + \mathbf{D}^{-1} \end{bmatrix} \begin{bmatrix} \hat{\gamma} \\ \hat{\psi} \end{bmatrix} = \begin{bmatrix} \lambda^T \Phi_h \mathbf{S} \\ \Gamma^T \Phi_h 2\mathbf{S} \end{bmatrix} \quad (15)$$

3.2 Theoretical Analysis of $\hat{\gamma}$ and $\hat{\psi}$

This section focuses on the theoretical analysis and the asymptotic property of the parameter estimations. On the basis of Semi-parametric Stochastic Mixed Models [11], we can derive the following approximate solutions to linear equations (10)

$$\hat{\gamma} \approx (\lambda^T \Phi_\gamma \lambda)^{-1} \lambda^T \Phi_\gamma \mathbf{S}$$

$$\hat{\psi} \approx (\Gamma^T \Phi_\psi \Gamma + \mathbf{D}^{-1})\Gamma^T \Phi_\psi \mathbf{S}$$

$$\Phi_\gamma = \Phi_h - \Phi_h \Gamma(\Gamma^T \Phi_h \Gamma + \mathbf{D}^{-1})^{-1}\Gamma^T \Phi_h \quad (16)$$

$$\Phi_\psi = \Phi_h - \Phi_h \lambda(\lambda^T \Phi_h \lambda)^{-1}\lambda^T \Phi_h$$

$\hat{\gamma}$ and $\hat{\psi}$ are asymptotically unbiased estimations of γ and ψ respectively [12]. Unbiased property is crucial for estimators. Based on Chebyshev's inequality, the probability of $\hat{\gamma}$ deviating from γ can be bounded. The proposed sensed data model can provide theoretical guarantee for prediction.

Theorem 2. *The probability of the deviation of estimator being larger than ε is $O(\varepsilon^{-2}N^{-1})$, where $N = m * n$.*

Proof. According to the previous work for mixed-effects model in regression analysis [12], for a general linear model, the variance function of $\hat{\alpha}(t)$ can approximated by $\eta^2(t)(2Nhf(t))^{-1}$. h is the bandwidth of kernel function and $f(t)$ is the probability density function of sensed data at time t. Then we have the following.

$$\eta^2(t) \approx \sum_{i=1}^{m} \psi_i^2(t)/m - 1 + \sum_{i=1}^{m} m^{-1}\sigma_i^2$$

In sensor networks, the domain of sensed data is bounded, therefore $\psi_i^2(t)$ and σ_i^2 are bounded. Let $\rho = \sum_{i=1}^{m} m^{-1}\sigma_i^2 + \max\{\psi_i^2(t)|i = 1, 2, ..., m\}$, then $Var[\hat{\alpha}(t)]$ is smaller than $\rho(2Nhf(t))^{-1}$.

Based on Chebyshev's inequality, the probability of the deviation from $\hat{\alpha}(t)$ being larger than ε is smaller than $var[\alpha(t)]\varepsilon^{-2}$. ρ and h are constants. $f(t)$ is the probability density function of sensed data at time t, where $t \geq 0$. Then $f(t)$ is discrete and bounded. Let ω denote $\min\{f(t_i)|i = 1, 2, ..., n\}$. Thus $var[\alpha(t)]\varepsilon^{-2}$ is smaller than $\rho(2Nhf(t))^{-1}$. We know that the probability of the deviation of estimator being larger than ε is $O(\varepsilon^{-2}N^{-1})$.

Corollary 1. *If N is larger than $\rho(2h\omega\delta\varepsilon^2)^{-1}$, the probability of the deviation between $\hat{\alpha}(t)$ and $\alpha(t)$ being larger than ε is smaller than δ, where $N = m * n$.*

Corollary 2. *The formulas of estimations for $\alpha(t)$ and $\varphi_i(t)$ are as follows.*

$$\hat{\alpha}(t) = e_1^T \left\{ \sum_{i=1}^{m} (I + G_i D_i)^{-1} G_i \right\}^{-1} \sum_{i=1}^{m} (I + G_i D_i)^{-1} c_i \qquad (17)$$
$$\hat{\varphi}_i(t) = e_1^T D_i (I + G_i D_i)^{-1} d_i$$

The predicted sensed value can be computed as following.

$$\hat{d}_i(t) = \hat{\alpha}(t) + \hat{\varphi}_i(t) + \Delta_i$$

Proof. For convenience, we introduce the following symbols. Let $g_{ir} = \sum_{j=1}^{n} \frac{K_h(t_j-t)*(t_j-t)^r}{\sigma_i^2}$, $c_{ir} = \sum_{j=1}^{n} \frac{K_h(t_j-t)*(t_j-t)^r s_{ij}}{\sigma_i^2}$, $\mathbf{g}_{ir} = (g_{ir}, ..., g_{ir+k})^T$, $d_{ir} = \sum_{j=1}^{n} \frac{K_h(t_j-t)*(t_j-t)^r(s_{ij}-\tau_j^T\hat{\gamma})}{\sigma_i^2}$.

Based on differential of $\hat{\gamma}$ and $\hat{\psi}$ in linear equations (10), when $j = 0$, we can obtain estimations of $\alpha(t)$ and $\varphi_i(t)$.

$$\hat{\alpha}(t) = e_1^T \{ \sum_{i=1}^{m} (I + G_i D_i)^{-1} G_i \}^{-1} \sum_{i=1}^{m} (I + G_i D_i)^{-1} c_i \qquad (18)$$
$$\hat{\varphi}_i(t) = e_1^T D_i (I + G_i D_i)^{-1} d_i$$

where $G_i = (\mathbf{g}_{i0}, \mathbf{g}_{i1}, ..., \mathbf{g}_{ik})$, $c_i = (c_{i0}, c_{i1}, ..., c_{ik})^T$, $d_i = (d_{i0}, d_{i1}, ..., d_{ik})^T$.

4   Model Based Adaptive Data Collection Algorithm

This section proposes Model based Adaptive Data Collection algorithm named MADC for short. According to the prediction error, MADC algorithm adaptively adjusts the frequency of data collection. MADC aims to reduce the data transmission and forward extreme value points as much as possible.

If the relative error between predicted value and true sensed value is larger than predefined ε, then the sensor node will collect sensed data at higher frequency. If the relative error gets smaller, the collection rate will be decreased. Unusual event or breakdown of sensor node will cause drastic variation of sensed data, which results in large error between predicted value and true sensed data. Therefore more collected data will be desirable for analysis. Small error implies that sensed data model can effectively reflect the physical world. Thus less collected data can meet the requirement. Based on above observation, prediction error is regard as the metric for adjusting cycle. In Sect. 5, experiment results validate the efficiency of the proposed approach.

Assuming a wireless sensor network consists of l disjoint clusters, member node communicates with head node in one-hop in a cluster. Each sensor node has r modes for data collection, where the i-th mode of data collection samples data every m_i unit time. We suppose that $m_1 < m_2 < \cdots < m_r$. The detail of MADC Algorithm for cluster C_j is described as following.

Step 1. We assume that C_j consists of one cluster head node and $m-1$ cluster member nodes. Based on latest historical information and empirical Bayesian estimation [13], sink achieves the estimation for covariance matrices D_j and R_j for cluster C_j. According to formula (17), sink computes the estimations for parameters and sends $\hat{\alpha}, \hat{\varphi}_1, ..., \hat{\varphi}_m$ and R_j to cluster head of C_j.

Step 2. When the cluster head of C_j receives the data packet from the sink, the cluster head broadcasts $\hat{\alpha}, \hat{\varphi}_1, ..., \hat{\varphi}_m$ and R_j to member nodes.

Step 3. Based on the estimations of parameters received from the head node, member node adaptively adjusts the frequency of data collection. It is supposed that the current sampling mode of sensor node is m_i. If the relative error is larger than $j * \varepsilon$, then the sensor node sends mode m_{i-j+1} to the cluster head. Then the node updates the predicted time at which extreme value will be collected.

Step 4. Based on the messages received from member nodes, the cluster head broadcasts new data collection mode to member nodes.

4.1   Distributed Adaptive Data Collection Algorithm

According to formula (17), MADC algorithm can provide more accuracy estimation for parameters, however it needs more communication, which is not suitable for energy-constrained sensor networks. For efficient energy utilization, this section proposes Distributed Adaptive Data Collection algorithm.

Corollary 3. *Based on formula (17), the approximate estimations for α and φ_i are as following.*

$$\hat{\alpha}_i(t) = \left\{\sum_{j=1}^{n} \frac{y_i(t_j)}{1 + x_i(t_j)}\right\} \bigg/ \left\{\sum_{i=1}^{n} \frac{x_i(t_j)}{1 + x_i(t_j)}\right\} \tag{19}$$

$$\hat{\varphi}_i(t) = \{y_i(t) - x_i(t)\hat{\alpha}(t)\}/\{1 + x_i(t)\}$$

The predicted sensed value can be computed as following.

$$\hat{d}_i(t) = \hat{\alpha}(t) + \hat{\varphi}_i(t)$$

Proof. Based on formula (17), estimation of parameters is more precise, but it requires high cost for transmission. Thus it is not suitable to sensor node with restricted capability under distributed architecture. In view of the consideration, we should design a less accuracy but lightweight computation algorithm for $\hat{\alpha}_i$ and $\hat{\varphi}_i$. Let k be zero, then D_i is degenerated a scalar function, which is denoted as δ_i. We can derive the following formula, which degenerates from formula (17).

$$\hat{\alpha}_i(t) = \left\{\sum_{j=1}^{n} \frac{y_i(t_j)}{1 + x_i(t_j)}\right\} \bigg/ \left\{\sum_{i=1}^{n} \frac{x_i(t_j)}{1 + x_i(t_j)}\right\}$$

$$\hat{\varphi}_i(t) = \{y_i(t) - x_i(t)\hat{\alpha}(t)\}/\{1 + x_i(t)\}$$

where $x_i(t)$ and $y_i(t)$ satisfy the following formula

$$x_i(t) = \delta_i \sum_{j=1}^{n} \frac{K_h(t_j - t)}{\sigma_i^2}, y_i(t) = \delta_i \sum_{j=1}^{n} \frac{K_h(t_j - t) * d_i(t_j)}{\sigma_i^2}$$

In sensed data model, $\alpha(t_j)$ models the mean of sensed data at time t_j in a specific monitoring area. In MADC algorithm, $\alpha(t)$ is the average value of sensed data within cluster. If the head node receives the data from member nodes, according to formula (17), the estimation of parameters can be obtained at head node. But this strategy also needs additional communication cost between head node and member nodes. To reduce communication cost, accuracy loss of estimation is desirable. Based on the observation, this section gives Distributed Adaptive Data Collection Algorithm, named DADC for short. DADC algorithm makes the tradeoff between communication cost and precision of estimation. The detail of DADC algorithm is described as follows.

Step 1. We assume that the latest sampled data of node i are $\{d_i(t_1), d_i(t_2), ..., d_i(t_n)\}$. Let $\alpha_i = \sum_{i=1}^{n} d_i(t_j)/n$. Due to the low computing capability of sensor node, node i regards subsample variance as estimation of δ_i, that is $\delta_i = \sum_{i=1}^{n} (d_i(t_j) - \sum_{i=1}^{n} d_i(t_j)/n)^2/n$.

Step 2. According to formula (19), member node can compute the estimations of $\alpha(t)$ and $\varphi_i(t)$ respectively.

Step 3. Based on the relative error between predicted value and true sensed data, node i adaptively adjusts frequency of data collection. Frequency adjustment of data collection is similar to the process of member node of MADC algorithm.

Step 4. According to the messages received from member nodes, the cluster head broadcasts new data collection mode to member nodes. It is similar to the process of cluster head node of MADC algorithm.

Both of the algorithms are based on topology of clusters. Because local similarity of sensed data exists among different sensor nodes of geographical adjacent area, thus the proposed methodology can be applied to other topology of sensor networks.

5   Performance Evaluation

To evaluate the performance of the proposed data collection algorithms, a series of experiments are conducted. Experiments have been carried out based on a real data set collected from a temperature and humidity monitoring application.

It is assumed that the sensor network consists of multiple clusters. A cluster is formed by the set of sensor nodes in a geographical area, where locality of sensed data exists among the sensor nodes. All sensor nodes in a cluster are called cluster members, including one cluster head. Sensed data set was collected from Intel Berkeley Research Lab during one month. The sensed data consists of environmental data regularly sampled by the sensor nodes, which were spread around the lab. Since the monitored physical world appears obvious periodic property, data collection frequency of different modes are set as 2, 4, 8, 16, 32, 64 min, respectively.

The first group of experiments are to investigate the accuracy of prediction based on sensed data model. To probe the accuracy of prediction with parameters, it is assumed that the sensor node is set by stationary frequency of data collection.

In Fig. 2, blue circles represent sampled data per 30 s. Red stars represent predicted value based on sensed data model. The estimation of parameters in the model is based on Corollary 2, and the frequency of data collection is two minutes. Figure 2 shows that the proposed model can achieve high accuracy. For example the maximal value of true temperature is 23.3 °C and the predicted value is 23.6 °C. The relative error is less 0.04. The sensed data model integrates mean sensed value of adjacent area and measurement error, which can avoid disturbance incurred by hardware and other factors.

In Fig. 3, blue circles represent the sampled data per minute. According to the sensed data model, red stars represent the predicted value The estimation of parameters is based on Corollary 3 and the frequency of data collection is 8 min. Estimation of parameters depends on the sensed data from individual node, then the proposed model is sensitive to noise and measurement error. For example, around 80 and 90 min, the predicted value is easily affected and the larger deviation between predicted value and true sensed data is brought. Though predicted value is based on local information, sensed data model still achieves

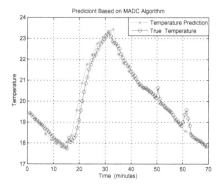

Fig. 2. Prediction based on MADC algorithm (Color figure online)

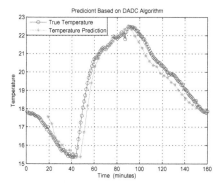

Fig. 3. Prediction based on DADC algorithm (Color figure online)

high accuracy. For example, predicted value is $17.45\,^\circ\mathrm{C}$ and true temperature is $17.83\,^\circ\mathrm{C}$. Experiment results verify the efficiency of the proposed model.

The second group of experiments investigates whether the strategy of adaptive data collection can collect extreme value points. In Fig. 4, blue curve represents sampled data per minute, which is regarded as true physical process. Green star represents the predicted value at the time of its x-coordinate and yellow circle represents collected data at the time of its x-coordinate. Process of MADC algorithm for cluster member node is simulated by C++. The figure demonstrates that the collected data based on MADC can reflect the physical process. And approximate extreme value points are collected. For example, at 106 min, temperature reaches the minimum value $15.91\,^\circ\mathrm{C}$, and the minimum value of collected data is $15.94\,^\circ\mathrm{C}$ at 102 min. Based on the error between predicted value and sensed data, sensor node adaptively adjusts the frequency of data collection. At the interval of 40 and 80 min, temperature gradually decreased and the period of data collection was 18 min. Between 116 and 140 min, temperature increased drastically, the frequency of data collection was 8 min. Figure 5, shows that based on DADC algorithm, collected data can reflect the physical process.

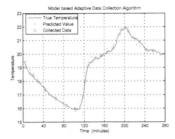

Fig. 4. Data sampled by MADC algorithm (Color figure online)

And approximate extreme value points are collected. Larger error of prediction leads to higher rate of data collection. Estimation of parameters in sensed data model is less accuracy than that of MADC algorithm, thus DADC algorithm yields higher frequency strategy of data collection.

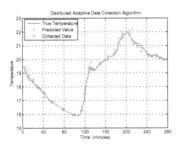

Fig. 5. Data sampled by DADC algorithm (Color figure online)

6   Conclusion

EFS based data collection approach is well known and widely applied in WSNs. Low frequency strategy may fail to collect extreme value points, and high frequency scheme produces massive sensed data, which consumes substantial energy for transmission. This paper proposes a novel sensed data model. Based on maximum likelihood estimation, the model can minimize measurement error. It is proved that the estimation is asymptotic approximate. Further, this paper proposes Model based Adaptive Data Collection algorithm and Distributed Adaptive Data Collection algorithm. Performance evaluation validates the efficiency of the given algorithms.

Acknowledgments. This work is supported in part by the National Natural Science Foundation of China (61602084, 61502099 61572104), the Post-Doctoral Science Foundation of China (2016M600202), the Doctoral Scientific Research Foundation of Liaoning Province (201601041).

References

1. He, Z., Cai, Z., Cheng, S., et al.: Approximate aggregation for tracking quantiles and range countings in wireless sensor networks. Theoret. Comput. Sci. **607**(3), 381–390 (2015)
2. Cheng, S., Cai, Z., Li, J., et al.: Extracting kernel dataset from big sensory data in wireless sensor networks. IEEE Trans. Knowl. Data Eng. **29**(4), 813–827 (2017)
3. Cheng, S., Cai, Z., Li, J.: Curve query processing in wireless sensor networks. IEEE Trans. Veh. Technol. **64**(11), 5198–5209 (2015)
4. Cai, Z., Goebel, R., Lin, G.: Size-constrained tree partitioning: approximating the multicast k-tree routing problem. Theoret. Comput. Sci. **412**(3), 240–245 (2011)
5. Cai, Z., Chen, Z., Lin, G.: A 3.4713-approximation algorithm for the capacitated multicast tree routing problem. Theoret. Comput. Sci. **410**(52), 5415–5424 (2009)
6. Li, J., Cheng, S., Gao, H., et al.: Approximate physical world reconstruction algorithms in sensor networks. IEEE Trans. Parallel Distrib. Syst. **25**(12), 3099–3110 (2014)
7. Wei, G., Ling, Y., Guo, B., et al.: Prediction-based data ag-gregation in wireless sensor networks: Combining grey model and Kalman Filter. Comput. Commun. **34**(6), 793–802 (2011)
8. Jiang, H., Jin, S., Wang, C.: Prediction of not? an energy-efficient framework for clustering-based data collection in wireless sensor networks. IEEE Trans. Parallel Distrib. Syst. **22**(16), 1064–1071 (2011)
9. Cheng, S., Cai, Z., Li, J., et al.: Drawing dominant dataset from big sensory data in wireless sensor networks. In: Proceedings of IEEE INFOCOM 2015, pp. 531–539. IEEE Computer Society, Hong Kong (2015)
10. Deshpande, A., Guestrin, C., Madden, S., et al.: Model-driven data acquisition in sensor networks. In: Proceedings of ACM VLDB 2004, pp. 588–599. ACM, Toronto (2004)
11. Wu, H., Zhang, J.: Local polynomial mixed-effects models for longitudinal data. J. Am. Stat. Assoc. **97**(459), 883–897 (2002)
12. Wolfinger, R.: Generalized linear mixed models: a pseu-do-likelihood approach. J. Stat. Comput. Simul. **48**(3), 233–243 (1993)
13. Saldju, T., Landgrebe, D.A.: Covariance estimation with limited training samples. IEEE Trans. Geosci. Remote Sens. **37**(4), 2113–2118 (1999)

An Energy Efficient Routing Protocol for In-Vehicle Wireless Sensor Networks

Chundong Wang, Zhentang Zhao[✉], Likun Zhu, and Honglei Yao

School of Computer Science and Engineering, Tianjin University of Technology,
Tianjin 300384, China
deviltangv@163.com

Abstract. In this paper, an advanced distributed energy-efficient clustering (ADEEC) protocol was proposed with the aim of balancing energy consumption across the nodes to achieve longer network lifetime for In-Vehicle Wireless Sensor Networks (IVWSNs). The algorithm changes the cluster head selection probability based on residual energy and location distribution of nodes. Then node associate with the cluster head with least communication cost and high residual energy. Simulation results show that ADEEC achieves longer stability period, network lifetime, and throughput than the other classical clustering algorithms.

Keywords: In-Vehicle Wireless Sensor Network · Routing protocol · Clustering · Energy efficiency

1 Introduction

Monitoring vehicular system and assuring safety of passengers have impelled the deployment of sensors inside cars. Examples include impact sensors, speedometers, knock sensors, and oxygen sensors. The sensors monitor critical system parameters and send their data to an Electronic Control Unit (ECU), which responds to abnormal conditions either automatically or manually. Since the sensors are critical to safety of driver and passengers, intra-vehicular data collection should be reliable and efficient. As the number of sensor nodes in a modern vehicle continues to grow rapidly, their wiring becomes á significant challenge to vehicle designers and manufactures. It is estimated that a modern sedan has in total more than 4 km of wires. Reducing wires by wireless technology means, for instance, can potentially reduce weight, ease manufacturing and design cars in a modular way.

The In-Vehicle Wireless Sensor Network (IVWSN) [1] consist of an vehicular base-station (BS) [2,3] and a number of vehicular wireless sensor nodes which is installable randomly to in-vehicle electronic components. They send and receive the operation commands and the vehicular sensor information through wireless communication to provide convenience/safety-related services such as parking assistance system, Tire Pressure Monitoring System (TPMS) and air-conditioner system. A base-station obtains and monitors the ECU status

© Springer Nature Singapore Pte Ltd. 2017
B. Zou et al. (Eds.): ICPCSEE 2017, Part II, CCIS 728, pp. 161–170, 2017.
DOI: 10.1007/978-981-10-6388-6_14

information transmitted from ECUs through the wired communication. Then, a base-station transmits operation commands to vehicular wireless sensor nodes according to ECU status information. A base-station receives and processes the sensor information from vehicular wireless sensor nodes. Potential wireless nodes in a IVWSN are required to operate under strict resource constraints. Specifically, power for transmitting data from sensor nodes to BS must be preserved so that battery life is extended and recharging is infrequent as possible. IVWSNs operate in a unique signal propagation environment including shadowing, multipath propagation and relatively short distance links. An efficient routing protocol is crucial for extending battery life peripherals. It follows that efficient routing protocols should specifically be tailored to the IVWSN environment and requirements.

To date, several communication technologies based on Zigbee, bluetooth, and ultra-wide band technologies, have been examined for in-vehicle wireless networking [4–6]. However, it is only recently that network layer aspects were considered. In [7] the authors study experimentally the Collection Tree routing Protocol for intra-vehicle peripheral communications, showing performance gains in terms of packet delivery rates as well as power consumption over the traditional star topology networks. Power consumption is crucial for the IVWSN application as removing the power wire to remote peripherals is very much desirable. The required long lifetime of the IVWSN poses stringent requirements on the energy consumption of each of the peripherals, and the network as a whole.

Routing protocols, specially cluster-based techniques, play an important role while achieving energy efficiency. According to this technique, members of the same cluster select a cluster head (CH) [8,9] and nodes belonging to that cluster send sensed data to the CH which forwards the aggregated data to the BS. Clustering can be implemented either in homogeneous or heterogeneous WSNs [10]. In homogeneous networks, nodes are equipped with the same energy level, and in heterogeneous networks, these levels differ.

The proposed protocol in Wireless Sensor Networks (WSNs) uses the optimal CH by the residual energy of nodes for balancing energy consumption and increasing network lifetime. Distributed energy-efficient clustering (DEEC) protocol is proposed in [11]. Authors in [12] proposed enhanced developed distributed energy-efficient clustering (EDDEEC) algorithm. This protocol changes the cluster head selection probability in an efficient and dynamic manner based on the absolute residual energy level. Authors in [13] proposed distance aware waiting based EDDEEC (DWEDDEEC) protocol. This protocol sets the threshold distance and have shown significant improvement over the existing EDDEEC.

In this paper, we proposed advanced distributed energy-efficient clustering (ADEEC), where the clustering is performed based on the residual energy and location distribution of node. The algorithm selects CH to make sure that nodes with relatively low remaining energy are avoided as relays for other nodes. We have also explored the feasibility and benefits of the WSNs modified DWEDDEEC routing protocol in a IVWSN environment. Experiments proved that the ADEEC protocol prolongs the network lifetime by heterogeneity-aware clustering algorithm.

2   Assumptions and Model

The proposed work consists of the following assumptions:

(1) Topology is static and each senor node has unique ID.
(2) All nodes are aware of their residual energy and location.
(3) The base station with a fixed position and sufficient energy has the enough communication capacity in the whole network.

2.1   Energy Level Model

ADEEC considers three-level heterogeneous networks that contains three different energy levels of nodes: normal, advanced, and super. The energy level of normal node is E_0, the energy level of every advanced node of fraction m with a times more energy than normal nodes is $E_0(1+a)$. Whereas, super nodes of fraction m_0 with b times greater power compared to the normal nodes is $E_0(1+b)$. As N is the total number of nodes in the network, then $Nmm_0, Nm(1-m_0)$, and $N(1-m)$ are the numbers of super, advanced, and normal nodes in the networks respectively. Here the sum of energy of whole network is specified by:

$$E_{super} = Nmm_0E_0(1+b) \tag{1}$$

$$E_{advanced} = Nm(1-m_0)E_0(1+a) \tag{2}$$

$$E_{normal} = N(1-m)E_0 \tag{3}$$

The total initial energy of three-level heterogeneous WSNs is therefore calculated as:

$$E_{total} = NE_0(1+m(a+m_0b)) \tag{4}$$

Thus, the three level heterogeneous networks have $m(a + m_0b)$ times greater energy as contrast to the heterogeneous WSNs.

2.2   Energy Consumption Model

The energy consumed by network should be as minimum as possible, therefore the radio dissipation model should be designed appropriately. We adopt the radio model used in [12], as the energy consumption of a node depends on its components (modules) for special purposes like sensing, processing, and wireless communication. Based on this assumption, the total energy consumption of a node E_T is given as follows:

$$E_T = E_S + E_P + E_W \tag{5}$$

where E_S is the energy consumed by the sensing module, E_P is the energy consumption cost of a processing module, and E_W is the energy consumption cost of a wireless communication module.

3   ADEEC Protocol

In this section, we present the details of the proposed ADEEC protocol.

3.1   Energy Efficient CH Election

In the existing EDDEEC, higher energy nodes are elected as CH to attain energy efficiency. To distribute the load uniformly among the nodes, node position distribution is considered. To reduce the energy consumption clustering process, each node is assigned with specific weight based on its residual energy and position distribution to become CH. The average energy of rth round from [11] is given as:

$$\bar{E}(r) = \frac{1}{N}E_{total}(1 - \frac{r}{R}) \tag{6}$$

where R denotes the total rounds during the network lifetime and is calculated as:

$$R = \frac{E_{total}}{E_{round}} \tag{7}$$

where E_{round} is the energy dissipated in a network during a single round and is calculated as:

$$E_{round} = l(2NE_{elec} + NE_{DA} + k\varepsilon_{mp}d_{toBS}^4 + N\varepsilon_{fs}d_{toCH}^2) \tag{8}$$

where E_{elec} is the energy dissipated per bit to ran transmitter or receiver circuit, and ε_{fs} and ε_{mp} are the radio amplifier types for free space and multipath respectively, E_{DA} is the data aggregation cost expended by CH, d_{toBS} is the average distance between the CH and the BS, and d_{toCH} is the average distance between cluster members and the CH.

$$d_{toCH} = \frac{M}{\sqrt{2\pi k_{opt}}}, \; d_{toBS} = 0.765\frac{M}{2} \tag{9}$$

By taking the derivative of E_{round} with respect to k and equating to zero, we can find the optimal number of clusters k_{opt} and is calculated as:

$$k_{opt} = \frac{\sqrt{N}}{\sqrt{2\pi}}\sqrt{\frac{\varepsilon_{fs}}{\varepsilon_{mp}}}\frac{M}{d_{toBS}^2} \tag{10}$$

At the start of each round, nodes decide on the basis of threshold whether to become CHs or not. The value of threshold is calculated as:

$$T(s_i) = \begin{cases} \frac{p_i}{1-p_i(\bmod{(r,\frac{1}{p_i})})} & if \; s_i \in G \\ 0 & otherwise \end{cases} \tag{11}$$

where G is the set of nodes eligible to become CHs for round r and p is the desired probability of the CH. The probabilities for three types of nodes for the CH selection by EDDEEC are given as:

$$p_i = \begin{cases} \frac{p_{opt} E_i(r)}{(1+m(a+m_0 b))\bar{E}(r)} & \text{for normal nodes, if } E_i(r) > T_{absolute} \\ \frac{p_{opt}(1+a)E_i(r)}{(1+m(a+m_0 b))\bar{E}(r)} & \text{for advanced nodes, if } E_i(r) > T_{absolute} \\ \frac{p_{opt}(1+b)E_i(r)}{(1+m(a+m_0 b))\bar{E}(r)} & \text{for super nodes, if } E_i(r) > T_{absolute} \\ c\frac{p_{opt}(1+b)E_i(r)}{(1+m(a+m_0 b))\bar{E}(r)} & \text{for all nodes, if } E_i(r) \leqslant T_{absolute} \end{cases} \qquad (12)$$

Where the best value of c from [12] as a variable controlling the clusters in number equals 0.025 for enhanced network efficiency, $T_{absolute}$ is the value of absolute residual energy level; $T_{absolute} = 0.7E_0$, p_{opt} denotes the optimal rate of clusters and is calculated as:

$$p_{opt} = \frac{k_{opt}}{N} \qquad (13)$$

Equation (12) primarily illustrations that the super and advanced nodes have more energy than the normal ones. So, the super and advanced nodes are largely preferred to be selected as CHs for the initial transmission rounds, and when their energy decreases to the same level as that of the normal ones, these nodes will have the same CH election probability like the normal nodes. However, the existing protocols either discuss residual energy of node or impact of the distance between the nodes. To balance the energy consumption in network, we focus on the consideration of both factors. If there are two nodes with equal distance from the BS, CH is selected based on higher residual energy. If there are two nodes with same residual energy of rth round, the node closer to the base station may results in selection as CH. Thus, we propose changes in the probability function defined by ADEEC. These changes are that nodes decide on both the residual energy as energy factor and the relative location as path factor whether to become CHs or not. The probabilities for the CH selection are given as:

$$p_i = \begin{cases} \frac{p_{opt}}{(1+m(a+m_0 b))}\left(w_1\frac{E_i(r)}{E(r)} + w_2\sqrt{\frac{d_{toBS}}{d_i}}\right) & \text{for normal nodes, } E_i(r) > T_{absolute} \\ \frac{p_{opt}(1+a)}{(1+m(a+m_0 b))}\left(w_1\frac{E_i(r)}{E(r)} + w_2\sqrt{\frac{d_{toBS}}{d_i}}\right) & \text{for advanced nodes, } E_i(r) > T_{absolute} \\ \frac{p_{opt}(1+b)}{(1+m(a+m_0 b))}\left(w_1\frac{E_i(r)}{E(r)} + w_2\sqrt{\frac{d_{toBS}}{d_i}}\right) & \text{for advanced nodes, } E_i(r) > T_{absolute} \\ c\frac{p_{opt}(1+b)}{(1+m(a+m_0 b))}\left(w_1\frac{E_i(r)}{E(r)} + w_2\sqrt{\frac{d_{toBS}}{d_i}}\right) & \text{for all nodes, } E_i(r) \leqslant T_{absolute} \end{cases}$$

$$(14)$$

where d_i is the distance between the node from BS, d_{toBS} is the average distance between the CH and BS, and w_1, w_2 is the function of nodes weight for energy factor and path factor respectively. Here w_1 and w_2 can be calculated as:

$$w_1 = \alpha \times e^{\left(-\frac{E_i(r)}{E_0^i}\right)}, \quad w_1 + w_2 = 1 \qquad (15)$$

where α as a variable adjust the weight of energy factor and path factor. Equation (14) illustrates that the path factor is more weight for original node at the start of operation. And soon after few round, with the decrease of residual energy, the node with more residual energy have more probability for CH selection. The detailed procedure of the proposed and selected algorithm is designed as follows:

Algorithm 1. Energy efficient CH selection

Input: The initial energy and location of every node
Output: The node is CH or cluster member for the current round

1 Initialize various parameters, such as radio parameters;
2 Calculate CH percentage p_{opt} and average energy of the network at present round $\bar{E}(r)$;
3 **for** *each node i* **do**
4   **if** $E_i(r) > 0.7E_0$ **then**
5   calculate p_i using specific probability function according to its type;
6   **else**
7   modify the probability of node based on $T_{absolute}$;
8   **if** *node has not been a cluster head in previous rounds* **then**
9   node belongs to set G, where G is set of nodes eligible to become a CH and node chose a random number between 0 and 1;
10   **if** *random number chosen is less than threshold fraction* **then**
11   node is CH for the current round;
12   **else**
13   node is cluster member and send data to their appropriate cluster head;
14 **final**;
15 **return** current node is CH or cluster member;

3.2   Node Association

When the node proclaims as CH, it advertises its node ID, residual energy and location information along with the proclamation. Based on this information, a non-CH node estimates CHs compatibility towards it. And a non-CH node need to determine whether to send data through cluster heads according to their own position and the residual energy.

The node selects CH with highest residual energy and lowest communication cost. The CHs aptness is the ratio between the residual energy of the CH and communication cost through this CH to BS. If a node receives multiple CH proclamations, the node estimates the weight of each CH as:

$$W_{ch}(n_i) = \frac{E_i(r)}{E_{Tx}(l, d)} \tag{16}$$

where $E_{Tx}(l, d)$ is the communication cost of the node through CH to BS. The node then associates with the highest weight CH in its vicinity.

4   Results and Discussion

In this section, we evaluate the performance of ADEEC protocol using MAT-LAB. IVWSN consists of $N = 100$ nodes which are randomly distributed in area of dimension $10 \times 2.5 \times 3.5\,\mathrm{m}^3$ with a centrally located BS. The proposed ADEEC protocol is compared with other energy efficient clustering protocols like IV_EDDEEC, and IV_DWEDDEEC. The performance metrics used to measure these protocols are: stability period, network lifetime, and number of packets sent to the BS.

Table 1. Simulation parameters

Parameter	Value
E_0	0.5 J
E_{elec}	5 nJ/bit
ε_{fs}	1 pJ/bit/m^2
ε_{mp}	0.013 pJ/bit/m^4
E_{DA}	5 nJ/bit/message
l	4000 bits

(1) Stability period: we mean the number of rounds from network initialization till the death of first node.
(2) Network lifetime: we mean the number of rounds from network initialization till the death of all nodes.
(3) Number of packets sent to the BS: we mean the total number of packets that are directly sent to BS either from CHs or non-CH nodes.

The radio parameters used in our simulations are shown in Table 1. Results along with discussions are provided in the following subsections.

Case 1: $m = 0.8$, $m_0 = 0.6$, $a = 1.5$, $b = 3.0$.

In this case, we set 20 normal nodes having E_0 energy, 32 advanced nodes having 1.5 times more energy than normal nodes, and 48 super nodes containing 3 times more energy than the normal nodes. Figure 1 depicts the number of dead nodes during the network lifetime. The first node for IV_EDDEEC, IV_DWEDDEEC and ADEEC dies at 1851, 1429, and 2156 rounds, respectively, and all nodes die at 7890, 7896, 7042 and 8343 round respectively. Figure 2 shows that the data sent to the BS is more for ADEEC as compared to the rest of the baseline protocols. ADEEC dynamically selects fittest CHs and adjusts the node association part. Thus, ADEEC consumes relatively less energy which leads not to only prolonged stability period but also prolonged network lifetime in comparison to the other protocols. Prolonged stability period and network lifetime means that the number of packets sent to BS are more in comparison to the other selected protocols.

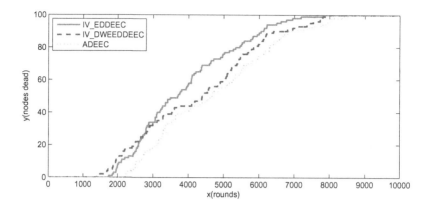

Fig. 1. Network lifetime (case 1)

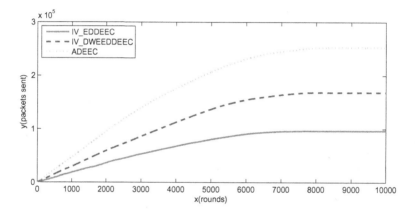

Fig. 2. Number of packets sent to BS (case 1)

Case 2: $m = 0.3, \quad m_0 = 0.2, \quad a = 1.0, \quad b = 2.0.$

In this case, we set 70 normal nodes having E_0 energy, 24 advanced nodes having 1 times more energy than normal nodes, and 6 super nodes containing 2 times more energy than the normal nodes. Figure 3 depicts the number of dead nodes during the network lifetime. The first node for IV_EDDEEC, IV_DWEDDEEC and ADEEC dies at 1565, 1431 and 1889 rounds, respectively, and all nodes die at 5868, 5994 and 6051 round respectively. Figure 4 shows that the data sent to the BS is more for ADEEC as compared to the rest of the protocols. It is obvious from the results that ADEEC is the most efficient among all protocols in terms of stability period, network lifetime, and number of packets sent to the BS, even in case of network containing less number of super and advanced nodes as compared to normal ones.

In this case 2, to illustrate the impact of different numbers of heterogeneous energy nodes on network performance, the number of normal nodes increased as compared to case 1. Similarly, the number of advanced and super nodes decreased

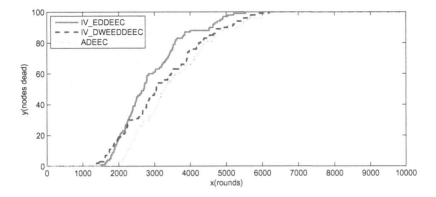

Fig. 3. Network lifetime (case 2)

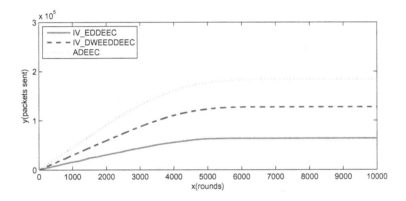

Fig. 4. Number of packets sent to BS (case 2)

as case 2 in comparison to case 1. The number of normal, advanced and super nodes can be adjusted according to actual application based on related parameters. Normal nodes have the least energy in comparison to advanced and super nodes. Thus, as a whole, the total energy of network is downscaled in this case as compared to the previous case. All the protocols are the same in this case as compared to case 1 but with less initial energy resources. Therefore, the stability period, network lifetime, and number of packets sent to BS are relatively on the lower side in this case as compared to the previous case.

5   Conclusions

In this paper, we proposed ADEEC clustering algorithm based on the impact of energy heterogeneity and the location of sensor nodes for the performance of the whole network. Our contribution involves a deterministic CH election that holds the energy cost factor and path factor. Secondly, the modified DEEC algorithm protocol originally designed for WSNs was shown to scale well to the IVWSN environment. The newly proposed protocol is implemented in MATLAB. The simulation results show that proposed the ADEEC protocol performs better for the selected performance metrics in IVWSNs. In near future we will research a extension that is to consider regenerated energy sources, i.e. energy harvesting nodes. In this approach each node has a different recharging rate on top of its current battery state. Estimating the recharging rate and including it into the routing scheme can potentially enhance network performance considerably.

References

1. Yun, D.S., Lee, S.J., Kim, D.A.: A study on the architecture of the in-vehicle wireless sensor network system. In: International Conference on Connected Vehicles and Expo, vol. 4, pp. 153–162. IEEE (2013)
2. Ahmad, A., Javaid, N., Khan, Z.A., Qasim, U., Alghamdi, T.A.: Routing scheme to maximize lifetime and throughput of wireless sensor networks. IEEE Sens. J. **14**(10), 3516–3532 (2014)
3. Heinzelman, W.B., Chandrakasan, A.P., Balakrishnan, H.: An application specific protocol architecture for wireless microsensor networks. IEEE Trans. Wirel. Commun. **1**, 660–670 (2002)
4. Reddy, A.D.G., Ramkumar, B.: Simulation studies on ZigBee network for in-vehicle wireless communications. In: International Conference on Computer Communication and Informatics, pp. 1–6. IEEE (2014)
5. Hillman, A.P.: On the potential of bluetooth low energy technology for vehicular applications. IEEE Commun. Mag. **53**(1), 267–275 (2015)
6. Bas, C.U., Ergen, S.C.: Ultra-wideband channel model for intra-vehicular wireless sensor networks beneath the chassis: from statistical model to simulations. IEEE Trans. Veh. Technol. **62**(1), 14–25 (2013)
7. Si, W.: Integrating wireless technologies into intra-vehicular communication. Dissertations and Theses - Gradworks (2016)
8. Anisi, M.H., Abdullah, A.H., Coulibaly, Y., Razak, S.A.: EDR: efficient data routing in wireless sensor networks. Int. J. Ad Hoc Ubiquit. Comput. **12**(1), 46–55 (2013)
9. Liao, Y., Qi, H., Li, W.: Load-balanced clustering algorithm with distributed self-organization for wireless sensor networks. IEEE Sens. J. **12**(1), 1498–1506 (2013)
10. Karalar, T.C., Yamashita, S., Sheets, M., Rabaey, J.: A low power localization architecture and system for wireless sensor networks. In: IEEE Workshop on Signal Processing Systems, pp. 89–94. IEEE (2004)
11. Qing, L., Zhu, Q., Wang, M.: Design of a distributed energy-efficient clustering algorithm for heterogeneous wireless sensor networks. J. Softw. **29**(12), 2230–2237 (2006)
12. Javaid, N., Rasheed, M.B., Imran, M., et al.: An energy-efficient distributed clustering algorithm for heterogeneous WSNs. EURASIP J. Wirel. Commun. Netw. **2015**(1), 1–11 (2015)
13. Kumar, R., Kaur, R., Bhardwaj, R.: DWEDDEEC: distance aware waiting based EDDEEC protocol for hetrogeneous WSNS (2015)

Energy-Conserving Transmission Network Model Based on Service-Awareness

Huyin Zhang[1,2], Chenghao Li[2], Tianying Zhou[2,3(✉)], Long Qian[2], and Jingcai Zhou[2]

[1] Shenzhen Institute, Wuhan University, Shenzhen 518057, China
zhy2536@whu.edu.cn
[2] Computer School, Wuhan University, Wuhan 430072, China
zhoutying@qq.com
[3] School of Computer Science and Technology,
Hubei University of Science and Technology, Xianning 437100, China

Abstract. In this paper, we propose a service-aware network model which is based on the traffic pattern in data center. First of all, we analyze the traffic model in data center networks. Then we use this model to make the net topology integration and classification through the software define network. In order to achieve the purpose of energy consumption optimization, we divide the hosts into same VLAN according to their interaction frequency to reduce the cross VLAN transmission consumption. Simulation results show that we get a great energy improvement in the fat tree net topology.

Keywords: Data center network · Transmission model · Energy-conserving

1 Introduction

With the development of cloud computing, some giant industry companies can offer the cloud services, such as Google, Amazon, Microsoft at overseas and Baidu, Alibaba at domestic. At the same time, with the rapid development of mobile internet, the user of searching, social networks and instant messaging is also booming. These both lead to the scale of cloud data center increasing, accompany the problem of data center energy consumption which has become the research hotspots.

The rapid development of distributed technology has led to the flow of data center networks [1] from traditional north-south-oriented models to east-west-oriented models. So we can save the energy by reducing the amount of redundant data traffic among hosts and decreasing the times of roundtrips to different switches.

In this paper, we propose a service-aware energy conservation model in the cloud data center. According to this model, we make the net topology integration and classification to achieve the purpose of energy consumption optimization. The main idea is that the less number of network equipments we used for data traffic transmission, the more energy we can save. Simulation shows that we can also save the energy of data center and improve the network performance.

The rest of the paper is organized as follows. In Sect. 2, we introduce the related work. In Sect. 3, we propose the mathematical model and define the problem.

© Springer Nature Singapore Pte Ltd. 2017
B. Zou et al. (Eds.): ICPCSEE 2017, Part II, CCIS 728, pp. 171–183, 2017.
DOI: 10.1007/978-981-10-6388-6_15

In Sect. 4, the service-aware network model and the algorithm are described in detail. In Sect. 5, the simulation results are given to show the improvement. In Sect. 6, we give the conclusions.

2 Related Work

Researching directions of green data center and energy-saving data center are mainly the following two. The first one is network-level's energy saving that using the integration of network link. The second is device-level's energy saving that using device's sleeping state. In this paper, our research is mainly focus on the network-level energy-saving technology.

The network-level energy-saving technology is mainly to consider that the energy consumption and link utilization on the link are independent of each other in the network topology. Even if the link is idle, its energy consumption is not change [2]. This is due to the link will sent the data frames to keep the synchronization [3]. So we can save energy by turning off the free link.

Fisher et al. [4] pointed out that the links between adjacent routers consume a lot of energy in the backbone network. Therefore, turning off the link can make great effect on energy saving in the network data flow when it is not the peak period. However, the dynamically remove of one or more links from the entire network can reduce the transmission capacity in a short period of time, and even to destroy the connectivity of the network. So, how to determine the best closure scheme is a multi-commodity flow (MCF) class problem [5], and also a NP hard problem. Vasic et al. have proposed a technique that takes energy consumption into account while taking the achievement of the same traffic rates as the energy-oblivious approaches [6]. Different with Vasic, Zhang et al. have proposed an intra-domain traffic engineering mechanism, GreenTE, which maximizes the number of links that can be put into sleep under given performance constraints such as link utilization and packet delay [7].

Another research put emphasis on the topology-aware energy-efficient routing. Cianfrani et al. [8, 9] proposed shortest path tree generation algorithm by modifying the OSPF protocol in some of the network node. It makes all the network nodes generated by the shortest path tree or a subset to provide routing services. It can achieve the energy saving by effectively reducing the links number in the network. Since the method has nothing to do with the traffic load, it can be applied into the scene where the network traffic changes frequently. Cuomo et al. proposed an ESACON (Energy Saving based on Algebraic Connectivity) algorithm [10]. The basic idea of this paper is to achieve energy savings by turning off the switch or link with the network connectivity as large as possible. Experiments show that ESACON has better performance than similar topology-aware algorithm. Amaldi et al. proposed a method of off-line IP traffic engineering. It allows the opening or closing of communication interfaces to accommodate the network traffic consumption with different daily traffic load scenarios [11]. The basic idea of this method is to change the switch state by modifying the link weight of OSPF to control the service quality and congestion control. These methods close the link directly, during which it generally takes tens of seconds when the network equipment convergence links from online to offline. To solve this problem,

Lee proposes a distributed loop-free routing update method, which allows the router to correctly update the routing table [12]. The experimental results show that this method makes great sense on energy-saving.

There is also an energy-efficient routing scheme based on elastic tree (ET) [13]. The basic idea is to solve the network topology with the constraints of network performance and fault-tolerant performance. The process of optimization solution can be a NP hard problem. In this paper, we introduced a dedicated interface counter, which increased the cost of devices. Xu et al. [14] proposed an energy-aware routing for general topology (ERGT), which is more widely used than elastic tree. Its main idea is, in the meanwhile of offering service by using minimal device, it also needs to meet the constraints of throughput of data center network. It can reduce the energy consumption, nevertheless, it also sacrifice system performance. ET and ERGT algorithms both can achieve good results when the traffic load is lower.

Recently, many experts and scholars have used the software define network to improve the data center network in order to achieve energy conservation [15–18].

3   Problem Analysis and Definition

3.1   Symbols and Basic Definitions

Cloud data center network energy-saving issues involve more symbols. For ease of description, this section gives all symbols and some basic definitions. Table 1 summarizes these symbols.

Table 1.   Commonly used symbol

Symbol	Basic definitions
P	Total energy consumption of data center network
W	Data set of data center switches
S	Data set of data center servers
W_i	Energy consumption of switch i
S_j	Energy consumption of server j
NIC_j	Energy consumption of NIC interfaces
CPU_j	Energy consumption of CPU's
$\Re i$	ith flow link
ℓ	Mapping table between business nodes
λ	Data set of VLAN partition
C	Energy consumption through the data transmission
$N(i,j)$	The number of switches between node i and j

3.2   Model Abstraction and Problem Definition

Energy Consumption Model. Total energy consumption of data center network P includes two parts. They are the interact consumption between servers and the consumption of the network equipment itself respectively.

$$P = \sum_{i \in I} W_i + \sum_{j \in J} S_j \qquad (1)$$

Here I and J represent data set of switches' interface and servers' interface respectively. W_i represents the energy consumption of switch i and S_j represents the energy consumption of server j.

The energy consumption of network device mainly depends on the hardware configuration of device and flow rate [19]. The hardware configuration includes: backplane type of switch, line card type and quantity, port configuration and so on. The energy consumption of S_j consist the consumption of NIC's interface (NIC_j) and the consumption of CUP (CPU_j).

$$S_j = NIC_j + CPU_j \qquad (2)$$

Then we can get this equation by Eqs. (1) and (2):

$$P = \sum_{i \in I} W_i + \sum_{j \in J} (NIC_j + CPU_j) \qquad (3)$$

Total energy consumption of data center network P is composed by the number of servers, the number of switches NIC and the number of switches' CPU.

Data Center Network Energy Consumption Problem. In the distributed architecture-based cloud data center, all kinds of business software (including databases, file systems, etc.) system architecture are naturally scattered in the business module on multiple servers, and they achieve coordination operations through the network. Sometimes an operation requires multiple network interactions to complete, which accounts for the transmission of the data center network traffic from the early 80% of the north-south traffic to 70% for the east-west flow. With the change of traffic model, the traditional network architecture can no longer meet the requirements of data center business, so the data center network architecture need to use non-blocking network technology such as fat tree. Moreover, due to the heavy-tailed distribution that the user's request for data tend to be, unicast transmission which based on end-to-end has caused a large number of duplicate data transmission on the network. Obviously, in this new business model and traffic characteristics, the traditional ways of achieving network energy saving method has been unable to get the target. So we can reduce the data center network energy consumption by reducing the amount of redundant data transmission and reducing the number of data transfers across the switch. Then we will analyze the workflow of HDFS to research the exist problem exposed of data center network.

HDFS has a master/slave architecture. A typical HDFS cluster consists of a single NameNode, a master server that manages the file system namespace and regulates access to files by clients. In addition, there are a number of DataNodes, usually one per node in the cluster, which manage storage attached to the nodes that they run on. HDFS exposes a file system namespace and allows user data to be stored in files. Internally, a file is split into one or more blocks and these blocks are stored in a set of DataNodes. The NameNode executes file system namespace operations like opening, closing, and renaming files and directories. It also determines the mapping of blocks to DataNodes. The DataNodes are responsible for serving read and write requests from the file system's clients. The DataNodes also perform block creation, deletion, and replication upon instruction from the NameNode. A typical data write operation is shown in Fig. 1, the data needs to be sent from the Client node to the first DataNode, then to the second DataNode, finally to the third DataNode. One same data has been forwarded three times in the whole data center network, which greatly expands the network device's energy consumption.

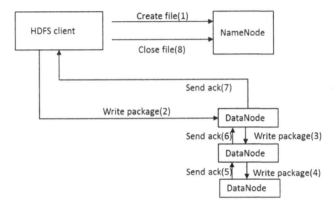

Fig. 1. The transmission model of HDFS write operation

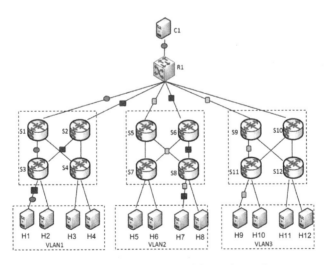

Fig. 2. Original VLAN partition (scheme 1)

Assume that there are one set DataNode (*H1, H7, H9*) in the HDFS and all the devices version is the same. We use *p* to represent power, *b* to represent bandwidth, and *n* to represent the total amount of data required to be stored. When *c1* start to write data in the Fig. 2 (scheme 1), the network should set up three links $\Re1$: *c1 -> H1*, $\Re2$: *H1 -> H7* and $\Re3$: *H7 -> H9*. The link $\Re1$ goes through the core router *R1*, core switch *S1* and *S3*. The link $\Re2$ goes through the core router *R1*, core switch *S3*, *S2*, *S6* and *S8*. The link $\Re3$ goes through *S8*, *S5*, *R1*, *S9* and *S11*. Under the condition of ignoring other data traffic, the duration of $\Re3$, $\Re2$ and $\Re1$ are $t3 = n/b$, $t2 = (t3 + n/b)$ and $t1 = (t2 + n/b)$. So the energy consumption of HDFS's writing a data block of size *n* is:

$$p * (n/b) * 5 + p * ((n/b) + (n/b)) * 5 + p * ((n/b) + ((n/b) + (n/b))) * 3 =$$
$p * (n/b) * 24$. We noted that in this scheme link $\Re1$ and $\Re2$ are idle at most of the time.

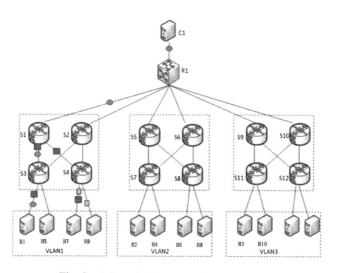

Fig. 3. Adjusted VLAN partition (scheme 2)

If we adjust the network topology by using VLAN partition, make the topology become to scheme 2 (Fig. 3). Then the energy consumption is:

$$p * (n/b) + p * ((n/b) + (n/b)) * 3 + p * ((n/b) + ((n/b) + (n/b))) * 3$$
$$= p * (n/b) * 16.$$

It follows that in different VLAN partition scheme, scenario 2 can save 33.3% more than scenario 1.

The HDFS energy consumption model in the network transmission is abstracted as:

$$\begin{cases} c_i = p * t_{(S,D_1)} * N_{(S,D_1)} \\ \quad + p * t_{(D_1,D_2)} * N_{(D_1,D_2)} \\ \quad + p * t_{(D_2,D_3)} * N_{(D_2,D_3)} \\ t_{(S,D_1)} = t_{(D_1,D_2)} + n/b \\ t_{(D_1,D_2)} = t_{(D_2,D_3)} + n/b \\ t_{(D_2,D_3)} = n/b \end{cases} \tag{4}$$

$$C = \sum_{\forall l \in \ell} \left(p * \sum_{j \in l} t_j * N_j \right) \tag{5}$$

Here we use c_i to represent the consumption of single mapping, C to represent the total energy consumption, $t_{(S,D_1)}$ to represent the duration between node S and node D_1, $N_{(S,D_1)}$ to represent the switches numbers of node S and node D_1. The mapping relationship between nodes is represented by ℓ. For each $\ell = \{S, D\}$ S is master node. D represents slave nodes set (The order is $D_j < D_{j+1} < D_{j+2}$). We can come to the conclusion from the Eq. (5) that we effectively reduce the energy consumption by reducing the number of switch that flow passes through and its duration.

4 Service-Aware Network Model

4.1 Service-Aware Network Architecture

In the service-aware network architecture (Fig. 4) each physical host has a virtual switch (vSwitch). Each vSwitch is connected with the virtual machine. The physical network card connected the vSwitch with SDN Controller, meanwhile vSwitch maintains a mapping table. vSwitch can find the corresponding virtual machine link according to the MAC address and then complete the data forwarding.

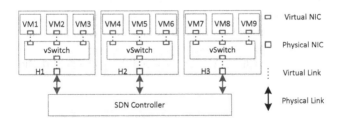

Fig. 4. Service-aware network architecture

Here is the data stream processing process:

(1) When the packet sent from the virtual machine, it would first pass through the virtual network card which is the virtual machine configured NIC. Virtual NICs determine how to process packets, such as release, blocking, or modification, based on established rules.

(2) Then data will be forwarded to vSwitch after being released by the network card. vSwitch matches the packets according to their own flow tables. If the match is successful, the operation is performed according to the corresponding instruction. If the match is not successful, the packet is sent to the controller to wait for the assignment and decrement of the relevant flow table.

(3) When the packet needs to be forwarded through the physical network card, it will be sent to the physical network card which is connected to the vSwitch, and then forwarded to the external network device.

In summary, the service-aware network consists of vSwitch. The whole network links and VLAN are controlled by the SDN Controller (can also be dynamically adjusted). When a network topology can't be completed through the SDN Controller, the system can drive virtual machine migration to achieve the network topology adjustment according to the intended target, to make sure that the entire network can be in accordance with the needs of business software dynamic distribution and adjustment.

4.2   Service-Aware VLAN Algorithm, SAVA

From our analysis given above, we can see that the key to influence the path's length is the VLAN partition. When VLAN partition is considered to be the optimal condition, the data flow involved in each transaction is forwarded only in this VLAN. Here is the basic idea of this algorithm. According to the network topology information G (W, H) of the data center, the distributed mapping relationship between the business software nodes ℓ and the constraints (Eq. 6), we adopt the greedy algorithm to divide the different hosts H into different VLAN. The final result is the constraint in the Eq. 6. Here are the constraints.

(1) The total number of network devices passing through each mapping item in the business software is the minimum.

(2) Each source node and destination node must belong to a unique VLAN and the total number must be equal to the host set (H).

(3) The VLAN number can't larger than old VLAN number.

(4) The hosts in each VLAN must conform to the constraints of the network topology G, that is to say, this means the hosts in the same VLAN must communicate with each other by using physical link.

(5) The host number in each VLAN can't exceed the vSwitch port number.

$$
\begin{cases}
Min \sum\limits_{i \in I} \Re i, I = size\ of\ (\ell) \\
\ell i = (s, d), s \in H, d \in H \\
\sum (s) + \sum d = H \\
size\ of\ (V) < MAX\_VLAN\_NUMBERS \\
h \subset V \subset G \\
H(V) < MAX\_PORTS
\end{cases} \tag{6}
$$

In this algorithm, firstly, we need to preprocess the topology of the input topology, construct directed graphs to reflect projected relationship and find out its weakly connected component. Set a weakly connected component equaling to a VLAN. Then, according to the physical link, the host migration can be feasible VLAN division results. Here is the algorithm pseudo code.

Algorithm 1. **SAVA**
Input:
    G : Data center network topology information
    ℓ : Distributed mapping between business nodes.
    H : Hosts set
Output: VLAN partition table
Time Complexity: $O(|\ell|+|H|)$

```
(1) def Planning_VLAN( ℓ )
(2)   graph=DiGraph( ℓ )
(3)   V=weakly_connected_components(graph).sort()
(4)   SE=switch.list
(5)   SF=NULL
(6)   while (SE==NULL and V==NULL)
(7)     v0=V[0]
(8)     if(v0.size > SE[i])# i means any switches
(9)       vclist=cut(v0).sort()
(10)      V. insert(0,vclist)#put the head inser into V
(11)      continue
(12)    else
(13)      if (SE[i] is not connect with hosts)
(14)        SE[i].vlan=v0.vlan
(15)      connect(SE[i],v0)
(16)      V.remove(v0)
(17)      if(SE[i]'s port is disabled)
(18)        SF.add(SE[i])
(19)        SE.remove(SE[i])
(20)  return SF
```

5   Evaluation

5.1   Methodology

In order to accurately evaluate the performance of a service-aware network algorithm in a network scenario, this paper uses mininet for simulation testing. The experimental environment information is shown in Table 2. Experimental hardware configuration is 16 G memory, 8-core Intel CPU. The topology is a fat tree structure, and experiments are carried out under k = 4, 6, 8 to verify the energy-saving efficiency, robustness, scalability and network performance of the algorithm. Figure 5 shows k = 4's fat tree structure. Table 3 shows the link bandwidth in the fat tree when k is different.

Table 2. Experimental environment table

Software	Version
Ubuntu	Ubuntu 14.04.4 LTS
Mininet	2.3.0d1
Ryu	4.7

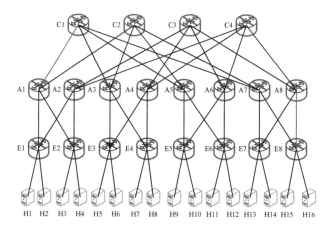

Fig. 5. Fat tree topology when k = 4

Table 3. Link bandwidth in the fat tree

K	Edge layer	Aggregation layer	Core layer
k = 4	2 M	4 M	8 M
k = 6	2 M	6 M	18 M
k = 8	2 M	8 M	32 M

Here is the experimental method.

Network topology is fat tree and the $k = [4, 6, 8]$.

(1) Energy consumption. We adopt the benchmark flow algorithm, ET (elastic tree) algorithm, ERGT algorithm and SAVA algorithm to construct the network flow table, meanwhile we carry out the experiments under different traffic load. Then we observe and measure the energy consumption, and do a detailed record.

(2) Network performance (delay information). We use 'iferf' command to test the network performance. We calculate the delay when the traffic is stable. By comparing the benchmark algorithm, ET algorithm, ERGT algorithm and SAVA algorithm, the performance of network transmission performance is evaluated.

5.2   Experimental Results and Analysis

In order to verify the energy-saving efficiency, this paper compares the ET algorithm, ERGT algorithm and SAVA algorithm with the baseline algorithm to get the percentage of energy saving, as shown in Fig. 6. These figures show that when the traffic load is low (load <10%), these three algorithms have better energy saving efficiency (energy efficiency >60%). But with the load increasing, energy-saving effect decreases significantly. Figure 6a (k = 4) shows that the energy saving effect is 1.8%, 2.3%, 4% (ET algorithm, ERGT algorithm, SAVA algorithm) when the traffic load is 100%. This is because both ET and ERGT are designed for idle link, when the link load is large, it can modify the number of links it can revise will become smaller, so will its energy efficiency is smaller too. The SAVA algorithm is designed for the communication model. If all the business is in the same VLAN, the energy consumption will be small. So when traffic load is 100%, the energy efficiency is better than ET and ERGT. SAVA energy saving effect: 4%, 12%, 11% with 100% traffic load and k = 4, 6, 8 respectively. With the switches number (k values)increase, the energy-saving efficiency is getting better and better. Because of the increasing of the switches and hosts, ratio of being divided into the same VLAN host node will increase, which can reduce the more cross-VLAN transmission to achieve the effect of energy saving. Especially when k is a multiple of 3, the host can be completely divided into the same VLAN. There is almost no cross-VLAN transmission. Therefore, the energy saving effect when k = 6 is better than k = 8. SAVA algorithm has better effect than the other two algorithms under any load condition when k value is a multiple of 3. Under the best condition, it can reach 8% higher than ERGT.

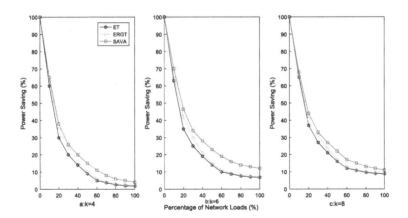

Fig. 6. Power saving with different network loads

In order to verify the influence of the algorithm on network performance, this paper is tested under the fat tree (k = [4, 6, 8]). The test results are shown in Fig. 7, where the horizontal axis means flow size (0.4–2 Mbit). As link bandwidth shows in Table 3, fat tree links are also different under different k. The edge layer is the link bandwidth between the switch and the host. The aggregation layer is the link bandwidth between

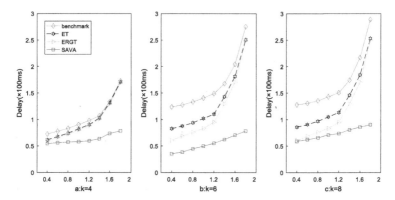

Fig. 7. Delay information with different k

the edge layer and the switch. The core layer is the link bandwidth between the aggregation layer and the core layer switch. We also use 'iperf' to test the performance. Here are the results.

In the three graphs, each of the algorithm increases with the flow rate. But the smallest increasing rate algorithm is SAVA. As mentioned above, the ET algorithm and ERGT algorithm are designed for idle links, which can achieve better results under low load and low flow. However, in the case of high load and high flow rate, it can't get the good results. The SAVA algorithm is designed for the business, which controls the host between each service in a smaller range (the same VLAN), so it can still be better in this test (lower delay). With the k value increase, the delay will increase. This is due to the switches number increase, the data flow go through the links increase too, result the delay's increase. We noted that, the delay of the SAVA algorithm is significantly lower than other two algorithms when k = 6. This because when k is 6, the hosts are in the same VLAN which have frequent business interactions, thus the number of cross-VLAN interaction get reduced. Due the lower cross-VLAN interaction a host communicates with other hosts don't go through with lots of switches. So the delay can get significantly reduced. Similarly, when k is three times, SAVA algorithm can achieve a better effect when k is a multiple of 3.

6 Conclusion

In this paper, we propose a service-aware network model and design a data flow programming algorithm based on this model. First, we analyze the business mode and find that we can change the data center link model to save energy. Then we use the programmable features of SDN and the dynamic migration function of virtual machine to programming the whole data center net topology. Therefore we realize that the path of data flow is shorter, the cross-VLAN data exchange is less. Based on these points, we build a service-aware network which can avoid redundant forwarding effectively and achieve the goal of energy consumption optimization in the data center. The simulation results demonstrate that SAVA save the energy of data center, compared with other methods.

Acknowledgments. This work was supported by the Shenzhen science and technology projects under Grant No. JCYJ20140603152449639 and the science and technology planning project of Guangdong province under Grant No. 2015B010131007.

References

1. Guo, C., Wu, H., Tan, K., et al.: Dcell: a scalable and fault-tolerant network structure for data centers. ACM SIGCOMM Comput. Commun. Rev. **38**(4), 75–86 (2008)
2. Chabarek, J., Sommers, J., Barford, P., et al.: Power awareness in network design and routing. In: INFOCOM, pp. 457–465 (2008)
3. Bianzino, A., Chaudet, C., Rossi, D., et al.: A survey of green networking research. IEEE Commun. Surv. Tutor. **14**(1), 3–20 (2012)
4. Fisher, W., Suchara, M., Rexford, J.: Greening backbone networks: reducing energy consumption by shutting off cables in bundled links. In: ACM SIGCOMM Workshop on Green Networking, pp. 29–34 (2010)
5. Even, S., Itai, A., Shamir, A.: On the complexity of time table and multi-commodity flow problems. In: Foundations of Computer Science Annual Symposium on IEEE, pp. 184–193 (1975)
6. Vasic, N., Kostic, D.: Energy-aware traffic engineering. In: Proceedings of the 1st International Conference on Energy-Efficient Computing and Networking, pp. 169–178. ACM (2010)
7. Zhang, M., Yi, C., Liu, B., et al.: Green TE: power-aware traffic engineering. In: Network Protocols (ICNP), pp. 21–30. IEEE (2010)
8. Cianfrani, A., Eramo, V., Listanti, M., et al.: An OSPF enhancement for energy saving in IP networks. In: INFOCOM Workshops, pp. 325–330 (2011)
9. Cianfrani, A., Eramo, V., Listanti, M., et al.: An energy saving routing algorithm for a green OSPF protocol. In: INFOCOM Workshops, pp. 1–5 (2010)
10. Cuomo, F., Abbagnale, A., Cianfrani, A., et al.: Keeping the connectivity and saving the energy in the internet. In: INFOCOM Workshops, pp. 319–324 (2011)
11. Amaldi, E., Capone, A., Gianoli, L.G., et al.: Energy management in IP traffic engineering with shortest path routing. In: World of Wireless, Mobile and Multimedia Networks, pp. 1–6 (2011)
12. Lee, S., Tseng, P., Chen, A.: Link weight assignment and loop-free routing table update for link state routing protocols in energy-aware internet. Future Gen. Comput. Syst. **28**(2), 437–445 (2012)
13. Cuomo, F., Cianfrani, A., Polverini, M., et al.: Network pruning for energy saving in the internet. Comput. Netw. **56**(10), 2355–2367 (2012)
14. Xu, M., Shang, Y., Li, D., et al.: Greening data center networks with throughput guaranteed power-aware routing. Comput. Netw. **57**(15), 2880–2899 (2013)
15. Shi, D., Ruixuan, L., Xiaolin, L.: Energy efficient routing algorithm based on software defined data center network. J. Comput. Res. Dev. **52**(4), 806–812 (2015)
16. Hongyu, P., Gang, C., Yinghai, Z., et al.: A new data center memory energy consumption optimization stragegy in SDN. J. Beijing Univ. Posts Telecommun. **38**(2), 78–82 (2015)
17. Yoon, M., Kamal, A.: Power minimization in fat-tree SDN datacenter operation. In: Global Communications Conference, pp. 1–7 (2015)
18. Rofoee, B., Zervas, G., Yan, Y., et al.: Griffin: programmable optical datacenter with SDN enabled function planning and virtualisation. J. Lightwave Technol. **33**(24), 5164–5177 (2015)
19. Mahadevan, P., Sharma, P., Banerjee, S., et al.: A power benchmarking framework for network devices. In: International Conference on Research in Networking, pp. 795–808 (2009)

A Multi-objective Optimization Data Scheduling Algorithm for P2P Video Streaming

Pingshan Liu[1,2(✉)], Xiaoyi Xiong[2(✉)], and Guimin Huang[2]

[1] Business School, Guilin University of Electronic Technology, Guilin, China
yai3xx@foxmail.com
[2] Guangxi Key Laboratory of Trusted Software,
Guilin University of Electronic Technology, Guilin, China

Abstract. In P2P video streaming, each peer requests its wanted streaming data from others and responses others' requests by its data scheduling algorithm. Recent years, some data scheduling algorithms are proposed either to optimize the perceived video quality, or to optimize the network throughput. However, optimizing the perceived video quality may lead to low utilization of the senders' upload capacity. On the other hand, optimizing the network throughput may lead to the degrading perceived quality, for some emergent data may not be transmitted in time. In this paper, to improve the two objectives simultaneously, we formulate the data scheduling problem as a multi-objective model. In the formulation, we not only consider the segment quality and emergency which affect the perceived video quality, but also consider the rarity of the segments, which influences the network throughput. Then, we propose a distributed data scheduling algorithm to solve the multi-objective problem in polynomial time. Through simulations, we show the proposed algorithm outperforms other conventional algorithms in perceived video quality and utilization of peers' upload capacity.

Keywords: Peer-to-Peer video streaming · Data scheduling · Throughput · Quality optimization

1 Introduction

Peer-to-Peer (P2P) video streaming has become popular in recent years. Compared with the traditional C/S (client/server) model, P2P network has more advantages such as scalability and high dynamics, since it can propagate content without relying on a particular network structure. P2P streaming system is mainly composed of two parts, the first part is overlay construction. In this part, the peers are organized to form an overlay network. Through the overlay network, participates can share video resources with each other. The second part and also the critical part is called data scheduling, which selects the senders for each video data.

Generally, the overlay can be classified into two types, constructed and unconstructed. Constructed overlay also names tree-based overlay, in which peers organize themselves into a tree structure where parents directly transmit the videos to the children according to the structure [1]. Obviously, this approach can achieve a low

© Springer Nature Singapore Pte Ltd. 2017
B. Zou et al. (Eds.): ICPCSEE 2017, Part II, CCIS 728, pp. 184–197, 2017.
DOI: 10.1007/978-981-10-6388-6_16

latency, and it doesn't need to take too much time to calculate the data schedule. However, it could spend extra overhead to maintain a structured overlay among peers and it is difficult to undergo the peer churn. The other type is mesh-based, or the gossip-style, in which peers connect each other to construct an unstructured network [2]. In this approach, the video to be spread is partitioned into some small segments which consist of a certain number of video frames. The peers in the mesh-overlay contribute the video segments to others as senders and request the video segments as receivers. To obtain a missing segment, a receiver must choose a neighbor as the sender of the segment from the multiple neighbors which have the segment. This is achieved by data scheduling and our research focuses on it in this paper.

As an important component of the P2P streaming system, data scheduling has been researched a lot in recent years [2–10], but it is still a valuable problem to study. Due to the dynamics of the overlay and the peers' heterogeneous condition, calculating the segment transmission schedule is computationally complex [10]. To meet the increasing demands of people on perceived video quality, a lot of researches have proposed some scheduling algorithm to improve the video quality [5, 6, 9]. However, maximizing the video quality could make the transmission requests concentrate on several peers, and that could lead to low utilization of the others' upload capacity. Although some other studies present algorithms to optimize the throughput of the system [10, 11], it may result in low perceived video quality because some of emergent segments can't be transmit in time.

In this paper, our purpose is to improve the perceived video quality and the system throughput simultaneously. Based on the two objectives, we formulate the data scheduling problem as a multi-objective optimization formulation by weighting the segments according to their different properties, such as rarity, emergency and segment quality. Then we design a data scheduling algorithm to solve the problem. Simulation results show that our algorithm is efficient and outperforms the other algorithms.

The rest of the paper is organized as follows. In Sect. 2, we summarize the related researches in literature. Section 3 gives the description of the P2P system model and the data scheduling model we study. Then we present our data scheduling algorithm in Sect. 4. Section 5 evaluates the proposed algorithm through the simulation. At the last, we conclude the paper and give a discussion in Sect. 6.

2   Related Work

Since the optimal data scheduling is a NP-complete problem, some researches proposed heuristic algorithm to get approximate optimal solutions to achieve a high perceived video quality, such as [4–6, 8, 9]. Chakareski et al. [4] use an iterative descent algorithm to solve the optimization problem which is formulated to improve the visual perception. The authors of [5, 9] construct an optimal model for the purpose of improving perceive video quality. They define the objective function by the quality of each segment, and then they propose an approximation algorithm to get an approximate optimal solution. Liu et al. [8] present an event-driven priority-based algorithm, not only take the response processing on receivers but also on senders into account. [6] considers the impact of each frame type, and proposes a receive side frame scheduling scheme as well as a sender selection scheme. But frames are interdependent in the video, taking the frame as the smallest transmit unit may result in decoding error due to the loosing of some frames, thus, the visual perception could be deteriorated accordingly. Most of them only aim to optimize the perceived video quality, and that may cause load imbalance in sender side. There also are some studies proposed data scheduling algorithms to improve the throughput of the system [10, 11]. Zhang et al. [10] transform the chunk scheduling problem into a min-cost problem to optimize the priority function, which is defined by considering the rarity and emergency of the video chunks. They optimize the throughput of peers in the overlay, but the algorithm is computationally expensive, and they don't enhance the perceived video quality. [11] proposes a load balancing strategy which uses a request migration algorithm to handle the load imbalance problem. However, most of them formulate the data scheduling problem as a single objective optimization, which could maximize the perceived quality with low utilization of the peers' upload capacity, or maximize the network throughput with degrading perceived video quality.

In this paper, we formulate the data scheduling problem as a multi-objective optimization formulation and then we propose a novel weighted segment first algorithm (WSF), which solves the scheduling model efficiently and achieves a high video quality with considerable utilization of senders' upload capacity simultaneously.

3 Statement of System Model and Data Scheduling Problem

In this section, we give the statement of the system model and describe the data scheduling problem in detail.

3.1 System Model

Due to the robustness and scalability and simplicity, the mesh-based overlay is widely used in P2P streaming systems such as [12]. In this paper, we employ a mesh-based overlay which is composed of a tracker, a video source and a lot of peers. The video source divides the video into segments, and each of segments includes a certain number of video coded frames. The size of frames is different due to the coding scheme so as to that the size of the segments is different. Every segment has a serial number for identification. When a peer starts to play a video, it firstly joins a swarm, in which the peers are watching the same video. The swarm in our model is dynamic as same as the real world and the peers can join and leave the swarm randomly. After participating the peer starts to receive video segments from other peers periodically and contributes its own upload bandwidth to transmit segments which are requested by its neighbors. To achieve this, the peers in the swarm periodically notify their upload bandwidth and buffer map, which is a binary string for indicating the availability of segments.

Most important of all, there could be more than one sender having the segment which is requested by a receiver. But because of the difference of the peer condition such as upload bandwidth and the transmit queue, the bad segment schedule could result in a lower perceived quality or overloading on a portion of senders. For instance, a peer chooses a neighbor as its sender for a missing segment, but the transmit queue is full of the segments to other receivers, so the segment becomes a late segment when the receiver receives it, which reduces the playback continuity. Therefore, it is challenging and important to design an efficient data scheduling algorithm for making an optimal or approximate optimal schedule.

3.2 Data Scheduling Problem Statement

In this paper, our goal is to calculate an approximate optimal schedule to optimize perceived video quality and the utilization of the senders' upload bandwidth simultaneously. In this part, we detailed describe the data scheduling problem in mesh-based P2P video streaming system.

Data scheduling problem: In mesh-based P2P video streaming system, we consider that each peer with its M neighbors forming a swarm Q to share a video. The frame rate of the video is V_f. The video is divided into N segments with $\mathcal{N}_{GOP}$ video frames and every segment has its own serial number $j(1 \leq j \leq N)$. The visual quality of each segment is different due to the encoder and the content of the segment. The peers in the swarm periodically exchange their upload bandwidth b_i and segments availability $a_{i,j}$. $a_{i,j} = 1$ means the segment j is available in the sender $i(1 \leq i \leq M)$ and $a_{i,j} = 0$ otherwise. When a peer starts to watch the video, it keeps a sliding window which contains δ seconds length of its missing segments, and we let S be the set of the missing segments' serial numbers. Compared to the current playback time, each segment in the sliding window has a decoding deadline calculated as following.

$$d_j = \frac{(j-1) \times \mathcal{N}_{GOP}}{V_f} \tag{1}$$

Therefore, the receiver should calculate the transmission schedule for each segment according to the above information. We use Q to present the sets of the schedules calculated by the scheduler. e.g. $\langle i, j, t \rangle$ is a schedule which means the receiver schedules the neighbor i to send the segment j at time t. For optimizing the perceived video quality and network throughput, the generated schedules should maximize the total quality of the on-time segments and the utilization of the senders' upload capacity.

In Sect. 4, we formulate the data scheduling problem and propose an efficient algorithm to solve the problem.

4 Solution of the Multi-objective Problem

In this section, we consider that segments have different impacts to perceived video quality and network throughput in different conditions. We design a segment weight calculation strategy to weight segments aim to optimize the two performance indexes. Then we propose a data scheduling algorithm to calculate the approximate segments schedules.

4.1 Segment Weight Calculation Strategy

To calculate the weight of a segment, we consider three properties for each segment: emergency, rarity and segment quality.

A segment must satisfy the following condition to be feasible to a receiver: (1) There must be a sender holding the segment. (2) The time when the receiver receives the segment must be before the deadline of the segment. First of all, the emergency is the most important property of a segment. The segments with the shorter time interval between the decoding deadline and the current playback time are more likely to be missing. So they should be transmitted preferentially. As a consequence, we define the function of emergency of segment j as follows:

$$\mathcal{P}^j_{emergency} = \frac{\delta - \left(d_j - T_{playback}\right)}{\delta} \tag{2}$$

We let $d_j - T_{playback}$ be the remaining time between the current playback time and the deadline of the segment j. δ is the total time of a scheduling window, it means that a receiver requires δ length of the video each scheduling period. In a transmission schedule, if the segments with higher $\mathcal{P}^j_{emergency}$ are transmitted firstly, receivers may miss fewer segments so as to achieve a better visual fluency.

We notice that the fewer the neighbors a receiver has, the higher the risk of a segment being lost. Thus, it is necessary to consider the rarity of a segment. A segment may have multiple senders, or only one sender even no sender. The segments with fewer senders are rarer compared with other segments. Hence, we define the rarity of a segment by counting the number of senders who holds this segment.

$$\mathcal{P}^j_{rarity} = \frac{\left(M - \mathcal{N}^j_{neighborCount}\right)}{M} \tag{3}$$

In (3), $\mathcal{N}^j_{neighborCount}$ is the number of the neighbors which has the segment j, and it is calculated as:

$$\mathcal{N}^j_{neighborCount} = \sum_{i \in Q} a_{i,j} \tag{4}$$

The third property we consider is segment quality. Many previous researches ignored the importance of the quality belongs to the segment itself. In this part, we define the quality of a segment by peak signal-to-noise ratio (PSNR) value [13]. A video segment consists of a certain number of video frames, and different video frames have different PSNR value. A video frame with a higher PSNR value has better visual quality than others. Here we use the average value of the PSNR values of video frames which are in a segment to present the quality index of the segment. Therefore, the quality property is defined as follows:

$$\mathcal{P}^j_{quality} = \frac{\left(PSNR_j - PSNR_{min}\right)}{\left(PSNR_{max} - PSNR_{min}\right)} \tag{5}$$

In (5), we let $PSNR_j$ present the quality of the segment j. $PSNR_{min}$ is the minimum value of the all segments' quality in a video, and the maximum value is $PSNR_{max}$.

Previous studies [3, 10] have showed that the rare segments have more impacts to the throughput of the network, and [5, 9] define the weight of each segment as the PSNR value to optimize the perceived video quality. Actually, emergency is also an important property contributed to quality. In this paper, we aim to optimize the network throughput and perceived video quality simultaneously. Therefore, we then have a discussion to define the weight of a segment synthetically. Notice that all the three property functions we defined before are normalized for unification, so we firstly define the weight function of segment j as follows:

$$\omega_j = \imath_q.\mathcal{P}^j_{quality} + \imath_e.\mathcal{P}^j_{emergency} + \imath_\imath.\mathcal{P}^j_{rarity} \tag{6}$$

Through this integrated formulation, the multi-objective optimization problem can be transformed to a single objective optimization problem. We let $\imath_q$, $\imath_q$ and $\imath_\imath$ be the ratio of the properties respectively, then we discuss about them in following three different conditions.

- Supplier missing

The peers in the system may leave randomly at any time for network failure or other personal reasons. That may cause $\mathcal{N}_{neighborCount} = 0$ which means none of the neighbors holds the requested segment. Therefore, it's useless to calculate the other property values in this situation, because no senders can send a segment which they don't have. So we set the weight of the segment j to 0 and the function becomes as follows:

$$\omega_j = 0 \tag{7}$$

- Rarity First

We hope that the rarity should account more proportions of the weight when the number of the neighbors who hold the segment is low. In our formulation, if $\mathcal{N}_{neighborCount}$ is not more than 3, then we dynamically set the ratio of the rarity to a linear representation:

$$\nabla_\nabla = 1 - \nabla_{\mathrm{II}} - 0.1 \times \mathcal{N}^j_{neighborCount} \tag{8}$$

In this case, the quality property has less impact than the other two properties. Since the last two properties are major to consider, the value of r_q is set to 0.1, then the weight of the segment is calculated according to (9):

$$\beth_| = 0.1 \cdot \mathcal{P}^j_{quality} + (0.9 - r_r) \cdot \mathcal{P}^j_{emergency} + \nabla_\nabla \cdot \mathcal{P}^j_{rarity} \tag{9}$$

In this way, the rare segments can spread wisely more rather than concentrate on one sender, so it's significant to avoid that a peer can't transmit a rare segment because of over loading. Furthermore, the segment loss rate also decreases as a result.

- Emergency and Quality First

If $\mathcal{N}_{neighborCount}$ is more than 3, we mainly consider emergency and quality property since there is no remarkable difference between the segments in rarity. For the quality property, we set ∇_q to 0.4 and ∇_e to 0.5. The higher proportion of emergency makes certain that the priority of the segments which are closer to current playback time is higher so as to maintain fluency. The second property to be considered in this case is segment quality. For some reasons such as the video degradation and encoding errors, there could be some segments with low PSNR value which could reduce the visual perception quality. Therefore, the weight of the segments which lead to a low quality should be decreased. We give the formulation in this part:

$$w_j = 0.4 \cdot P^j_{quality} + 0.5 \cdot P^j_{emergency} + 0.1 \cdot P^j_{rarity} \tag{10}$$

Therefore, the scheduling problem is changed to maximize the total weights of the segments in each request period τ for receiving a best visual perception and an idealized network throughput simultaneously. Thus, we formulate the data scheduling problem as a weighted segment optimization function for a receiver:

$$
\begin{aligned}
Maximize : \mathcal{Z} &= \sum_{i \in Q} \sum_{j \in S} w_j x_{i,j} \\
\text{s.t.} (a) &\sum_{i=1}^{M} x_{i,j} \leq 1, \forall j \in S, \\
(b) &\sum_{j \in S} x_{i,j} s_j \leq b_i \tau, \forall i \in Q, \\
(c) &x_{i,j} \leq a_{i,j}, \forall i \in Q, j \in S, \\
(d) &x_{i,j} \in \{0,1\}, \forall i \in Q, j \in S.
\end{aligned} \tag{11}
$$

In formula (11), constraint (a) guarantees that each segment shouldn't be scheduled with more than one sender. Constraint (b) ensures that each sender avoids being excessive. Here s_j is the size of the segment j, and τ is the request period. Next, constraint (c) means the sender only sends the segment it holds. Last, constraint (d) limits the value range of variables.

4.2   Proposed Algorithm for the Data Scheduling Problem

To compute an efficient schedule for the weighted segment problem in real time, the condition of the senders must be calculated first. As mentioned before, there could be more than one sender for a segment. So the receiver must choose a sender for the segment and send request of the segment to that receiver. Upon the sender receives a request, it put the requiring segment to the transmit queue, in which the segments will be transmit by the sender in transmit time order.

However, a sender may receive too many requests in a schedule period and some of segments could not be sent in time due to the limitation of the sender's upload bandwidth. On the one hand, the late segments could lead to interruptions, and that results in a low visual perception quality on receiver side. On the other hand, some senders undertake too many sending tasks which are excessive for their capacity, which actually causes load unbalanced on the senders. Therefore, the upload bandwidth's utilization rate of the senders should be considered when a receiver calculates segments schedule.

We calculate the utilization rate by the ratio of transmit time in a scheduling period. The accumulated transmission time is calculated in (12)

$$T = \frac{s_{queue}}{b_i} \tag{12}$$

where s_{queue} is the total size of the segments in the transmit queue of the sender, and b_i presents the upload bandwidth of the sender i. From (12), we can infer that the time a receiver receives segment j is calculated by the following formula:

$$T_{receiver} = T + \frac{s_j}{b_i} + T_{tpd} \tag{13}$$

In the formula, s_j/b_i is the transmit time of the segment j. T_{tpd} is the propagation delay. It means the time it takes for a segment to be transmitted from the sender to the receiver in the medium. We ignore the propagation delay in this paper for convenience since it is a small value and it can't influence the schedule calculation.

In the next, when the weight of the requiring segments and the receive time for every neighbors are calculated, the receiver starts to calculate the schedules by our proposed distributed scheduling algorithm and sends the schedules to its neighbors to ask for the segments. The pseudo code of the algorithm is in following:

Algorithm 1 The proposed distributed scheduling algorithm

1: **Procedure**
2: Calculate all the w_j for each segment in the sliding window of the receiver, and sort the segments in a descending order with w_j
3: **For** each segment j in the sliding window of the receiver **do**
4:      Calculate the $T_{receiver}$ for each sender, and sort the senders by ascending order with $T_{receiver}$
5:      **For** each sender i **do**
6:          Boolean $available_1$ = availability of segment j in the sender i
7:          Boolean $available_2 = \tau - T_{receiver} > 0$? 1:0
8:          Boolean $available_3 = d_j - T_{receiver} > 0$? 1:0
9:          **If** ($available_1$ && $available_2$ && $available_3$)
10:             Add $<i,j,T>$ to the schedule
11:             Update T for the sender i
12:             **Break for**
13:         **End if**
14:     **End for**
15: **End for**
16: **End procedure**

The idea of the algorithm can be detailed as follows. First, it calculates the weights of the segments in the sliding window of the receiver and sorts the segments in descending order by the value of the weight in line 2. Second, it iterates the segments in order and does the following steps for each segment in line 3: (1) Calculate the $T_{receiver}$ for all the senders and sort those in ascending order in line 4. (2) Iterate all the senders in line 5 to 14, if the sender is able to send the segment to the receiver in time, (that is, the time when the segment received is before the deadline of the segment, and the sending time is before the end of the scheduling time.) then it is scheduled to send the segment and updates the value of T in the line 10 and 11. Finally, a transmission schedule is calculated through the algorithm then the senders send the segments to the receiver according to the schedule.

5 Evaluation

In this section, we evaluate the performance of the proposed algorithm by simulation. We firstly introduce the setup of the simulator we used and the metrics, and then we compare our algorithm with the other classic data scheduling algorithms by the results from the simulation.

5.1 Simulation Setup

We employ an event-driven simulator which is implemented by Shen in [9]. The simulator is coded by Java, and we extend it by adding the proposed algorithm in this paper. The simulation platform provides a mesh model for overlay construction and implements 5 algorithms: SSTF and WSS in [9], LRF [3], Min-cost [10], and the proposed WSF. WSS is an approximation algorithm which considers the quality of the segments. SSTF implements an unweighted data scheduling algorithm, which preferentially schedules the segments with shortest transmission time. LRF is a general scheduling algorithm widely used in real systems such as PPTV [12] and CoolStreaming [3]. It preferentially schedules the segment with fewest holders and chooses a holder with higher bandwidth as its sender. Lastly, Min-cost is a heuristics algorithm which forms the data scheduling problem as a min-cost flow model and solves it.

In our simulation, we deploy a tracker, a video source server, and 3000 peers in the simulator, and the simulator randomly selects 1% of the peers as the propagation peers, which we called as the seeds. The upload bandwidth of peers follows the distribution given in Table 1 as recommended in [14]. And the upload bandwidth and download bandwidth of the tracker are both set to 10 Mb. For each peer, we set the number of the neighbors to 10. And we set that the simulation runs 24 h for each algorithm. To simulate the dynamics in real system, the peers randomly join the system individually and a fraction of them leave during the simulation time. The peers in the system exchange the buffer map to inform segment availability and upload bandwidth periodically. While a peer starts to watch the video, it holds a sliding window which contains its missing segments, and we set the time of the sliding window to 10 s in the simulation. To obtain the missing segments in the sliding window, the peer runs data scheduling algorithm to calculate a transmit schedule and sends it to its neighbors. Then, the neighbors send the segments to receivers according to the schedule.

We run the simulator on a normal PC, with a 2.3-GHz Intel CPU and 8 GB of memory. We choose a high-resolution video trace from the Arizona State University video trace library for simulation [13, 15]. The parameter of the video trace is showed in Table 2.

We consider four important performance metrics as used in [3, 9]:

Average Perceived Video Quality α:

$$\alpha = \sum_{n=1}^{N} \frac{q_n u_n}{N} \tag{14}$$

where q_n is the PSNR value of the segment n, and u_n indicates that whether segment n arrives on time.

Continuity Index:

$$\beta = \sum_{n=1}^{N} \frac{u_n}{N} \tag{15}$$

Continuity index is the value of the proportion of the on-time segments over the total segments. It indicates the fluency of the perceived video. The higher value of the continuity index brings a better watching experience for the viewers.

Table 1. Peer upload bandwidth distribution

Distribution (%)	10.0	14.3	8.6	12.5	2.2	1.4	6.6	28.1	16.3
Total bandwidth (kbps)	256	320	384	448	512	640	768	1024	>1500
Contributed bandwidth	150	250	300	350	400	500	600	800	1000

Table 2. Video parameters

Video name	From Mars to China
Resolution	HDTV (1920 * 1080)
Frame rate (fps)	30
Number of frames	51715
Group of Pictures (GOP)	12
Quantization Parameter (QP)	28-28-30
Frame size (bits): min/max/mean	80/326905/20207.12
PSNR value (dB): min/max/mean	43.314/68.458/47.94675
Mean bitrate (kbps)	4849.71

Load of the Sender i:

$$\ell = \sum_{j \in Schedule_i} \frac{s_j}{\delta b_i} \tag{16}$$

We use the utilization of upload bandwidth of the sender i to measure its load. In the formula (16), $\sum_{j \in Schedule_i} s_j$ is the total size of the segments scheduled to the sender i.

Load Balancing Factor:

$$r = \sqrt{\frac{\sum_{p=1}^{P} \left(\ell_p - \bar{\ell}\right)^2}{P}} \tag{17}$$

The load balancing factor r is the standard deviation of ℓ over the all scheduling periods. The smaller the value of r, the better the performance of the load balancing and the network throughput. In formula (17), we use P to present the number of the scheduling periods for the sender i.

In addition, we use execution time to indicate the simplicity of the algorithm. If the algorithm takes too much time to calculate a segment schedule, it may slow down the peer which doesn't have enough capacity.

5.2   Simulation Results

We run the simulator for each algorithm and log the performance metrics we use above for every sender. To compare the performance metrics of each algorithm clearly, we compute and plot the cumulative distribution function curves (CDF) for each metric. Finally we summarize the results in the following figures.

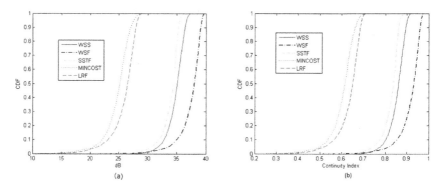

Fig. 1. CDF of PSNR (a) and continuity index (b)

The CDF of PSNR value is showed in Fig. 1(a). We can see WSF outperforms the others obviously. More than 90% of peers can obtain more than 35 dB quality at least while the WSS can offer 33 dB, and the next is SSTF with 32 dB. That means WSF is able to ensure higher visual perception quality for most of receivers. Furthermore, we observe that WSF can achieve 39 dB at most whereas the WSS also can obtain 37 dB. Next to WSF are WSS and SSTF, they can achieve 33 dB and 32 dB for 90% peers respectively. The result shows that the MINCOST and LRF can't take an acceptable visual perception to the receivers due to the max quality of MINCOST is only 28 dB, and it can only achieve 23 dB for less than 90% peers.

The same as the PSNR value, the continuity index is plotted in the Fig. 1(b) for each algorithm respectively. The simulation data shows WSF still achieves the best performance over the 5 algorithms. From the Fig. 1(b), we see that the continuity index with the WSF algorithm is 0.8 for 98% peers, whereas the continuity index of WSS is 0.75, which means the WSF can provide 0.8 continuity index for at least 98% peers. For SSTF and WSS, they perform similarly and also can obtain 0.8 continuity index for more than 80% peers. Compared to them, LRF and min-cost can't produce high continuity index, they only achieve 0.71 continuity index at most among all the peers.

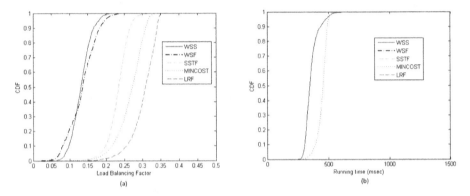

Fig. 2. CDF of load balancing factor (a) and execution time (b)

We next plot the CDF curves of the load balancing factor in the Fig. 2(a). We observe that MINCOST and SSTF as well as LRF have the much higher load balancing factor value compared with the other two algorithms. The highest value of the load balancing factor is only 0.23 in the curve of WSF. And more than 90% of scheduling periods' value is controlled to less than 0.2. In this metric, performance of WSS is approximate the same as WSF.

The Fig. 2(b) shows the CDF curve of the execution time of the algorithms over the receiver. We can only see the execution time of two heuristic algorithms. This is because the execution time of WSF and SSTF are both enough small so as to be invisible in the figure. Therefore, WSF is able to be deployed in the most of real-time P2P streaming systems due to the low overhead.

In summary, we can arrive at the following conclusion through the simulation: (1) Proposed WSF improves the perceived video quality of the receivers and the throughput of the network simultaneously. (2) WSF is a light-weight algorithm which could run fast at a commodity PC with low performance.

6   Conclusions and Future Work

In the paper, we have proposed a multi-objective optimization-based distributed data scheduling algorithm for mesh-based P2P video streaming systems. We formulate the data scheduling problem as a multi-objective problem. To solve the problem efficiently, a segment calculation strategy is proposed, which calculates the weight of the segments according to segment rarity, segment emergency and segment quality. Following, we present a distributed algorithm to schedule the segment on the receiver side. To demonstrate the performance of the proposed algorithm, we have developed an event-driven mesh-based P2P video streaming simulator with other four implemented classical algorithms. Through simulations, we show that WSF can achieve a significant improvement in visual perception quality and network throughput. Furthermore, our algorithm is also efficiency in execution time.

Besides the work of this paper, there are still a lot of works to do in the future to extend the research. First, we haven't deployed our algorithm in the real systems such as [3]. Second, the research in this paper only focuses on the single layer video streaming system, and we will consider the formulation and rebuild the algorithm to run in the multi-layer video streaming system.

Acknowledgments. This research was supported by Guangxi Natural Science Foundation under Grant No. 2016GXNSFAA380011, the Science and Technology Research Program of Guangxi University (No. KY2015ZD047).

References

1. Chu, Y., Rao, S.G., Seshan, S., Zhang, H.: A case for end system multicast. IEEE J. Sel. Areas Commun. **20**(8), 1456–1471 (2002)
2. Pai, V., Kumar, K., Tamilmani, K., Sambamurthy, V., Mohr, A.E.: Chainsaw: eliminating trees from overlay multicast. In: Castro, M., van Renesse, R. (eds.) IPTPS 2005. LNCS, vol. 3640, pp. 127–140. Springer, Heidelberg (2005). doi:10.1007/11558989_12
3. Zhang, X.Y., Liu, J.C., Li, B., Yum, T.: Cool streaming/DONet: a data-driven overlay network for peer-to-peer live media streaming. In: Makki, K., Knightly, E. (eds.) IEEE INFOCOM SERIES2102-2111 (2005)
4. Chakareski, J., Frossard, P.: Utility-based packet scheduling in P2P mesh-based multicast. In: Visual Communications and Image Processing 2009, San Jose, CA (2009)
5. Hsu, C., Hefeeda, M.: Quality-aware segment transmission scheduling in peer-to-peer streaming systems, 2010, pp. 169–180. ACM (2010)
6. Bideh, M.K., Akbari, B., Sheshjavani, A.G.: Adaptive content-and-deadline aware chunk scheduling in mesh-based P2P video streaming. Peer Peer Netw. Appl. **9**(2), 436–448 (2016)
7. Efthymiopoulou, M., Efthymiopoulos, N., Christakidis, A., Athanasopoulos, N., Denazis, S., Koufopavlou, O.: Scalable playback rate control in P2P live streaming systems. Peer Peer Netw. Appl. **9**(6), 1162–1176 (2016)
8. Liu, P., Huang, G., Feng, S., Fan, J.: Event-driven high-priority first data scheduling scheme for p2p vod streaming. Comput. J. **56**(2), 239–257 (2013)
9. Shen, Y., Hsu, C., Hefeeda, M.: Efficient algorithms for multi-sender data transmission in swarm-based peer-to-peer streaming systems. IEEE Trans. Multimedia **13**(4), 762–775 (2011)
10. Zhang, M., Xiong, Y., Zhang, Q., Sun, L., Yang, S.: Optimizing the throughput of data-driven peer-to-peer streaming. IEEE Trans. Parall. Distr. **20**(1), 97–110 (2009)
11. Huang, G., Li, C., Liu, P.: Load balancing strategy for P2P VoD systems. KSII Trans. Internet Inf. Syst. **10**(9) (2016)
12. PPTV. http://www.pptv.com. Accessed 1 Mar 2017

13. Van der Auwera, G., David, P.T., Reisslein, M.: Traffic and quality characterization of single-layer video streams encoded with the H. 264/MPEG-4 advanced video coding standard and scalable video coding extension. IEEE Trans. Broadcast **54**(3), 698–718 (2008)
14. Liu, P., Feng, S., Huang, G., Fan, J.: Bandwidth-availability-based replication strategy for P2P VoD systems. Comput. J. **57**(8), 1211–1229 (2014)
15. Seeling, P., Reisslein, M., Kulapala, B.: Network performance evaluation using frame size and quality traces of single-layer and two-layer video: a tutorial. IEEE Commun. Surv. Tutor. **6**(3) (2004)

A Novel Range-Free Jammer Localization Solution in Wireless Network by Using PSO Algorithm

Liang Pang[1,2]([⊠]), Xiao Chen[1,2], Zhi Xue[1,2], and Rida Khatoun[1,2]

[1] School of Electronic Information and Electrical Engineering,
Engineering Research Center of Network Information Security Management and
Service, Ministry of Education, Shanghai Jiao Tong University, Shanghai, China
`cyclone0000@163.com`, {`chenxiao,zxue`}`@sjtu.edu.cn`
[2] Telecom ParisTech, Paris, France
`rida.khatoun@telecom-paristech.fr`

Abstract. In wireless networks, jamming attacks are easy to launch and can significantly impact the network performance. The technique which localizes the jamming attacker is useful to address this problem. Some range-based localization schemes depend on the additional hardware of wireless nodes too much, and they can not work in resource-constrained wireless networks. Solutions in range-free localization are being pursued as a cost-effective alternative to more expensive range-based approaches. In this paper, we propose a novel range-free algorithm to localize the source of the attacker. We show that our approach only relies on the positions of each jammed or no-jammed node in the network, PSO algorithm is used to get the minimum covering circle of jammed positions and the circle center is the estimated jammer location. We compare our work with some existing range-free solutions via extensive simulations in two models, which are wireless sensor network (WSN) and vehicular ad hoc network (VANET) respectively. The experimental results suggest that our proposed algorithm achieves higher accuracy than the other solutions, and the localization error goes down with larger number of recorded jammed positions. In additional, when the recorded jammed positions are distributed in a specific constrained area, the localization error goes higher, we also propose an improved PSO algorithm to deal with this issue.

Keywords: Jamming attack · Jammer localization · Vehicular ad hoc network (VANET) · Wireless sensor network (WSN) · PSO · Minimum covering circle

1 Introduction

Due to the openness of the wireless transmission medium, wireless communications are particularly vulnerable to radio interference. Adversaries can easily purchase low-cost jamming devices and use these commonly available platforms

© Springer Nature Singapore Pte Ltd. 2017
B. Zou et al. (Eds.): ICPCSEE 2017, Part II, CCIS 728, pp. 198–211, 2017.
DOI: 10.1007/978-981-10-6388-6_17

to launch a Denial-of-Service (DoS) attack with little effort. Denial-of-Service attack can eliminate a network's capacity to execute its normal function [1] by affecting its throughput, network load, end to end delay [2]. For example in WSN, Denial-of-Service attack makes nodes can not exchange data in jammed region, it will seriously affect the original function of WSN or even disable the whole network; And in VANET, Denial-of-Service attack cause the congestion of control channel which is used for transmitting safety information, it will further impact the communication among vehicles and increase the probability of on road accidents.

According to Xu et al. [3], jammer can be categorized as constant jammer, deceptive jammer, random jammer and reactive jammer. Some methods of detecting the jamming were widely researched in [4–9]. In order to cope with this kind of Denial-of-service attacks, a lot of sophisticated physical layer strategies and techniques have been developed such as Frequency-Hopping Spread Spectrum (FHSS) and Direct Sequence Spread Spectrum (DSSS), Ultrawide Band Technology (UWB), Antenna Polarization, directional transmission methods [10]. And many kinds of evasion strategies also have been researched, such as wormhole-based anti-jamming techniques [11], channel surfing [12] and timing channel [13]. However, these solutions are overwhelmingly reliant on cryptography as basis, and the additional infrastructural overhead is also too big to be used in some resource-constrained networks.

Recently, new approaches utilizing localizing algorithm associated with range-based properties to combat attackers in wireless networks have been proposed. But, the additional cost of hardware required by range-based solutions is also unacceptable in resource-constrained networks, and sometimes the range-based solutions can not reach the required location precision. Furthermore, when the size of wireless network grows larger or the nodes in network are high autonomy, the interferences produced by inner nodes should make the measurements of range-based data inaccurate, it also increase the localization error. Thus, an alternate range-free solution to the jamming localization problem is needed. We focus on using range-free algorithms to find the location of the jamming attacker in wireless networks, which are great important for eliminating the attackers from the network and guiding the routing protocol to avoid the jammed region. And the existing range-free solutions are still not enough accurate and the localization error goes high when the reference point set is not uniform. This work is proposed to address these problems. We assume the jammer is static and it has an isotropic effect that the jammed region can be modeled as circular region centered at the jammer's location. Then, we use PSO algorithm to get the minimum circle which covers all jammed positions, and this circle is regarded as an approximate jammed region, hence the center of this circle is treated as the estimated position of the jammer. The main contributions of our work are:

Firstly, we use particle swarm optimization (PSO) algorithm to find the minimum covering circle of all jammed positions.

Secondly, We improve the PSO algorithm to deal with the non-uniform distributed condition which means the recorded jammed positions are limited to locate in a local area of jammed circle.

Finally, We make simulations with both moving nodes and static nodes in models of WSN and VANET. The experimental results show that our approach is more effective than the other existing rang-free solutions.

The rest of the paper is organized as follows. We place our work in the context of related research in Sect. 2. Our network assumption is presented in Sect. 3. We formulate the problem of localizing the jamming attacker as an optimization problem of using PSO algorithm in Sect. 4. We introduce our simulation experiments and do some analytical discussions in Sect. 5. Finally, we conclude our work in Sect. 6.

2   Related Work

The localization techniques are researched widely [14,15]. Dealing with ranging methodology, range-based algorithms involve distance estimation to landmarks using the measurement of various physical properties such as Received Signal Strength (RSS) [16,17], Time Of Arrival (TOA) [18], Time Difference Of Arrival (TDOA), and Direction Of Arrival (DOA) [19]. In [20], a heterogeneous network containing powerful nodes with established location information is considered. In this work, anchors beacon their position to neighbors that keep an account of all received beacons. Using this proximity information, a simple centroid localization (CL) model is applied to estimate the listening nodes' location. An enhance version of CL is weighed centroid localization [21], which adds weight value into the process of estimating target node position. [22] proposed the virtual force iterative localization (VFIL) algorithm which outperforms better than CL and WCL in terms of localization accuracy. VFIL first estimates the jammer's location by using traditional coarse localization scheme like CL, and then iteratively localizes the jammer by defining a virtual force. Both of them can be used to locate the jammer. However, the localization accuracy of these solutions is unsatisfactory. And they even perform incorrectly when the testbed is non-uniform distributed. In [23,24], they study the Packet Delivery Ratio (PDR), [23] uses it to detect the jamming attack, and [24] use the relationship between the distance to jammer and the value of PDR to further localize the jammer. The works [25,26] are most closely related to us. [25] proposed a range-free APIT scheme in a heterogeneous network which includes anchors equipped with high-powered transmitters and GPS. APIT isolated the environment into triangular regions between beaconing nodes, and a node can estimate its location by narrowing down the triangular region in which it can potentially reside. [26] proposed Minimum-covering-circle based (MCCL) jammer localization algorithm in wireless sensor networks (WSN). He used the plane geometry knowledge to form an approximate jammed region. But this approach is also inappropriate when the testbed is non-uniform distributed. Our work differs from the previous study in that our PSO-based localization algorithm does well in both uniform and non-uniform data set, and the performance grows up with larger data set, it proofs that our method is applicable for big data analysis in wireless networks.

3   Network Assumptions

3.1   Network Model

We consider two network models: one of them is wireless sensor network (WSN), the other is vehicular ad hoc network (VANET), and they are all part of internet of things. In WSN model, we chose an open map, that means any pairwise coordinate (x, y) is accessible for any node in the network. We research two kinds of node model according to the characteristics of WSN, they are static node model and moving node model respectively. We also assume that the static nodes in WSN know their location coordinates. And the moving nodes know their original location coordinates, they can calculate their next position coordinates through their present velocity and moving direction. This is a reasonable assumption, because the movement of each moving node can be considered as a Markov process, and the gravity sensor and speed-sensing device are widely used in WSN; In VANET model, we chose an urban traffic map, it is an part of Shanghai traffic map. All nodes in this network model are moving model, and they are strictly restricted to run on the road. At last, we focus on localizing a jammer after some jammed regions have been detected. So we do not consider the process of detecting jamming attack (A lot of researches are focus on this technique, like [4–9]).

3.2   Jamming Model

As shown in Fig. 1, nodes can perceive two different categories under jamming condition, jammed and unaffected-jammed. Nodes under unaffected-jammed can broadcast and receive messages as usual. It is important for the static node model in WSN, because the nodes under jammed condition can not transmit their position coordinates to other nodes. And the unaffected-jammed region is also necessary to the algorithms like WCL [21], which is used to compare with our algorithm. However, the moving node model can record their position coordinates

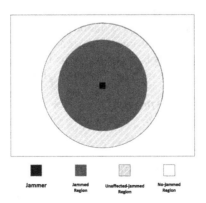

Fig. 1. Jamming model

in jammed region, and transmit them to other nodes or server in unaffected-jammed or no-jammed region. Thus, nodes record a triple $(x, y, if\text{-}jammed)$ every fixed time interval, where the x and y are the coordinates and if-jammed equals to 1 under jamming condition, otherwise if-jammed equals to 0. Furthermore, we assume a static jammer which has an isotropic effect, in other words, the jammed region can be modeled as a circular region centered at the jammer's location.

4 Method

We can divide the positional information recorded by nodes into two data set through the if-jammed signal. They can be expressed as $X = \{x_k\}, k = 1, 2, 3, ..., p$ and $Y = \{y_j\}, j = 1, 2, 3, ..., n$ respectively, where x_k is a 2-dimensional sample represents the coordinate of the kth point in jammed region or unaffected-jammed region, y_j is a 2-dimensional sample represents the coordinate of the jth point in no-jammed region. We formulate the problem of localizing the jammer source as an optimalization problem. It can generally be described as: The data set X is given; We need to produce the minimum circle area which can cover all points in X. To address this optimalization problem, we use Particle Swarm Optimization (PSO) algorithm.

4.1 Particle Swarm Optimization (PSO) Algorithm

There are two insights of a process of optimization of swarm particles [27–29]: (1) being assumed as particles in space, the points will tend to move toward and influence one another, with the objective of seeking agreement with their neighbors; (2) the space in which the particles move is heterogeneous with respect to evaluation: some regions are better than others. Some points in the parameter space result in greater fitness than others. With this architecture, the process of finding better regions can be viewed as solving the problems of optimization.

PSO algorithm suggests that individuals moving through a socio-cognitive space should be influenced by their own previous behavior and by the successes of their neighbors. As the system is dynamic, each individual is presumed to be moving at all times. The direction of movement is a function of the current position and velocity, the location of the previous best success of individual, and the best position found by any member of the neighborhood. It shows as Eq. 1.

$$\begin{cases} v_i(t) = v_i(t-1) + \varphi_1(p_l - x_i(t-1)) + \varphi_2(p_g - x_i(t-1)), \\ x_i(t) = x_i(t-1) + v_i(t). \end{cases} \quad (1)$$

In Eq. 1, x_i is the algebraic symbol of the position of a particle i; v_i is called for velocity, it is a vector that are added to the position coordinates in order to move the particle through iteration; φ_1 and φ_2 are the effective parameters; p_l is the best solution of particle i in advance; p_g is the best solution of all particles in advance.

4.2   PSO-Based Localization Algorithm

We assume a is any point on open map or urban traffic map. And the minimal covering circle area of X with the circle center a can be expressed as Eq. 2, where x_i is the ith element in X, $|X|$ is the total number of elements in X and d_i is the Euclidean distance between point x_i to the circle center a. Then, we transform the jammer localization problem to search the most appropriate a which has the minimum S_a. In PSO algorithm, particles can be seen as potential solutions of the optimal a which is a point on map, and the optimal circle center a is obtained when all particles reach an agreement at $S_{optimal}$ through several iterations.

$$\begin{cases} S_a = \max\{\pi d_i^2\}, i = 1, 2, 3, ..., |X| \\ d_i = ||x_i - a|| = \sqrt{(x_{ix} - a_x)^2 + (x_{iy} - a_y)^2} \\ S_{optimal} = \min_{a \in D}\{S_a\}, D \text{ is the set of all points on map} \end{cases} \tag{2}$$

In order to use PSO algorithm, the self-optimal circle center p_l and the global optimal circle center p_g are required to be determined. If we use n particles to search the most appropriate a, the particle h is represented as a_h and S_{a_h} represents the minimum covering circle area of X at a_h, where $h = 1, 2, 3..., n$. Then, the self-optimal circle center of particle h named p_{hl} can be defined easily as the value of a_h when the minimum S_{a_h} in the previous iterations is obtained. And the global optimal circle center p_g is defined as the value of p_{hl} when the minimum $S_{p_{hl}}$ in the previous iterations is obtained. Hence, every particle in PSO algorithm iterates with the following Eq. 3.

$$\begin{cases} v_h(t) = v_h(t-1) + \varphi_1(p_{hl} - a_h(t-1)) + \varphi_2(p_g - a_h(t-1)), \\ a_h(t) = a_h(t-1) + v_h(t). \end{cases} \tag{3}$$

When PSO algorithm falls in a stable state and almost all particles are equal to the global optimal centroid p_g. That means the agreement is reached by all particles. Then the criteria for stopping iteration of algorithm can be defined as Eq. 4, where $a_h(k)$ is the value of a_h at kth iteration, ε is the condition of termination, it equals to a very small positive value. The algorithm is listed in Algorithm 1.

$$V_{stop} = \frac{\sum_{h=1}^{n}(a_h(k) - p_g)}{n} < \varepsilon \tag{4}$$

In addition, as shown in Fig. 2, sometimes the distribution of the points in X is not uniform. That is because there exists some inaccessible areas for nodes, and these inaccessible areas are caused by many reasons like obstacles, nodes are placed on road strictly, and so on. This situation will lead to a incorrect result that some no-jammed positions will be included in jammed region or unaffected-jammed region. To address this issue, we proposed to use the data set Y to improve PSO algorithm. If one circle is given, we chose the point which has the minimum distance to the circle center in the point set of outside this circle. It is easy to see that the circle with the same center and above-mentioned minimum distance as the radius is infinitely close to the original circle. In our work, we

Algorithm 1. PSO-based Jammer Localization Algorithm

Ensure: Data set X;
Require: JammerLocation;
1: Const ParticleSize $= n$;
2: Var $h = 1 : n$;
3: Init iteration number $m = 1$, particle value $a_h(0)$;
4: **for** $h = 1; h < n; h + +$ **do**
5:     $p_{hl} = a_h(0)$, $S_{p_{hl}} = S_{a_h}(0)$
6: **end for**
7: $p_g = p_{kl}$, where $S_{p_{kl}} = \min_{h \in [1,n]}\{S_{p_{hl}}\}$
8: **repeat**
9:     **for** $h = 1; h < n; h + +$ **do**
10:         Calculate $\boldsymbol{v}_h(m)$ and $a_h(m)$
11:         **for all** x_i in X **do**
12:             $d_i = \|x_i - a_h(m)\|$
13:             $S_{a_h}(m) = \pi(\max\{d_i\})^2$
14:         **end for**
15:         **if** $S_{a_h}(m) \le S_{p_{hl}}$ **then**
16:             $p_{hl} = a_h(m)$;
17:         **end if**
18:     **end for**
19:     **for** $h = 1; h < n; h + +$ **do**
20:         **if** $S_{p_{hl}} \le S_{p_g}$ **then**
21:             $p_g = p_{hl}$;
22:         **end if**
23:     **end for**
24:     $m = m + 1$
25: **until** $V_{stop} < \varepsilon$
26: $JammerLocation = p_g$;

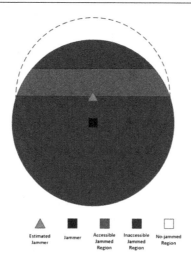

Fig. 2. Situation of the data set X is not uniform distributed

call it reference circle for short, it is produced by data set Y as shown in Eq. 5, where a is still the potential solution and $|Y|$ is the total number of elements in Y. And the evaluation function is also changed as Eq. 5, it is a process of searching the circle which is most closely to its reference circle. But the improved PSO algorithm is different from PSO algorithm: The improved PSO use the absolute value of $S_{reference} - S_a$, when the improved PSO algorithm is performed to find the circle which is closest to the reference circle, some different values of a will reach the same evaluation value of $|S_{reference} - S_a|$, that is because $S_{reference} - S_{a_1} = S_{a_2} - S_{reference}$. And these special potential solutions like a_1, a_2 makes all particles not sure which can be agreed with, they will oscillate between a_1 and a_2, they therefore can not converge to a point $a_{optimal}$. Instead, the distance between a_1 and a_2 can be used as the acceptable error, many pairs of a_1, a_2 form an area which can be seen as a circle range, so it converges to a very small circle region which has the same center $a_{optimal}$. Hence, the error generated by improved PSO is larger, and it is only used with the not uniform distributed X.

$$\begin{cases} S_{reference} = \min\{\pi||y_i - a||^2\}, i = 1, 2, 3, ..., |Y| \\ S_{optimal} = \min_{a \in D}\{|S_{reference} - S_a|\}, D \text{ is the set of all points on map} \end{cases} \quad (5)$$

5 Experiments and Analytical Discussions

5.1 Experimental Environment

We implement our approach in Matlab. We choose a 500×500 open map for simulating WSN network model, the static node models on it are randomly distributed, and the moving node models follow Markov movement which can be expressed as $X(t+1) = X(t) + v\Delta t$, the location of any node at $t+1$ is only related to its location at t; The urban traffic map is generated by the SUMO simulator in our work, it is a part of Shanghai traffic map, and we use it for simulating VANET network model. Nodes in this traffic map are limited to run on the road and choose a random direction to move at each corner. The parameters used in our experiments are shown as follows: The time interval $\Delta t = 1$, the particle size $m = 50$, and the PSO effective parameters $\varepsilon = 0.1, \varphi_1 = 0.0001 \times rand(1)$ and $\varphi_2 = 0.0003 \times rand(1)$, where $rand(1)$ represents a random number from 0 to 1.

5.2 Evaluation Criterion

We define localization error as shown in Eq. 6: The Euclidean distance between the estimated location of the jammer and the true location of the jammer in the network. And the localization error can be seen as an important metric to evaluate the accuracy of our PSO algorithm with other existing rang-free solutions.

$$e = ||a_{true} - a_{estimate}|| = \sqrt{(a_{true\,x} - a_{estimate\,x})^2 + (a_{true\,y} - a_{estimate\,y})^2} \quad (6)$$

5.3 Experiments of Static Node Model

As mentioned above, the input data of our work is only the data set X or $\{X, Y\}$. Thus, in this section we investigate both the impact of jammer's transmission range R and node number n which decide the size of our data set. In this experiment, CL [20] and MCCL [26] are performed to make a comparison of the localization performance with our proposed PSO algorithm (WCL is same as CL here, because the positions in jammed region can not be recorded). First we keep node number the same $n = 600$ and change the jammer's transmission range during four standard value 50 ft, 80 ft and 120 ft. Each localization algorithm was used to obtain the estimated position of the jammer, and the jammer with the same transmission range was deployed at 10 different locations, it means that each algorithm run 10 times. The experimental results are shown in Fig. 3(a)(b)(c), we observe that our proposed PSO algorithm has achieved the best localization accuracy, and it performs stably while others produce large localization errors occasionally. We also observe that the localization error becomes smaller with the jammer's transmission range increases. The reason is that bigger transmission range means more unaffected-jammed positions will be recorded, so does the various kinds of particles. And the differences among particles are the key point of PSO to evolve all particles into the optimal solution. Then, we keep jammer's transmission range the same $R = 80$ ft and change the node number during four standard value 600, 800 and 1000. Figure 3(d)(e)(f) shows the experimental results. In general, our method still performs better than others, and the localization error reduces with the bigger node number, that is because the node density is proportional to node number, that means more recorded jammed positions will be included.

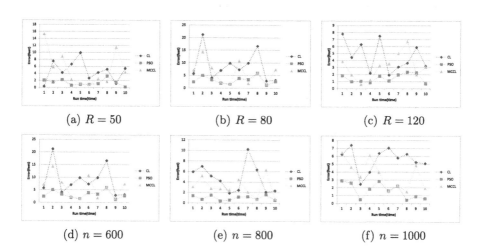

Fig. 3. (a)(b)(c) Impact of jammer's transmission range R when node number is fixed by using static node model (d)(e)(f) Impact of node number n when jammer's transmission range is fixed by using static node model

5.4   Experiments of Moving Node Model

The same as last section, we investigate both the impact of jammer's transmission range R and node number n here, but the only difference is that the positions in jammed region will be recorded by moving node model. The first experiment is performed on the open map, and the simulation time $t = 200\,\text{s}$ is fixed. On this basis, we choose a fixed node number $n = 30$ combine with three different transmission range $R = 50, 100, 150\,\text{ft}$ at first, furthermore we choose a fixed transmission range $R = 100\,\text{ft}$ combine with three different node number $n = 20, 30, 50$. Each case still runs ten times with different jammer location and the experimental results are shown in the following Fig. 4. As we have discussed before, our method is also the most outstanding one and the localization error is inversely proportional to node number and jammer's transmission range. Moreover, the localization error is obviously smaller than the static node model, because moving nodes can record more jammed or unaffected-jammed positions who increase the diversity among particles.

The similar experiments are performed on the urban traffic map too. For the urban traffic map is much bigger than the open map, we keep the simulation time the same at $t = 300\,\text{s}$, a fixed $n = 60$ combine with $R = 50, 150, 200\,\text{ft}$ and a fixed $R = 100\,\text{ft}$ combine with $n = 40, 60, 100$ are selected to run ten times respectively. Figure 5 shows the experimental results, PSO algorithm is performed well as usual, and the localization error is also inversely proportional to node number and jammer's transmission range. However, the localization errors generated with open map are lower than errors generated with urban traffic map. Because the urban traffic map has inaccessible areas, it makes the detected jammed positions are only distributed on a lot of fixed roads in the jammed circle, and the diversity among particles reduces; Furthermore, when

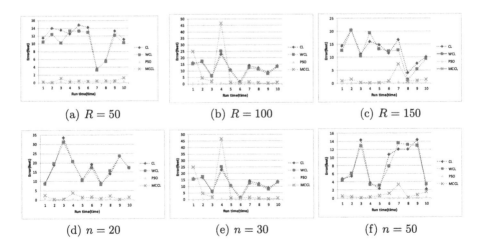

(a) $R = 50$ \qquad (b) $R = 100$ \qquad (c) $R = 150$

(d) $n = 20$ \qquad (e) $n = 30$ \qquad (f) $n = 50$

Fig. 4. (a)(b)(c) Impact of jammer's transmission range R when node number is fixed by using moving node model on the open map (d)(e)(f) Impact of node number n when jammer's transmission range is fixed by using moving node model on the open map

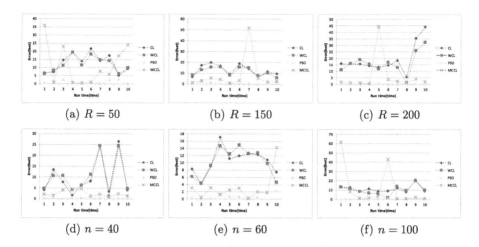

(a) $R = 50$ (b) $R = 150$ (c) $R = 200$

(d) $n = 40$ (e) $n = 60$ (f) $n = 100$

Fig. 5. (a)(b)(c) Impact of jammer's transmission range R when node number is fixed by using moving node model on the urban traffic map (d)(e)(f) Impact of node number n when jammer's transmission range is fixed by using moving node model on the urban traffic map

the transmission range R or the node number n is small, like $R = 50$ ft, $n = 40$ the errors are unstable, it is caused by the uniform distribution of X. Actually, the location error is significantly impacted by the detected positions whether locate in most areas of the jammed circle, larger X just means detected points has a big chance to be located in more areas.

5.5 Experiments of Improved PSO Algorithm

We have introduced the applying scene of our improved PSO above. In this experiment, we create a corresponding jammed region which allows moving nodes to access partial area of it. Our experiment is only performed on the open map. The transmission range of the jammer is fixed at 100 ft, and the node number is fixed at 50. Each location algorithm is ran 10 times to obtain the estimated position of different jammers. Figure 6 shows the experimental results. Overall,

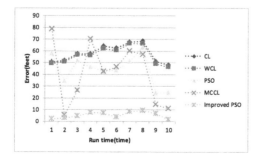

Fig. 6. Validity experiment of effectiveness about improved PSO algorithm

we observed that all the other algorithms include PSO are failed into an unacceptable lower precision except improved PSO. Because it was performed to find the circle which is not include any element of Y, it means no-jammed region was not allow to be located in a jammed circle. However, the errors were also higher, it clearly validates that the improved PSO algorithm is only suitable for special situations.

5.6   The Relationship Between Iterations and Particle Size

In this experiment, we discuss the relationship between the number of iterations and particle size, it also represents the execution time of PSO algorithm. We choose ten different data set X generated by the fixed n, t, R, and then we use PSO algorithm to localize the jammer with various particle size $m = 10, 20, 30, 40, 50, 60$. One advantage of PSO algorithm is that the increased particle size improves implementation efficiency which alternatively means the number of iterations decreases. From Fig. 7, it can be clearly seen that the mean value of iterations decreases while the particle size increases. So it can be inferred that larger particle size leads to less number of iterations. And every particle in PSO is performed independently in every iteration, so it is suitable for distributed parallel computing.

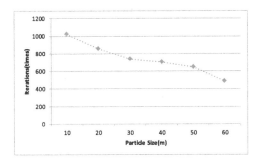

Fig. 7. The relationship between the number of iterations and the particle size in PSO

6   Conclusion

In this work, we proposed to use a range-free solution to locating jamming attacker in vehicular ad hoc network or in wireless sensor network. Both static node model and moving node model are chosen to detect jammed or unaffected-jammed positions on the open map or the urban traffic map. We use PSO algorithm to localize the jammer source on the collected data set, and an improved PSO algorithm is also developed to estimate the jammer source on the not uniform distributional data set. To validate our approach, we conducted a lot of simulation experiments in Matlab. The experimental results provided strong evidence of the effectiveness of our proposed approach. We found that PSO can always achieve a higher accuracy than the other existing range-free algorithms, and the improved PSO can still make a more effective estimate compared to the other algorithms when the distribution of data set is not uniform. At the end, we observe that PSO algorithm executes better with the larger data set, and it does well in distributed parallel computing.

References

1. Wood, A.D., Stankovic, J.A., Son, S.H.: JAM: a jammed-area mapping service for sensor networks. In: 24th IEEE Real-Time Systems Symposium, RTSS 2003, pp. 286–297 (2003)
2. Jrme Hrri, F.F.: Mobility models for vehicular ad hoc networks: a survey and taxonomy. IEEE Commun. Surv. Tutorials **11**, 19–41 (2009)
3. Xu, W., Trappe, W., Zhang, Y., Wood, T.: The feasibility of launching and detecting jamming attacks in wireless networks. In: ACM International Symposium on Mobile Ad-hoc Networking and Computing, MobiHoc 2005, p. 46 (2005)
4. Schmitt, J.B., Giustiniano, D.: Detection of reactive jamming in DSSS-based wireless networks detection of reactive jamming in DSSS-based wireless networks. IEEE Trans. Wirel. Commun. **13**, 1593–1603 (2013)
5. Sasikala, E., Rengarajan, N.: An intelligent technique to detect jamming attack in wireless sensor networks (WSNs). Int. J. Fuzzy Syst. **17**, 76–83 (2015)
6. Siddhabathula, K.: Fast jamming detection in wireless sensor networks. In: IEEE International Conference on Communications, pp. 934–938 (2010)
7. Misra, S., Singh, R., Mohan, S.V.R.: Information warfare-worthy jamming attack detection mechanism for wireless sensor networks using a fuzzy inference system. Sensors **10**, 3444–3479 (2010)
8. Muraleedharan, R., Osadciw, L.A.: Jamming attack detection and countermeasures in wireless sensor network using ant system. In: Proceedings of the SPIE - International Society for Optical Engineering, vol. 6248, pp. 62480G–62480G-12 (2006)
9. Liu, Q., Yin, J., Yu, S.: A bio-inspired jamming detection and restoration for WMNs: in view of adaptive immunology (2013)
10. Pickholtz, R.L., Schilling, D.L., Milstein, L.B.: Revisions to "theory of spread-spectrum communications - a tutorial". IEEE Trans. Commun. **32**, 211–212 (1984)
11. Agalj, M., Apkun, S., Hubaux, J.P.: Wormhole-based antijamming techniques in sensor networks. IEEE Educational Activities Department (2007)
12. Xu, W.: Channel surfing: defending wireless sensor networks from interference. In: 2007 6th International Symposium on Information Processing in Sensor Networks, pp. 499–508 (2007)
13. Xu, W., Trappe, W., Zhang, Y.: Anti-jamming timing channels for wireless networks. In: Proceedings of the first ACM Conference on Wireless Network Security - WiSec 2008, p. 203 (2008)
14. Vadlamani, S., Eksioglu, B., Medal, H., Nandi, A.: Jamming attacks on wireless networks: a taxonomic survey (2016)
15. Hussein, A.A., Rahman, T.A., Leow, C.Y.: A survey and open issues of jammer localization techniques in wireless sensor networks. J. Theor. Appl. Inf. Technol. **71**, 293–301 (2015)

16. Bahl, P., Padmanabhan, V.N.: RADAR: An in-building RF based user location and tracking system. In: Proceedings IEEE INFOCOM 2000, Conference on Computer Communications, Nineteenth Annual Joint Conference of the IEEE Computer and Communications Societies (Cat. No. 00CH37064), pp. 775–784 (2000)
17. Elnahrawy, E., Li, X., Martin, R.P.: The limits of localization using signal strength: a comparative study. In: 2004 First Annual IEEE Communications Society Conference on Sensor and Ad Hoc Communications and Networks, IEEE SECON 2004, pp. 406–414 (2004)
18. Misra, P., Enge, P.: Global Positioning System: Signal, Measurements, and Performance. Ganga-Jamuna Press, Lincoln (2006)
19. Yang, Z., Ekici, E., Xuan, D.: A localization-based anti-sensor network system. In: Proceedings of IEEE INFOCOM, pp. 2396–2400 (2007)
20. Bulusu, N., Heidermann, J., Estrin, D.: GPS-less low-cost outdoor localization for very small devices. IEEE Pers. Commun. **7**, 28–34 (2000). Special Issue on Smart Spaces and Environment
21. Blumenthal, J., Grossmann, R., Golatowski, F., Timmermann, D.: Weighted centroid localization in Zigbee-based sensor networks. In: IEEE International Symposium on Intelligent Signal Processing, pp. 1–6 (2007)
22. Liu, H., Xu, W., Chen, Y., Liu, Z.: Localizing jammers in wireless networks. In: 7th Annual IEEE International Conference on Pervasive Computing and Communications, PerCom 2009, pp. 1–6. IEEE (2009)
23. Mokdad, L., Ben-Othman, J., Nguyen, A.T.: DJAVAN: detecting jamming attacks in vehicle ad hoc networks. Perform. Eval. **87**, 47–59 (2015)
24. Pelechrinis, K., Koutsopoulos, I., Broustis, I., Krishnamurthy, S.V.: Jammer localization in wireless networks: an experimentation-driven approach. Comput. Commun. **86**, 75–85 (2016)
25. He, T., Huang, C., Blum, B.M., Stankovic, J.A., Abdelzaher, T.: Range-free localization schemes for large scale sensor networks. In: Proceedings of the 9th Annual International Conference on Mobile Computing and Networking (2003)
26. Sun, Y., Wang, X., Zhou, X.: Jammer localization for wireless sensor networks. Chin. J. Electron. **20**, 735–738 (2011)
27. Poli, R., Kennedy, J., Blackwell, T.: Particle swarm optimization. In: Swarm Intelligence Symposium, SIS 2007, pp. 120–127. IEEE (2007)
28. Kennedy, J.: The particle swarm: social adaptation of knowledge. In: Proceedings of 1997 IEEE International Conference on Evolutionary Computation, ICEC 1997, pp. 303–308 (1997)
29. Eberhart, R., Kennedy, J.: A new optimizer using particle swarm theory. In: Proceedings of the Sixth International Symposium on Micro Machine and Human Science, pp. 39–43 (1995)

An Algorithm for Hybrid Nodes Barrier Coverage Based on Voronoi in Wireless Sensor Networks

Xiaochao Dang[1,2], Rucang Ma[1(✉)], Zhanjun Hao[1,2], and Meixiu Ma[3]

[1] College of Computer Science and Engineering,
Northwest Normal University, Lanzhou 730070, China
984301633@qq.com
[2] Gansu Province Internet of Things Engineering Research Center,
Lanzhou 730070, China
[3] College of Educational Technology, Northwest Normal University,
Lanzhou 730070, China

Abstract. In order to make up for the deficiencies and insufficiencies that wireless sensor network is constituted absolutely by static or dynamic sensor nodes. So a deployment mechanism for hybrid nodes barrier coverage (HNBC) is proposed in wireless sensor network, which collaboratively consists of static and dynamic sensor nodes. We introduced the Voronoi diagram to divide the whole deployment area. According to the principle of least square method, and the static nodes are used to construct the reference barrier line (RBL). And we implemented effectively barrier coverage by monitoring whether there is a coverage hole in the deployment area, and then to determine whether dynamic nodes need limited mobility to redeploy the monitoring area. The simulation results show that the proposed algorithm improved the coverage quality, and completed the barrier coverage with less node moving distance and lower energy consumption, and achieved the expected coverage requirements and objectives.

Keywords: Wireless Sensor Network (WSN) · Voronoi diagram · Hybrid node · Barrier coverage · Reference Barrier Lines (RBL) · The least square method

1 Introduction

Wireless Sensor Network (WSN) is distributed Ad hoc network that are deployed in the monitoring area with static or mobile micro sensor nodes with limited perceptual, computational, data processing, storage and wireless communication capabilities [1, 2].

The coverage is an important topic in the field of wireless sensor network. The barrier coverage [3] model is also widely used in wireless sensor network. Its motivation is to guarantee that the mobile target can be efficiently monitored by the network when the target tries to pass through the area where the wireless sensor network is deployed [4].

© Springer Nature Singapore Pte Ltd. 2017
B. Zou et al. (Eds.): ICPCSEE 2017, Part II, CCIS 728, pp. 212–229, 2017.
DOI: 10.1007/978-981-10-6388-6_18

The barrier coverage technology [5] is proverbially applied to military, national security, marine pollution, forest fires, military march and other fields. For example, the wireless sensor nodes are deployed on the border line to monitor illegal immigrants. They are deployed in the vicinity of forests or chemical sites to prevent the occurrence of fire and other dangerous situations. Around the enemy camp, the wireless sensor network can detect the enemy's troops and weapons and so on [6, 7].

In this paper we construct the RBL, use Voronoi diagram to divide the monitoring area, and study the HNBC strategy. After sensors deployment, monitor whether there is a barrier hole, if there is and then repairs the barrier holes by mobile sensor nodes. Our main contributions are summarized as follows:

There are many factors that affect the performance of the barrier coverage, such as sensor nodes density and radius, monitoring area length, the relationship between the number of RBL and k-barrier coverage, and the comprehensive impact above factors. Our analysis show that the barrier coverage rate increases with varying degrees of sensor nodes density and/or radius, and the total moving distance, average moving distance and energy consumption are decreasing. In the k-barrier coverage, the value of k has a great effect on the coverage performance. With the increase of k value, the total moving distance of sensor nodes under HNBC increases linearly, and when k increases to some a value, the average moving distance remains basically unchanged, but fluctuate up and down at a certain value. As the length of the monitoring area, the total moving distance, energy consumption and coverage rate of sensor nodes increase under the HNBC, but the average moving distance is basically the same, it shows that the HNBC has good network scalability. Our results show that the sensor nodes deployment strategy and the sensor nodes mobility have a certain connection. In addition, if the node mobility performance is used efficiently, the barrier coverage will be improved. And the deployment strategy of hybrid sensor nodes can provide useful guidance for wireless sensor network barrier coverage.

2 Related Work

At present, wireless sensor network barrier coverage research has acquired great achievements. Among them, Kumar et al. [8] first proposed the concept of strong k-barrier coverage and weak k-barrier coverage. Mostafaei and Meybodi [9] proposed an energy efficient barrier coverage algorithm, and they researched strong k-barrier coverage construction by static wireless sensor network. Tian et al. [10] proposed a two-dimensional k-barrier coverage scheme, and the deployment area is divided into multiple sub-regions to build the barrier. Li et al. [11] studied the coverage rate of weak barrier coverage. In [12], the mobile model is introduced, which studied the probability of the intruder passing through the barrier area. He et al. [13] studied the problem of constructing barrier coverage for multiple mobile sensor nodes; Saipulla et al. [14] took the limitation of node mobility into consideration the barrier construction. Ban et al. [7] studied the k-barrier coverage of omni-directional mobile sensor network, and proposed an energy efficient barrier coverage construction algorithm. Balister et al. [13] studied a reliable node deployment density estimation method such that randomly deployed sensor network based on the density estimate form 1-barrier coverage and is s-t connected. Kumar et al. [14] proposed Optimal Sleep-Wakeup algorithm that studied barrier constructing algorithm under the two conditions that is the sensor nodes' survival time is same and different, and researched how to build k-barrier by scheduling limited barrier to prolong the network life cycle. In [15], barrier coverage control algorithm was proposed. The probabilistic sensing model and the data fusion technology were utilized to build virtual nodes to increase coverage area. In order to reduce communication overhead between sensors, the divide and conquer strategy was applied to construct barrier. At the same time, redundant sensor nodes were scheduled to sleep for reducing energy consumption and achieving the purpose of extending the network life cycle. In [16], in order to reduce the energy consumption of the strong k-barrier, it was proved that the minimum energy consumption of the strong k-barrier is NP problem, and then a heuristic energy-saving algorithm with node-aware power adjustable is proposed.

In the above all, there is blindness that deployed the sensor nodes to build the barrier, the sensor nodes are redundantly deployed so as to realize effective coverage, and there are such shortcomings as a crossing path can not be detected, high energy consumption, high complexity, lack of feasibility, poor economy, which are all notable features of dynamic algorithm. In order to compensate for the deficiencies and insufficient that wireless sensor network is constituted absolutely by static or dynamic sensor nodes, we combine with the advantages of the algorithms in the above, and introduced the Voronoi diagram to divide the whole deployment area, to select RBL using stationary sensor nodes as the reference in accordance with the least square method. Finally, a hybrid network deployment mechanism consisting of static and dynamic sensor nodes is proposed.

3 Network Model and Definition

3.1 Network Model

n hybrid wireless sensor nodes will be randomly deployed in a narrow belt region with two horizontal boundaries. The length is L, the width is W, satisfies $A = L \times W$, the number of static sensor nodes is n_s, and the number of dynamic sensor nodes is n_d, where $n = n_s + n_d$. The sensor node set is $S = \{S_1, S_2, \ldots, S_n\}$, meanwhile, Voronoi diagrams are constructed with the center of the sensor nodes.

Assume that the sensor node has the following properties:

(1) All sensors nodes are isomorphic, i.e. node has the same sense radius Rs, communication radius Rc, and initial energy E_0.
(2) Each node has its own positioning system that can obtain the location information of itself and neighbor's nodes, which can be converted into plane coordinates by some a mean.
(3) Dynamic sensor nodes have limited mobility. We allow nodes to move only once, with moving mark id, initially $id = 0$, if $id = 1$, then the node can not move again.
(4) The sense area $A = \pi Rs^2$, the communication range $C = \pi Rc^2$.
(5) The deployed sensor nodes meet requirements of network connectivity, i.e. $Rc \geq 2 \times Rs$.

Research has shown that sensor nodes sense quality is exponentially decayed with the increase of the distance between the target and the sensor node in practical application. Therefore, we adopt probabilistic sensing model, as shown in Eq. (1).

$$p_{x_P, y_P}(S_i) = \begin{cases} 0, r + r_e \leq d(S_i, P) \\ e^{-\lambda \alpha^\beta}, r - r_e \leq d(S_i, P) < r + r_e \\ 1, r - r_e \geq d(S_i, P) \end{cases} \tag{1}$$

where $p_{x_P, y_P}(S_i)$ is the probability that arbitrary point P in the area is covered by the sensor node S_i, $d(S_i, P)$ represents the Euclidean distance between the sensor node S_i and arbitrary node P, where $d(S_i, P) = \sqrt{(x_P - x_{S_i})^2 + (y_P - y_{S_i})^2}$, $\alpha = d(S_i, P) - (r - r_e)$, $r_e(r_e < r)$ is a parameter that it is determined by the sensor hardware that reflects the node's sense ability. If $d(S_i, P) > r_e$ is satisfied, $p_{x_P, y_P}(S_i)$ is affected by the parameter λ and β, where λ, β are determined by the physical characteristics of the sensor node.

3.2   Problem Description

Coverage performance is one of the key indicators of random deployment of wireless sensor network. Currently, we analyzed others' papers, most of the coverage mechanism is concerned about the network is constituted through the entire static nodes or the complete dynamic nodes. However, the former, its network coverage area that deploys randomly by static nodes composed is stationary. There may be some barrier holes in the network, or there are crossing paths that can not be detected, or throw sensor nodes in excess to increase node deployment density, and to improve coverage rate, and to ensure effective coverage, but it lead to redundancy and waste. For the latter, although it can monitor the crossing paths, the sensor nodes need to be moved for several times, or to perform multi-step approximation calculation to build barrier. Therefore, there are some such problems as the algorithm complexity is bigger, the energy consumption is higher, the network cost is higher, the feasibility is not enough, and it is difficult in the practical application promotion. In view of this, in this paper, we introduce the Voronoi diagram to divide the whole deployment area, to select RBL by using stationary sensor nodes as the reference according to the least square method. Finally, a hybrid network deployment mechanism consisting of static and dynamic sensor nodes is proposed, and realizes the effective barrier coverage.

3.3   Definitions

Definition 1 (Voronoi diagram). It is a continuous polygon consisting of a vertical bisector of a straight line of connecting two adjacent points, which is an important part of calculating geometry. Its formal definition is as follows: Given a finite set $P = \{p_1, p_2, \ldots, p_n\} \subset R^2$, $n > 2$, and $\forall i, j \in I_n = \{1, 2, \ldots, n\}$, if $i \neq j$ then $p_i \neq p_j$. Where $d(p_i, q)$ denote the Euclidean distance between point p_i and point q. Then

$B(p_i, p_j) = \{q \in R^2 | d(p_i, q) = d(p_j, q), i \neq j\}$ is the vertical bisector between points p_i and p_j.

Definition 2 (Crossing path). In the monitoring area A, it is an arbitrary curve segment that the starting point and ending point are located on its upper and lower borders, and dividing the area into any of the left and right sub-regions.

Definition 3 (Coverage rate). It is a ratio of the acreage covered by the sensor nodes that can be deployed in the monitoring area to the total acreage of the monitoring area. That is to say, the ratio of the acreage that n sensor nodes covered effectively to the acreage of the monitoring area.

$$Cov = \frac{\sum_{i=1}^{n} \left(\pi Rs_i^2 \right)}{L \times W} \tag{2}$$

Definition 4 (Node density). It refers to the ratio of the number of sensor nodes to the acreage of the monitoring area, which has an important effect on the formation of the barrier.

$$\rho = \frac{n}{L \times W} \tag{3}$$

Definition 5 (k-barrier coverage [4, 7]). If the moving object along arbitrary crossing paths, it traverses the monitoring area that deployed the sensor nodes, at least k sensor nodes sense it, then the area is k- barrier coverage.

4 HNBC Algorithm

Initially, the sensor nodes are randomly deployed in the monitoring area, so the sensor nodes failed to effectively cover the monitoring area due to the nodes distributed unreasonably, or malfunction, or energy exhausted and so on, and coverage holes appear.

4.1 Select the Reference Barrier Line

In order to reduce the energy consumption, it is necessary to reduce the moving distance of the sensor nodes as much as possible. According to the least square method, the appropriate RBL is selected by the static sensor nodes in the monitoring area. It is impossible that the band-shaped area is a straight line in the actual environment. Therefore, we divided the barrier area in accordance with differential thought, each of small section of the barrier area is approximately regard as a straight line, and then using the least square method to determine the location of the RBL. As shown in Fig. 1, the dashed line is the selected RBL.

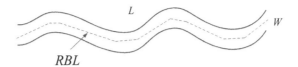

Fig. 1. Reference barrier line

Assume the fitting equation of the RBL is $\hat{y} = ax + b$, make the sensor nodes as close as possible to the line, then variance $\sum_{i=1}^{n}(y_i - \hat{y})^2$ between the measured value y_i and the calculated value $\hat{y}$ of

$$
\begin{cases}
a = \dfrac{n\sum\limits_{i=1}^{n} x_i y_i - \sum\limits_{i=1}^{n} x_i \sum\limits_{i=1}^{n} y_i}{n\sum\limits_{i=1}^{n} x_i^2 - \left(\sum\limits_{i-1}^{n} x_i\right)^2} \\[4mm]
b = \dfrac{1}{n}\sum\limits_{i=1}^{n} y_i - \dfrac{a}{n}\sum\limits_{i=1}^{n} x_i
\end{cases}
\tag{4}
$$

the fitting equation is the smallest. Therefore, as shown in Formula (4) by the least square method.

Substitute the value of a, b in Eq. (4) into the fitting equation $\hat{y} = ax + b$ to obtain the RBL. Among them, the number of RBL and k-barrier coverage is closely related. This article does not research this relation.

4.2 Monitor Barrier Holes

In the target area, it is possible to determine whether there is a coverage hole based on the maximum distance $d(S, V)$ that is between the sensor nodes and the vertices of the Voronoi polygons and the sense radius Rs.

The distance between the sensor node and the vertex of the Voronoi polygon where they are located is $d(S, V)$, then

$$
d(S, V) = max\left(\sqrt{(S_x - V_x)^2 + (S_y - V_y)^2}\right)
\tag{5}
$$

where $d(S, V)$ represent the distance between the sensor nodes and each vertex of the Voronoi polygon, S is the sensor nodes, and V is the Voronoi vertex.

It is known from the network model that each sensor node can obtain position information of its own and the neighbor nodes in the Voronoi diagram. Compared the size between distance $d(S, V)$ and the sense radius Rs, it is known that there are three possible positional relationship between the sense radius of the sensor node and the Voronoi polygon. As follows:

(1) The sense region may have one or more intersections with the Voronoi polygon vertices, as shown in Fig. 2(a).

(2) Voronoi polygons may be completely surrounded by the sense area, on the contrary, likewise, the sense area may be perfectly surrounded by the Voronoi polygon, while the latter does not exist coverage hole, but there will be overlapping, node deployment redundancy, as shown in Fig. 2(b).
(3) When the distance $d(S, V)$ between the center of the adjacent sensor nodes and the vertex of the Voronoi region is greater than the sense radius Rs, then where there is/are hole(s) in the Voronoi region, as shown in Fig. 2(c).

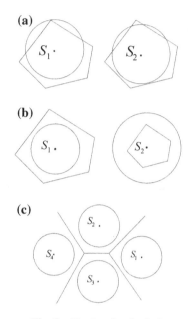

Fig. 2. Monitor barrier hole

4.3 Deploy the Mobile Nodes

After the sensor nodes are randomly deployed, to monitor whether there is hole in the BRL near, if there is, it is necessary to move the dynamic sensor nodes to repair the hole. Our provisions is that the average number of times that the sensor nodes are moved can not exceed the total number of dynamic nodes 1/5, and the average moving distance can not exceed 15 m, and coverage rate reach to 80%–90% so that the nodes involved in the repair process as much as possible to reduce the number of mobile sensor nodes, to cut down the node energy consumption, to extend the network life cycle. There are some blemishes that the stored energy of the sensor nodes is small and it is almost impossible to replenish energy. It is necessary to minimize the energy consumption and extend the network life cycle in the wireless sensor network deployment. Therefore, we must clear the direction and distance of movement, to the utmost extent, to minimize the moving distance, to reduce energy consumption.

As shown in Fig. 3, we move the dynamic sensor nodes to hole area involving the vicinity of the RBL to determine the direction of the sensor node movement so that reduce the hole acreage and repair hole, where, S_1A and S_3C are the sense radius R_s, S_1 and S_3 are the static sensor nodes, S_2 is a dynamic sensor node for repairing hole. Drawing the tangent $\overrightarrow{S_2A}$ and $\overrightarrow{S_2C}$ of the dynamic nodes S_1 and S_3, respectively, where $\overrightarrow{S_2A} + \overrightarrow{S_2C} = \overrightarrow{S_2B}$.

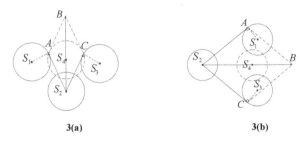

3(a)   3(b)

Fig. 3. The vector angle is acute

1. As shown in Fig. 3, when the vector angle $\left\langle \overrightarrow{S_2A}, \overrightarrow{S_2C} \right\rangle$ (denoted as $\angle AS_2C$) is an acute angle, $\overrightarrow{S_2B}$ is the direction in which the dynamic node moves to the barrier hole area, that is, moving dynamic node S_2 along the $\overrightarrow{S_2B}$ so that reduces the hole acreage and effectively repairs the hole, and dynamic node S_4 is the location after the dynamic node S_2 mending the coverage hole.

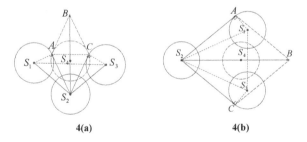

4(a)   4(b)

Fig. 4. Determining moving distance

$$|\overrightarrow{S_2B}| = |\overrightarrow{S_2A} + \overrightarrow{S_2C}| = \sqrt{|\overrightarrow{S_2A}|^2 + 2 \times |\overrightarrow{S_2A}| \times |\overrightarrow{S_2C}| cos\angle AS_2C + |\overrightarrow{S_2C}|^2} \quad (6)$$

It can be seen from Fig. 4, repair coverage hole, the sensor node moving maximum distance is $1/2|\overrightarrow{S_2B}|$, $|\overrightarrow{S_2B}|$ is as follows:

Among them, $\angle AS_2S_1$, $\angle S_1S_2S_3$, $\angle CS_2S_3$ can be obtained, shown in Fig. 4(a), angle $\angle AS_2C = \angle S_1S_2S_3 - \angle S_1S_2A - \angle S_3S_2C$; Fig. 4(b) shows the angle $\angle AS_2C = \angle S_1S_2S_3 + \angle S_1S_2A + \angle S_3S_2C$.

Let $|\overrightarrow{S_2B}|$ be the minimum, the value of $cos\angle AS_2C$ is analyzed:

(1) If $\angle AS_2C = 0°$, point A and point C coincide, two static nodes tangent, there is no coverage hole.

(2) If $\angle AS_2C = 90°$, then $|\overrightarrow{S_2B}| = \sqrt{|\overrightarrow{S_2A}|^2 + |\overrightarrow{S_2C}|^2}$, at the same time, the dynamic node move $1/2|\overrightarrow{S_2B}|$ along the $\overrightarrow{S_2B}$ direction to repair hole.

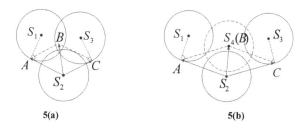

5(a)   5(b)

Fig. 5. The vector angle is obtuse

2. As shown in Fig. 5, when the vector angle $\angle AS_2C$ is obtuse, if the static node coverage area is intersected, then, as shown in Fig. 5(a), there is no coverage hole. Therefore, we studies the position where the vector angle is obtuse angle when there is coverage hole shown in Fig. 5(b), and dynamic node S_4 is the location after the dynamic node S_2 mending the coverage hole.

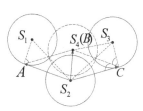

Fig. 6. Determining moving distance

$$|\overrightarrow{S_2B}| = |\overrightarrow{S_2A} + \overrightarrow{S_2C}| = \sqrt{|\overrightarrow{S_2A}|^2 + 2 \times |\overrightarrow{S_2A}| \times |\overrightarrow{S_2C}|cos\angle AS_2C + |\overrightarrow{S_2C}|^2} \quad (7)$$

It can be seen from Fig. 6, the maximum distance is $|\overrightarrow{S_2B}|$ that node moves to repair coverage hole, at same time, point B and point S_4 coincide, the solution is as follows:

Among them, $\angle AS_2S_1$, $\angle S_1S_2S_3$, $\angle CS_2S_3$ can be obtained, the angle $\angle AS_2C = \angle AS_2S_1 + \angle S_1S_2S_3 + \angle CS_2S_3$.

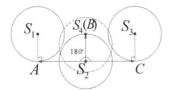

Fig. 7. The vector angle is flat angle

In particular, as shown in Fig. 7, if $\angle AS_2C = 180°$, $cos\angle AS_2C = -1$ is minimum value, we stipulate where the direction of the vector $\overrightarrow{S_2B}$ is vertical line through the point S_2, and point to the RBL, the size is the sense radius Rs.

The dynamic sensor node moves in the direction of $\overrightarrow{S_2B}$, and the dynamic sensor node stop moving when it can sense to coverage area on the left side of the hole and on the right side, the mark $id = 1$, and then to judge whether there is hole area nearby, to repair holes and to redeploy the monitoring area to reach the desired coverage requirements with the above algorithm.

4.4 Describe Algorithm

In this section, we describe the implementation process of an algorithm for hybrid sensor nodes barrier coverage based on Voronoi in wireless sensor network in a band region $L \times W$, as follows:

Step 1: Determine the target area. The barrier area is determined as a band target area $L \times W$.

Step 2: Initialize the node to construct the Voronoi diagram. The hybrid sensor nodes are deployed in the target area, and the Voronoi diagram is constructed to each node as the center.

Step 3: Select the RBL. To be based on the distribution of static sensor noes and the least square method select the RBL.

Step 4: Monitor barrier coverage holes. In accordance with Sect. 4.2, to monitor whether there is coverage hole in the vicinity of the RBL. If there is, then repair the hole; otherwise, go to **Step 7**.

Step 5: Repair barrier coverage holes. With different strategies to determine the dynamic sensor nodes moving direction and distance when repair the coverage hole in accordance with the vector angle for the acute angle or obtuse angle.

Step 6: Redeploy the sensor nodes to construct the Voronoi diagram, to select the RBL, and to judge the coverage hole. The holes have been repaired completely, jump to **Step 4** to monitor barrier coverage hole once again, if there is hole, then implement **Step 5**; otherwise, quit the algorithm.

Step 7: Repeat **Step 4–Step 6**, until meet the barrier coverage requirements of the target area.

The pseudo-code of the algorithm is showed in Algorithm 1.

Algorithm 1 HNBC Algorithm for monitoring and repairing the barrier holes

Input
1: Area length L and width W.
2: The number of sensor nodes n, and static node is n_s, dynamic node is n_d.
3: Sensor node radius is Rs and initial energy E_0.
4: All sensor nodes' location coordinates :(si.x, xi.y) for $1 \leq i \leq n$

Initialization
1: The first dynamic node's coordinates :(0, 0).
2: The flag id=0.

Main loop
1:    for $d(S,V) > Rs$
2:        if $d(S,V) > Rs$
2:            if vector angle is acute angle
3:            Calculate $\overline{S_2B}$ and $1/2|\overline{S_2B}|$
4:            end if
5:            if vector angle is obtuse angle or flat angle
6:            Calculate $\overline{S_2B}$ and $1/2|\overline{S_2B}|$
7:            end if
8:        id=1
         end if
9:        else
10:            quit algorithm
11:        end else
12:    end for

Output
The barrier coverage, the total moving distance, the average moving distance and network energy consumption.

In summary, in this paper, the algorithm complexity consists of the initial deployment sensor nodes to construct the Voronoi diagram, the construction of the RBL, the holes monitoring, the dynamic sensor nodes movement to repair hole and the redeployment sensor nodes, and details as follows:

The algorithm complexity that construct Voronoi diagram by Delaunay triangulation is $O(n^2)$. The complexity of construction the RBL is related to the number of static nodes, it is $O(s)$. For n hybrid sensor nodes, the distance $d(S, V)$ is compared with the sense radius Rs, the complexity is $O(n)$. In the worst case, if there is a barrier holes, and moved each dynamic node to repair the hole, then the complexity is $O(n_d^2)$. Therefore, the algorithm complexity has polynomial complexity, which is $O(n^2)$.

5 Simulation Experiment and Analysis

Our algorithm is simulated in the *MATLAB R2010b* programming environment, and research the impact of parameters change from different dimensions on barrier coverage quality. Assuming that the wireless sensor network is deployed in the rectangular area $L \times W = 1000 \text{ m} \times 100 \text{ m}$, all experimental data take the average over 1000 times independent experiments. The simulation parameters are shown in Table 1:

Table 1. Simulation parameters

Parameter name	Value
Simulation area A	100000 m^2
Area length L	1000 m
Area width W	100 m
Simulation experiments time	1000
Number of nodes N	50
Sense radius Rs	15 m
Communication radius Rc	30 m
Energy consumption E_0	3.6 J/m
Initial deployment density ρ	$4 \times 10^{-3}/m^2$

5.1   Sensor Nodes Density and Coverage Performance

The sensor node density has an important effect on the barrier formation and coverage performance. With the increase of node density, the efficiency of barrier coverage will be improved to different degrees. After performing HNBC algorithm, the total moving distance, average moving distance and network energy consumption are obviously lower than the other three algorithms [7, 24, 27], and the coverage rate is better than that of other three, As shown in Fig. 8(a), (b), (c) and (d).

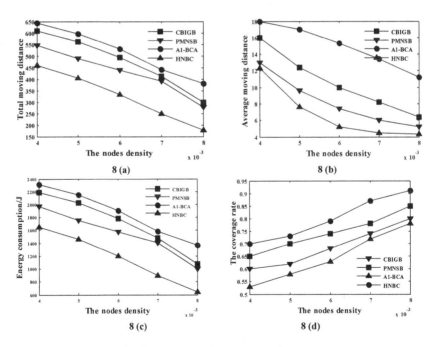

Fig. 8. Density and coverage performance

In [24], PMNSB algorithm has better performance than CBIGB algorithm [7]. However, PMNSB algorithm compares with HNBC algorithm, whose algorithm complexity is bigger under the same conditions. If you build a k-coverage, then move more sensor nodes, as a consequence of larger total moving distance and greater energy consumption. Therefore, the HNBC algorithm is superior to the other three algorithms with the increase of the node density.

5.2   Sensor Node Radius and Coverage Performance

In the wireless sensor network, the sensor node radius is one of the main factors affecting the barrier coverage. The simulation results show that the four algorithms' the total moving distance, the average moving distance and the network energy consumption decrease and the coverage rate increase in different degrees with the radius of the node increasing with a gradient of 5 m. However, the HNBC algorithm performance is obviously better than the other three algorithms, as shown in Fig. 9(a), (b), (c) and (d).

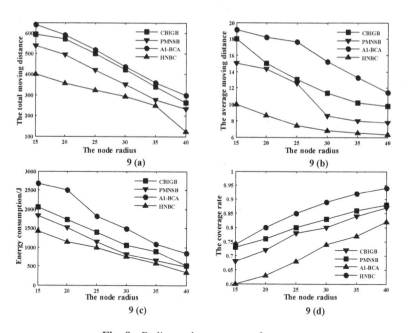

Fig. 9. Radius and coverage performance

According to the simulation results, it can be seen that node average moving distance and the total moving distance rapidly decrease. They demonstrate a large negative correlation with the increasing of the sensor node density and the node radius, at the same time, the coverage rate shows a large positive correlation. As a result of the comprehensive analysis, the probability of overlapping of the sensing range becomes large when the sensor nodes radius increase and/or the number of sensor nodes grow. Therefore, with the increasing of the sense radius and/or the node density in the experiment, the probability for constructing barrier to repair hole becomes smaller than original deployment. Furthermore, the less the total distance the node to move, and then reduce the average moving distance of the sensor node, reduce the network energy consumption and increase the coverage rate rapidly.

5.3   K-Barrier Coverage Formation

Due to the equivalent relationship between the number of BRL and the number of the value k, we research the relationship between k value, node moving distance and [7, 24, 26] in this section. The simulation results show that the impact of the k value on the barrier coverage as is shown in Fig. 10. With the increasing of the number of barriers, it can be seen that the two indexes of the four algorithms are increased from the figure. So as to construct respectively the 1-barrier and 6-barrier, the total moving distance is 150 m, 989 m, the average distance of 4.2 m, 6.5 m of HNBC algorithm, while the total moving distance is 178 m, 1400 m, the average distance of 6 m, 9 m of PMNSB algorithm, the total moving distance is 400 m, 1800 m, the average distance of 8 m, 12 m of GBICB algorithm, and the total moving distance is 270 m, 1650 m, the average distance of 5 m, 10.5 m of kMCBarrier algorithm.

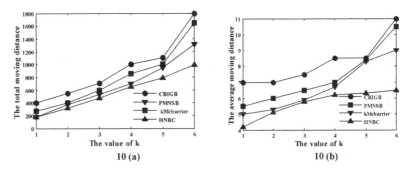

Fig. 10. K value and moving distance

In order to construct a k-barrier coverage that satisfies the requirements, it is necessary to move a larger number of nodes. Nevertheless, the total distance of the nodes will increase accordingly, and the different algorithms have their own corresponding solutions. However, the HNBC algorithm is based on the least squares method to select the RBL makes the total moving distance and average moving distance get smaller. So our scheme solves effectively the problem because of larger the node distance causes the network energy consumption too large.

5.4   Different RBL Comparisons

The impact of different RBL selection on coverage quality and performance is not the same. The HNBC algorithm is compared with the BGBS3 algorithm in [7], the PMNSB algorithm in [24], and the A1-BCA algorithm in [27]. The simulation experiment, as shown in Figs. 11 and 12.

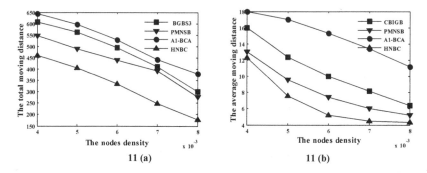

Fig. 11. Different density and moving distance

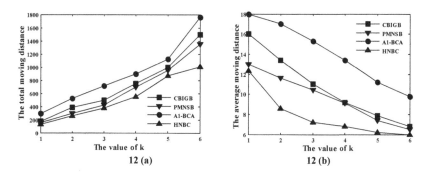

Fig. 12. Different k and moving distance

As can be seen from the figure that the total moving distance is respectively 300 m, 386 m, 523 m and the average moving distance is individually 11 m, 13 m, 17 m of the PMNSB, BGBS3, A1-BCA and HNBC algorithms to construct 2-barrier when the node density is 0.004. Due to sensor node energy is limited, however, the moving distance involving the sensor nodes make up for the barrier holes must consumes a lot of energy. Accordingly, In terms of moving distance, regardless of node density or k value is the same, the simulation results show that the total distance and average distance of the node movement is the smallest to execute the HNBC algorithm in the above four algorithms. That is, HNBC algorithm performance is better than PMNSB, BGBS3, and A1-BCA algorithm.

5.5   The Impact of Area Length

The sensor nodes are deployed in monitoring areas with length of 1000 m and width of 250 m, and node sense radius is 15 m. As is shown that the simulation experimental results of the monitoring zone length and coverage performance under different node density when the sensor nodes are deployed in a monitoring area of 1000 m * 250 m and node density is 0.004, 0.005, 0.006, 0.007, 0.008 in Fig. 13, respectively.

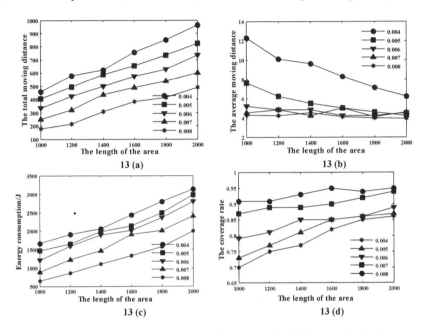

Fig. 13. Area length on coverage performance under different node density

It can be seen from the figure that the total distance, coverage and energy consumption of the nodes are increased in different degrees when the area length increases from 1000 m to 2000 m with the gradient of 200 m, and there is a large positive correlation, especially the correlation of the total distance of the nodes. The average moving distance is reduced significantly with the increase of the area length when the node density is 0.004. At the same time, the average moving distance of the nodes is in the interval [4, 5] with the increasing of the area length when the node density increases. And the average moving distance does not increase with the increase of the area length, that is, the algorithm stability and expansibility is better, can be applied to large-scale sensor network.

6   Conclusion

It is found that the network is composed by all completely static sensor nodes, it is necessary to redundantly deploy sensor nodes to achieve effective coverage. Simultaneously, the network is composed through all perfectly dynamic sensor nodes, its feasibility is not enough, with high complexity, high energy consumption, high cost and where there are some characteristics that wireless sensor network topology is easy to change. Therefore, we proposed an algorithm for hybrid node barrier coverage based on Voronoi to dynamically build wireless sensor network so as to solve problem of coverage holes to effectively cover the monitoring area. The simulation result shows that the proposed algorithm can solve the barrier holes through selecting the RBL. At last, compared with the algorithm of A1-CBA, BGBS2, CGICB, kMCBarrier, PMNSB from different dimensions, we achieve the barrier coverage requirements and objectives. In the future, we research the relationship of the number of RBL and the value k of the k-barrier.

Acknowledgement. This work was supported by the National Natural Science Foundation of China under Grant No. 61363059 and No. 61662070, Science and Technology Support Program of Gansu Province under Grant No. 1604FKCA097, the Young Teachers' Research Ability Improvement Program for Northwest Normal University (No. NWNU-LKQN-13-24).

References

1. Tong, X.J.: The novel block encryption scheme based on hybrid chaotic maps for the wireless sensor networks. Acta Physica Sinica **61**(3), 030502-379 (2012) ·
2. Zhu, W., Qi, W., De Bao, W., Ling, W.: Relay node placement and addition algorithms in wireless sensor networks. Acta Physica Sinica **61**(12), 855–865 (2011)
3. Saipulla, A.: Barrier coverage in wireless sensor networks. Diss. Theses-Gradworks **3**(6), 298–305 (2010)
4. Chen, A., Kumar, S., Lai, T.H.: Local barrier coverage in wireless sensor networks. IEEE Trans. Mob. Comput. **9**(4), 491–504 (2010)
5. Li, L., Zhang, B., Shen, X., et al.: A study on the weak barrier coverage problem in wireless sensor networks. Comput. Netw. Int. J. Comput. Telecommun. Netw. **55**(3), 711–721 (2011)
6. Atzori, L., Iera, A., Morabito, G.: The internet of things: a survey. Comput. Netw. **54**(15), 2787–2805 (2010)
7. Ban, D.S., Wen, J., Jiang, J., et al.: Constructing k-barrier coverage in mobile wireless sensor networks. J. Softw. **22**(9), 2089–2103 (2011)
8. Kumar, S., Lai, T.H., Arora, A.: Barrier coverage with wireless sensors. In: International Conference on Mobile Computing and Networking, pp. 626–630. ACM (2011)
9. Mostafaei, H., Meybodi, M.R.: An energy efficient barrier coverage algorithm for wireless sensor networks. Wirel. Pers. Commun. **77**(3), 2099–2115 (2014)
10. Tian, J., Zhang, W., Wang, G., et al.: 2D k-barrier duty-cycle scheduling for intruder detection in wireless sensor networks. Comput. Commun. **43**(5), 31–42 (2014)
11. Li, L., Zhang, B., Shen, X., et al.: A study on the weak barrier coverage problem in wireless sensor networks. Comput. Netw. **55**(3), 711–721 (2011)
12. Yen, L.H., Cheng, Y.M.: Range-based sleep scheduling (RBSS) for wireless sensor networks. Wirel. Pers. Commun. **48**(3), 411–423 (2009)
13. Balister, P., Bollobas, B., Sarkar, A., et al.: Reliable density estimates for coverage and connectivity in thin strips of finite length. In: ACM International Conference on Mobile Computing and Networking, pp. 75–86. ACM (2007)

14. Kumar, S., Lai, T.H., Posner, M.E., et al.: Optimal sleep-wakeup algorithms for barriers of wireless sensors. In: International Conference on Broadband Communications, Networks and Systems, BROADNETS, pp. 327–336. IEEE (2007)
15. Luo, Q., Lin, Y., Wang, L., Yin, B., et al.: Barrier coverage control based on data fusion for wireless sensor network. J. Electron. Inf. Technol. **34**(4), 825–831 (2012)
16. Guo, X.: Energy-efficient algorithm of k-barrier coverage in wireless sensor network. J. Comput. Appl. **33**(8), 2104–2107, 2111 (2013)
17. He, S., Chen, J., Li, X., et al.: Cost-effective barrier coverage by mobile sensor networks. Proc. IEEE INFOCOM **131**(5), 819–827 (2012)
18. Saipulla, A., Liu, B., Xing, G., et al.: Barrier coverage with sensors of limited mobility. In: ACM Interational Symposium on Mobile Ad Hoc NETWORKING and Computing, MOBIHOC 2010, Chicago, IL, USA, pp. 201–210, September 2010
19. Ma, H., Yang, M., Li, D., et al.: Minimum camera barrier coverage in wireless camera sensor networks, vol. 131, no. 5, pp. 217–225 (2012)
20. Tao, D., Tang, S., Zhang, H., et al.: Strong barrier coverage in directional sensor networks. Comput. Commun. **35**(8), 895–905 (2012)
21. Wang, Y., Cao, G.: Barrier coverage in camera sensor networks. In: Twelfth ACM International Symposium on Mobile Ad Hoc NETWORKING and Computing, pp. 3967–3974. ACM (2011)
22. Wang, Z.B., Liao, J.L., Cao, Q., et al.: Barrier coverage in hybrid directional sensor networks. IEEE Trans. Mob. Comput. **13**(7), 222–230 (2013)
23. Wang, Z.B., Liao, J.L., Cao, Q., et al.: Achieving k-barrier coverage in hybrid directional sensor networks. IEEE Trans. Mob. Comput. **13**(7), 1443–1455 (2014)
24. Chao, W., Xinggang, F., Heng, W., et al.: An effective realization scheme for strong K-barrier coverage in WSN. Chin. J. Sens. Actuators **2**, 227–233 (2015)
25. Chen, Y., Zetong, X.: Algorithm for barrier coverage of limited mobile WSNs. Comput. Eng. Des. **11**, 3804–3807 (2014)
26. Shuai, L., Keqing, L., Huan, D., et al.: Research on k-barrier coverage of mobile sensor. Transducer Microsyst. Technol. **33**(5), 52–54 (2014)
27. Zhiqiang, S., Xianzhong, Z., Huaxiong, L.: On barrier coverage based on mobile wireless sensors nodes. Comput. Appl. Softw. **31**(9), 122–124 (2014)

Measurement Analysis of an Indoor Positioning System Based on LTE

Jiahui Qiu[✉], Qi Liu, Wenhao Zhang, and Yi Chen

China Unicom Company, Beijing 100048, People's Republic of China
qiujh@dimpt.com

Abstract. Location information is playing an increasing important role in people's life. Specially, indoor positioning systems have been sufficiently researched to provide location information for persons and devices. This paper focuses on the research and development of the indoor positioning system based on LTE uplink signal, which consists of transmitter, Location Measurement Unit (LMU) and server. The Sounding Reference Signal (SRS) is employed as the excitation signal for the estimation of TDOA because of its good autocorrelation. The LMU is used for signal processing and Time-Difference of Arrival (TDOA) calculation, which is connected with four segregate received antennas. Synchronization is not necessary between any two LMUs. Therefore, they can be flexibly deployed on Base Stations (BSs). Additionally, a spatial filtering algorithm without the prior information is proposed to optimize the route tracking. The test results demonstrate that the fixed point positioning error is less than 3 m, and the estimated trajectory is basically accordant with the real path in the dynamic test.

1 Introduction

There are surging demands for the positioning services with the development of the society and technologies. Positioning applications in outdoor scenario include navigation, tracking and traffic management, etc. Global Positioning System (GPS) has been extensively developed and can basically meet the daily needs of people. There are much more widespread availabilities in indoor scenarios. According to the research of Strategy Analytics Corporation, people take 80–90% of their time in indoor environments, and 70–80% of the communication also occurs in indoor environments [1]. For commercial demands, common use cases such as shopping in malls or supermarkets, warehouse management, game development and so on, are continuously improved. There are more new applications in 5G era. Internet of Vehicles (IoV), for example, needs the positioning information for smart parking, autopilot and navigation. Another typical application of positioning in the all interconnected era is the smart home, which needs the accurate positioning information of people to control the switch of the home appliances. More applications even can be found in Industry 4.0, Virtual Reality (VR) and so on.

© Springer Nature Singapore Pte Ltd. 2017
B. Zou et al. (Eds.): ICPCSEE 2017, Part II, CCIS 728, pp. 230–240, 2017.
DOI: 10.1007/978-981-10-6388-6_19

However, GPS can not be deployed for indoor positioning service because there is not line of sight (LOS) transmission between the transceiver, which is prerequisite for satellite communication. Furthermore, the indoor environments are much more complex compared with that of the outdoor. Different solutions for indoor positioning are employed according to the different demands and environments. Some articles such as [2,3] have given an overview of various available technology options, as well as the comparisons including accuracy, complexity, robustness, cost, etc.

Whatever the positioning system is, there is a universal system architecture of positioning system which is shown in Fig. 1. The four parts included in the architecture are signal receiving, parameter estimation, positioning calculation and optimization. Details of the four parts are as following.

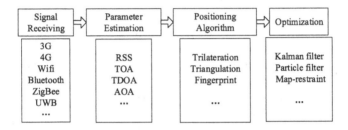

Fig. 1. General architecture of positioning system.

(1) Signal receiving. Possible signals that can be used for positioning include the cellular signals (3G/4G) and WLAN signals involving WiFi, Bluetooth, UWB and ZigBee and so on, which can provide various communication services in different application scenarios. Considering the wide coverage of cellular networks and the no need for updating of mobile terminal hardware, cellular positioning has attracted much attention. However, restricted by the positioning accuracy (generally based on cell-ID), cellular signal-based positioning is usually used in outdoor environments.

(2) Parameter estimation. The distance information would be obtained through the channel parameters estimation consisting of Received Signal Strength (RSS), Time of Arrival (TOA), Time Difference of Arrival (TDOA), Angle of Arrival (AOA) and even the Channel Impulse Response (CIR). The basic principle for the RSS is that the attenuation of emitted signal strength is related to the propagation distance. TOA is the propagation delay that can be used to calculate the distance between the transceiver. In order to get the precise delay, it is required that the transceiver should be strictly synchronized. To deal with this issue, the TDOA measurement is proposed, transforming the absolute distance to the relative distance by measuring the difference in time at which the signal arrives from multiple transmitters or anchors. AOA-based system, which refers to detecting the arrival angle of

the signal at the receiver, estimates the position by the intersection of several angle direction lines. The advantages are that the time synchronization is not required, and that the position can be determined using as few as two measurement units for 2-D positioning.

(3) Positioning calculation. The final position is what the channel parameter measurements mentioned above are mapped to through the positioning algorithm. The common method is to calculate the position by using the geometric properties, for example, distance and angle, which corresponds to two classic positioning algorithms named trilateration and triangulation. The algorithms mentioned above are classified as geometry positioning algorithm. There are some other algorithms which are based on the probabilities, such as pattern matching algorithm (e.g. fingerprint) and proximity algorithm (e.g. cell-ID). Fingerprint, a process of signal collection and matching, is another common method used for positioning. In this scheme, signal properties such as received strength [4] and CIR [5] are compared to a database of properties previously collected at a variety of locations. The closest match is returned as the estimated position.

(4) Optimization. The positioning points obtained from the positioning algorithm are usually dispersive and irregular, which need to be filtered to make the positioning route more smooth and precise. The common filters include Kalman filter, particle filter and so forth. Inertial sensors, such as accelerator and gyroscope, which can provide the step length and direction information, are often utilized in the filtering. Another optimization method is map-restraint. For example, in the hallway, the terminal can only move forward or background, and is not possible penetrate the walls or other obstacles.

For telecommunication operator, it is significant to develop the indoor positioning. On one hand, it is easy and convenient to implement the positioning system in the current cellular networks. On the other hand, it is helpful to optimize the networking such as call drop distribution, signal strength distribution and data service destiny distribution, etc. Therefore, in this paper, to sufficiently exploit the operator's network vantages and reduce the complexity and cost of the network deployments for indoor positioning, a high-accuracy indoor positioning system based on LTE is proposed, employing the LTE uplink signal and accomplishing the positioning processing at the BS side. By using the Location Measurement Unit (LMU) connected with the received antenna cluster, synchronization is just necessary among antennas in one cluster, but not among clusters. The positioning accuracy is less than 3 m in the test.

The rest of the paper is organized as follows. In Sect. 2, the indoor positioning system based on LTE uplink signal is introduced. The positioning processing including signal processing, positioning algorithm and filtering is described in Sect. 3. The test results are presented and analyzed in Sect. 4. Finally, Sect. 5 concludes the paper.

2 System Architecture

The positioning system is based on the LTE uplink signal. The Sounding Reference Signal (SRS), which has good autocorrelation and cross correlation characteristics, is employed as the excitation signal for the estimation of TDOA in the positioning system.

According to the basic system architecture shown in Fig. 2, four antennas, grouping one antenna cluster and using the same synchronization clock, are connected to one LMU. In the LMU, the SRS received from the four antennas are extracted and correlated with the local SRS independently. By selecting one of the four links as the reference, three TDOAs can be calculated. It should be noted that the synchronization among the four links plays an important role in the system and impacts the accuracy of TDOA calculations. Therefore, it is essential to compensate for the transmission bias of the four links to make sure that the bias is less than 1 ns (corresponding to 30 cm distance offset). Additionally, the isolation between the adjacent two links also affects the TDOA calculation. It is requested that the channel isolation is not less than 80 dBm. The TDOAs obtained from the LMU are transmitted to the server through the router or in a wireless way. In the server, the positioning calculation and filtering are accomplished. More antenna clusters can make the positioning more accurate, and there is no need for synchronization among clusters. Therefore, the antenna arrangements in units of clusters make the network deployment more flexible. The operator can add or diminish the LMUs for the accuracy and cost requirements.

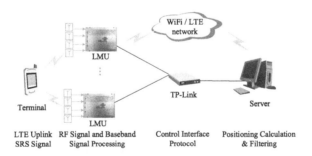

Fig. 2. Architecture of LTE-based positioning system.

3 Positioning Processing

3.1 Signal Processing

The overview of LMU is shown as Fig. 3. In the LMU, the received signals are sampled at the intermediate frequency. The radio frequency (RF) signal is transferred from $f_m = 1.755\,\text{GHz}$ to $f_m = 138.24\,\text{MHz}$ through the filter and mixer, and sampled with the frequency of 184.32 MHz through analog to digital converter (ADC) [6].

Fig. 3. Overview about the LMU.

The main tasks of Field Programmable Gate Array (FPGA) (Xilinx XC7K70T 160T [7]) include I/Q signal extraction, digital down-conversion and TDOA calculation. Firstly, the signals in each link are transferred from 138.24 MHz to 30.72 MHz. Then, the correlations of the received signals in the four RF links with the local SRS are carried out separately. According to the peak-value detection principle, four TOA values can be obtained, which are not absolute but relative. The signals after the correlation are up-sampled to 2 GHz to increase the detection accuracy of TOAs. 20 groups of TOAs are outputted by the FPGA per second. At last the TDOAs can be calculated by choosing one of the four links as the reference anchor (for example, choosing the link with the maximum power as the reference).

3.2 Positioning Algorithm

To achieve the high-accuracy positioning, there are much more requirements to the algorithm performances such as reliability and rate of convergence. Therefore, the Non-Linear Least Square (NLLS) algorithm [8,9] is employed in this paper, which can calculate the positioning rapidly by using Taylor series expansion and inner iteration.

Assume that $\boldsymbol{\theta} = [x, y, z]$ is the target position to be estimated, and $\mathbf{v}_m, m = 1, 2, \ldots, M$ is the positions of M anchors, and $d_m(\boldsymbol{\theta}) = \| \mathbf{v}_m - \boldsymbol{\theta} \|$ is the true distance between the mth anchor and the target. If the first anchor ($\mathbf{v}_1$) is chosen as the reference, the TDOA values are written as following without considering the measurement offset caused by non-line of sight (NLOS),

$$r_{m,1} = d_m(\boldsymbol{\theta}) - d_1(\boldsymbol{\theta}) + n_m - n_1, m = 2, 3, \ldots M \quad (1)$$

where $n_m \sim N(0, \sigma_m^2)$ is the range measurement error. Function $d_m(\boldsymbol{\theta})$ is neither linear nor quadratic, which can be linearized by Taylor series expansion, i.e.,

$$d_m(\boldsymbol{\theta}) \approx d_m(\boldsymbol{\theta}_0) + \tilde{d}_m(\boldsymbol{\theta}_0)(\boldsymbol{\theta} - \boldsymbol{\theta}_0), \quad (2)$$

where $\boldsymbol{\theta}_0$ is the known position and

$$\tilde{d}_m(\boldsymbol{\theta}_0) = [\frac{\partial d_m(\boldsymbol{\theta})}{\partial x} \frac{\partial d_m(\boldsymbol{\theta})}{\partial y} \frac{\partial d_m(\boldsymbol{\theta})}{\partial z}]_{\theta=\theta_0}. \tag{3}$$

Then the $M - 1$ TDOA values are written as

$$\tilde{\mathbf{r}} = [\tilde{r}_{2,1}, \tilde{r}_{3,1}, \dots, \tilde{r}_{M,1}]^T = \mathbf{H}\boldsymbol{\theta} + \mathbf{n}, \tag{4}$$

where

$$\tilde{r}_{m,1} \approx h_{m,1}\boldsymbol{\theta} + n_m - n_1,$$

$$h_{m,1} = \tilde{d}_m(\boldsymbol{\theta}_0) - \tilde{d}_1(\boldsymbol{\theta}_0),$$

$$\mathbf{H} = [h_{2,1}, h_{3,1}, \dots, h_{M,1}]^T,$$

$$\mathbf{n} = [n_2 - n_1, n_3 - n_1, \dots, n_M - n_1]^T.$$

The covariance matrix of the noise vector is expressed as

$$\mathbf{C} = \frac{1}{N}(\mathbf{A} + \sigma_1^2 \times \mathbf{1} \times \mathbf{1}^T), \tag{5}$$

where $\mathbf{A} = diag([\sigma_2^2, \sigma_3^2, \dots, \sigma_M^2])$ is a diagonal matrix and $\mathbf{1}$ is a $(M - 1) \times 1$ vector whose elements are all equal to 1.

The least squares estimator is expressed as

$$\tilde{\boldsymbol{\theta}} = (\mathbf{H}^T\mathbf{C}^{-1}\mathbf{H})^{-1}\mathbf{H}^T\mathbf{C}^{-1}\tilde{\mathbf{r}}. \tag{6}$$

The iteration is applied to minimize the error, i.e.,
1. Initialize $\boldsymbol{\theta}_0$;
2. **Repeat**
3.   Estimate $\tilde{\boldsymbol{\theta}}^k$ according to (6);
4.   If $\|\tilde{\boldsymbol{\theta}}^k - \tilde{\boldsymbol{\theta}}^{k-1}\| < \varepsilon$, then,
5.   Break.
6.   End if.
7. $\boldsymbol{\theta}_0 = \tilde{\boldsymbol{\theta}}^k$, go to 3.
8. **Until** the maximum number of iteration is reached or until convergence.

Considering that the algorithm is sensitive to the initial position, some other methods are essential to provide a coarse estimation as the initial position, at the cost of increasing the overall computational complexity. Additionally, to further improve the positioning accuracy, it needs to eliminate some unconscionable TDOAs by setting a threshold.

The positioning results obtained from the positioning algorithm are dispersive and irregular, which need to be smoothed by filtering. The common filters mainly involve Kalman filter and particle filter [10]. Both of them rely on the prior information such as step length and direction based upon the empirical models or inertial sensors. In this subsection, a spatial filter is proposed, which can get the smooth track without the prior information. More details about the spatial filtering algorithm are as following.

Step 1. Accumulate the N estimated position inputs $\mathbf{p}_j$ in matrix $\mathbf{P}^i = [\mathbf{p}_1, \mathbf{p}_2, \ldots, \mathbf{p}_N]$, where $\mathbf{p}_j = (x_j, y_j, z_j)^T$, $j = 1, 2, 3, \ldots$ is the index of the estimated positions, and $i = 1, 2, 3, \ldots$ is the index of the filtering output.

Step 2. Calculate the median position $\mathbf{P}^i_{med} = (X^i, Y^i, Z^i) = med(\mathbf{P}^i)$, where $med(A)$ means keeping the median value of each row of A.

Step 3. Put $\mathbf{P}^i_{med}$ into the position matrix $\mathbf{P}_{matrix} = [\mathbf{P}_{matrix}, \mathbf{P}^i_{med}]$. Initially, $\mathbf{P}_{matrix}$ is null.

Step 4. Filter the position matrix $\mathbf{P}_{matrix}$ with $\mathbf{a} = [a_i, a_{i-1}, \ldots, a_{i-u}]$ to get the ith positioning output $\mathbf{P}^i_f$, where a_k is the impact fact of the kth positioning result to the ith positioning result, and $a_k \in (0, 1]$, i.e.,

if $i \leq u + 1$,
    for t=1 : T
        $\mathbf{R} = \mathbf{P}_{matrix}(t, :) \otimes \mathbf{a}$
        $\mathbf{P}^i_f(i, t) = \mathbf{R}/i$
    end
else
    for t=1 : T
        $\mathbf{R} = \mathbf{P}_{matrix}(t, :) \otimes \mathbf{a}$
        $\mathbf{P}^i_f(i, t) = \mathbf{R}/(u + 1)$
    end
end

where t is index of the dimensions of the location coordinate.

Step 5. Keep the last k positions in $\mathbf{P}^i$ and continue to accumulate $\mathbf{p}_j$ until the length of $\mathbf{P}^i$ is N under the condition that the distance between $\mathbf{p}_j$ and $\mathbf{P}^i_f$ is less than L. Then, $i = i + 1$, and go back to Step 2.

4   Test Results

In this section, we present numerical results for the proposed LTE-based positioning system. The measurements are carried out in the office room and meeting room. Fixed point test is implemented in the office room while both the fixed point test and dynamic test are carried out in the meeting room. LTE uplink signals are configured and generated by the signal generator with the bandwidth of 20 MHz at the carrier frequency of 1.755 GHz. Details about the test scenario descriptions and result analysis are as following.

4.1   Office Room Test

We consider a typical office room as shown in Fig. 4 with dimensions of 5.4 m × 8 m × 3 m. SMU200A is employed as the transmitter and connected with the omni antenna. The transmitting power is −15 dBm. Two LMUs are used in the test, each of which is connected with four directional receiving antennas.

11 fixed points distributed along the yellow line "7" are chosen as the test targets, and at each point, 100 measurements are carried out to obtain the average the results.

Fig. 4. Fixed point test environment in the office room. (Color figure online)

Figure 5 shows the positioning results on the map. The points marked by "□" on the blue line are the target points, and the points marked by "★" on the red line are the estimated results of the targets. Table 1 shows the positioning errors of the 11 targets. As observed, the positioning errors are very small for most of the targets, which are all less than 2 m. The average positioning error is about 0.84 m.

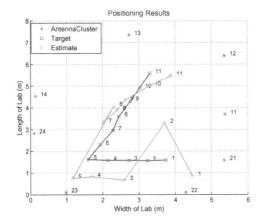

Fig. 5. Fixed point test results in the office room. (Color figure online)

4.2   Meeting Room Test

The dynamic test is conducted in a ladder meeting room with the length of 12 m and width of 13 m. For the implementation of position measurements, the underlying geometric construction system is essential with regard to the positioning accuracy. Figure 7 shows the antenna setup which has been established based on the Geometric Dilution of Precision (GDOP) [11], and provides an overview of the test area.

Figure 6 demonstrates the probabilities of the positioning accuracies within 2 m, 3 m and 4 m at 5 random-selected locations, respectively. As observed, the

Table 1. Positioning errors in the office room test.

Target index	Positioning error (m)
#1	1.00
#2	1.80
#3	0.91
#4	1.00
#5	0.97
#6	1.80
#7	0.43
#8	0.30
#9	0.15
#10	0.29
#11	0.59
Average	0.84

probability of positioning accuracy within 3 m and 4 m is larger than 80% and 90% in most cases, respectively. When the user is moving along the red line at a walking speed, holding the hand-held signal generator with the transmitting power of -10 dBm, the route tracking result is shown as Fig. 7. The estimated trajectory is represented as the green line, which has been optimized by the spatial filtering algorithm. According to the statistic, the positioning error is less than 3 m.

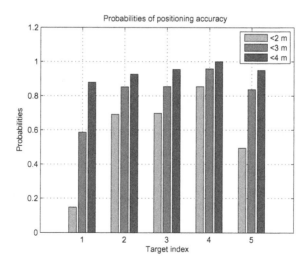

Fig. 6. Probabilities of positioning accuracy in the meeting room. (Color figure online)

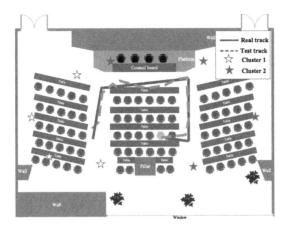

Fig. 7. Dynamic test results in the meeting room.

5   Conclusion

Considering that the WLAN-based positioning systems have some disadvantages such as discontinuous coverage, additional terminals and high cost, etc., the LTE-based indoor positioning system is investigated in this paper, which would be implemented in current cellular networks without the repetitive network deployment. The key unit of the system named LMU is researched and developed, and the spatial filtering is proposed. The fixed point test and dynamic test are conducted, respectively. The test results show that the positioning error is less than 3 m.

References

1. FCC 14–13: Wireless E911 Location Accuracy Requirements. http://www.fcc.gov/document/proposes-new-indoor-requirements-andrevisions-existing-e911-rules
2. Liu, H., Darabi, H., Banerjee, P., Liu, J.: Survey of wireless indoor positioning techniques and systems. IEEE Trans. Syst. Man Cybern. Part C Appl. Rev. **37**(6), 1067–1080 (2007)
3. Deng, Z., Yu, Y., Yuan, X., Wan, N., Yang, L.: Situation and development tendency of indoor positioning. China Commun. **10**(3), 42–55 (2013)
4. Ding, G., Zhang, J., Zhang, l., Tan, Z.: Overview of received signal strength based fingerprinting localization in indoor wireless LAN environments. In: 2013 5th MAPE, pp. 160–164 (2013)
5. Khanbashi, N.A., Alsindi, N., Al-Araji, S., Ali, N., Aweya, J.: Performance evaluation of CIR based location fingerprinting. In: 2012 IEEE 23rd PIMRC, pp. 2466–2471 (2012)
6. Texas Instruments ADS62C17 Data Sheet. http://www.ti.com/product/ads62c17
7. Xilinx Kintex-7 FPGAs Data Sheet. http://www.digikey.com/product-detail/en/XC7K160T-2FFG676C/122-1836-ND/3671575

8. Ye, R., Liu, H.: UWB TDOA localization system: receiver configuration analysis. In: 2010 ISSSE, vol. 1, pp. 1–4 (2010)
9. Kim, W., Lee, J.G., Jee, G.I.: The interior-point method for an optimal treatment of bias in trilateration location. IEEE Trans. Veh. Technol. **55**(4), 1291–1301 (2006)
10. Trees, H.L.V., Bell, K.L.: A Tutorial on Particle Filters for Online Nonlinear/NonGaussian Bayesian Tracking. Wiley-IEEE Press, pp. 723–737 (2007)
11. Sharp, I., Yu, K., Guo, Y.J.: GDOP analysis for positioning system design. IEEE Trans. Veh. Technol. **58**(7), 3371–3382 (2009)

Urban Trace Utilizing Mobile Sequence

Yukun Ma$^{(\boxtimes)}$, Bin Xu, and Qi Li

Department of Computer Science and Technology, Tsinghua University,
Beijing, China
yukunma@aliyun.com, xubin@tinghua.edu.cn, zhongguoliqi@163.com

Abstract. Urban trace mining and modeling of large scale active mobile phone holders is of great significance as it can effectively reduce traffic jam and hit-and-run incidents as well as other urban problems. Traditionally, people monitors traffic flows by GPS devices installed on buses or cars and traces hit-and-run incidents by road cameras, which turns out to be inefficient in that the data are sparse to represent a massive of people traces. In this paper, we propose a novel approach called mobile sequence to describe urban roads and moving object trajectories. Firstly, we use real urban roads information and base stations location information to generate urban roads mobile sequence. Then we extract people's mobile sequence from massive mobile network logs. Finally, we match roads' mobile sequence with people's mobile sequence to obtain human traces. As a validation, we use application installed on mobile phone to gather some real urban roads and base stations' information and then calculate their real mobile sequences to match the theoretical mobile sequences and gain 94.8% covering rate. Also, we use people's mobile sequences extracted from mobile network logs with theoretical mobile sequences and gain 90.6% matching accuracy.

Keywords: Urban trace · Mobile sequence · Urban real roads · Mobile network logs

1 Introduction

With wider and deeper urbanization of the world, big cities emerge in a large number, which have engendered many great issues involving environmental pollution, energy wasting and many transportation problems. Facing with the kinds of great challenges, governments and researchers proposed a myriad of creative solutions. Among all of the novel tackles, methodologies based on big data are repeatedly mentioned because of the advent of the era of big data. Additionally, more conceptions and assumptions to solve modernization issues are put forward, such as smart cities [3], urban computing [20], etc.

When people use intelligent phones to access the Internet, user behavior data are recorded in logs by operator. If we know a majority of people's traces in advance, then some urban issues, such as traffic congestion and hit-and-run incidents, can be eased and reduced efficiently. At present there are mainly three

© Springer Nature Singapore Pte Ltd. 2017
B. Zou et al. (Eds.): ICPCSEE 2017, Part II, CCIS 728, pp. 241–255, 2017.
DOI: 10.1007/978-981-10-6388-6_20

kinds of methods to detect a road traffic jam. The first method is based on GPS of the mobile terminal. When related applications in mobile holders are started, they will collect GPS location information to upload to the server and the server will compute the terminal speed so that we can judge if there is a traffic congestion. The second method is to cooperate with special industries, such as bus companies, taxi companies. In the industries, cars will be installed GPS sensor, then the company can collect a huge of real-time GPS data. The data will be combined with the historical big data at the same period to compute the speed of the car in this road section. The last method is based on the detecting roadside cameras or speed radar equipment so as to detect and monitor the number and the rough speed of vehicles and to estimate the trafficability of a road. The forementioned methods are mainly based on GPS data or depend on some particular devices to collect vehicles' precise location information, which is not so convenient and economical. In contrast, we propose a novel method to detect the traffic anomalies based on mobile big data from mobile operator. The main challenge, when we apply mobile network data to modeling and analysis, is that we need extract spatio-temporal data as accurately as possible from people's Internet behavior traces. One reason is that the granularity of the base station coverage ranges from dozens of meters to hundreds of meters, which is determined by the stations' density in that area. If we can compute the average speed of the mobile devices, then we can determine a person's transportation ways, such as walking or by car, and we can also judge if the person is on the highway. In this paper, we propose an approach only based on people's mobile phones to solve the challenge. In our approach, urban roads and people's traces are represented by mobile sequence. The approach's effectiveness was demonstrated by real urban roads information collected by specialized applications installed in mobile phones. When we make a strict matching between the theoretical mobile sequence and the practical mobile sequence, we obtain about 74% correctly matching rate. However, almost 95% practical mobile sequences are covering by theoretical value, which means we just change the matching strategy then we can mine almost all the mobile users' trace based on the approach. In summary, the contributions of our work are as follows:

- We extract all of the users' daily traces from their mobile network logs and solved traces mining problem more conveniently and massively;
- We proposed a novel approach called mobile sequence to depict urban roads and people's urban traces;
- We provide a more convenient and efficient method to solve traffic issues.

The rest of the paper is organized as follows. In Sect. 2, we describe related work on urban trace data mining based on network data generated from mobile phones. Section 3 describes some preliminaries and some definitions. The novel approach we proposed to describe moving object traces in this paper will be demonstrated in Sect. 4. Section 5 shows the mobile sequences matching procedures and gives detail analysis of the results. Finally, we conclude this paper in Sect. 6 and also some future work will be illustrated in this section.

2   Related Work

Until now, lots of research on the city computing of urban trace mining remains emerge in endlessly [19]. On one hand it benefits from the convenience of the advance in location-acquisition technologies, which results in a myriad traces data including people, cars or other objects' mobility. Actually, the unprecedented traces data foster a broad range of novel thought patterns on smart cities, especially in location-based social networks [21] and urban computing [20]. On the other hand, in smart cities and urban computing, urban trace data mining has become an increasingly hot and productive research orientation.

Generally speaking, urban traces, after some data preprocessing, may lead us to discovery some predictable patterns or trends of urban dynamics [2,6,7,13,14]. Though a myriad of data categories in urban trace, a user carrying a mobile phone unintentionally generates traces recorded by a sequence of cell tower IDs. Thus, a phone naturally becomes a better sensor of a person's daily whereabouts [9]. If a person keeps open the phone flow everywhere, his Internet behavior traces are tracked by network logs. Even he just keeps his phone power on, his locations are passively tracked because his phone must communicate with the cellular network. All of the traces data are mined from big data and modeled to infer people's behavior [4]. Also, mobility trace data are applied to the population movement and patterns minning [12], but some universal features should be taken into consideration. Some researches pay more attention on the presence of preferential locations of one person's trace [13,14]. As for urban trace model, modeling is even more fine-grained over time. In 2002, White and Wells [15] considered moving objects' start points and end ports and proposed OD (Origin-Destination) Matrix to describe moving objects' traces to avoid more accurate traces problem only based on mobile data. According to the OD matrix, Iqbal et al. [8] combines the base stations with the traffic flows to monitor people movement flow. In 2007, POI (Point of Interest) [5] was proposed to discovery a part of people's similar interests. Because of different cover areas of the base stations, much more fine-grained traces are difficult to gather only from cellular network and mobile data. In 2015, Chao et al. [16,17] explored the location relationship between three base stations and the person who is covered simultaneously by three stations. They evaluate the person's position by triangulation method and the error distance reaches hundreds meters. Some researcher [10,18] take the spatio-temporal data mining as the frequent pattern problems to solve the trace mining problem.

In academia, two main trajectory technology methods of wireless location are network-based and handset-based. Network-based technology refers to analyze people's network behavior data when they are moving in a city and keep accessing the Internet. One problem is that people's Internet duration is limited in the past but now the data are intensive and dent because of the real-time applications' prevalence, such as Wechat, QQ. People usually keep their real-time applications on even they are driving, walking or eating. Thus, the task of depicting a user's trace from the network data is enough because high accuracy trace of a person may violate privacy, which is illegal. Based on the fact, handset-based technology, on one hand, we can use sensors in the mobile, such as GPS receiver to collect peoples' locations and have a relatively higher accuracy. On the other hand, almost all of the people are not willing to expose their location data unconditionally to other people or institutions. That means we can gather people's network data instead of their accurate location data, which is unobtrusive and continual.

Different from the methods above, in this paper, we will introduce a novel representation method based on a series of mobile base stations. A person moves from a base station to another base station and his position changeover will be covered by a new network called Thiessen polygon [1,11] network, composed of many Thiessen polygon instead of simple cellular network. The replacement reason will be illustrated in Sect. 3. We use this method to represent every moving objects' traces, and to match new traces so as to tackle the traffic congestion and other track problems.

3   Preliminaries and Definitions

Cellular network or mobile network is a mobile communication architecture, which is divided into analog cellular network and digital cellular network. In the real world, standard cellular network is an idealized model, which is influenced by some other geological and climatological conditions, such as buildings, humidity and temperatures. Therefore, the coverage of the base stations is not a standard cellular network and the standard network is unfeasible to indicate the coverage of one base station. An enlightment is from meteorology, then we alternatively choose Thiessen polygon [1,11] to divide the service coverage of all the base stations, which works fine practically. Figure 1 shows the real-world base stations network divided by Thiessen polygon.

In this paper, we probe into urban traces based on mobile network logs, which lead us give some definitions about human spatial-temporal trace and our

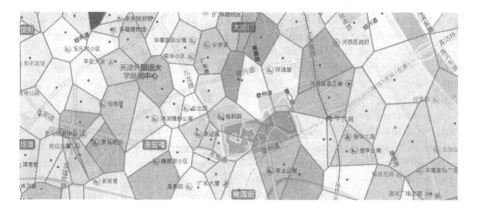

Fig. 1. Actual service coverage divided by Thiessen polygon in cellular network. (Color figure online)

new method called mobile sequence. As a rule, mobile base stations are always built along the roads. The whole road is always covered by some base stations, so we define RMS (road mobile sequence) to represent all the urban roads based on Thiessen polygon. Besides, after we extract human traces from network logs, we need define PMS (people mobile sequence) to stand for human traces in our mining algorithm. So the definitions are as follows:

Definition 1: *point sequence of spatial-temporal trace.* *It refers to a tuple* $T_p = \{p_0, p_1, \cdots, p_i, \cdots, p_n\}$, *where* $p_i = \{x_i, y_i, t_i\}$ *is an coordinate of the space.* x_i *is longitude,* y_i *is latitude and* t_i *is the time.*

Definition 2: *human trace.* *It refers to a tuple* $T_e = \{p_0, p_1, \cdots, p_j, \cdots, p_m\}$, *where* p_j *is the same as definition 1.*

Definition 3: *mobile sequence.* *It refers to a trace-around base stations' sequence which are generated by the base stations that cover the trace. Mobile sequence (MS) is a sorted set contained a series of base stations. So we define* $MS = \{BS_1, BS_2, \cdots, BS_i, \cdots, BS_n\}$, *where* BS_i *is the i-th base station in the sequence.*

Definition 4: *RMS.* *It refers to a mobile sequence generated from urban real roads based on Thiessen polygon algorithm.*

Definition 5: *PMS.* *It refers to a mobile sequence generated from a human trace extracted from network logs.*

4   Method

In this section, we will generally describe our method. Firstly, we use urban real roads network information and mobile base stations information to generate RMS (see definition RMS) based on Thiessen polygon algorithm. Secondly, we give the approach to generate PMS based on human moving sequences extracted from mobile network logs. Finally, we match PMS with RMS and obtain all the urban human traces.

4.1   Generating Road Representation Base on Mobile Sequence (RMS)

In this subsection we provide a simple RMS generating algorithm in Algorithm 1. After generating RMS, how does the forementioned RMS represent the urban road? Now we give the details of the representative method according to the Fig. 2. In Fig. 2, *road_A*, *road_B*, *road_C* and *road_D* are four roads and the cross of every two roads are assigned by the two road labels. For instance, the cross of *road_A* and *road_B* is labeled as *AB* and the rest can be done in the same manner. Now every road can be represented by the labels. (For example, [$A1, AC, AD, A2$] represents *road_A*). In Fig. 2, only the base station correlated with road network are labeled but the irrelevant ones are not.

Algorithm 1. Generate RMS Algorithm

Require: Urban Real Roads Set: $RS=\{rs_1, rs_2, \ldots, rs_n\}$
Require: Base Stations Set: $BS=\{bs_1, bs_2, \ldots, bs_m\}$
1: $TN \Leftarrow [];CC \Leftarrow [];TP \Leftarrow [];RMS \Leftarrow \{\};$
2: **for** bs_i in BS **do**
3:     Construct Delaunay triangulation network;
4: **end for**
5: **for** bs_i in BS **do**
6:     Find all the triangles numbers adjacent to bs_i save in TN;
7:     Sort every triangle adjacent to bs_i clockwise or anticlockwise;
8: **end for**
9: **for** tn_i in TN **do**
10:     Compute all the centers of triangles' circumscribed circle save in CC;
11: **end for**
12: **for** cc_i in CC **do**
13:     Connect all the centers of triangles' circumscribed circle according to every neighbouring triangle to generate Thiessen polygons in TP;
14: **end for**
15: **for** rs_i in RS **do**
16:     **for** tp_j in TP **do**
17:         **if** tp_j covers rs_i **then**
18:             $RMS[rs.id].add(tp_j)$
19:         **end if**
20:     **end for**
21: **end for**

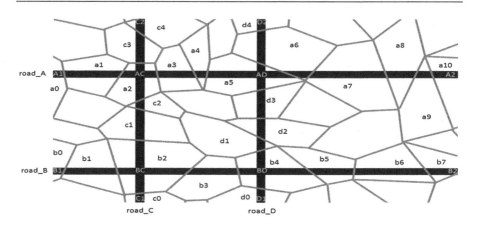

Fig. 2. Urban road network and its coverage base stations.

Know about the generation of the RMS, we demonstrate some road mobile sequences (RMS) in Fig. 2. Obviously, all the urban roads can be represented by mobile sequence. Figure 3 is an example which replaces roads with mobile sequences.

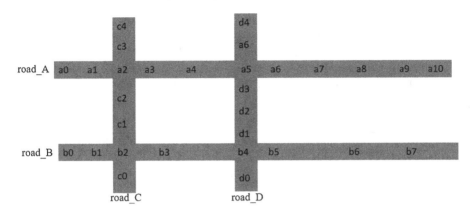

Fig. 3. RMS represents road network.

4.2 Extracting Human Trace Representation Based on Mobile Sequence (PMS)

Assume that base stations' identifiers are as a set BS_SET and the one which communicates with the users is identified by p ($p \in BS\_SET$). In some specified time span (t_s, t_e) ($t_s < t_1 < t_2 < \cdots < t_i < \cdots < t_n < t_e$), where, $i = 1, 2, 3, \cdots, n$. A person's trace can be represented as (t_s, t_e). Now we take Fig. 2 as an example to describe how to generate PMS from mobile network logs. An assumption demonstrates like this, if a mobile terminal moves along $[A1, AC, AD, BD, D1]$ and time is ignored temporarily, then a possible log record is $[a0, a1, a1, a1, a2, a2, a3, a3, a3, a4, a5, a5, a5, d3, d3, d2, d1, b4, b4, b4, d0]$. Based on the log record, a mobile sequence can be obtained by merging the adjoining labels. Figure 4 illustrates the generation of mobile sequence.

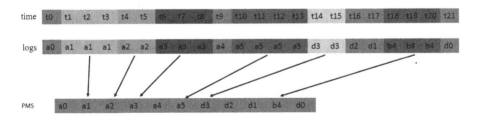

Fig. 4. The generation of PMS.

4.3 Obtain Human Trace by Matching PMS with RMS

As for the entire urban road networks, we can generate all the RMS and replace the representation form of the urban road networks with RMS network. Most of the roads can be described by RMS and most part of the people's urban traces can also be depicted by PMS. As a result, people's urban traces are converted to mobile sequence network traces. How strong the capability of the representation method will be expressed in Sect. 5. In this subsection, we solve the last problem in our method, that is how to obtain human traces by matching PMS with RMS in urban real roads network. Here is the algorithm (Algorithm 2). Find all parts of the RMS via traversing each point in PMS, then filter the points that has been added because of the intersection of two roads.

Algorithm 2. Obtain human trace by matching PMS with RMS Algorithm

---

Require: $RMS=\{\{ID_1,[rms_1,rms_2,...,rms_{n_1}]\},\{ID_2,[rms_1,$
    $rms_2,...,rms_{n_2}]\},...,\{ID_k,[rms_1,rms_2,...,rms_{n_k}]\}\}$
Require: $PMS=\{pms_1,pms_2,...,pms_m\}$
 1: $HT \Leftarrow \{\}$; {//save human traces}
 2: $threshold \Leftarrow 2$;
 3: **for** pms_i in PMS **do**
 4:    **for** rms_j in RMS **do**
 5:       **if** rms_j contains pms_i **then**
 6:          $HT[rms_j.id].add(rms_j)$;
 7:       **end if**
 8:    **end for**
 9: **end for**
10: **for** ht_j in HT **do**
11:    **if** $length(ht_j) < threshold$ **then**
12:       $HT[rms_j.id].remove()$;
13:    **end if**
14: **end for**

---

5   Experiment

In previous section, we propose the RMS to represent urban real roads network and extract PMS from mobile network logs. However, the capability of the method should be measured and the data source we extract human traces also should be described. So the next subsection data source will be shown and some measurement will be defined to make sure the approach's representative capability. Therefore, in our matching procedure, an experiment will be undertaken. In our experiment, we define a measurement called CMR (correctly matching rate) and CR (covering rate) to measure the representative ability of the mobile sequences. Also, we use application to collect urban real roads information to validate if the RMS representation is effective in urban human trace mining.

5.1   Data Source

In our data analysis and experiment, three datasets will be used and some of them are from China Mobile Company. The first dataset is called mobile network logs, which is originated from users' network log records. The time span of the data is 7 days from 2016-05-03 to 2016-05-09 and the total size is 4.9 TB. Mobile network data are a kind of XDR data. A XDR record is generated by a session and uniquely indicated by a XDR identifier. When A session is transferring in the same base station, it generates a XDR record. When the session switches to a new base station, it generates a new XDR record. As for the long sessions, a time threshold is set to ensure a certain recording frequency. When a session time exceeds the threshold, the current record reaches its end and a new record starts. In XDR data, the threshold is 5 min, which is enough for traces.

Table 1. Partial fields of mobile network logs

Fields	Description
XDR_ID	Record id, the unique indicator of the record (coded)
IMSI	User id, the unique indicator of one user (coded)
IMEI	Phone id, the unique indicator of one phone (coded)
CELL_ID	A base station's id
PROCEDURE STARTTIME	A request start time, format: yyyy-MM-dd HH:mm:ss.S
PROCEDURE ENDTIME	A request start time format: yyyy-MM-dd HH:mm:ss.S
CELL_ID	A base station unique identifier
LONGITUDE	The base station longitude
LATITUDE	The base station latitude

The second dataset is called base station information, which contains 24905 records. All the location information describes Heping district, Tianjin City. The third dataset is urban real road network information. Every road is illustrated by a sequence and every point in a sequence have three variables: longitude, latitude and the road width at that point. The detail of the data is described in Table 1.

5.2 Measurement Definition

Now we define two measurements called CMR (correct matching rate) and CR (covering rate) to measure the representative ability of the theoretical RMS.

Definition 6: *CMR (correctly matching rate)*. *CMR is used to describe the matching degree between the theoretical RMS and the practical RMS. We define the former as $RMS_t = \{BS_{t1}, BS_{t2}, \cdots, BS_{ti}, \cdots, BS_{tn}\}$, where BS_{ti} is the i-th base station in the theoretical sequence. The latter is defined as $RMS_p = \{BS_{p1}, BS_{p2}, \cdots, BS_{pj}, \cdots, BS_{pm}\}$, where BS_{pj} is the j-th base station in the practical sequence. If we find n_1 elements in the sorted set RMS_p and the found elements have the same order with RMS_t, then we define $CMR = n_1/n$. The larger CMR is, the stronger matching ability of the theoretical RMS and the practical RMS.*

Definition 7: *CR (covering rate)*. *CR is used to demonstrate the covering ability of the theoretical mobile sequence. If a practical mobile sequence is covered by a theoretical mobile sequence perfectly, then we maintain that theoretical mobile sequence has strong representation ability to depict human trace. We define two sorted sets RMS_t and RMS_p as Definition 4. If we find m_1 elements in the sorted set RMS_p and the found elements have the same order with RMS_t, then we define $CR = m_1/m$. The larger the CR is, the stronger covering ability of the theoretical RMS.*

From the definitions above, CMR and CR are used to describe the match degree of RMS_p and RMS_t. If n, n_1 in definition CMR is equal with n, m_1 in definition CR, then it indicates that RMS_p and RMS_t match completely. As a matter of fact, our final purpose is to find out people's trace according to PMS, therefore, we define matching accuracy (MA) to describe the matching degree of PMS and RMS_t.

Definition 8: *MA (matching accuracy)*. *MA is used to demonstrate the matching ability of the PMS and RMS_t. If a PMS is covered by a RMS_t perfectly, then we indicate that RMS_t has strong representation ability to depict PMS. We define two sorted sets RMS_t as Definition 4 and PMS (defined as $PMS = \{BS_{p1}, BS_{p2}, \cdots, BS_{pk}, \cdots, BS_{pl}\}$, where BS_{pk} is the k-th base station in the practical sequence). If we find l_1 elements in the sorted set PMS and the found elements have the same order with RMS_t, then we define $MA = l_1/l$. The larger the MA is, the stronger matching ability of the theoretical RMS.*

5.3   Real Roads Matching Theoretically Analysis

The relationship between the base station coverage area and roads (Fig. 1, some urban roads are marked as yellow) is analyzed as follows.

– Most of the base stations are usually distributed on both sides of the road. The road is normally covered by more than one base stations.
– In urban areas, base stations' density is relatively higher, the coverage of each base station is smaller (coverage radius is around 100 m).
– Also, to ensure running vehicle communication quality, the base station will be built along the main road.

Based on the above analysis, the road matching method can be obtained as follows:

– A user is moving a longer distance and produces a longer mobile sequence that can be matched with the road network. And the elements in the sequence (represented by the base stations) are located in both sides of the road.
– If the mobile sequence generated by the movement of a known user can matches the road network and the incremental mobile sequence still matches the road network when the user continues moving, then we judge that the user is on the road. If the generation of mobile sequence stops for a long time, then we stop matching.

– Because of some factors (weather, alternating day and night, etc.), a very small part of base stations' coverage will slightly get changed. So some threshold should be added in matching. If the distance is less than a threshold value we provide, then it can be thought correctly matched.

5.4   Using Application to Collect Real Roads Data

To determine whether the real world practical RMS matches the theoretical RMS generated by Thiessen polygon dividing the real roads, we developed an application program to collect the real world urban roads data and base stations information along the roads. We install our application on our mobile phone and keep running on some main urban roads. We collect the roads longitude and latitude sequences and the base stations' identifiers. The data fields we collected are shown in Table 2.

Table 2. Application collection data fields

Fields	Description
TIMESTAMP	Time, format: yyyy-MM-dd HH:mm:ss.S
GPS_LONGITUDE	Road updating longitude
GPS_LATITUDE	Road updating latitude
CELL_ID	Base station unique identifier
LAC	Location area code

5.5   The Theoretical RMS and Practical RMS in Real Urban Roads

As illustrated in Fig. 1, the base stations are distributed along the two sides of the roads. We assign some uppercase letters to represent the roads for convenience. Hereinafter, roads will be described by letters. To demonstrate every road theoretical RMS (RMS_t) and practical RMS (RMS_p), we take one road as an example (Fig. 5).

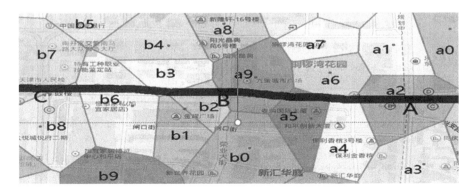

Fig. 5. An example to generate RMS_t and RMS_p in real roads.

First of all, to generate RMS_t, we visualized the dataset in Sect. 5.1 based on Algorithm 1, and part of the visualization is like Fig. 1. Then for all of the roads we assign some identifiers for them, respectively. Now we take Fig. 5 as an example to describe the case. In this figure, we use the combination of lowercase letters a–z and single digits 0–9 to identify a Thiessen polygon. Therefore, RMS_t is $\{a2, a4, a6, a5, a9, b2, b3, b6, b8, b7\}$ according to the definition of RMS. Meanwhile from our collected data, RMS_p is $\{a2, a6, a9, b3, b6, b8, b7\}$, which is generated based on the method in previous section.

5.6 Matching Results of RMS_t and RMS_p

After generating all of the experimental roads RMS_t and RMS_p, we demonstrate all the sequences in Table 3. From Table 3, we can see that RMS_p is shorter than RMS_t and almost all of the elements in RMS_p are contained in RMS_t, which means RMS_t has an excellent covering ability. In the perspective of the definition of CMR, we obtain an average 73.6%. But in practical trace mining

Table 3. Matching results of RMS_t and RMS_p

Road	RMS_t	RMS_p	CMR	CR
ABC	a2, a4, a6, a5, a9, b2, b3, b6, b8, b7	a2, a6, a9, b3, b6, b8, b7	0.700	1.000
BJ	a1, a3, a5, b5, b4, c4, c5	a1, a3, a5, b5, c4, c5	0.857	1.000
CDEF	a1, a3, a6, a7, a8, b2, b4, b5, b7, b9, c0, c1, c3, c4, c6, d0, c9, d5, d6	a1, a3, a7, a8, b2, b4, b7, b9, c0, c1, c4, c6, d0, d5, d6	0.789	1.000
DJK	a0, a2, a3, a4, a6, a8, b0, b4, b7, b9, c3, c4, c2, c5, c8	a0, a4, a6, a8, b0, b4, b7, c4, c2, c8	0.667	1.000
FOXWV	a0, a2, a3, a5, a7, a8, b0, b2, b3, b5, b6, b7, b9, c1, c0, c3, c4, c6, c8, d1, d2, d3, d7, d6, d9, e0, e2, e3, e4, e7	a0, a2, a3, a5, a7, b0, b2, b3, b7, c0, c3, c6, c8, d1, d4, d3, d7, e0, e2, e3, e7	0.700	0.952
KLM	a2, a6, a9, b3, b7, b9, b8, c2, c7	a2, a6, a9, b3, b9, b8, c2, c7	0.889	1.000
LNOPH	a0, a6, a5, a9, b3, b5, c1, c4, c7, c8, c9, d1, d2	a0, a5, b0, b3, b5, c1, c4, c7, d0, d1, d2	0.692	0.818
MYXQ	a0, a1, a6, a7, a8, c3, c4, d0, d1, d7, d6, e0, e1, e3, e7, f1, f4, g3, g2, g1, g7, h1	a0, a1, a7, a8, c3, c4, d1, d6, e1, e3, e7, f5, f4, g2, g1, h1	0.682	0.933
QR	a8, b2, b5, b4, b3, b9	a8, a7, b5, b4, b9	0.667	0.800
SW	a5, a6, a8, b1, b4, b6, b8, c2, c6, c7, c9, d1	a5, a6, a8, b4, b6, c2, c6, c7, c9, d1	0.833	1.000
TUV	a3, a6, a7, a9, b1, b4, b8, c4, c5, c6, c7, c8, d4, d7, d9	a3, a6, a7, b1, b4, b8, c5, c6, c8, d4, d5, d9	0.733	0.917
GPQST	a2, a3, a4, a6, a7, a8, b8, b7, c2, c3, c4, c5, c6, e5, e9, f6, f2, f3, f7, f8, g1, g2, g5	a2, a3, a4, a7, a8, b7, c2, c3, c5, c6, e5, e9, f6, f8, g2, g5	0.696	1.000
GHI	a2, a1, a3, a4, b1, b6, b5, b4, c7, c8, d0, c9, d5, d4, d7	a2, a3, a4, b1, b5, b4, b3, c8, c9, d5, d7	0.667	0.909
Mean			0.736	**0.948**

system or platform, we just need to consider whether the RMS_p is covered by RMS_t and whether the order of the sequences are consistent. So we provide a new measurement CR and obviously we get a decently higher CR. That is to say, mobile sequence is an effective approach to mine persons' mobile traces.

5.7   Matching Results of PMS and RMS_t

Now we use PMS to match RMS_t. PMS is extracted from mobile network logs utilizing the method described in Sect. 4.2 and RMS_t is the same as the Sect. 5.5. We use the same labeling method in Sect. 5.5 and give the points of PMS and RMS_t. Then the results is described in Table 4. From the results, we can see, mean MA is high, which means PMS matches RMS_t well. That is to say we can find a person's real roads according to PMS. Because PMS and RMS_t have mapping relationship and RMS_t and real roads network also have mapping relationship. From the mappings, human traces can be obtained.

Table 4. Matching results of RMS_t and PMS

Num	RMS_t	PMS	MA
1	a0, a1, a3, a4, a6, a8, b0, b2, b4, b7, b9, c3, c4, c5, c8	a1, a4, a8, b0, b4, b7, c2, c4, c8	0.889
2	a1, a2, a3, a5, a8, b0, b2, b3, b5, b8, b9, c0, c2, c3, c5, c7, d0, d1, d2	a1, a3, a5, b0, b3, b8, c0, c3, c7, d1	1.000
3	a0, a3, a5, b2, b3, b4, b6, c0, c2	a0, a3, b2, b4, b5, b6, c2	0.857
4	a0, a2, a1, a4, b1, b2, b4, b5, b8, b9, c0, c1, c3, c4, c7, c8, c9, d1, d2	a0, a1, b0, b2, b5, b9, c1, c4, c8, d1, d2	1.000
5	a0, a1, a2, a3, a4, a6, a8, a9, c1, c2, c3, c4, c6, c8, c9	a0, a1, a4, a8, c0, c2, c4, c6, c9	0.889
6	a2, a3, a9, b0, b3, b5, b8, b9, c1, c3	a2, a3, a4, a9, b0, b3, b8, c0, c3	0.778
7	a3, a4, a7, b0, b1, b3, b5, c0, c3, c5, c6, c8, d1, d3, d4, d6	a3, a4, a6, b1, b5, c5, c6, d1, d3, d5, d6	0.818
8	a2, a3, a5, a8, b0, b2, b9, c0, c3, c4, c7, c9, d2, d3, d4, d6, d7, d9	a2, a5, b0, b2, b9, c3, c9, d2, d4, d6, d9	1.000
9	a0, a2, a4, a5, a7, a9, b2, b7, c1, c2, c3, c4, c5, d0, d3, d4, d6, d7, d8, e2, e4, e5, e7, f0, f3, f4, f5, f7, g1, g2, g4, g5, g7	a0, a4, a7, a8, b2, c2, c5, c6, d3, d5, e2, e5, e9, f4, f7, g2, g5	0.824
10	a2, a3, a4, b0, b1, b3, b4, b6, b8, d0	a2, a3, b1, b4, b8, d0	1.000
Mean			**0.906**

6   Conclusion and Future Work

In this paper, we proposed a novel representative method called mobile sequence to depict urban roads and then human's mobility traces can be described and mined more easily and conveniently. Then we collect real urban roads information and urban base stations identifier to validate the effectiveness of our method, as a result, they gain a very high CMR and MA, which means the method maintain a strong capability of representation urban traces. That is to say, on one hand we just need to extract PMS from mobile network logs and establish the mapping relationship with RMS_t. On the other hand, we also build the mapping relationship between RMS_t and urban real roads network. From the analysis and experimental results, the method we proposed to represent urban traces is excellent in mining urban trace.

In mobile sequence, we have a bottleneck. That is if two mobile traces or urban roads share a common mobile sequence, it is hard to determine which road the person is on. On one hand, this case happens because of some tiny roads locate along the main urban roads. The efficient areas of mobile sequence are generated by Thiessen polygon, so to solve tiny roads problem is a little difficult. But on the other hand this problem can be partially solved based on trace frequency and real urban roads network in practical urban trace query platform in that we have human's network data as well as the urban roads network data. From the distinct data source, we will probe a more fine-grained urban trace mining approach in the future work. Meanwhile, the complexity of the proposed algorithm is high and there is much space to improve the matching algorithm to reduce the computing complexity though we can handle this problem on spark platform.

Acknowledgments. This work is supported by Ministry of Education-China Mobile Research Fund under grant MCM20150507 and Tsinghua University Initiative Scientific Research Program (No. 20131089190). Beijing Key Lab of Networked Multimedia also supports our research work.

References

1. Aurenhammer, F.: Voronoi diagrams—a survey of a fundamental geometric data structure. ACM Comput. Surv. **23**, 345–405 (1991)
2. Brockmann, D., Hufnagel, L., Geisel, T.: The scaling laws of human travel. Nature **439**(7075), 462–465 (2006)
3. Caragliu, A., Del Bo, C., Nijkamp, P.: Smart cities in Europe. J. Urban Technol. **18**, 65–82 (2011)
4. Giannotti, F., Nanni, M., Pedreschi, D., Pinelli, F., Renso, C., Rinzivillo, S., Trasarti, R.: Unveiling the complexity of human mobility by querying and mining massive trajectory data. VLDB J. Int. J. Very Larg. Data Bases **20**(5), 695–719 (2011)
5. Giannotti, F., Nanni, M., Pinelli, F., Pedreschi, D.: Trajectory pattern mining. In: Proceedings of the 13th ACM SIGKDD International Conference on Knowledge Discovery and Data Mining, pp. 330–339. ACM (2007)
6. Gonzalez, M.C., Hidalgo, C.A., Barabasi, A.L.: Understanding individual human mobility patterns. Nature **453**(7196), 779–782 (2008)
7. Hasan, S., Schneider, C.M., Ukkusuri, S.V., González, M.C.: Spatiotemporal patterns of urban human mobility. J. Stat. Phys. **151**(1–2), 304–318 (2013)
8. Iqbal, M.S., Choudhury, C.F., Wang, P., González, M.C.: Development of origin-destination matrices using mobile phone call data. Transp. Res. Part C: Emerg. Technol. **40**, 63–74 (2014)

9. Lane, N.D., Miluzzo, E., Lu, H., Peebles, D., Choudhury, T., Campbell, A.T.: A survey of mobile phone sensing. IEEE Commun. Mag. **48**(9), 140–150 (2010)

10. Luo, W., Tan, H., Chen, L., Ni, L.M.: Finding time period-based most frequent path in big trajectory data. In: Proceedings of the 2013 ACM SIGMOD International Conference on Management of Data, pp. 713–724. ACM (2013)

11. Okabe, A., Boots, B., Sugihara, K.: Spatial Tessellations: Concepts and Applications of Voronoi diagrams. Wiley, Hoboken (1992)

12. Simini, F., González, M.C., Maritan, A., Barabási, A.L.: A universal model for mobility and migration patterns. Nature **484**(7392), 96–100 (2012)

13. Song, C., Koren, T., Wang, P., Barabási, A.L.: Modelling the scaling properties of human mobility. Nat. Phys. **6**(10), 818–823 (2010)

14. Song, C., Qu, Z., Blumm, N., Barabási, A.L.: Limits of predictability in human mobility. Science **327**(5968), 1018–1021 (2010)

15. White, J., Wells, I.: Extracting origin destination information from mobile phone data. In: 2002 Eleventh International Conference on Road Transport Information and Control (Conf. Publ. No. 486), pp. 30–34. IET (2002)

16. Wu, C., Xu, B., Li, Q.: Parallel accurate localization from cellular network. In: Wang, Y., Xiong, H., Argamon, S., Li, X.Y., Li, J.Z. (eds.) BigCom 2015. LNCS, vol. 9196, pp. 152–166. Springer, Cham (2015). doi:10.1007/978-3-319-22047-5_13

17. Wu, C., Xu, B., Shi, S., Zhao, B.: Time-activity pattern observatory from mobile web logs. Int. J. Embed. Syst. **7**(1), 71–78 (2014)

18. Zhang, C., Han, J., Shou, L., Lu, J., La Porta, T.: Splitter: mining fine-grained sequential patterns in semantic trajectories. Proc. VLDB Endow. **7**(9), 769–780 (2014)

19. Zheng, Y.: Trajectory data mining: an overview. ACM Trans. Intell. Syst. Technol. (TIST) **6**(3), 29 (2015)

20. Zheng, Y., Capra, L., Wolfson, O., Yang, H.: Urban computing: concepts, methodologies, and applications. ACM Trans. Intell. Syst. Technol. **5**, 38 (2014). http://research.microsoft.com/apps/pubs/default.aspx?id=211950

21. Zheng, Y., Zhou, X.: Computing with Spatial Trajectories. Springer Science & Business Media, New York (2011)

An Extension to ns-3 for Simulating Mobile Charging with Wireless Energy Transfer

Ping Zhong[1], Yating Li[1], Weile Huang[1], Xiaoyan Kui[1(✉)],
Yiming Zhang[2], and Yingwen Chen[2]

[1] School of Information Science and Engineering, Central South University,
Changsha 410083, China
xykui@csu.edu.cn
[2] College of Computer, National University of Defense Technology,
Changsha 410073, China

Abstract. Many theoretical derivation of the energy model requires extensive simulation in Internet of Things (IoT). Network Simulator 3 (ns-3) provides a simulation platform for various experimental studies including energy harvest. However, the function of charge schedule and wireless energy transfer model is not yet implemented. To address this problem, in this paper we propose an extension to ns-3 for simulating mobile charging with wireless energy transfer. First, we utilize a WET Harvest Class to harvest energy from the environment and a Charge Schedule Class for the mobile charger to choose the optimal node charging in the charging request queue in ns-3. Second, we use Charge Energy Model to judge what the mobile charger will do next when the energy of current node is higher or lower than energy threshold. Evaluation results show that our improvements are feasible and helpful with charge schedule and energy model in ns-3.

Keywords: Network Simulator 3 · Energy harvest · Charge schedule · Wireless energy transfer model

1 Introduction

The future Internet of Things (IoT) networks are expected to be composed of a large population of low cost and intelligent devices with wireless energy and data transfer. In the past years, a lot of research has focused on Wireless Sensor Networks (WSNs) to prolong the network lifetime with finite batteries [1, 2]. And the batteries do not offer sufficient energy density to achieve long lifetime, high cost for WSNs nodes in the current technology [3]. Recently, there have been a growing number of applications that power wireless sensor networks by wireless charging technology. The utilization of the mobile energy chargers provides a more reliable energy supply than the systems that harvested dynamic energy from the surrounding environment [4]. Wireless energy transfer (WET) [5] technology is now commercially available. Especially, it provides another way for solving the finite batteries problem and extends the lifetime of sensors to maintain network continued operation. Generally speaking, WET can be summarized into two modes, namely, the "near-field" techniques including coupling and magnetic coupling, and the "far-field" electromagnetic radiation [6].

© Springer Nature Singapore Pte Ltd. 2017
B. Zou et al. (Eds.): ICPCSEE 2017, Part II, CCIS 728, pp. 256–270, 2017.
DOI: 10.1007/978-981-10-6388-6_21

Extensive researches in the literature have been modeling the mobile charging with WET. In [7, 8], nodes send charging request to mobile charger when lower than the minimum threshold. Then the mobile charger moves to the request node for charging with WET technology. In [9, 10], each mobile charger using WET charging is responsible for a part of the network. However, every theoretical derivation needs the realization of the simulation to determine whether it is right. Simulation of network protocols and algorithms over realistic device operation is known as a necessary task before implementation or production. It provides a flexible and fast way to present the test result. Moreover, simulations have played a principal role in modeling, analyzing and evaluating new technologies in order to validate the algorithms, protocols or system performance before real implementation and deployment [11]. In addition, simulations save a lot of unnecessary spending compared to real world.

Network Simulator 3 (ns-3) [12] provides a simulation platform for experimental studies on various communication technologies and network topologies, and it is already used by many researchers around the world. ns-3 is not perfect to support energy consumption and energy harvest simulation [13] because of lacking of energy source and consumption models. However, the current ns-3 energy framework allows simulating the energy consumption at a node as well as simulating the harvest energy from surrounding environment. There are three main parts for energy harvest and charging, namely, Energy Source, Device Energy Model and Energy Harvester.

In order to achieve better charging way for nodes and scheduling method for charging request nodes, in this paper we propose an extension of the energy framework currently released with ns-3 in order to realize the wireless energy transfer. It is different from the current energy harvester, e.g., in the form of solar, motion vibration, temperature difference, wind, coupled magnetic resonance and Radio Frequency [14, 15]. The wireless energy transfer is based on the demand of energy replenishment but not on periodical charging. Besides, nodes need to judge whether the energy of nodes is lower or higher than the minimum or maximum threshold, and then send request to the mobile charger for the next action. To quantify the priority of a charging task, we integrate energy consumption rate, residual energy and distance between the mobile charger and nodes into a charging priority. Every time, the charging request with the highest priority will be served first. There are two typical methods for on-demand charging, namely, Nearest-Job-Next with Preemption (NJNP) and First-Come-First-Served (FCFS). If a new charging is closer to the mobile charger, NJNP allows to charge the closer one first. In contrast, FCFS only considers the time factor and always serves the first recharging nodes according to the charging time sequence. At last, we utilize comprehensive simulations to verify the new energy framework for energy charging and scheduling.

The rest of paper is organized as follows. In Sect. 2, we give the related work. Section 3 introduces the energy charging framework and present the design objective, assumption and overall structure. Section 4 presents the detail of implementation in Network Simulator 3. Validation and Results are presented in Sect. 5. Finally, Sect. 6 concludes our work.

2 Related Work

As we all know, ns-3 is an open source discrete event simulator for simulating including complex network topologies, virtual network environment and network scenario. A variety of systems such as the attribute and automatic memory system make ns-3 provide a wide range platform for experimental studies [12]. In [13], the author presented an energy framework with ns-3 that allows users to simulate energy harvest as well as energy consumption to prolong the network lifetime. Tapparello et al. [16] proposed an extension of energy framework based on ns-3 for introducing energy predictor and energy harvester. Modeling the harvest capabilities of a device based on ns-3 is studied in [17]. With the progress of the times, ns-3 is a powerful and useful tool to simulate energy consumption and energy harvest in wireless networks.

Sensors can harvest energy from the surrounding environment and be recharged periodically with ns-3. As mentioned before, the ns-3 energy framework includes Energy Source, Device Energy Model and Energy Harvester. In addition, the Energy Source represents each node's energy provision. A node can have one or more energy sources and each energy source can be connected to multiple device energy models. Besides, when nodes do not sufficient energy, the energy source will give an energy warning to nodes and achieve the batteries' loss and recovery parameters. And the Device Energy Model is an energy consumption model serviced for a node. It is designed to be a node's state such as transmitting, receiving, idle and sleep. Each state has different consumption value for energy. The Energy Source will recalculate the nodes' energy and consumption rate of energy when nodes' state is changed. The Device Energy Model is the link between the node and the Energy Source. In addition, the Energy Harvester can harvest energy from the environment and change the Energy Source when nodes need to be charged.

In literature, there are three schemes for energy charging in wireless rechargeable sensor networks such as periodical charging, collaborative charging and optimizations for charging performance [4].

The periodical charging [8, 18, 19] used by ns-3 usually changes the energy charging problem into a Traveler Salesman Problem (TSP) [20] according to the rate of energy consumption and distance between the mobile charger and node. There are two types in the periodical charging, for example, single-node charging [8] and multi-node charging [21]. There are many constrains in the periodical charging such certainty and periodicity [22–25]. Therefore, there is no doubt collaborative charging is necessary for such constrains. There are many charging schedule ways for this such as based on tree structure [26] and combined routing protocol [27] to reduce charging consumption rate and charging delay. As for optimizations for charging performance in networks, predecessors have done a lot of contribution, for example, analyzing schedule problem in random topology [28, 29] and establishing the basic of Quality of Monitoring (QoM) [30–32] in networks. They usually optimize the system performance in terms of the mobile charger's speed, charge delay and nodes' lifetime and ignoring others.

From what has been discussed above, we will propose a new schedule algorithm considering the rate of energy consumption, residual energy and the distance between the mobile charger and nodes to improve the overall performance on networks.

3   Energy Charging Framework for ns-3

This section introduces the design objective and overall structure of energy charging framework. The network is assumed a multi-hop sensor networks and the node enabled WET module. The application scenario is non-contact charging scenario. As long as the mobile charging resource to the position of static sensor nodes, the charging process can successfully proceed. The mobile charging car can get the position and energy state of all nodes in the area.

3.1   Design Objective

The objective of energy framework includes the monitoring of node states, the optionality of charging scheme, and the controllability of charging.

The monitoring of node states is implementation the tracing of node energy variation. The node will take different action under different energy threshold and inform different models. At the same time, it monitors the location of mobile charging car. There is an ordered queue of charging strategies about the distance between nodes and charging car, the remaining energy of node.

The optionality of charging scheme is based on the monitoring information to realize an optimized charging strategy.

The controllability of charging is to selectively charge for a given node and control the charging energy for the given node. After the charging requirement and charging scheme, the node who meets the condition will be charged. Further, when the ability of mobile charging node is strong and the charging scheme is better, the node only needs to try its best completing own task without consider energy constrain. At this point, the WSNs both sensors and mobile charging sensor achieve the ideal situation.

3.2   Overall Structure

The overall structure of energy framework is based on the basic structure of ns-3. According to the realization of different subclasses, we achieve the corresponding energy charging and harvesting functions according to the design goals. As shown in the following Fig. 1, method of energy collection by wireless energy transfer is added in the Energy Harvester module. The dashed line is our contribution.

In addition, there are some assumptions in energy charging framework.

First, in reality, the energy can be transmitted directly to the sensor node through the magnetic coupling resonance or other wireless charging technology, for example, electro-magnetic induction, which can be achieved as long as the distance between mobile charging source and the sensor is relatively close. However, in the simulation scenario, it is assumed that mobile charging source can charge node only when it reaches the position of charging sensor.

Then, the energy status and positions of all sensors can be transferred to the mobile charging sources through wireless technology. Nevertheless, the time and cost needed for the process are relatively smaller compared to charging process and mobile process of charging source. Thus, we ignore it in simulation.

The last but not least, the energy consumption of sensors is more complex in real environment. However, we assume all nodes are based on four states, for example, receiving, transmitting, idle and sleep. At the same time, different status of energy consumption is different.

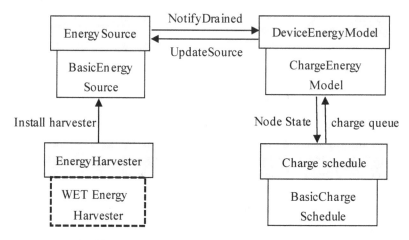

Fig. 1. The overall structure of energy module

4   Implementation

In this section, we mainly talk about some important classes with respect to method of energy collection by wireless energy transmission. It principally includes WET Harvester class, Charge schedule Class and Charge Energy Model.

4.1   WET Harvester Class

At the beginning of this section, what do we mainly introduce is the different classes that are constructed in WET Harvester class compared to original module in ns-3. As depicted in Fig. 2, "+" indicates that the method belongs to public, "–" means the method is private, "#" shows the method is protected.

The original subclass in ns-3 is BasicEnergyHarvester merely. We build a new subclass called WET EnergyHarvester which is a new subclass of Energy Harvester in this paper. The main difference between WET EnergyHarvester and BasiEnergyHarvester is the method of UpdateChangePower and UpdateHarvestedPower. The function of UpdateHarvestedPower is to obtain a certain range of random voltage value. Besides, it can transmit vibrational, wind, heat, light and electromagnetic energy into electric energy which sensor can use directly. The simulation in ns-3 managed and scheduled by Simulator class can schedule the periodic events and collect energy from the environment continuously. However, as is known to all, the energy density from environment is relatively low and the conversion efficiency is not very high with the existing technology. What we need Charging way is on-demand charging not periodical charging, thus, we need to modify the UpdateHarvestedPower method in Basic Energy Harvester to achieve on-demand charging.

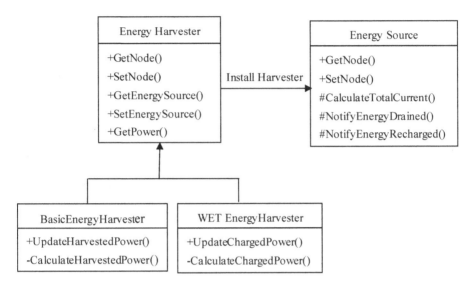

Fig. 2. WET harvester class diagram

First of all, the method of UpdateHarvestedPower can obtain the voltage values and it can be added to the node's EnergySource. Although we not use periodical charging way, the periodic update charging values method can be retained. The obtained voltage value updates to 0 when nodes do not need to be charged, which is equivalent to not replenish energy to the sensors. Thus, when nodes call the method of UpdateHarvestedPower, the following two conditions needed to judge:

- Mobile charger arrives at the location of the node.
  The current time that mobile charger confirms to charge a node is T_1, and the mobile charger arrives at the node the time of T_2. Therefore, the time T when node gets energy supplement can get.

$$T_2 = d/v \tag{1}$$

$$T = T_1 + T_2 \tag{2}$$

Where d means the distance between the mobile charger and the request node and v means the mobile charger's speed. Hence, how to determine whether the mobile charger arrived, that is, whether the time is greater than or equal to T.

- The node is optimal choice in the charging request queue.
  The optimal node is the node that gives highest priority to energy supplement by out algorithm.

If these conditions are satisfied, the node can be recharged through wireless energy transfer technology. Otherwise, the obtained voltage value updates to 0, which means the node does not get charged.

4.2   Charge Schedule Class

The next section is mainly about charge schedule. There is a charge schedule class getting the optimal node in the charging request queue through it. When multiple nodes send energy request to the mobile charger, these nodes will be placed in a charging request queue firstly. Then the Charge Schedule Class will determine the charging order of the queue which makes the charging efficiency achieve higher for the whole networks. At the same time, there are three factors affecting the charging efficiency, for example, the residual energy values of current node, the energy consumption rate of the current node and the distance between the mobile charger and request node. So, the Charge Schedule Class corresponds to these three program modules which is EnergyRemin, ConsumeRate and Distance. The MixSort module is the synthesis of the three modules, as shown in Fig. 3.

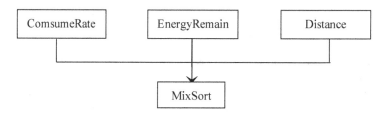

Fig. 3.   Three factors affecting the charging efficiency

We have previously assumed that each node's energy state and position state are known to the mobile charger. Therefore, a charging request queue is buffered in the mobile charger according to these state information in our simulation. In addition, the priority queue node varies with each node's energy, so we need to focus on the method of UpdateEnergySouce in Basic Energy Source Class. Besides, when UpdateEnergy-Souce is executed to the end, we detect the current energy state of the node. If the energy state is lower than the minimum charging threshold, we add the node into the charging request queue. In the contrast, if the energy state is higher than the maximum, the node will be removed from the charging request queue, which means charging process completed. if there are other charging request after removing process, we need to select a node to charge through a choosing approach. And the approach is the integration of EnergyRemin, ConsumeRate and Distance in the following.

- EnergyRemain

  The EnergyRemin means the residual energy of the sensor. It is clear that the node cannot keep it working for a long time with less energy and will die because of energy depletion. Thus, it is urgent needed to be charged and has higher priority under the same conditions. Besides, it is necessary to ensure the real-time and reliability of network transmission. So, if the energy of node is depleted, the mobile charger will not charge it and see it as death node. Therefore, there is no doubt that ensuring the nodes to survive is much important. And the number of survival nodes affects the quality of wireless sensor networks directly. Therefore, we give a node with less energy a higher priority based on the above discussion.

- ConsumeRate

  The rate of nodes energy consumption, which means the speed of current nodes energy expenditure, are different based on transmitting and receiving data packets under the different environment. Nodes with less energy may consume less energy lately, so, they may live longer according to the remaining energy. And some nodes with more energy may transmit or receive more data packets recently, thus its energy consumption rate is much higher than others. Hence, we give nodes with higher rate of energy consumption a higher priority for charging.

- Distance

  The distance between the mobile charger and the node is also a main factor for charging efficiency. And if the distance is relatively longer, which shows needing more time to move to node for the mobile charger, the node may not be able to charge timely. So, we need to choose a shortest distance to charge for the most optimal charging efficiency.

From what has been discussed above, each scheme will get a charge queue for requested nodes. Assuming that three queues of EnergyRemain, ConsumeRate and Distance are ranked in the order of A, B and C.

$$Q = A\alpha + B\beta + C\gamma \tag{3}$$

$$\alpha + \beta + \gamma = 1 \tag{4}$$

where α, β, γ are weight of the three schemes, Q means the final order of charging queue. We can select the best coefficient ratio for the optimal solution according to the needs of the actual environment.

4.3 Charge Energy Model

We have assumed that all sensors are based on four states in My Device Model, for example, transmitting, receiving, idle and sleep. As shown in Fig. 4, the Wifi Radio Energy Model original device model in ns-3 has two functions as follows.

- HandleEnergyDepletion: executed when the energy of current node lower than the minimum threshold.
- HandleEnergyRecharged: executed when the energy of current node higher than the maximum threshold.

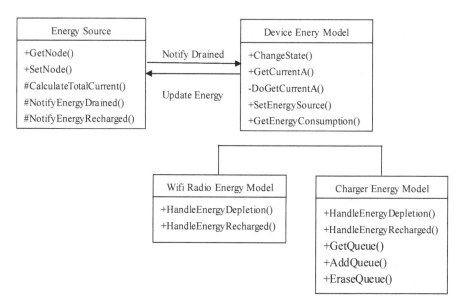

Fig. 4. The energy model class diagram

Here is difference between the original ns-3 and our work. First, in original ns-3, these two functions are empty, which means when node is changed only changing working state, working current and the energy consumption rate of node. And they do not realize how to choose and change to another node. However, we have realized it in this paper.

Look at the Charger Energy Model in Fig. 4, it has four methods used for charging. The first is HandleEnergyDepletion that indicates the energy of current node lower than the minimum threshold. The node needs to be charged immediately to prevent running out of energy. The process of the entire function is as follows. If the charging request queue is not empty, which indicates that the mobile charger is charging for other nodes, we just put the node into the charging request queue. And after completion of the current charging process, the mobile charge calculates all charging request nodes' priority to select the next node to charge according to the residual energy and the rate of energy consumption and so on. However, if the charging request queue is empty at this time, so the node will be charged immediately. Apart from join this node in the charging request queue, the mobile charger needs to calculate the distance and charging time and so on to prepare this charging process.

The next method is HandEnergyRecharged that executed when the residual energy of this node is higher than the maximum threshold. Then the charging request queue will remove the node because of energy higher than the maximum threshold. Next the mobile charger will detect the charging request queue to judge if charging request node does exist. There are three chooses for the mobile charger to execute. Firstly, the charging request is empty, which means no node need to be charged and nodes' energy are sufficient for them to transmit or receive data packets. At this time, this method is over and waiting for the next energy request node. Secondly, if the charging request queue is not empty and only has one charging request node, there is no doubt that select this node to charge and calculate the distance and charging time and so on to prepare this charging process. Thirdly, if the charging request queue has more than one request node, what we should do is use the previous three charging module the residual energy and energy consumption rate and distance to decide which node is best choose to charge and keep the networks continues operation at the same time.

5   Performance Evaluation

This section is to test the mobile charging with wireless energy transfer implemented by our energy framework. We use a network including 60 wireless rechargeable sensors, distributed uniformly over a 100 m × 100 m area for initial conditions. Nodes are not considering the impact of obstacles and aging problem collect information from environment. Besides, the data generation process can be modeled as a Poisson process with average rate λ. The initial energy for battery is 80–100%, which can prevent nodes charging requests on a period of time. In addition, the operating voltage for nodes is 5 V and the parameter settings are summarized in Table 1.

Table 1. Parameter settings for mobile charging

Parameter	Value
Battery capacity of sensor nodes	300 J
Node current consumption in idle	0.273A
Node current consumption in transmit	0.38A
Node current consumption in sleep	0.033A
Node current consumption in receive	0.313A
Moving speed of the mobile charger	4 m/s

5.1   Charging Threshold

The charging threshold refers to the number of survival nodes and receiving packets. When residual energy is lower than charging threshold, node switched into sleep state cannot transmit packets to the base station. Nodes can transmit packets normally while receiving energy supply. Otherwise, nodes will be nonfunctional after exhausted. We should determine energy threshold first to ensure nodes alive as much as possible.

As show in Fig. 5, setting different charging threshold have big influence on the throughput which means packets collected by base station. We can observe that base station receives more packets successfully while setting charging threshold equal 30% of battery capacity. At this point, the mobile charger's charging efficiency is highest and less nodes are exhausted. When charging value is 45% of nodes capacity, lots of nodes with sufficient residual energy frequently send energy charge to mobile charger. Meanwhile, nodes will switch into sleep state reducing packets numbers. Such case will increase the numbers of charge, movement of mobile charger and maintenance costs of mobile charger. Moreover, nodes are able to prolong the working time while charging threshold is equal to 15% of nodes capacity, but nodes will be nonfunctional because the residual energy is too small and multiple charging requests arrive at the same time. When threshold equal to 15% and time is between 1200 s and 1800 s, we can draw from the picture that a large number of nodes are death due to run out of energy. From what has been discussed above, in order to ensure no death nodes and best network performance, letting charging threshold is 30%.

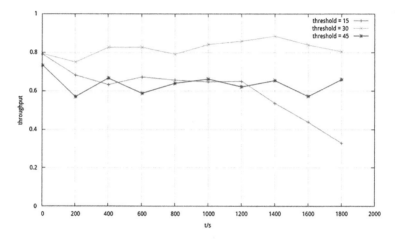

Fig. 5. The charging threshold comparison

5.2   The Number of Nodes

There are lots of factors effecting the throughout such as the speed of mobile charger, network size and utility weight on three energy efficiency elements. After determining the charging threshold, we change the number of nodes to find optimal network size. By adjusting the nodes number, base station can collect more packets with appropriate number of nodes.

Figure 6 shows the throughout rate, throughout rate indicates whether generated data packets can be collected by base station, when number of nodes are 30, 60 and 90. First, we can see that throughout rate increases with lower nodes number. Then, the packets are not successfully received by base station nearly half of packets generated by nodes, which means a mobile charger is not enough to charge nodes in area. Thus, the large network size requires multiple mobile charger to achieve higher throughout rate.

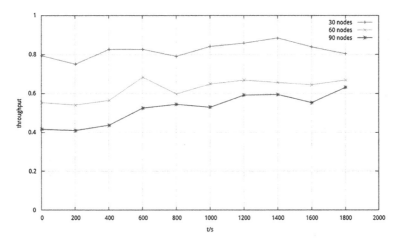

Fig. 6. The network size comparison

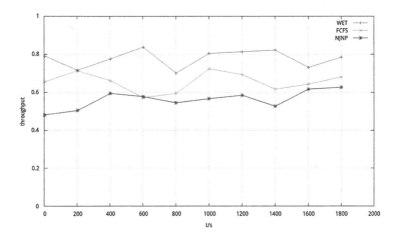

Fig. 7. The charging schedule algorithm comparison

5.3 Performance Evaluation Under Different Charging Schedule Algorithm

After determining the charging threshold and network size, we next discuss the charging schedule. There are three schedule algorithms, namely, WET proposed in this paper, FCFS and NJNP. As shown in Fig. 7, it is obvious that the WET is the best in three charging schedules. Then, when nodes nearing the mobile charger generate more packets, the NJNP may have better performance. However, data generation rate of each node is same and the network size is relatively large, so, nodes away from mobile charger will not be charged and the network performance is the worst. Moreover, when the speed of mobile charger is suitable, the FCFS will get high network performance relatively. But the mobile charger will continue to move and cost increasing when charging nodes widely distributed. Therefore, the WET considering main factor is significantly better than the NJNP and FCFS.

6 Conclusion

In this paper, we presented an extension energy framework with wireless energy transfer in ns-3. The framework supports to simulate the mobile charging with different on-demand charging schedule and charging schedule under different charging algorithm. We first design a WET Harvest class to collect energy from the mobile charger. Then, we propose a Charge schedule class used for node selection in request queue and Charge Energy model to judge what is the next action when residual energy is higher or lower than the maximum or minimum threshold. At last, several simulations are conducted to show that the proposed charging schedule is more outperformed than others. And the scheme improves network performance in throughout, request response time and charging threshold.

Acknowlegements. The work described in this paper was supported by the grant from the National Natural Science Foundation of China (Nos. 61402542, 61502540 and 61672539); National Science Foundation of Hunan Province (No. 2015JJ4077).

References

1. Wang, C., Li, J., Yang, Y., et al.: A hybrid framework combining solar energy harvesting and wireless charging for wireless sensor networks. In: IEEE INFOCOM 2016 - IEEE Conference on Computer Communications, pp. 1–9. IEEE (2016)
2. Zhao, M., Li, J., Yang, Y.: A framework of joint mobile energy replenishment and data gathering in wireless rechargeable sensor networks. IEEE Trans. Mob. Comput. **13**(12), 2689–2705 (2014)
3. Wang, C., Li, J., Ye, F., et al.: Recharging schedules for wireless sensor networks with vehicle movement costs and capacity constraints. In: Eleventh IEEE International Conference on Sensing, Communication, and NETWORKING. pp. 468–476. IEEE (2014)
4. Lin, C., Wang, Z., Han, D., et al.: TADP: enabling temporal and distantial priority scheduling for on-demand charging architecture in wireless rechargeable sensor Networks. J. Syst. Architect. **70**, 26–38 (2016)
5. Krikidis, I., Timotheou, S., Nikolaou, S., et al.: Simultaneous wireless information and power transfer in modern communication systems. IEEE Commun. Mag. **52**(11), 104–110 (2014)
6. Zeng, Y., Zhang, R.: Optimized training design for wireless energy transfer. IEEE Trans. Commun. **63**(2), 536–550 (2014)
7. Peng, Y., Li, Z., Zhang, W., et al.: Prolonging sensor network lifetime through wireless charging. In: IEEE Real-Time Systems Symposium, RTSS 2010, San Diego, California, USA, 30 November–December, pp. 129–139. DBLP (2010)
8. Xie, L., Shi, Y., Hou, Y.T., et al.: Making sensor networks immortal: an energy-renewal approach with wireless power transfer. IEEE/ACM Trans. Netw. **20**(6), 1748–1761 (2012)

9. Dai, H., Wu, X., Xu, L., et al.: Using minimum mobile chargers to keep large-scale wireless rechargeable sensor networks running forever. In: International Conference on Computer Communications and Networks, pp. 1–7. IEEE (2013)
10. Hu, C., Wang, Y.: Minimizing the number of mobile chargers in a large-scale wireless rechargeable sensor network. In: Wireless Communications and NETWORKING Conference, pp. 1297–1302 IEEE (2015)
11. Gholami, K.E., Elkamoun, N., Hou, K.M., et al.: A new WPAN Model for NS-3 simulator. In: Nicst (2013)
12. The ns-3 network simulator. http://www.nsnam.org/
13. Wu, H., Nabar, S., Poovendran, R.: An energy framework for the network simulator 3 (ns-3). In: Proceedings of the 4th International ICST Conference on Simulation Tools and Techniques, pp. 222–230. ICST (Institute for Computer Sciences, Social-Informatics and Telecommunications Engineering) (2011)
14. Lu, X., Wang, P., Niyato, D., et al.: Wireless networks with rf energy harvesting: a contemporary survey. IEEE Commun. Surv. Tutor. **17**(2), 757–789 (2014)
15. Bi, S., Ho, C., Zhang, R.: Wireless powered communication: opportunities and challenges. Commun. Mag. IEEE **53**(4), 117–125 (2014)
16. Tapparello, C., Ayatollahi, H., Heinzelman, W.: Extending the energy framework for network simulator 3 (ns-3). eprint arXiv arXiv:1406.6265v1 (2014)
17. Benigno, G., Briante, O., Ruggeri, G.: A sun energy harvester model for the network simulator 3 (ns-3). In: Workshop on Smart Wireless Access Networks for Smart City, pp. 49–54. IEEE (2015)
18. Xie, L., Shi, Y., Hou, Y.T., et al.: Wireless power transfer and applications to sensor networks. IEEE Wirel. Commun. **20**(4), 140–145 (2013)
19. Xie, L., Shi, Y., Hou, Y.T., et al.: On traveling path and related problems for a mobile station in a rechargeable sensor network. In: Fourteenth ACM International Symposium on Mobile Ad Hoc NETWORKING and Computing, pp. 109–118. ACM (2013)
20. Lin, S., Kernighan, B.W.: An effective heuristic algorithm for the traveling-salesman problem. Oper. Res. **21**(2), 498–516 (1973)
21. Lin, C., Wu, G., Obaidat, M.S., et al.: Clustering and splitting charging algorithms for large scaled wireless rechargeable sensor networks. J. Syst. Softw. **113**(C), 381–394 (2015)
22. He, L., Cheng, P., Gu, Y., et al.: Mobile-to-mobile energy replenishment in mission-critical robotic sensor networks. In: 2014 Proceedings of IEEE INFOCOM, pp. 1195–1203. IEEE (2014)
23. He, L., Gu, Y., Pan, J., et al.: On-demand charging in wireless sensor networks: theories and applications. In: IEEE International Conference on Mobile Ad-Hoc and Sensor Systems, pp. 28–36. IEEE (2013)
24. Madhja, A., Nikoletseas, S., Raptis, T.P.: Distributed wireless power transfer in sensor networks with multiple mobile chargers. Comput. Netw. **80**, 89–108 (2015)
25. Madhja, A., Nikoletseas, S., Raptis, T.P.: Hierarchical, collaborative wireless energy transfer in sensor networks with multiple mobile chargers. Comput. Netw. **97**, 98–112 (2016)
26. Guo, S., Wang, C., Yang, Y.: Mobile data gathering with wireless energy replenishment in rechargeable sensor networks. In: 2013 Proceedings of IEEE INFOCOM, pp. 1932–1940. IEEE (2013)
27. Li, Z., Peng, Y., Zhang, W., et al.: J-RoC: a joint routing and charging scheme to prolong sensor network lifetime. In: IEEE International Conference on Network Protocols, ICNP 2011, Vancouver, BC, Canada, October, pp. 373–382. DBLP (2011)
28. Jiang, F., He, S., Cheng, P., et al.: On optimal scheduling in wireless rechargeable sensor networks for stochastic event capture. In: IEEE International Conference on Mobile Adhoc and Sensor Systems, MASS 2011, Valencia, Spain, October, pp. 69–74. DBLP (2011)

29. Zhang, Y., He, S., Chen, J.: Data gathering optimization by dynamic sensing and routing in rechargeable sensor networks. In: Sensor, Mesh and Ad Hoc Communications and Networks, pp. 273–281. IEEE (2013)
30. Cheng, P., He, S., Jiang, F., et al.: Optimal scheduling for quality of monitoring in wireless rechargeable sensor networks. IEEE Trans. Wireless Commun. **12**(6), 3072–3084 (2013)
31. Dai, H., Jiang, L., Wu, X., et al.: Near optimal charging and scheduling scheme for stochastic event capture with rechargeable sensors. In: IEEE International Conference on Mobile Ad-Hoc and Sensor Systems, pp. 10–18. IEEE (2013)
32. Dai, H., Wu, X., Xu, L., et al.: Practical scheduling for stochastic event capture in wireless rechargeable sensor networks. In: Wireless Communications and NETWORKING Conference, pp. 986–991. IEEE (2013)

Design and Implementation of Distributed Broadcast Algorithm Based on Vehicle Density for VANET Safety-Related Messages

Wei Wu[1], Zhijuan Li[2(✉)], Yunan Zhang[2(✉)], Jianli Guo[1], and Jing Zhao[2(✉)]

[1] Science and Technology on Communication Networks Laboratory,
Shijiazhuang, China
[2] Harbin Engineering University, Harbin, Heilongjiang, China
{lizhijuan,zhangyunan,zhaoj}@hrbeu.edu.cn

Abstract. Vehicle ad hoc network (VANET) is a research hotspot in industrial and academic fields now and after. Dedicated short-range communication (DSRC) is a key technology of vehicular safety services and most research adhere to IEEE 802.11p standard. The safety-related services face channel congestion, message collision, and hidden terminal problem in different traffic conditions. This paper focuses on the broadcast of safety-related message under different vehicle densities. In this paper, we firstly divide the safety-related messages into three categories and assign them different broadcast priorities. Secondly, we design different distributed broadcast algorithms for the three type messages. Then, we propose a method used to evaluate the vehicle density and present the relationship between the vehicle density and the transmit power. Then the safety module selects proper transmit power according to the relationship before the message is sent. Finally, we conduct the simulation experiment using NS3 software in the Linux environment. Simulation results show that the broadcast scheme can effectively ensure that the emergency message is correctly received in the 200 m range for high vehicle density. Compared with the algorithm without considering the vehicle density, the performance has been greatly improved.

Keywords: Distributed broadcast · Vehicle density · Transmit power · Transmission distance

1 Introduction

There are more traffic accidents and fatalities on the road every day. Once the vehicle ad hoc network technology is mature and put into use, not only can make the traffic more convenient, but also greatly reduce the harm to people's lives and property. As we all know, vehicles usually have a higher speed, their speed is at least 30 km/h on the urban traffic roads, even up to 120 km/h on highways. Then traffic accidents take place in a few seconds or even in a second. If drivers can predict the danger ahead of time, many lives and property will

© Springer Nature Singapore Pte Ltd. 2017
B. Zou et al. (Eds.): ICPCSEE 2017, Part II, CCIS 728, pp. 271–285, 2017.
DOI: 10.1007/978-981-10-6388-6_22

be protected from danger. Vehicle ad-hoc networks (VANETs) are mobile ad hoc networks that communicate with each other between vehicles in a traffic environment. We call the communication method as vehicle-to-vehicle (V2V) communication. The goal is to build a self-organizing, low cost, easy to build and access wireless communication network in a variety of traffic environments. Therefore, the vehicle ad hoc network can provide drivers safety-related services, for example, traffic accident warning, assist driving, traffic information query function, vehicular communication and so on, which is realized by broadcasting safety-related message to the surrounding vehicles.

Dedicated short-range communication (DSRC) radio technology with a 75-MHz bandwidth at the 5.9-GHz band [1] is projected to support low-latency wireless data communications among vehicles and from vehicles to roadside units. IEEE802.11p standard defines specifications of the physical layer and the medium access control (MAC) layer of the vehicular wireless communication networks based on DSRC, has been created and distributed for discussions. V2V communications form a basis for decentralized active safety-related applications, which is expected to reduce accidents and their severity [2].

In general, when a vehicle equipped with a DSRC device and a GPS positioning system travels on a road, traffic information and beacon information are often exchanged with other vehicle nodes. The transmit power of message determines its transmit distance. In contrast to the general message, the safety-related messages have higher delay requirements. Rebroadcasting message by the source node is the common method that can resolve broadcast failure in some situation but can bring broadcast storm if there exists traffic jam. We consider another situation. If the vehicle on the road travels slowly, the probability of terminal package collision is higher if the communication distance of message is long. And if the running vehicles on the road are very few, wasting energy is not advocated. Therefore, the paper adopts the distributed broadcast algorithm and designs different distributed broadcast algorithm based on vehicle density for safety-related messages.

The remainder of this paper is organized as follows: Sect. 2 briefly reviews the work related to the design and modeling of safety-related message broadcast in VANETs. Section 3 firstly divides the safety-related message into three types, and introduces the distributed broadcast algorithm of safety-related message. Section 4 presents the vehicle density assessment scheme and analyses the relationship between the transmission power and the vehicle density and applies to the algorithms. In Sect. 5, the proposed algorithm is validated by extensive software simulations. This paper is concluded in Sect. 6.

2   Related Work

The safety-related services face channel congestion, message collision, and hidden terminal problem in different traffic conditions. And there have been lots of active researches on the analysis of broadcast strategy for safety services in VANETs. In summary, we divide these broadcast schemes into the following four categories according to the different influence factors.

(1) Control Transmit Power

The transmit power is higher, the impact range of the message is larger, and the message can achieve more farer. But in the case of high vehicle density, the more serious the information loss if the transmit power keep the same because of channel congestion problem. The scheme on controlling the transmission power of message has been studied.

Torrent-Moreno et al. [3,4] proposed a distributed vehicle transmission power adaptive algorithm which calculated the minimum transmit power to meet the communication within the perceptual range.

All vehicles in the perceived range would use the same transmit power. However, the coordinated transmit power doesn't necessarily meet all vehicle requirements. And the method also requires infrastructure support, poor feasibility. Gozalvez et al. [5] proposed an opportunity-driven adaptive radio resource management scheme (OPRAM) that increases transmission power at a crossroads to avoid vehicle collisions. But when the crossroads congestion is more serious, the package collision rate is relatively high, even the special signal of the channel is cleared, collision will happen. Hafeez et al. [6] analyzed the influence range of single-hop broadcasts and established a mathematical model that can assess the interference range of the transmission distance. The mathematical model can be used to calculate the maximum communication distance available for single-hop broadcast.

(2) Control Information Sending Frequency

The channel load of VANETs is larger if the vehicle density is higher. Fixed message transmission frequency is easy to cause broadcast storm, which leads to channel congestion and information loss. In this case, the probability is very small that the information can be received correctly by the surrounding node. By controlling the transmission frequency of the information, especially the transmission frequency of the general message, it is possible to reduce the collision event of the packet.

In [7,8], Christoph et al. proposed a new urban vehicle ad hoc network communication protocol, which is prone to broadcast storms due to the fixed periodic transmission of beacon message in the vehicle information system. Congestion and loss of message, the system uses a dynamic cycle to send beacons.

In [9], Tielert et al. proposed a design principle for designing collision control in an on-board ad hoc network, including distributed principles, participatory principles, and fairness principles, and based on this design principle, a method called PULSAR rate control protocol. In [10,11], He Jianhua et al. designed an adaptive rate control vehicle network security application for DSRC. The application can control the message transmission rate so that the emergency information can quickly access the channel so that the emergency message can be sent as soon as possible.

(3) Control the Carrier Listening Threshold

It is difficult for the sending node to judge the status of the current receiving node, and it is difficult for the transmitting node to control the hardware of the receiving node to change the communication parameter by using the forwarding policy which center is transmitting node. Therefore, some scholars have proposed a forwarding strategy based on the receiving node as the central control carrier sense threshold, and the strategy can be realized by the network simulator. In [12], Robert et al. proposed a method for controlling the transmission of information by controlling the perceived threshold to ensure that the security information can be received normally for the loss of information caused by high load on the vehicle network.

(4) Combine the Above Several Factors

The transmission power, the transmission frequency and the carrier sense threshold have an impact on the network congestion. Some scholars have proposed a new congestion control forwarding strategy. In [13, 14], the control of the channel congestion was realized by controlling the transmission power, the transmission frequency, the carrier sense threshold, and the competitive back-off window. But these methods are still not very good to adapt to urban traffic and highway traffic environment under different vehicle density.

In summary, a series of studies have been carried out on broadcast of the safety-related message in VANETs, and vehicle density is an important factor. But vehicle density doesn't be considered to broadcast of safety-related message in VANETs now.

3   Distributed Broadcast Algorithm of Three Type Safety-Related Messages

As mentioned above, there are safety-related services and non-safety-related services in VANETs, So each vehicle node here is equipped two modules: the non-safety-related application module and the safety-related application module. In this paper, we only discuss the processing of the safety-related module. Different type safety-related application requires different response time. Once an emergency situation occurs, it is critical to inform surrounding vehicles as soon as possible. Because the driver reaction time to traffic warning signals can be on the order of 700 ms or longer, the update interval of safety messages should be less than 500 ms (we refer to it as the lifetime of safety messages) [15], and according to the requirements in [15], the probability of message delivery failure in a vehicular network should be less than 0.01. Firstly we classify the safety-related messages in a VANET into three categories.

- The Class-one messages are emergency warning messages, mainly includes: the front vehicle braking message, the front vehicle collision message and other hazardous road conditions, such message has the highest priority. For example, when the front vehicle emergency brakes, emergency braking message will be send to the surrounding vehicles, the surrounding vehicles can make the right judgment and avoid traffic accidents.

- The Class-two messages are lower in priority than Class-one messages. This type of messages is a long-distance notification, mainly on environmental warning and traffic warning, help drivers know the road conditions ahead and select the appropriate traffic routes. Although this type of messages is sent at a lower frequency, it must be sent prior to the general message.
- The Class-three messages stand for periodic beacon messages which inform the location of the vehicle, the movement direction of the vehicle, the speed of the vehicle and so on. The running vehicle periodically broadcasts its own beacon message and receives beacon message from other nodes around it, and updates its own adjacent node table (ANT).

3.1 Priority Mechanism of Three Type Safety-Related Messages

According to the definition of safety-related messages, their priority is descending from class-one to class-three. The traditional IEEE802.11 adpots distinct ranges of backoff window sizes and AIFS duration to distinguish different services. Now we design preemptive priority scheme as Fig. 1. we set the backoff window of emergency message as zero, and the backoff window of beacon message as nonzero. And we divide a DIFS interval into a number of minislots. The length of minislot l_m and the number of minislots n_m can be calculated as Eqs. (1) and (2). The adopt of minislot [16] ensures the emergency message broadcast prior to beacon message broadcast when their backoff window are zero.

$$l_m = 2\sigma + SIFS \tag{1}$$

$$n_m = \lfloor DIFS/L \rfloor \tag{2}$$

where σ is the maximum propagation delay at the farthest communication distance, $SIFS$ is the switching time between the receiving mode and the sending mode. In this paper, we set wait1, and wait2 respectively for class-one messages and class-two messages and $L \leq Wait1 \leq Wait2 \leq DIFS$. Figure 1 shows the broadcast priority for different type safety-related messages.

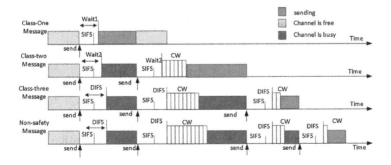

Fig. 1. The broadcast priority for different type messages

3.2 The Distributed Broadcast Algorithm of Class-One Safety-Related Message

One or more vehicle node will receive the message after the source vehicle node transmits the emergency message. Then some of these vehicles who receive message successfully in first cycle will be selected as the relay node to rebroadcast the message. And the selected node activates a delay Timer (AD timer). The value of the Timer derived from the following expression.

$$t_{AD} = T_{Max} \left(1 - \frac{d}{R} \right) \tag{3}$$

where t_{AD} is the value of a timer, d is the distance between the vehicle node and the sending node, R is the maximum communication distance of the vehicle node, T_{Max} is the maximum delay time and is usually less than the life cycle of the message. According to the above formula, we can find that the value of timer is smaller if the distance between the receiving vehicle node and the sending node is larger, so that the remote nodes can quickly repeat the emergency message broadcasting.

Please remember that the vehicle node who receive message successfully can calculate the distance from the sending node according to the adjacent node table (ANT). Otherwise, each receiving node can distinguish a copy of broadcast packets and new generation according to 12 serial number in the MAC header by IEEE802.11 [12] design. All selected nodes will rebroadcast in the manner described above until the number of copies of the emergency message reaches a certain number N_c.

The sending node will monitor the forward situation of other nodes after broadcasting the message. If the number of copies of the emergency message doesn't reach the expected number when T_{max} finishes, then the source node will rebroadcast the emergency message. Note that the receiver will inform the sender about the error if an error occurs in the forward processing. Each sender assigns an identification serial number and the sender identification serial number for the message or the copy of the message. Even if there are some errors caused by message-self or broadcast collision, the source node can record the number of copies of the emergency message. The following Algorithm 1 is the process of broadcast of class-one safety-related messages.

3.3 The Distributed Broadcast Algorithm of Class-Two Safety-Related Messages

The class-two messages broadcast selects the farthest distance receiving node in the direction of the source node in the maximum communication range as the relay node. The distance between the node and the sending node is defined as the projection of the Euclidean distance between nodes, and the node moves along the direction of movement of the source node. The distance between the relay node and the source node is shown in Fig. 2.

Algorithm 1. Broadcast algorithm of class-one safety-related messages

```
 1: if messageCountOfNode ≤ N_c then
 2:    if lifeOfMessage ≠ 0 then
 3:       if nodeID == messageNodeID then
 4:          if messageCountOfNode ≠ 1 then
 5:             messageCountOfNode ← messageCountOfNode+1
 6:             return
 7:          else
 8:             Send tone
 9:             info
10:             Start monitor module
11:          end if
12:       else
13:          Start t_AD timer
14:          Send tone
15:          messageCountOfNode ← messageCountOfNode+1
16:       end if
17:    else
18:       return
19:    end if
20:    return
21: end if
```

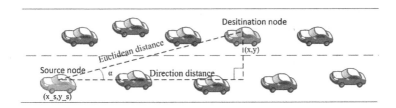

Fig. 2. The direction distance between the forwarding node and the source node

Assuming the position of the relay node is (x, y), the position of the source sending node is (x_s, y_s). And we assume that the direction distance is d_d and the Euclidean distance is d_e, the following relationship can be obtained:

$$d_d = d_e cos\alpha \tag{4}$$

Among them, the European distance can be obtained by the following formula:

$$d_e = \sqrt{(x - x\_s)^2 + (y - y\_s)^2}$$
$$\alpha = arctan(\frac{y - y\_s}{x - x\_s})$$

So the timer t_{AD} and the direction distance have the following relationship:

$$t_{AD} = T_{Max}(1 - \frac{d_d}{R}) \tag{5}$$

where $= T_{Max}$ is the maximum delay time and R is the communication distance of the sending node. When the vehicle receives the class-two messages, the value of the timer will be set as shown in Eq. (5). Equation (5) shows that nodes far from the source node can send message faster and the sending frequency may be higher.

Note that when the timer of the vehicle node is running, the success of correctly received emergency message from other nodes, then the node will stop the timer, and cancel the rebroadcast action. Because of radio interference caused by collision or noise information broadcast failure, leading to other nodes have received an error message, there may even be the farthest node or partial node does not receive emergency message. This kind of situation is likely to exist, so when this happens, other candidate nodes will continue to maintain the timer count, until a certain candidate node as relay node of the new rebroadcast of the emergency message. The process will continue until the emergency message is relayed by the relay node at least once in the maximum communication range. Algorithm 2 describes the process in detail.

Algorithm 2. Broadcast of class-two safety-related messages

1: **if** message in messageListOfNode **then**
2:     **if** lifeOfMessage $= 0$ **then**
3:         **return**
4:     **else**
5:         **if** NodeID $\neq$ nodeIDOfMessage **then**
6:             Start t_{AD} timer
7:         **end if**
8:         Send tone
9:         Send message
10:        Start monitor module
11:    **end if**
12: **else**
13:     Stop t_{AD} timer
14:     **return**
15: **end if**

3.4   The Distributed Algorithm of Class-Three Safety-Related Message

According to the above definition, the class-three safety-related message about the position, direction and speed of the broadcast vehicle are periodically given to the surrounding vehicle nodes. If the beacon message comes from other vehicle node, the program will update its adjacent node table and send the message, if the beacon message derives from its own, then directly send it.

3.5   Monitor Module

The sending node monitors the status of the class-one and class-two safety-related messages. For the class-one safety-related messages, the source node will record the broadcast times of the message after the message is be broadcasted, if the sending delay time of the message finishes but the life of the message is not over, and the number of copies of the message doesn't reach N_c, then the monitor module will rebroadcast the message. For the class-two safety-related messages, the sending node listens to the channel after the message is sent. If the sending delay time of the message finishes but the life of the message is not over, and the message is not be rebroadcasted with the presence of other nodes around, then the sending node will rebroadcast the message. Finally, the sender will rebroadcast the message if the sender has been told that there are some errors in the message. The callback function of the monitor module is described in Algorithm 3.

Algorithm 3. The callback function of the listener module

1: **if** message.T_{Max} == 0 **and** lifeOfMessge $\neq$ 0 **then**
2:    **if** ANT.IsEmpty() **then**
3:        **return**
4:    **else**
5:        **if** (typeOfMessage = 1 **and** countOfMessage $\leq N_c$) **or** (typeOfMessage = 2 **and** message in messageListOfNode = 0) **then**
6:            **if** nodeIDOfMessage $\neq$ NodeID **then**
7:                Start T_{AD}
8:            **end if**
9:            Send tone
10:            Send message
11:        **end if**
12:    **end if**
13: **else**
14:    **return**
15: **end if**

There is no manager in the distributed broadcast algorithm of safety-related message in VANETs, so the entire protocol does not add additional hardware overhead to the vehicle node except that each vehicle node need install a hardware device for sending safety-related message hints. The hardware device that sends the safety-related message hints sends an unmodulated low frequency signal, so the bandwidth is very small.

4   Distributed Broadcast Algorithm Based on Vehicle Density for Safety-Related Messages

4.1   Vehicle Density Assessment

The traffic condition varies over time, we defined the number of vehicles within the unit distance as the vehicle density. When the vehicle density is high, if the maximum communication distance is large, it will affect the communication of other nodes around the node, and even cause broadcast collision. When the vehicle density is low, if the maximum communication distance is short, emergency message may not be received by other nodes. Therefore, under different traffic conditions, the vehicle nodes should use different transmission power according to the different vehicle density. Assuming that the maximum communication distance of the vehicle node is R, the sensing range of the vehicle node is 2R. Setting the perceived distance of the vehicle to twice the communication distance can suppress the broadcast collision caused by the hidden terminal and improve the reliability of the emergency message broadcast. Then the node density in the area around the vehicle node is:

$$\rho = \frac{N}{2R} \tag{6}$$

where N is the number of nodes in the currently adjacent node table and R is the communication distance. Assuming the distance between the two junctions is 1000 m, each car is 5 m length, then a lane will have up to 200 cars. If there are four lanes on the road, then 1000 m on the road there will be up to 800 cars. Assuming that the communication distance is 500 m, then the sensing distance is 1000 m, so the maximum vehicle density is 0.8 in the sense range of the vehicle node, and the minimum vehicle density is 0.

4.2   The Transmission Power Based on the Vehicle Density

Assuming that the maximum communication distance is 500 m and the sensing distance is 1000 m. Taking into account the impact of vehicle density on vehicle communications, the message transmission power based on vehicle density is as follows:

$$P_{tx} = P_{min} + (P_{max} - P_{min})(0.2C - \rho) \tag{7}$$

where P_{tx} is the transmission power of the vehicle node, P_{min} is the minimum transmission power (this paper refers to the transmission power of 100 m), P_{max} is the maximum transmission power (this paper refers to the transmission power of 500 m), C is the number of lane (C is at least 1).

4.3   Distributed Broadcast Algorithm Based on Vehicle Density

Now we know how to judge the vehicle density, and the relationship between signal transmission power and signal transmission distance. Section 3 has given the whole process of the distributed broadcast algorithm without vehicle density. Now, we call a function CalTranPower() before send message and get the current best transmit power. Firstly, the vehicle calculates the vehicle density within the minimum communication distance range based on the ANT. As the assumption, minimum communication distance is 100 m, and the vehicle is 5 m length, then the single lane up to 40 vehicles on the sense range of the vehicle, therefore, four lanes will have 160 vehicle nodes, It can be concluded that the vehicle density is at most 0.8. If the vehicle density is above 0.4(include 0.4), the vehicle will use the minimum transmission power. When the vehicle density is less than 0.4, the transmission power will be calculated using Eq. (7) (see Algorithm 4).

Algorithm 4. CalTranPower()

1: **if** n/(2*dmin) ≥ 0.4 **then**
2: **return** pmin
3: **else**
4: **return** pmin+(pmax-pmin)*(0.2C-n/(2*dmin))
5: **end if**

---

5   Experimental Setup and Simulation Results

In this paper, two reliability indexes are used to evaluate the performance of the safety-related message broadcast strategy, which are the packet delivery ratio (PDR) and the packet reception ratio (PRR).

– PDR refers to the possibility that the packet send by source node is received successfully by all the receiving nodes at the maximum communication distance range [17, 18].
– PRR refers to the ratio of nodes that successfully receive packets to all receivers within the maximum communication distance after the source sends the packet [19, 20].

The definition showed that PDR is the sender-centric evaluation criteria, and PRR is the receiver-centric evaluation criteria. In comparison, PDR is more stringent and more sensitive to a variety of factors. For example, channel attenuation, channel noise and other factors will have some impact on the message broadcast, PDR fluctuations can reflect the impact of these factors.

5.1   Simulation Settings

The cost of actual experiment about VANETs is very expensive, especially when the vehicle density is high, thousands ofvehicles and vehicle drivers are needed, so most scholars tend to use the simulation software to validate theory. At present, the more commonly used simulation software includes NS2, NS3, OPNET, OMNET++ and so on. This paper is executed under NS3 experimental simulation. Table 1 is the parameters of the simulation experiment in this paper.

In the simulation experiment, the channel decay model uses Nakagami attenuation model, the packet transmission frequency is 0.1–0.2, the path loss intensity is 3–5, the scene area is set to a 5000-m-long freeway and the positions of vehicles spatially form a Poisson process. Assuming all vehicle nodes perceive the default range of 1000 m, is twice the max communication distance, used to suppress hidden terminal problems.

Table 1. Experimental parameters for simulation

Parameter name	Parameter value	Parameter value	Parameter value
Communication distance	500 m	CWMin	15–1024
Slot	20 μs	Channel bandwidth	10 MHZ
DIFS	50 μs	Modulation	BPSK\QPSK\ 16-QAM\64-QAM
SIFS	10 μs	Delay	1 μs

5.2   Performance Analysis

In this paper, we carried out three group simulation experiments. First is the broadcast of non-safety messages with IEEE802.11p under different vehicle density. Second is the distributed broadcasting strategy of class-one safety-related messages without considering the vehicle density. The last is the distributed broadcasting strategy based on the vehicle density for the class-one safety-related messages. R represents transmit distance, and λ represents the vehicle density in all figures.

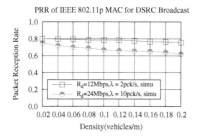

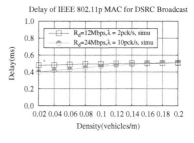

Fig. 3. PRR of non-safety services ($R = 500$ m)

Fig. 4. Delay of non-safety services ($R = 500$ m)

Figures 3 and 4 show that the PRR and delay of non-safety messages under common 802.11p. The PRR is usually under 0.8 which doesn't meet the safety-related messages' requirements. And the fast data rate leads the PRR to drop. Figure 4 shows that the delay is under 0.6 ms.

Figures 5 and 6 show that the PRR and delay of class-one safety messages under broadcast algorithm with different vehicle density. We find that the PRR of one-hop safety services can reach up to 0.9 when the number of rebroadcast is five, and the delay time can meet the requirments of the emergency message.

Figures 7 and 8 demonstrate that the PRR of the class-two safety-messages broadcast algorithm is better than RAD broadcast scheme [22, 23]. And the end-to-end delays on the 5000-m long freeway of our algorithm is smaller than RAD scheme. Because the average rebroadcast distance of our algorithm is much longer than that of the RAD scheme.

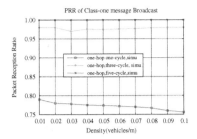

Fig. 5. PRR of safety services ($R = 500\,\text{m}$)

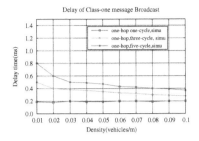

Fig. 6. Delay of safety services ($R = 500\,\text{m}$)

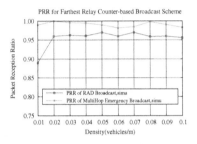

Fig. 7. PRR of class-two messages

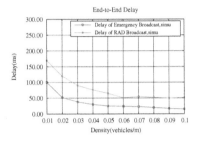

Fig. 8. Delay of class-two messages

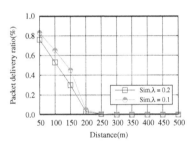

Fig. 9. PDR of different receive distance with $R = 500\,\text{m}$

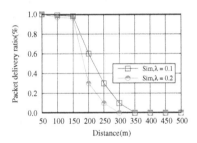

Fig. 10. PDR of different vehicle density with varying R

As shown in Fig. 9, if we adopt the fixed transmit power, the algorithm can ensure that emergency message can be received successfully within 100 m in the case of high vehicle density. The PDR of the safety-related message is higher in the case of low vehicle density, but the PDR of the safety-related message is lower in the case of high vehicle density. As contrast, from Fig. 10, we can see that the PDR in the range of 200 m is better, but the PDR descends more severe out of 200 m. This mainly due to the transmit power will be reduced when the vehicle density is high resulting in the nodes away from the sending node can't receive the message. And the PDR of the safety-related message is higher in the case of low vehicle density, the PDR of the safety-related message is lower in the case of high vehicle density.

6 Conclusion

In this paper, we design the distributed broadcast algorithm based on vehicle density of safety-related messages in VANETs. Generally speaking, the algorithm includes three kinds of distributed broadcast algorithm of different safety-related messages, a monitor module and a transmission power adjustment scheme based on the vehicle density. Then, we evaluate our scheme using NS3 simulation software in Linux environment. Simulation results show that the scheme can meet the requirements of the safety-related messages, and the optimization algorithm can effectively ensure that the emergency message in the 200 m range is correctly received in the case of high vehicle density. Compared with the algorithm without considering the vehicle density, the performance has been greatly improved. Our algorithm is a feasible scheme for the urban traffic environment which always changes with time, and it provides a new idea and method for the emergency message transmission mechanism in VANETs.

References

1. Standard Specification for Telecommunications and Information Exchange Roadside and Vehicle Systems-5GHz Band Dedicated Short Range Communications (DSRC) Medium Access Control (MAC) and Physical Layer (PHY) Specifications, April 2009
2. Ma, X., Zhang, J., Yin, X., et al.: Design and analysis of a robust broadcast scheme for vanet safety-related services. IEEE Trans. Veh. Technol. **61**(1), 46–61 (2012)
3. Mittag, J., Schmidteisenlohr, F., Killat, M., et al.: Analysis and design of effective and low overhead transmission power control for VANETs. In: Ad Hoc Networks (2008)
4. Torrent-Moreno, M., Mittag, J., Santi, P., et al.: Vehicle-to-vehicle communication: fair transmit power control for safety-critical information. IEEE Trans. Veh. Technol. **58**(7), 3684–3703 (2009)
5. Gozalvez, J., Sepulcre, M.: Opportunistic technique for efficient wireless vehicular communications. IEEE Veh. Technol. Mag. **2**(4), 33–39 (2007)
6. Hafeez, K.A., Zhao, L., Liao, Z., et al.: A new broadcast protocol for vehicular ad-hoc networks safety applications. In: Global Communications Conference (2010)
7. Sommer, C., Tonguz, O.K., Dressler, F.: Adaptive beaconing for delay-sensitive and congestion-aware traffic information systems. In: Vehicular Networking Conference (2010)
8. Sommer, C., Tonguz, O.K., Dressler, F.: Traffic information systems: efficient message dissemination via adaptive beaconing. IEEE Commun. Mag. **49**(5), 173–179 (2011)
9. Tielert, T., Jiang, D., Chen, Q., et al.: Design methodology and evaluation of rate adaptation based congestion control for Vehicle Safety Communications. In: Vehicular Networking Conference (2011)
10. He, J., Chen, H., Chen, T.M., et al.: Adaptive congestion control for DSRC vehicle networks. IEEE Commun. Lett. **14**(2), 127–129 (2010)
11. Guan, W., He, J., Bai, L., et al.: Adaptive rate control of dedicated short range communications based vehicle networks for road safety applications. In: Vehicular Technology Conference (2011)

12. Schmidt, R.K., Brakemeier, A., Leinmuller, T., et al.: Advanced carrier sensing to resolve local channel congestion. In: Ad Hoc Networks (2011)
13. Huang, C., Fallah, Y.P., Sengupta, R., et al.: Adaptive intervehicle communication control for cooperative safety systems. IEEE Netw. **24**(1), 6–13 (2010)
14. Stanica, R., Chaput, E., Beylot, A.: Congestion control in CSMA-based vehicular networks: do not forget the carrier sensing. In: Sensor, Mesh and Ad Hoc Communications and Networks (2012)
15. Xu, Q., Mak, T., Ko, J., et al.: Vehicle-to-vehicle safety messaging in DSRC. In: International Workshop on Vehicular Ad Hoc Networks, Philadelphia, PA, USA, October 2004, pp. 19–28. DBLP (2004)
16. Shan, H., Zhuang, W., Wang, Z.: Distributed cooperative MAC for multihop wireless networks. IEEE Commun. Mag. **47**(2), 126–133 (2009)
17. Ma, X., Yin, X., Trivedi, K.S.: On the reliability of safety applications in VANETs. Int. J. Perform. Eng. **8**(2), 115–130 (2012)
18. Zhao, J., Govindan, R.: Understanding packet delivery performance in dense wirelesssensor networks. In: International Conference on Embedded Networked Sensor Systems, SENSYS 2003, Los Angeles, California, USA, pp. 1–13, November 2003
19. Ye, F., Yim, R., Roy, S., et al.: Efficiency and reliability of one-hop broadcasting in vehicular ad hoc networks. IEEE J. Sel. Areas Commun. **29**(1), 151–160 (2011)
20. Ma, X., Zhang, J., Wu, T.: Reliability analysis of one-hop safety-critical broadcast services in VANETs. IEEE Trans. Veh. Technol. **60**(8), 3933–3946 (2011)
21. Ma, X., Butron, G.: On the reliability in d-dimensional broadcast wireless networks. In: International Conference on Computing, NETWORKING and Communications, pp. 957–961 (2015)
22. Williams, B., Mehta, D.P., Camp, T., et al.: Predictive models to rebroadcast in mobile ad hoc networks. IEEE Trans. Mob. Comput. **3**(3), 295–303 (2004)
23. Zhang, H., Jiang, Z.P.: Modeling and performance analysis of ad hoc broadcasting schemes. Perform. Eval. **63**(12), 1196–1215 (2006)

Prediction of Cell Specific O-GalNAc Glycosylation in Human

Yuanqiang Zou[1], Kenli Li[1], Taijiao Jiang[2,3(✉)],
and Yousong Peng[4(✉)]

[1] College of Computer Science and Electronic Engineering, Hunan University,
Changsha 410082, China
[2] Center of System Medicine, Institute of Basic Medical Sciences, Chinese
Academy of Medical Sciences and Peking Union Medical College,
Beijing 100005, China
taijiao@ibms.pumc.edu.cn
[3] Suzhou Institute of Systems Medicine, Suzhou 215123, Jiangsu, China
[4] College of Biology, Hunan University, Changsha 410082, China
pys2013@hnu.edu.cn

Abstract. Glycosylation is one of the most extensive post-translation modifications of proteins. Although lots of computational models have been developed to predict the glycosylation sites, none of them considered the tissue and cell specificity of glycosylation. Here, we built a two-step computational method GlycoCell to predict the cell-specific O-GalNAc glycosylation, the most complex type of O-glycosylation reported so far, in 12 human cell types. The first step predicted whether a site had the potential to be O-glycosylated. The model achieved an accuracy of 0.83. The second step predicted whether a potential glycosite would be O-glycosylated in the given cell type. For 12 cell types, a model was built for each cell type. The accuracies for these models ranged from 0.78 to 0.87. To facilitate the usage of GlycoCell for the public, a web server was built which is available at http://www.biomedcloud.com.cn/GlyoCell/main.htm. It could be useful for investigating the cell-specific O-glycosylation in human.

Keywords: O-glycosylation · Cell-specific · SVM · Computational · Prediction · Word vector

1 Background

Glycosylation is one of the most extensive post-translational modifications (PTMs) of proteins. It is reported to take part in various biological roles, including protein structure maintenance, protein refolding regulations, enzymatic activities, immunogenicity, pathogenecity, and so on [1–3]. Based on the linkage between the amino acid and the sugar, five types of glycosylation were defined in eukaryotes: N-linked, O-linked, C-linked, P-linked and G-linked, among which the N- and O-linked

Electronic supplementary material The online version of this chapter (doi:10.1007/978-981-10-6388-6_23) contains supplementary material, which is available to authorized users.

B. Zou et al. (Eds.): ICPCSEE 2017, Part II, CCIS 728, pp. 286–292, 2017.
DOI: 10.1007/978-981-10-6388-6_23

glycosylation are most commonly observed in cells [4]. O-linked glycosylation occurs in the Golgi apparatus, during which the O-glycan were attached to Serine (Ser), Threonine (Thr) and to less extent, Tyrosine (Tyr) [4]. No consensus sequences were found for the O-linked glycosylation in eukaryotes [4, 5]. The GalNAc-type O-glycosylation is by far the most complex-regulated type of protein O-glycosylation, during which up to 20 GalNAc-transferase isoenzymes take part in [6, 7].

As is reported in many previous studies, the tissue or cell specific PTMs play important roles in regulating the activities of cells [8–10]. Recently, by mapping the human O-GalNAc glycoproteome using the SimpleCell technology, Steentoft et al. found that the distribution of glycoprotein and glycosite varied substantially in 12 human cell lines [7], which suggests that O-glycosylation of glycosite is cell-specific. This discovery greatly expanded our view of O-glycosylation, suggesting that more precise mapping of the site-specific PTMs at tissue or cell level was needed to further understand the PTMs.

The experimental detection of glycosylation sites is still a challenging task, which often requires extensive laboratory work and considerable expense. As the rapid development of sequencing technology, many computational approaches based on protein sequence were developed to predict the glycoproteins and their respective glycosites, and achieved considerable success in glycosylation site prediction, such as the classic NetOGlyc [7, 11], and the recently developed GlycoEP [4] and GlycoMine [5]. However, none of them try to predict the tissue or cell-specific glycosites. Here, by using the data in Steentoft's work about the human O-GalNAc glycoproteome in 12 cell types [7], we developed a two-step method for predicting the cell-specific O-glycosylation (see Fig. 1). The _rst step predicted whether an amino acid residue was a potential glycosite and the second step further predicted whether the potential glycosite would be O-glycosyated in a given cell type.

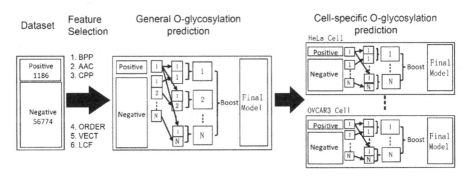

Fig. 1. The flowchart of this work.

2   Methods

Data of O-glycosylation Sites

The experimentally determined O-GalNAc glycosylation (in short as O-glycosylation in the remaining manuscript) sites were extracted from Table S2 (see Additional file (1) of Steentoft's work [7]. The corresponding protein sequences of these glycosites were downloaded from SwissProt [12]. To build the general O-glycosylation prediction model in the first step, all the Serines (Ser) and Threonines (Thr) in the proteins mentioned above which were O-glycosylated in any cell type were used as positives, while the remaining Sers or Thrs in these protein were used as negatives. To build the cell-specific O-glycosylation prediction models, for a given cell type, the glycosites which were O-glycosylated in this cell type were used as positive, while the remaining glycosites were used as negatives. The number of positives and negatives for the general and cell-specific O-glycosylation prediction models were listed in Table S1 (see Additional file 1).

To extract the local sequence features around the potential O-glycosylation sites, a local sliding window which comprised 21 residues were used, where the potential O-glycosylation site was located at the center with ten neighboring residues upstream and downstream of the central site.

Sequence Features

Six kinds of sequence feature set were used in this study, which were described as follows.

Binary Profile of Patterns (BPP) was adapted from Chauhan's work [4], during which each residue was represented by a vector of 21 dimensions, which contains 20 amino acids and one dummy amino acid "X". For example, Ala was encoded by 1,0,0,0,0,0,0,0,0,0,0,0,0,0,0,0,0,0,0,0,0 and Cys was encoded by 0,1,0,0,0,0,0,0,0,0,0, 0,0,0,0,0,0,0,0,0,0,0. In total, each potential glycosite with a window of 21 residues would be transformed into a vector of 442 dimensions.

Amino Acid Composition (AAC) is the fraction of each amino acid in a protein sequence. Each glycosite could be represented by a vector of 21 dimensions.

Central Position Pattern (CPP) is the weighted frequency of amino acids in a protein sequence. The weight is position-dependent, with the central residues in the window defined as 0, and the rest position from the middle to the sides defined as1 10.

Protein-Vector (VECT) is the distributed representation for a protein sequence. According to Asgari's work [13], the 3-gram word consisting of 3 amino acids is represented as a 100 dimensional vector. Then, the protein sequence within the window mentioned above is represented as the summation of the vector representation of overlapping 3-grams. Thus, each potential glycosite could be represented as a vector of 100 dimensions.

Order Information (ORDER) is the frequency of 3-nucleotides word in the DNA sequences transformed by a protein sequence, which is adapted from Hu's work [14]. Firstly, a protein sequence is transformed into three DNA sequences, each position of which contains the Kth (K = 1 $\sim$ 3) nucleotide of a codon encoding the corresponding amino acid. Then, the frequency of all the 3-nucleotides word (in total 64) were calculated in each DNA sequence. Thus, a protein sequence could be transformed into a 192 dimensional vector.

Lambda Correlation Factor (LCF) was adapted from Hu's work [14]. Firstly, each amino acid in a protein sequence is encoded by one of six letters according to their physicochemical property. Then, all the two-letter words with lambda (lambda = 1 $\sim$ 6) letters apart in the newly encoded sequences were extracted. The frequencies for all the two-letter words (in total 36 words) in the given Lambda value were calculated. Thus, for lambda equaling 1 ~ 6, a protein sequence could be transformed into a 216 dimensional vector.

Prediction of O-glycosylation Sites

Support Vector Machine (SVM), which is one of the most popular classifier with excellent performances, was used to predict the O-glycosylation sites in this work. Other classifiers including Naïve Bayes and decision tree were also used and performed inferior to SVM in this work (Table S2). Considering that in all the datasets the number of negatives is much larger than that of positives (Table S1), the boosting strategy was employed in the modeling as follows (see Fig. 1): firstly, the negatives were randomly separated into multiple parts, each of which has equal size to that of the positives; then, each part of negatives and all the positives were combined to form a new dataset, which results in multiple new datasets; then, each new dataset was used to build a SVM classifier; finally, to incorporate these SVM classifiers in prediction, a simple voting strategy with equal weight on each classifiers were adopted.

Five-fold cross-validations were used to evaluate the performance of SVM classifiers. The average and standard deviations of the performances in five-fold cross-validations were reported for better judgment on performance stability of the classifiers. The Area Under receiver operating characteristic Curves (AUC), accuracy, sensitivity and specificity were used to measure the performance of SVM classifiers in cross-validations.

3   Results

Distribution of Amino Acids Around the O-glycosylation Site

Based on the derived datasets, we firstly analyzed the site specificity of the O-glycosylation sites in human. The frequencies of the amino acids at each position were calculated for all the O-glycosylation sites. The sequence logos were used to visualize and analyze the sequence-level site specificity of all the O-glycosylation sites (see Fig. S1). Consistent with previous reports, no specific sequence motifs around the O-glycosylation site were observed, which suggests that the sequence-derived features that describe the local site specificity might be useful for predicting O-glycosylation.

Prediction of General O-glycosylation Sites with Six Kinds of Feature Set

Then, we attempted to predict the general O-glycosylation sites with six kinds of feature set using the SVM classifier. Different combinations of these kinds of feature set, from 1 to 6 kinds of features, were used to build the SVM model (Table S3). Table 1 listed the feature set combinations which performed best in the corresponding number of feature set combinations. The feature set "CPP" was observed to perform best among six kinds of feature set (Tables 1 and S3). The combination of top four feature sets, including the CPP, AAC, ORDER and VECT, performed best among all the feature combinations, with AUC (1 indicates the perfect match with the experimental data), Accuracy, Sensitivity and Specificity equaling to 0.849, 0.830, 0.869 and 0.829, respectively (Table 1).

Table 1. The best performance of SVM models with the N (1 ~ 6) feature sets combinations. "AUC", area under the receiver operating characteristics curves; "Acc", accuracy; "Sen", Sensitivity; "Spe", Specificity.

Num. of feature sets	Best feature combination	AUC	Acc	Sen	Spe
One	CPP	0.821 ± 0.014	0.808 ± 0.005	0.835 ± 0.030	0.807 ± 0.006
Two	CPP+ORDER	0.838 ± 0.013	0.816 ± 0.005	0.862 ± 0.024	0.814 ± 0.005
Three	AAC+CPP +ORDER	0.846 ± 0.011	0.822 ± 0.004	0.870 ± 0.022	0.821 ± 0.004
Four	AAC+CPP +ORDER+VECT	0.849 ± 0.008	0.830 ± 0.003	0.869 ± 0.018	0.829 ± 0.004
Five	AAC+CPP+VECT +ORDER+BPP	0.847 ± 0.01	0.821 ± 0.004	0.875 ± 0.017	0.819 ± 0.004

Prediction of Cell Specific O-glycosylation Sites

Steentoft's study [7] shows that the O-glycosylation sites differed much in cells. Our analysis also shows that weak correlations exist between the O-glycosylation sites in 12 cell types (Table S4). Therefore, we next attempted to predict whether or not a potential O-glycosite would be O-glycosylated in a given cell type. The same feature sets (CPP, AAC, ORDER and VECT) were used to build the cell specific O-glycosylation prediction model. Twelve models were built for predicting the cell specific O-glycosylation in twelve cell types, with each model corresponding to one model. The accuracies for these models range from 0.78 to 0.87 (Table 2). The models in OVCAR3 and HeLa cells performed best, with accuracies greater than 0.86.

Table 2. Performance of cell-specific O-glycosylation prediction models in 12 cell types in human. "AUC", area under the receiver operating characteristics curves; "Acc", accuracy; "Sen", Sensitivity; "Spe", Specificity.

Cell type	AUC	Acc	Sen	Spe
Capan1	0.849 ± 0.046	0.846 ± 0.033	0.855 ± 0.091	0.842 ± 0.028
Colo205	0.805 ± 0.026	0.786 ± 0.022	0.828 ± 0.062	0.782 ± 0.027
K562	0.795 ± 0.041	0.783 ± 0.022	0.81 ± 0.092	0.78 ± 0.027
HepG2	0.79 ± 0.036	0.786 ± 0.044	0.797 ± 0.049	0.784 ± 0.051
IMR32	0.817 ± 0.041	0.812 ± 0.03	0.828 ± 0.077	0.807 ± 0.032
T47D	0.826 ± 0.036	0.823 ± 0.026	0.831 ± 0.059	0.821 ± 0.025
Hek293	0.836 ± 0.016	0.853 ± 0.016	0.773 ± 0.036	0.899 ± 0.026
HaCaT	0.79 ± 0.042	0.792 ± 0.029	0.786 ± 0.09	0.793 ± 0.036
MDA231	0.795 ± 0.024	0.787 ± 0.016	0.812 ± 0.045	0.779 ± 0.015
HeLa	0.851 ± 0.038	0.863 ± 0.033	0.83 ± 0.064	0.872 ± 0.037
OVCAR3	0.855 ± 0.022	0.866 ± 0.019	0.807 ± 0.044	0.904 ± 0.018
MCF7	0.804 ± 0.037	0.801 ± 0.032	0.81 ± 0.063	0.798 ± 0.033

4   Conclusion

A Web Server for Predicting Cell Specific O-glycosylation Sites

To make the computational models developed above public available, a web server named GlycoCell was built, which is available at http://www.biomedcloud.com.cn/GlyoCell/main.htm. It includes two steps, both of which take the protein sequence as input. The first step is to predict whether a site has the potential to be O-glycosylated; if a site is a potential O-glycosite, GlycoCell would continue to predict whether the glycosite would be O-glycosylated in a given cell type.

5   Discussion

Steentoft et al. [7] reported that O-glycosylation differed much in different cell types in human. Here, based on their work, we built a two-step method to predict the cell specific O-glycosylation in 12 cell types. As far as we know, this is the first attempt to predict the cell-specific glycosylation. The O-glycosylation is a very complex process in cells. Many factors influenced the O-glycosylation process, such as the site specificity of enzymes [7], the physicial-chemical state of proteins, the flanking sequences around the glycosites [15], and so on. Our work shows that most of the models in cells predict the O-glycosylation with accuracies greater than 0.8, suggesting that sequence features contributed significantly to the cell-specific O-glycosylation. Among the feature sets used, the feature VECT is firstly used to predict the PTM. This work demonstrated its strength in representing the sequence information (Table S3). It may be useful in predicting other kinds of PTMs.

A limitation of this work is that only 12 cell types in human are involved in the cell-specific O-glycosylation prediction. More data about the cell-specific O-glycosylation are needed to predict the O-glycosylation in more cell types. Besides, since the O-glycosylation happened on Ser and Thr in most of the time, the methods in this study only predicted the O-glycosylation on Ser and Thr, ignoring those on other amino acids. Finally, more factors associated with glycosite O-glycosylation mentioned above should be considered to further improve the cell-specific O-glycosylation prediction models.

Acknowledgements. This study was supported by the National Natural Science Foundation (31500126 and 31371338), the National Key Plan for Scientific Research and Development of China (2016YFD0500300 and 2016YFC1200200) and the International Scientific and Technological Cooperation project (2014DFB30010).

There are no conflicts of interest.

References

1. Wright, A., Morrison, S.L.: Effect of glycosylation on antibody function: implications for genetic engineering. Trends Biotechnol. **15**(1), 26–32 (1997)

2. Arnold, J.N., et al.: The impact of glycosylation on the biological function and structure of human immunoglobulins. Annu. Rev. Immunol. **25**, 21–50 (2007)
3. Moremen, K.W., Tiemeyer, M., Nairn, A.V.: Vertebrate protein glycosylation: diversity, synthesis and function. Nat. Rev. Mol. Cell Biol. **13**(7), 448–462 (2012)
4. Chauhan, J.S., Rao, A., Raghava, G.P.S.: In silico platform for prediction of N-, O-and C-glycosites in eukaryotic protein sequences. PloS one **8**(6), e67008 (2013)
5. Li, F., et al.: GlycoMine: a machine learning-based approach for predicting N-, C-and O-linked glycosylation in the human proteome. Bioinformatics **31**, 1411–1419 (2015). btu852
6. Bennett, E.P., et al.: Control of mucin-type O-glycosylation: a classification of the polypeptide GalNAc-transferase gene family. Glycobiology **22**(6), 736–756 (2012)
7. Steentoft, C., et al.: Precision mapping of the human O-GalNAc glycoproteome through SimpleCell technology. The EMBO J. **32**(10), 1478–1488 (2013)
8. Müller, S., Hanisch, F.-G.: Recombinant MUC1 probe authentically reflects cell-specific o-glycosylation profiles of endogenous breast cancer mucin. High density and prevalent core 2-based glycosylation. J. Biol. Chem. **277**(29), 26103–26112 (2002)
9. Romanova, J., et al.: Distinct host range of influenza H3N2 virus isolates in Vero and MDCK cells is determined by cell specific glycosylation pattern. Virology **307**(1), 90–97 (2003)
10. Christensen, B., et al.: Cell type-specific post-translational modifications of mouse osteopontin are associated with different adhesive properties. J. Biol. Chem. **282**(27), 19463–19472 (2007)
11. Hansen, J.E., et al.: NetOglyc: prediction of mucin type O-glycosylation sites based on sequence context and surface accessibility. Glycoconjugate J. **15**(2), 115–130 (1998)
12. UniProt. (2016). http://www.uniprot.org/
13. Asgari, E., Mofrad, M.R.K.: Continuous distributed representation of biological sequences for deep proteomics and genomics. PLoS ONE **10**(11), e0141287 (2015)
14. Hu, Q.: The Research on Protein Sequence Feature Extraction and Its Application on Protein Subcellular Location. Hunan University, Changsha (2013)
15. O'Connell, B.C., Hagen, F.K., Tabak, L.A.: The influence of flanking sequence on the O-glycosylation of threonine in vitro. J. Biol. Chem. **267**(35), 25010–25018 (1992)

Supervised Learning for Gene Regulatory Network Based on Flexible Neural Tree Model

Bin Yang[✉] and Wei Zhang

School of Information Science and Engineering, Zaozhuang University,
Zaozhuang, China
batsi@126.com

Abstract. Gene regulatory network (GRN) inference from gene expression data remains a big challenge in system biology. In this paper, flexible neural tree (FNT) model is proposed as a binary classifier for inference of gene regulatory network. A novel tree-based evolutionary algorithm and firefly algorithm (FA) are used to optimize the structure and parameters of FNT model, respectively. The two *E.coli* networks are used to test FNT model and the results reveal that FNT model performs better than state-of-the-art unsupervised and supervised learning methods.

Keywords: Gene regulatory network · Flexible neural network · Binary classifier · Firefly algorithm

1 Introduction

Transcriptional regulation is a basis of many crucial molecular processes such as oscillator, differentiation and homeostasis, and the correct inference of gene regulatory networks (GRN) is a helpful and essential task to understand the intricacies of the complex biological regulations and gain insights into biological processes of interest in systems biology for researchers [1–3]. With the development of microarray and next generation sequencing technology, a large amount of gene expression data and gene annotation information have been generated [4–8].

However gene regulatory network is a complex and nonlinear dynamics system, inference of gene regulatory network is still a big challenge. Many machine-learning methods have been developed to infer gene regulatory network ranging from unsupervised to supervised methods. The unsupervised learning methods contain Boolean network [9], Bayesian network [10], Petri network, differential equation [11] and Information theory model [12]. The supervised learning methods require gene expression data and a list of known regulation relationships. Inference of gene regulatory network is considered as a binary classification problem [13, 14]. For each regulatory factor, the targeted genes which could be regulated by regulatory factor are set as positive samples, while the targeted genes which could not be regulated by regulatory factor are set as negative samples. The tradition classification methods have been successfully applied to the inference of gene regulatory network, especially Support Vector Machines (SVM). Gillani et al. developed a tool (CompareSVM) based on SVM to compare different kernel methods for inference of GRN [15].

© Springer Nature Singapore Pte Ltd. 2017
B. Zou et al. (Eds.): ICPCSEE 2017, Part II, CCIS 728, pp. 293–301, 2017.
DOI: 10.1007/978-981-10-6388-6_24

Neural networks (NN) have been widely applied in classification problem, because that NN could incorporate both statistical and structural information and achieve better performance than the simple minimum distance classifiers. Liu et al. proposed a new supervised approach based on radial basis function (RBF) neural network for inference of gene regulatory network. However NN model has some disadvantages, such as slow convergence and over-fitting phenomenon. Flexible neural tree (FNT) was proposed to solve these problems. Compared with neural networks, a FNT model has two advantages. (1) The model can be seen as a flexible multi-layer feedforward neural network with over-layer connections and free parameters in activation functions, so it is powerful and flexible model to model complex systems. (2) The model can select the proper input variables for constructing a model automatically, so as to select the input variables or features automatically [17].

In this paper, flexible neural tree (FNT) model is proposed to solve binary-class classification problem for inferring gene regulatory network. In order to search the optimal FNT model, a hybrid evolutionary method based on tree-based evolutionary algorithm and firefly algorithm (FA) is proposed to optimize the structure and parameters of FNT model. The gene express data from *E.coli* network are used to test the performance of FNT model.

The paper is organized as follows: Sect. 2 gives the introduction about FNT model. Section 3 presents two experiments for construction of the small and large gene regulatory networks. Some concluding remarks are presented in Sect. 4.

2 Methods

2.1 Flexible Neural Instructor

Flexible neural tree (FNT) model was proposed by Chen in 2005. The used function set F and terminal instruction set T for creating a FNT model are described as follows:

$$S = F \cup T = \{+_2, +_3, \ldots +_N\} \cup \{x_1, x_2, \ldots, x_n\}. \tag{1}$$

where $+_i$ denotes non-leaf node's instruction taking i arguments and x_i is leaf node's instruction taking no arguments. The output of a non-leaf node $+_i$ is calculated as a flexible neural operator (Fig. 1), which could be calculated as follows.

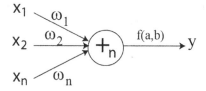

Fig. 1. A flexible neuron operator.

$$net_i = \sum_{j=1}^{i} w_j x_j,$$

$$out_i = f(a_i, b_i, net_i) = \frac{1}{1 + e^{-(\frac{net_i - a_i}{b_i})^2}}.$$

(2)

where w_j is weight, x_j is the input to node $+_i$, $f(\cdot)$ is the flexible activation function, two adjustable parameters a_i and b_i are randomly created as flexible activation function parameters.

In the FNT, every node is selected randomly from the predefined instruction/ operator sets S. If a leaf node is selected, this branch is terminated. If a non-leaf node $+_i$ is selected, i children are created in the next layer (do not exceed pre-defined maximum depth of FNT). A typical flexible neural tree model is shown as Fig. 2.

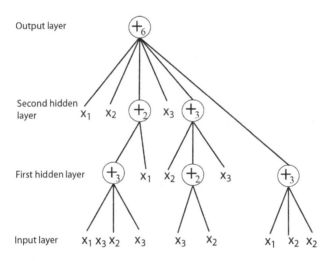

Fig. 2. An example of flexible neural tree model.

2.2 Structure Optimization

In this paper, a novel tree-based evolutionary algorithm is proposed, which contains three neural tree variation operators: mutation, crossover and selection.

(1) Mutation: We choose four mutation operators to generate offsprings from the parents.

- Change one terminal node: randomly select one terminal node in the tree and replace it with another terminal node, which is generated randomly.
- Change one subtree node: randomly select one subtree and replace it with another submit, which is generated randomly.

- Grow: select a random leaf in a hidden layer of the neural tree and replace it with a newly generated subtree.
- Prune: randomly select a function node in the neural tree and replace it with a terminal node selected in the set T.

(2) Crossover: First two neural trees are selected according to the predefined crossover probability P_c. One node is randomly selected for each neural tree. And swap them.

(3) Selection: A stochastic universal sampling selection is applied to select the parents for the next generation.

2.3   Parameters Optimization

Firefly algorithm (FA) is an efficient optimization algorithm which was proposed by Yang in 2009 [18]. It is very simple, has few parameters and easy to apply and implement, so this paper uses firefly algorithm to optimize the parameters of FNT model.

Firefly algorithm is the random optimization method of simulating luminescence behavior of firefly in the nature. The firefly could search the partners and move to the position of better firefly according to brightness property. A firefly represents a potential solution. In order to solve optimization problem, initialize a firefly vector $[x_1, x_2, \ldots, x_n]$ (n is the number of fireflies). As attractiveness is directly proportional to the brightness property of the fireflies, so always the less bright firefly will be attracted by the brightest firefly.

The brightness of firefly i is computed as

$$B_i = B_{i0} * e^{-\gamma r_{ij}} \tag{3}$$

where B_{i0} represents maximum brightness of firefly i by the fitness function as $B_{i0} = f(x_i)$. γ is coefficient of light absorption, and r_{ij} is the distance factor between the two corresponding fireflies i and j.

The movement of the less bright firefly toward the brighter firefly is computed by

$$x_i(t+1) = x_i(t) + B_i(x_j(t) - x_i(t)) + \alpha \varepsilon_i \tag{4}$$

where t is time point, B_i is brightness of firefly i, α is step size randomly created in the range [0, 1], and ε_i is Gaussian distribution random number.

2.4   Flowchart of Our Method

(1) Data preparation
   According to known regulation information from public databases, gene expression data need to be preprocessed. For each regulatory factor, the targeted genes which could be regulated by regulatory factor are set as positive samples, while the targeted genes which could not be regulated by regulatory factor are set as negative samples. Thus gene expression data could be divided into two classes.

We adopt a cross-validation procedure to measure the performance of classifier during the model training step.

(2) Construct the classifier

According to the processed gene expression data, search the optimal FNT model. The process is described as follows.

    (1) Create the initial population (flexible neural trees and their corresponding parameters).

    (2) Structure optimization by neural tree variation operators as described in Sect. 2.2.

    (3) If the better structure is found, then go to step 4), otherwise go to step 2).

    (4) Parameter optimization is achieved by FA as described in Sect. 2.3. In this stage, the tree structure or architecture of flexible neural tree model is fixed, and the best tree is taken from the end of run of the tree-based evolutionary method. All the parameters used in the best tree formulated a parameter vector to be optimized by FA.

    (5) If the maximum number of iterations of FA algorithm is reached, or no better parameter vector is found for a significantly long time (100 steps) then go to step 6); otherwise go to step 4).

    (6) If satisfactory solution is found, then stop; otherwise go to step 2).

3 Experiments

Five criterions (sensitivity or true positive rate (TPR), false positive rate (FPR), positive predictive (PPV), accuracy (ACC) and F-score) are used to test the performance of the method. Firstly, we define four variables, i.e., TP, FP, TN and FN are the number of true positives, false positives, true negatives and false negatives, respectively. Five criterions are defined as followed.

$$
\begin{aligned}
TPR &= TP/(TP+FN),\\
FPR &= FP/(FP+TN),\\
PPV &= TP/(TP+FP),\\
ACC &= (TP+TN)/(TP+FP+TN+FN),\\
F-score &= 2PPV*TPR/(PPV+TPR)
\end{aligned}
\tag{5}
$$

To evaluate the performance of our method, we compare it with state-of-the-art unsupervised methods (CLR [19], and GENIE [20]) and supervised methods (SVM [15], NN, and RBF [16]).

3.1 Small-Scale Gene Regulatory Network

The expression data generated from sub network from *E.coli* network using three different experimental conditions (knockout, knockdown and multifactorial) are used to test FNT model [15]. This sub network contains 150 genes and 202 true regulations. Each gene has 151 sample points.

Through several runs, the averaged results are listed in Table 1. From the results, we can see that supervised learning methods (SVM, NN, RBF and FNT) perform better than unsupervised learning methods (CLR and GENIE) except that CLR has the highest sensitivity (TPR) with multifactorial data. Among four supervised learning methods, FNT model performs best, which means that the inferred network achieves the optimal balance in terms of sensitivity and positive predictive rate (more true regulations and less false positive regulations).

Table 1. Comparison of six methods on *E.coli* subnetwork with different experimental conditions.

		TPR	FPR	PPV	ACC	F-score
Knockout data	CLR	0.4356	0.3478	0.0114	0.6444	0.0222
	GENIE	0.4010	0.2515	0.0145	0.7387	0.0279
	SVM	0.4554	**0.0076**	0.3552	0.9786	0.3991
	NN	0.4703	0.0081	0.348	0.9783	0.4
	RBF	0.485	0.0076	0.363	0.9787	0.4152
	FNT	**0.5010**	0.0076	**0.3826**	**0.9788**	**0.4388**
Knockdown data	CLR	0.4406	0.3602	0.0111	0.6323	0.0217
	GENIE	0.4009	0.2925	0.0125	0.6984	0.0242
	SVM	0.5198	**0.0073**	0.3962	0.9796	0.4497
	NN	0.5347	0.0087	0.36	0.9782	0.4303
	RBF	0.5495	0.0082	0.3827	0.9796	0.4512
	FNT	**0.5704**	0.0078	**0.4021**	**0.9796**	**0.4734**
Multifactorial data	CLR	**0.8168**	0.3355	0.0219	0.6600	0.0427
	GENIE	0.3366	0.2931	0.1046	0.6973	0.0203
	SVM	0.5445	0.0076	0.3971	0.9795	0.4593
	NN	0.5495	0.0085	0.3725	0.9786	0.444
	RBF	0.5842	0.0078	0.4069	0.9796	0.4797
	FNT	0.5863	**0.0075**	**0.4114**	**0.9798**	**0.4854**

In addition, to assess the effectiveness of two supervised learning methods (SVM and FNT), the ROC curves obtained by SVM and FNT on *E.coli* network with three experimental conditions are shown in Figs. 3, 4 and 5 respectively. The results show that FNT model performs better than the popular supervised learning methods (SVM).

3.2   Large-Scale Gene Regulatory Network

In this part, the expression data generated from *E.coli* network are used to test our method. *E.coli* network is from a well-know dataset DREAM5, which contains 334 transcription factors (TFs), 4511 target genes, 2066 true regulations and 805 chips [21].

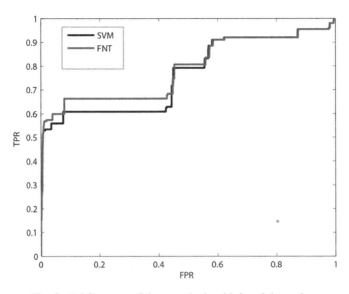

Fig. 3. ROC curves of three methods with knockdown data.

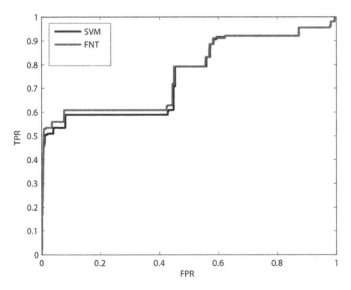

Fig. 4. ROC curves of three methods with knockouts data.

Through several runs, the results are listed in Table 2. From the results, we can see that supervised learning methods (NN and our method) perform better than unsupervised learning method (CLR). From the results of supervised learning methods it can be seen that FNT model is superior to the NN with respect to five indexes.

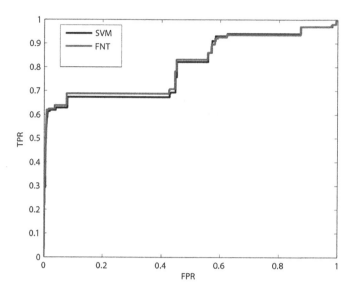

Fig. 5. ROC curves of three methods with multifactorial data.

Table 2. Comparison of several methods on *E.coli* network.

	TPR	FPR	PPV	ACC	F-score
CLR	0.37222	0.017301	0.001517	0.98266	0.003022
NN	0.4342	4.03E-04	0.0707	0.9995	0.121652
FNT	**0.450145**	**0.000269**	**0.10579**	**0.9997**	**0.171318**

4   Conclusions

In order to improve the accuracy of gene regulatory network, flexible neural tree is proposed for inference of gene regulatory network. According to the regulation prior knowledge, gene expression data could be divided into two classes. A novel hybrid evolutionary method is proposed to construct FNT model. Two gene regulatory networks from *E.coli* network are used to validate our method. The results reveal that FNT model performs better than unsupervised learning methods (CLR, GENIE) and supervised learning methods (SVM, NN and RBF).

Acknowledgments. This work was supported by the PhD research startup foundation of Zaozhuang University (No. 2014BS13), Zaozhuang University Foundation (No. 2015YY02), and Shandong Provincial Natural Science Foundation, China (No. ZR2015PF007).

References

1. Wu, J., Zhao, X., Lin, Z., Shao, Z.: Large scale gene regulatory network inference with a multi-level strategy. Mol. BioSyst. **12**(2), 588–597 (2016)

2. Mandal, S., Khan, A., Saha, G., Pal, R.K.: Reverse engineering of gene regulatory networks based on S-systems and Bat algorithm. J. Bioinform. Comput. Biol. **4**, 1650010 (2016)
3. Omranian, N., Eloundou-Mbebi, J.M., Mueller-Roeber, B., Nikoloski, Z.: Gene regulatory network inference using fused LASSO on multiple data sets. Sci. Rep. **6**, 20533 (2016)
4. Ellwanger, D.C., Leonhardt, J.F., Mewes, H.W.: Large-scale modeling of condition-specific gene regulatory networks by information integration and inference. Nucleic Acids Res. **42**(21), e166 (2014)
5. Vera-Licona, P., Jarrah, A., Garcia-Puente, L.D., McGee, J., Laubenbacher, R.: An algebra-based method for inferring gene regulatory networks. BMC Syst. Biol. **8**, 37 (2014)
6. Xie, Y., Wang, R., Zhu, J.: Construction of breast cancer gene regulatory networks and drug target optimization. Arch. Gynecol. Obstet. **290**(4), 749–755 (2014)
7. Penfold, C.A., Millar, J.B., Wild, D.L.: Inferring orthologous gene regulatory networks using interspecies data fusion. Bioinformatics **31**(12), i97–i105 (2015)
8. Baur, B., Bozdag, S.: A canonical correlation analysis-based dynamic bayesian network prior to infer gene regulatory networks from multiple types of biological data. J. Comput. Biol. **22**(4), 289–299 (2015)
9. Yang, M., Li, R., Chu, T.: Construction of a Boolean model of gene and protein regulatory network with memory. Neural Netw. **52**, 18–24 (2014)
10. Adabor, E.S., Acquaah-Mensah, G.K., Oduro, F.T.: SAGA: a hybrid search algorithm for Bayesian network structure learning of transcriptional regulatory networks. J. Biomed. Inform. **53**, 27–35 (2015)
11. Sun, M., Cheng, X., Socolar, J.E.: Causal structure of oscillations in gene regulatory networks: Boolean analysis of ordinary differential equation attractors. Chaos **23**(2), 025104 (2013)
12. Wang, J., Chen, B., Wang, Y., Wang, N., Garbey, M., Tran-Son-Tay, R., Berceli, S.A., Wu, R.: Reconstructing regulatory networks from the dynamic plasticity of gene expression by mutual information. Nucleic Acids Res. **41**(8), e97 (2013)
13. Maetschke, S.R., Madhamshettiwar, P.B., Davis, M.J., Ragan, M.A.: Supervised, semi-supervised and unsupervised inference of gene regulatory networks. Brief. Bioinform. **15**(2), 195–211 (2014)
14. Cerulo, L., Elkan, C., Ceccarelli, M.: Learning gene regulatory networks from only positive and unlabeled data. BMC Bioinform. **11**, 228 (2010)
15. Gillani, Z., Akash, M.S., Rahaman, M.D., Chen, M.: CompareSVM: supervised, Support Vector Machine (SVM) inference of gene regularity networks. BMC Bioinformatics **15**, 395 (2014)
16. Liu, S., Yang, B., Wang, H.: Inference of gene regulatory network based on radial basis function neural network. In: Pardalos, Panos M., Conca, P., Giuffrida, G., Nicosia, G. (eds.) MOD 2016. LNCS, vol. 10122, pp. 442–450. Springer, Cham (2016). doi:10.1007/978-3-319-51469-7_39
17. Chen, Y.H., Yang, B., Dong, J., Abraham, A.: Time-series forecasting using flexible neural tree model. Inf. Sci. **174**(3/4), 219–235 (2005)
18. Yang, X.-S.: Firefly algorithms for multimodal optimization. In: Watanabe, O., Zeugmann, T. (eds.) SAGA 2009. LNCS, vol. 5792, pp. 169–178. Springer, Heidelberg (2009). doi:10.1007/978-3-642-04944-6_14
19. Butte, A.J., Tamayo, P., Slonim, D., Golub, T.R., Kohane, I.S.: Discovering functional relationships between RNA expression and chemotherapeutic susceptibility using relevance networks. Proc. Natl. Acad. Sci. USA **97**(22), 12182–12186 (2000)
20. Huynh-Thu, V.A., Irrthum, A., Wehenkel, L., Geurts, P.: Inferring regulatory networks from expression data using tree-based methods. PLoS ONE **5**, e12776 (2010)
21. Marbach, D., Costello, J., Kuffner, R., Vega, N., Prill, R., Camacho, D., Allison, K., the DREAM5 Consortium, Kellis, M., Collins, J., Stolovitzky, G.: Wisdom of crowds for robust gene network inference. Nat. Methods **9**(8), 796–804 (2012)

Predicting the Antigenic Variant of Human Influenza A(H3N2) Virus with a Stacked Auto-Encoder Model

Zhiying Tan[1], Beibei Xu[1], Kenli Li[1], Taijiao Jiang[2,3(✉)],
and Yousong Peng[4(✉)]

[1] College of Computer Science and Electronic Engineering,
Hunan University, Changsha 410082, China
[2] Center of System Medicine, Institute of Basic Medical Sciences,
Chinese Academy of Medical Sciences and Peking Union Medical College,
Beijing 100005, China
taijiao@ibms.pumc.edu.cn
[3] Suzhou Institute of Systems Medicine, Suzhou 215123, Jiangsu, China
[4] College of Biology, Hunan University, Changsha 410082, China
pys2013@hnu.edu.cn

Abstract. The influenza virus changes its antigenicity frequently due to rapid mutations, leading to immune escape and failure of vaccination. Rapid determination of the influenza antigenicity could help identify the antigenic variants in time. Here, we built a stacked auto-encoder (SAE) model for predicting the antigenic variant of human influenza A(H3N2) viruses based on the hemagglutinin (HA) protein sequences. The model achieved an accuracy of 0.95 in five-fold cross-validations, better than the logistic regression model did. Further analysis of the model shows that most of the active nodes in the hidden layer reflected the combined contribution of multiple residues to antigenic variation. Besides, some features (residues on HA protein) in the input layer were observed to take part in multiple active nodes, such as residue 189, 145 and 156, which were also reported to mostly determine the antigenic variation of influenza A(H3N2) viruses. Overall, this work is not only useful for rapidly identifying antigenic variants in influenza prevention, but also an interesting attempt in inferring the mechanisms of biological process through analysis of SAE model, which may give some insights into interpretation of the deep learning models.

Keywords: Stacked auto-encoder · Antigenic variation · Influenza · Machine learning

1 Introduction

Influenza A viruses have not only caused tremendous economic loss, but also caused much mortality and morbidity to human society [1]. Vaccination is currently the most effective way to fight against it. Owing to rapid mutations on HA protein, the virus could change its antigenicity frequently [2], leading to immune escape and failure of vaccination. Rapid determination of the influenza antigenicity could help identify the antigenic variants in time.

© Springer Nature Singapore Pte Ltd. 2017
B. Zou et al. (Eds.): ICPCSEE 2017, Part II, CCIS 728, pp. 302–310, 2017.
DOI: 10.1007/978-981-10-6388-6_25

Traditional methods for determining the antigenicity of influenza viruses, mainly the hemagglutination inhibition (HI) assay and the micro-neutralization assay, are reported to be lack of consistency [3]. Given the rapid development of DNA sequencing technology, determination of HAs of influenza viruses in influenza surveillance has become routine work. The sequence-based computational methods for predicting antigenic variation of influenza virus have been developed and demonstrated to be helpful in influenza prevention [4–7]. Most of them used the models with simple structures, such as the linear/logistic regression or the naive bayes models. For example, Du et al. used a naive bayes model with 12 features to predict the antigenic variant of human influenza A(H3N2) viruses [6]. To further improve the accuracy of predicting antigenic variants, this work developed a novel method based on Stacked Auto-Encoder (SAE), a kind of deep learning model. Deep learning is the most up-to-date method in the field of machine learning [8]. It was reported to be able to capture the high-level abstractions from data, which is especially suitable for the biological data with large complexity and clear hierarchy. Actually, the deep learning methods have been used extensively in bioinformatics, such as the residue contact prediction [9], gene expression prediction [10], alternative splicing of genes [11], and so on. They were demonstrated to perform better than other machine learning methods did. However, few of them attempted to infer the biological mechanisms through the analysis of the deep architecture model. Here, we not only built a SAE-based method for predicting the antigenic variant of human influenza A(H3N2) virus, which achieved a predictive accuracy of 0.95, but also obtained interesting insights of antigenic variation through analysis of the model.

2   Materials and Methods

2.1   Data

The antigenic data, which included 31878 pairs of viruses with antigenic distances, were derived from Smith's work [12]. The antigenic distances were discretized as one (antigenic different) or zero (antigenic similar) with a cutoff of four according to Lee and Chen's work [4]. The HA1 protein (the immunodominant part of HA) sequences for the viruses in the dataset were downloaded from Influenza Virus Resources [13]. Each residue was considered as a feature. Those residues with complete conservation were removed. 154 residues were kept. Then, the differences between the HA1 protein sequences of a pair of viruses were converted to binary data based on amino acid identity. Therefore, for each pair of viruses, a vector of binary data with 154 dimensions was derived.

As is shown in Du's work [6], the pairs of viruses with over nine amino acid differences on HA1 protein has a probability of 0.99 to be antigenic different. Therefore, it is unnecessary to make prediction for these pairs of viruses. Besides, for the pairs of viruses with the same data vector but different labels, only those with the major label were kept. Finally, a total of 8097 pairs of viruses were used in this study, including 2723 and 5374 pairs of antigenically similar and distinct virus, respectively.

2.2   SAE Model

To predict the probability of antigenic variant for a pair of viruses with 154 dimensions, a Stacked Auto-Encoder (SAE) model was built (Fig. 1) with the sigmoid function as the activation function. The 154 nodes of the input layer refer to the 154 dimension of the input data. The number of hidden layers and the number of nodes in each hidden layer were selected with cross-validations as described in the Result sections. Here, only one hidden layer with 100 nodes was used in the SAE model. Since there are only two classes, either antigenic similar or distinct, only one node in the output layer was used. When the output for the node is greater than or equal to 0.5, it is considered as antigenic distinct, otherwise antigenic similar. The codes for SAE modeling were adapted from Rasmus's work [14]. Five-fold cross validations were used to test the model.

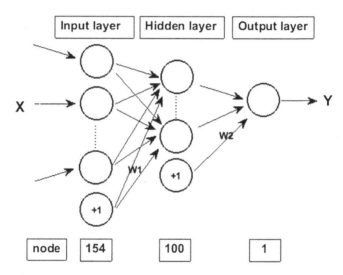

Fig. 1. Schema for the SAE model

2.3   Logistic Regression, Support Vector Machine and Decision Tree Models

Similar to those used in the SAE model, all these models took the data of 150-dimensions as input and output the labels of antigenic similar or distinct. The logistic regression model was achieved with the function of glm() in R; The SVM model was achieved with the functions of svmtrain() and svmclassify(); the Decision Tree model was achieved with the functions of classregtree() and treeval() in Matlab.

3   Results

3.1   Determination of Hyper-Parameters for SAE

The choice of hyper-parameters has an important influence on SAE model. There are six hyper-parameters for the SAE model in this study. The hyper-parameter of learning rate was set to be 0.3 and declined as the training process; the momentum was set to be 0.5, as it has little influence in the modeling (data not shown); the sparse penalty was set as 0.05 according to Yoshua's work [8]. For the remaining three hyper-parameters, the weight-decay, the number of nodes in the hidden layer and the number of hidden layer, they were determined using the cross-validations, as were shown in Fig. 2. The hyper-parameter of weight-decay and number of nodes in the hidden layer were set to be 0.00005 and 100 respectively (Fig. 2A and B). Surprisingly, the model with one hidden layer was observed to perform best (Fig. 2C), suggesting that a shallow neural network would be the best choice for predicting the antigenic variant of influenza viruses in our study.

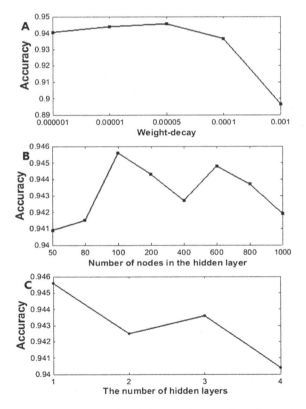

Fig. 2. Selection of hyper-parameters of weight-decay (A), number of nodes in the hidden layer (B) and the number of hidden layers (C). When selecting each of these hyper-parameters, the other two hyper-parameters were fixed to their "best value" (shown in the figure) according to our experiences.

3.2   Performances of SAE in Prediction of Antigenic Variants

After the selection of hyper-parameters for the SAE model, we tested its performance with five-fold cross-validations (Table 1). The accuracy, sensitivity and specificity were 0.95, 0.95 and 0.93, respectively. For comparison, we also built and tested the logistic regression model, Support Vector Machine (SVM) and Decision Tree model for predicting the antigenic variants of influenza A(H3N2) viruses. As is shown in Table 1, their performances were inferior to the SAE model in the cross-validations.

Table 1. Performances of SAE and logistic regression, SVM and Decision Tree models in cross-validations.

Model	Accuracy	Sensitivity	Specificity
SAE	0.95	0.95	0.93
Logistic regression	0.93	0.94	0.91
SVM	0.92	0.92	0.92
Decision tree	0.93	0.95	0.89

3.3   SAE Model Reflected the Mechanisms of Antigenic Variation

Since the SAE model was observed to perform well for predicting the antigenic variants of influenza viruses, we next attempted to analyze the model in details. Firstly, the weight matrix between the input layer and hidden layer (W1 with dimension of 154 * 100, see Fig. 3A), and those between the hidden layer and output layer (W2 with dimension of 100 * 1, see Fig. 3B) were visualized and analyzed. Gray cells refer to weights close to zero. As is shown in Fig. 3, in both W1 and W2, most weights were close to zero.

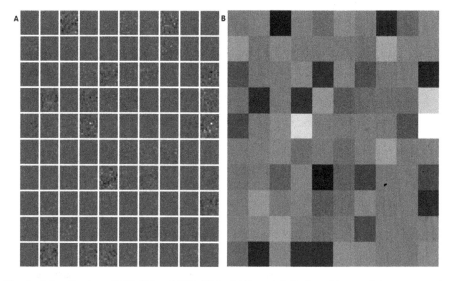

Fig. 3. Visualization of W1 (A) and W2 (B) in SAE model. Each cell in panel A represent the weights of input layer to a node in hidden layer, with dimension of 14 * 11.

Then, the weights from hidden layer to output layer were sorted decreasingly according to their absolute values. 20 nodes with top weights were selected. They contributed most to the antigenic variation in the SAE model. For each of these nodes, the weights from the input features (residues on the HA protein) to them were normalized. The input features with normalized weight greater than 0.5 were supposed to contribute large to the node selected. They were listed in Table 2. A total of 38 residues were observed to contribute large to antigenic variation, while 90% of them was located on the five canonical antigenic epitopes, especially for the epitopes A and B which were reported to contribute most to antigenic variation of influenza viruses.

Table 2. Residues contributing large to the top nodes in hidden layer. They were sorted based on the absolute of weights in W2. "A-E" refers to five canonical antigenic epitopes A to E, while "O" refers to other residues on HA1 protein.

Node	A	B	C	D	E	O
21	138	156, 189, 186	–	247, 226, 214	94	269
95	145	–	–	–	–	–
47	–	156, 189	–	227, 229, 214, 201, 208	–	231
93	135, 124	158, 189	–	172	–	289
20	138	156, 189	–	247, 214	–	–
71	145	189, 156	278	–	–	–
35	145	–	–	247	–	–
98	145, 138	189, 157	–	226	–	–
50	133, 145, 135	189, 155, 158	276	172, 214, 167	–	289
14	135, 122	156, 189, 159, 193	278	226	262	–
40	135, 138	189, 156	275	–	62	231
43	–	156, 189, 196, 157, 190, 159	275	247, 226	–	–
94	138	196, 156, 158	54	213	62, 80	3
34	145	156	–	–	–	–
67	145	193	–	–	262	–
72	145	193	–	247	–	–
97	133, 124	156, 189, 158, 193	–	214	62	–
5	145, 133	189	276	–	–	–
69	145, 133	156	–	–	–	–
76	–	189, 155, 159	–	–	94	–

For the 20 nodes in the hidden layer which contributed most to the antigenic variation, nearly all of them were composed of multiple residues which were located in multiple antigenic epitopes, suggesting the complex interactions between residues in antigenic variation. For example, the 14th nodes in the hidden layer were composed of 9 residues (Table 2) which were located in five epitopes (Fig. 4), including two in epitope A, four in epitope B, one in epitope C to E each.

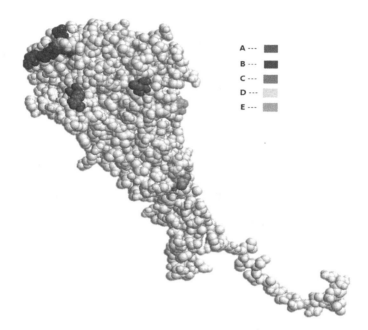

Fig. 4. Residues distribution on HA1 protein for the 14th node in the hidden layer.

Some residues were observed to play a decisive role in the hidden layer nodes, such as residue 145. We further summed the number of nodes contributed for each residue (Fig. 5). Surprisingly, the residues 189, 145 and 156 all took part in 10 nodes. In Koel's studies [15], they identified seven residues (145, 155, 156, 158, 159, 189, 193) on HA protein which mostly determine the antigenic variation of influenza A(H3N2) viruses, five of them were included in the top ten residues contributed most to antigenic variation (Fig. 5).

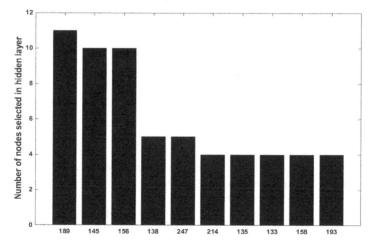

Fig. 5. Top ten residues which contribute large to the nodes selected in the hidden layer.

4   Discussion

This work built a SAE model for predicting antigenic variant of influenza A(H3N2) virus. It achieved a predictive accuracy of 0.95, which was better than the logistic regression model. Based on the model, we found that multiple residues cooperated to influence the antigenic variation. Some residues were observed to contribute much larger to antigenic variation than other residues did, such as 189, 145 and 156, which were also reported to play important roles in antigenic variation in Koel's studies [15].

Although we attempted to build the deep learning method, surprisingly, the model with a shallow architecture (only one hidden layer) was found to be more suitable for the data. This may reflect the direct influence of residue interactions on antigenic variation. Based on the shallow model, we obtained some interesting insights for helping us understand the antigenic variation of influenza viruses. The nodes in the hidden layer could be viewed as the first abstraction of information from input data. Most nodes included the contributions of multiple residues, which reflected the cooperation of residues in antigenic variation. Although only one hidden layer was used in the model, they captured the important features in antigenic variation. For example, most residues involved in the top ranked nodes in hidden layer were located in five canonical antigenic epitopes, and five of seven most important residues for antigenic variation were also captured by the model.

Overall, this work is not only useful for rapidly identifying antigenic variants in influenza prevention, but also is an interesting attempt in inferring the biological mechanisms through the analysis of deep learning model, which may help for interpreting the deep learning models.

Acknowledgement. This study was supported by the National Natural Science Foundation (31500126 and 31371338), the National Key Plan for Scientific Research and Development of China (2016YFD0500300 and 2016YFC1200200), and the Chinese Academy of Medical Sciences (2016-I2 M-1-005).

There are no conflicts of interest.

References

1. Thompson, W.W., Shay, D.K., Weintraub, E., Brammer, L., Cox, N., Anderson, L.J., et al.: Mortality associated with influenza and respiratory syncytial virus in the United States. JAMA **289**, 179–186 (2003)
2. Taubenberger, J.K., Kash, J.C.: Influenza virus evolution, host adaptation, and pandemic formation. Cell Host Microbe **7**, 440–451 (2010)
3. Stephenson, I., Heath, A., Major, D., Newman, R.W., Hoschler, K., Junzi, W., et al.: Reproducibility of serologic assays for influenza virus A(H5N1). Emerg. Infect. Dis. **15**, 1252–1259 (2009)
4. Lee, M.S., Chen, J.S.E.: Predicting antigenic variants of influenza A/H3N2 viruses. Emerg. Infect. Dis. **10**, 1385–1390 (2004)
5. Wu, A., Peng, Y., Du, X., Shu, Y., Jiang, T.: Correlation of influenza virus excess mortality with antigenic variation: application to rapid estimation of influenza mortality burden. PLoS Comput. Biol. **6**(8), e1000882 (2010)

6. Du, X., Dong, L., Lan, Y., Peng, Y., Wu, A., Zhang, Y., et al.: Mapping of H3N2 influenza antigenic evolution in China reveals a strategy for vaccine strain recommendation. Nat. Commun. **3**, 709 (2012)

7. Li, H., Peng, Y., Zou, Y., Huang, Z., Wu, A., Li, K., et al.: PREDAC-H5: a user-friendly tool for the automated surveillance of antigenic variants for the HPAI H5N1 virus. Infect. Genet. Evol. **28**, 62–63 (2014)

8. Bengio, Y.: Learning deep architectures for AI. Found. Trends® Mach. Learn. **2**(1), 1–127 (2009)

9. Eickholt, J., Cheng, J.: Predicting protein residue-residue contacts using deep networks and boosting. Bioinformatics **28**, 3066–3072 (2012)

10. Chen, Y., Li, Y., Narayan, R., Subramanian, A., Xie, X.: Gene expression inference with deep learning. Bioinformatics **32**, 1832–1839 (2016)

11. Leung, M.K., Xiong, H.Y., Lee, L.J., Frey, B.J.: Deep learning of the tissue-regulated splicing code. Bioinformatics **30**, i121–i129 (2014)

12. Smith, D.J., Lapedes, A.S., de Jong, J.C., Bestebroer, T.M., Rimmelzwaan, G.F., Osterhaus, A.D., et al.: Mapping the antigenic and genetic evolution of influenza virus. Science **305**, 371–376 (2004)

13. Bao, Y.M., Bolotov, P., Dernovoy, D., Kiryutin, B., Zaslavsky, L., Tatusova, T., et al.: The influenza virus resource at the national center for biotechnology information. J. Virol. **82**, 596–601 (2008)

14. Palm, R.B.: Prediction as a candidate for learning deep hierarchical models of data, Master, Informatics and Mathematical Modelling. Technical University of Denmark (2012)

15. Koel, B.F., Burke, D.F., Bestebroer, T.M., van der Vliet, S., Zondag, G.C., Vervaet, G., et al.: Substitutions near the receptor binding site determine major antigenic change during influenza virus evolution. Science **342**, 976–979 (2013)

A Novel Statistical Power Model for Integrated GPU with Optimization

Qiong Wang$^{(\boxtimes)}$, Ning Li, Li Shen, and Zhiying Wang

College of Computer, National University of Defense Technology,
Changsha, China
wangqiong@nudt.edu.cn

Abstract. GPUs are of increasing interests in the multi-core era due to their high computing power. However, the power consumption caused by the rising performance of GPUs has been a general concern. As a consequence, it is becoming an imperative demand to optimize the GPU power consumption, among which the power consumption estimation is one of the important and useful solutions. In this work, we present a novel statistical model that is capable of dynamically estimating the power consumption of the AMD's integrated GPU (iGPU). Precisely, we adopt the linear regression for power consumption modeling and propose a mechanism called kernel extension to lengthen the kernel execution time so that we can sample system data for model evaluation. The results show that the median absolute error of our model is less than 3%. Furthermore, to reduce the latency of power consumption estimation, we conduct a study to explore the possibility to simplify our statistical model. The results suggest that the accuracy and stability is still acceptable in the simplified model. This provides a desirable option to reduce our model latency when it is applied to the iGPU power consumption optimization in the real world.

Keywords: Regression model · Integrated GPU architecture · Performance · Power estimation

1 Introduction

In recent years, graphics processing units (GPUs) are obtaining substantial attentions due to its efficient parallel computing abilities. Particularly, they have been widely use to accelerate the solving of large scale computation problems because of the development of parallel programming languages, e.g., CUDA [1] and OpenCL [2]. However, the rising performance of GPUs is usually at the cost of increasingly higher power consumption, especially when GPUs are integrated with massive transistors (e.g., NVidia GeForce GTX 280 contains 1.4 billion transistors). For example, an Nvidia GTX 280 could consume as high as 236 watts power [3], while it is usually less than 150 W [4] for a typical multi-core CPU. As a result, more complex cooling solutions are needed to reduce the

© Springer Nature Singapore Pte Ltd. 2017
B. Zou et al. (Eds.): ICPCSEE 2017, Part II, CCIS 728, pp. 311–324, 2017.
DOI: 10.1007/978-981-10-6388-6_26

system temperature. This could compensate the benefits gained from the performance improvement. Therefore, it is highly necessary to reduce the GPU power consumption, among which analyzing and modeling the power consumption of GPUs is clearly a priority.

In this work, we aim at the Accelerated Processing Unit (APU) [5], which is a revolutionary processor released by AMD in 2011. APU combines CPU and integrated GPU (iGPU) into one single chip on which CPU and GPU can share on-chip resources. In order to improve the efficiency of APU and figure out bottlenecks of kernels, determining the iGPU power plays a significant role. Consequently, there have been several statistic models, such as statistic models [7,8], work flow graph [9], static code analysis [10] and machine learning technique [11]. Diop et al. [12] introduced a power model for heterogeneous processors. Zhang et al. [13] and Karami et al. [14] also proposed models that are primarily built for discrete GPUs rather than iGPUs in APUs. Therefore, one can note that most models in the literature mainly focus on either the whole APU or the discrete APU, while the iGPU in the APU has received relatively little attentions. Since iGPUs and discrete GPUs differ from each other on some important architectural characters, we believe that studying iGPUs will provide us new insights.

Our Contributions. In order to precisely study the power consumption of iGPU, we employ a rigorous statistical model to facilitate our analysis. We aim at conducting a comprehensive investigation on the iGPU performance and its power consumption, and more importantly, providing a accurate predication for reducing the power consumption of iGPUs. In general, the main contributions of this work could be summarized as follows:

– We found that the widely-adopted profiling tool, i.e., AMD CodeXL [6] cannot be directly used to collect the energy consumption of OpenCL kernels. Precisely, the power sampling period of CodeXL is at lease 100 ms while most OpenCL kernel execution time is less than 10 ms. By lengthening kernel extension time, we introduce a novel solution to sample data in one kernel execution. We insist that this mechanism is generally applicable for collecting data from OpenCL-based benchmarks.
– Relying on the multivariable linear regression model, we successfully build a statistical power model for the iGPU. We then adopt the proposed kernel extension for data collecting to evaluate our built model. The results show that our model is of higher accuracy than most existing models. Moreover, unlike previous models that collect data from GPU simulators, our experiment results are for real iGPUs and hence more meaningful in practice.
– Low latency is a desirable property for a power model especially when it is used to estimate power consumption. For this issue, we explore the possibility of simplifying our built model. Specifically, we generate a sensitivity analysis called single-simplification model to evaluate the importance of each performance counter. Based on the above analysis, we simplify our statistic model and reduce the number of performance counters without significantly reducing the model accuracy.

Organization. The rest of this paper is organized as follows. We give a brief background on APU architecture and OpenCL execution model in Sect. 2. In Sect. 3, we build the power model for iGPU and evaluate its accuracy. Section 4 provides a detailed analysis regarding the importance of each performance counter and the model simplification. Finally, we draw the conclusion in Sect. 5.

2 Preliminaries

2.1 APU and iGPU Architecture

We choose the AMD A10-7850 as our target APU, which is an important product of AMD's Kaveri APUs and has four Steamroller CPU cores and 8 GCN R7 graphics. Here Graphics Core Next (GCN) is the codename for a series of AMD graphics processor micro-architecture. It is an RISC SIMD microarchitecture which gains higher execution efficiency. Generally speaking, A10-7850k is equipped with 8 GCN computing units and 512 shade processors. Each GCN computing unit consists of a branch & message unit, a CU scheduler, 4 SIMD vector units (each 16-lane wide), 4 texture filter units, 16 texture fetch load/store units, 4 64 KB VGPR files, 1 scalar unit, a 4 KB GPR file, a 64 KB local data share and a 16 KB L1 Cache. Four compute units are wired to share a 16 KB instruction cache and a 32 KB scalar data cache. It also has 8 asynchronous compute engines for independent scheduling and work item dispatching.

2.2 OpenCL Execution Model

In this work, we use OpenCL as the programming tool. OpenCL [2] is an open programming framework for heterogeneous devices and platforms, such as CPU, GPU, DSP, FPGA and other accelerators. It provides a programming language for running kernels and application programming interfaces (APIs). These APIs are used to control the platform and run different kernels on OpenCL devices. OpenCL can provide task-based and data-based parallel computing capabilities to meet the needs of different applications. An OpenCL program usually consists of the host program and kernels. The host program defines the context and controls the kernel execution. A kernel is defined as a function executed on an OpenCL device. A running kernel is called a work item. Several work items on the same computing unit are organized into a work group. Each work item is tagged with a local ID and each work group is tagged with a global ID. With the local ID and global ID, we can exclusively identify each work item in the global space. When a kernel is executing on an OpenCL device, the host program creates a context which includes a device list, kernels, program objects and memory objects. The context defines the operating environment of kernels. Then several work groups are launched to form many wavefronts. These wavefronts are submitted to the command queues and executed on the GPU, in-order or out-of-order.

3 Statistical Power Modeling

3.1 Model Description

In this work we rely on the linear regression to build the power prediction model for iGPU. Noting that the selection of regression factors (i.e., explanatory variables in the regression model) would significantly affect the accuracy of the built prediction model, we carefully analyze the architecture of iGPU to figure out those on-chip resources that are closely related to its power consumption. Particularly, these on-chip resources mainly consist of **Arithmetic Logic Unit (ALU), Local Memory** and **Global Memory**. The main reason is that arithmetic operations in iGPU incur frequent data access with the memory and hence usually consume much power. That is, hardware events occurring in these operations account for the majority of power consumption. In our work, we utilize *performance counters* to record these events. Performance counters are generally used to count the number of hardware events and finish the application profiling.

Table 1. Performance counters description

Counters	Descriptions
VALUInsts	Number of vector ALU instructions
SALUInsts	Number of scalar ALU instructions
VFetchInsts	Number of vector fetch instructions
SFetchInsts	Number of scalar fetch instructions
FlatVMemInsts	Number of flat instructions
VALUUtilization	Percentage of active vector ALU threads in a wave
VALUBusy	Percentage of time vector ALU instruction are processed
SALUBusy	Percentage of time scalar ALU instruction are processed
LDSInsts	Number of LDS read or LDS write instructions
LDSBankConflict	Percentage of time LDS is stalled by bank conflicts
CacheHit	L2 cache hit rate
MemUnitBusy	Percentage of time memory unit is active
MemUnitStalled	Percentage of time memory unit is stalled
WriteUnitStalled	Percentage of time write unit is stalled
FetchSize	Total kilobytes fetched from the memory
WriteSize	Total kilobytes written to the memory

We choose the AMD CodeXL v1.9 as the profiler to sample the performance counter information during the program execution. We found that there are 15 types of performance counters collected by the CodeXL. Details are provided in Table 1. One can note that all the hardware events recorded by these performance counters are indeed closely related to the arithmetic operations and data access in iGPU. Precisely, **ALU** operations include VALUInsts, SALUInsts, VFetchInsts, SFetchInsts, VALUUtilization, VALUBusy and SALUBusy. **Local Memory** operations include LDSInsts and LDSBankConflict while **Global Memory** operations include FetchSize, WriteSize, CacheHit, MemUnitBusy, MemUnitStalled and WriteUnitStalled.

Note that the power consumption of an iGPU consists of two parts. The first part (also the main one) is the dynamic power consumption while the other part is the static one. More precisely, the dynamic part occurs due to the computation task, whose volume is accurately reflected by the number of hardware events captured by the performance counters listed in Table 1. That is, the number of activated hardware events would be significant when many computation tasks occur, which of course results in high power consumption. Regrading the static one, it is also known as basic power consumption as it is mainly caused by the bare operating system without any workload running. It is normally related to the processor architecture and also the outside environment such as the temperature.

Based on the aforementioned analysis, we formally set the model equation as follow.

$$P_{\mathsf{iGPU}} = \sum_{i=1}^{i=15} A_i \cdot \mathsf{E_i} + C \tag{1}$$

In the Eq. (1), P_{iGPU} denotes the total power consumption of the targeted iGPU. $\mathsf{E_i}$ represents the i-th performance counter listed in Table 1 and A_i is its corresponding coefficient. C is a constant value that denotes the static power consumption.

3.2   Model Evaluation

Experiment Setup. We choose A10-7850K as our experimental platform which contains 4 Steamroller CPUs and 8 Radeon R7 series GPUs. All the experiments are performed on an Ubuntu 12.04 LTS machine running the AMD Catalyst driver version 13.30. AMD APP SDK v2.9 is installed as the OpenCL implementation. We choose Rodinia 3.2 Benchmark Suite for the model evaluation.

As mentioned above, we choose the AMD CodeXL v1.9 as the profiler to sample the performance counter information and also the power during the program execution. Unfortunately, the smallest sampling period is 100 ms for CodeXL while most of the kernel execution time in Rodinia 3.2 are less than 10 ms. That is, we cannot get a power sample within one kernel execution because of time issue.

Kernel Extension. In order to overcome the above problem, we propose an approach called **kernel extension** to lengthen their execution time so that the power could be sampled within one kernel execution. Figure 1 illustrates the necessity of kernel extension.

Particularly, we rewrite kernels and make the original functions repeat for thousands of times. The rewritten benchmarks run longer time producing enough

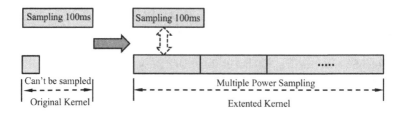

Fig. 1. Illustration of Kernel extension

hardware events and profiling data which is an necessary input to the power model. In order to reach higher accuracy, we should collect at least 10 samples from CodeXL within one kernel execution, which means that the kernel execution time should be longer than 1 second. By kernel extension, we can fill in the gap between kernel execution time and the sampling period of CodeXL.

The extended kernel execution time for each application in Rodinia is depicted in Table 2. It is worth mentioning that simply repeating functions may occurs memory error. One should adaptively adjust the repeating procedure according to the reported issue.

Table 2. Extended Kernel execution in Rodinia

Applications	Kernel	iT(ms)	eT(ms)
Backprop	*bpnn_layerforward_ocl_k1*	2.01	1699.45
	bpnn_adjust_weights_ocl_k2	3.10	1988.48
BFS	*BFS_1_k1*	0.10	2188.54
	BFS_2_k2	0.09	1674.26
B+tree	*findRangeK_k1*	2.15	1819.42
	findK_k2	8.19	2114.41
Gaussian	*Fan1_k1*	0.01	2049.41
	Fan2_k2	0.20	1429.43
Heartwall	*kernel_gpu_opencl_k1*	131.71	1530.07
Hotspot	*hotspot_k1*	0.38	3526.27
Kmeans	*kmeans_swap_k1*	88.92	2213.84
	kmeans_kernel_c_k2	31.12	2182.16
Leukocyte	*GICOV_kernel_k1*	14.29	1440.90
	dilate_kernel_k2	7.55	1641.12
	IMGVF_kernel_k3	42.17	2432.67
LavaMD	*kernel_gpu_opencl_k1*	155.93	1556.52
NN	*NearestNeighbor_k1*	0.10	1553.09
PF-naive	*particle_kernel_k1*	5.53	1007.59

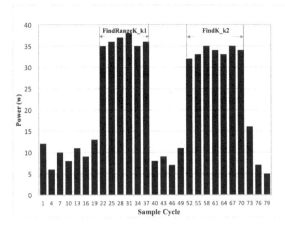

Fig. 2. Power for kernels in B+Tree

Power Measurement. It is worth noting that the power sampled by the CodeXL is the global consumption and not necessarily always related to the kernel execution. For example, each time before a program start to execute, there always exists a setup phase which usually does not involve the kernel. If we straightforwardly use the sample power from the CodeXL for model evaluation, it may occur errors and thus reduce the model accuracy. Therefore, we need to pick the kernel-related power from the recorded data to achieve high model accuracy.

To provide a clear picture, we choose the B+tree to illustrate this point. As depicted by Table 2, there are two kernels for B+tree: *findRangeK_k1* and *findK_k2*. We show the details in Fig. 2. One can note that the first 19 sample cycles is mainly for execution environment setup, and hence the power consumption is quite small, e.g., 5–15 W. After that, the kernel of *findRangeK_k1* begin to execute and meanwhile the power rapidly increases to around 35 W correspondingly. When it comes to the 38 cycle, the power recorded decreases sharply, which means the end of *findRangeK_k1* execution. Therefore, the power sampled from 19 cycle to 38 cycle are particularly the total power consumption for *findRangeK_k1*. As shown in Table 2, the running time of *findRangeK_k1* is about 1800 ms, and hence is consistent with the sample period. One can also note that the follow up power recorded appears similarly and is mainly for the *findK_k2* execution.

Model Evaluation. We choose the Statistical Product and Service Solutions (SPSS) to evaluate our regression model.

The statistical result of our model is shown in Table 3 and the regression coefficient of each performance counter is listed in Table 4. Precisely, the result achieves an adjusted R-square 95.9% which means that the linear relationship between the independent and dependent variables is pretty convincing and

Table 3. Statistical result

Multiple	0.997615
R square	0.995235
Adjusted R square	0.959498
Standard error	4.898391

reliable. Moreover, the median absolute error is 2.12%, indicating a high predicting precision. We also test the predication error of each kernel in the Rodinia benchmark. The result is given in Table 5. Overall speaking, our model is of high accuracy as each kernel error is small. e.g., only 0.25% for *kernel_gpu_opencl_k1*.

Table 4. Regression coefficients of the model

PC	Coefficient	PC	Coefficient
VALUInsts	−1.5E-05	WriteSize	1.54E-05
SALUInsts	3.03E-05	CacheHit	−0.76
VFetchInsts	0.000187	MemUnitBusy	0.75
SFetchInsts	−1.9E-06	MemUnitStalled	−1.30
VALUUtilization	−1.03153	WriteUnitStalled	226.03
VALUBusy	1.500774	LDSInsts	0.00022
SALUBusy	−1.43736	LDSBankConflict	37.90
FetchSize	2.55E-06		

Table 5. Predication errors for all Kernels

Kernel	Error	Kernel	Error
bpnn_layerforward_ocl_k1	2.63%	*hotspot_k1*	0.48%
bpnn_adjust_weights_ocl_k2	0.64%	*kmeans_swap_k1*	0.31%
BFS_1_k1	2.36%	*kmeans_kernel_c_k2*	0.62%
BFS_2_k2	3.60%	*GICOV_kernel_k1*	0.80%
findRangeK_k1	4.88%	*dilate_kernel_k2*	3.91%
findK_k2	1.06%	*IMGVF_kernel_k3*	4.27%
Fan1_k1	1.52%	*kernel_gpu_opencl_k1*	0.25%
Fan2_k2	3.86%	*NearestNeighbor_k1*	2.21%
kernel_gpu_opencl_k1	0.36%	*particle_kernel_k1*	6.02%

4   Model Simplification

4.1   Motivations for Model Simplification

Our goal here is to reduce the latency of the power model. Particularly, we ask whether it is possible to reduce the number of performance counters involved in the power model while the model accuracy remain high. In this work, we propose the **Single-Simplification** and **Multi-Simplification** power model to tackle with the aforementioned dilemmas. The single-simplification power model mainly explores the relationship between the type of performance counters and the standard error of power models. We then build multi-simplification power models to explore the connection among the number of performance counters and the stability and prediction accuracy of power models.

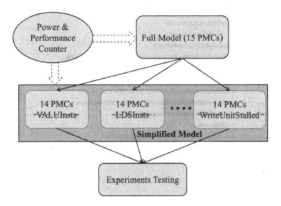

Fig. 3. Single-simplification model building

4.2   Single-Simplification Model

As depicted by Fig. 3, for each testing, one performance counter is removed from the full model which is of 15 performance counters. We then run the benchmark and sample the power and performance counter to reconstruct the regression model. Formally, for j-th ($j \in [1, 15]$) simplification, the model equation is as follows.

$$P_{\text{iGPU}} = \sum_{i=1}^{j-1} A_i \cdot \mathsf{E_i} + \sum_{i=j+1}^{15} A_i \cdot \mathsf{E_i} + C$$

The goal of each simplified model is to evaluate the importance of the removed performance counter, based on which we can simplify the model by removing those performance counters that are of less importance and hence the simplified one can still achieve high accuracy.

Results and Analysis. The accuracies, shown in Figs. 4 and 5, are compared to our full model with 15 performance counters. The accuracy variation is a direct reflection of the importance of performance counters. If the accuracy significantly descends, the missing performance counter is regarded as an important component. Precisely, we have the following conclusions.

– Roughly speaking, most of the general operation performance counters are the major arguments as all the simplified models have higher errors and accuracy variations. Particularly, VALUBusy is the most crucial factor among all the general performance counters. It is worth noting that the SFetchInsts is an exceptional, without which the median absolute error of model is 2.21% and quite close to the original one.

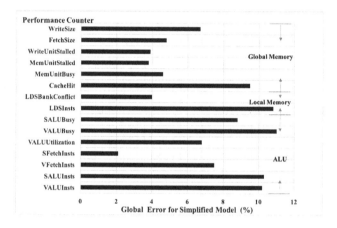

Fig. 4. Global error for performance counter

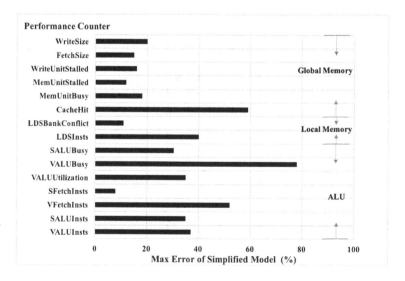

Fig. 5. Maximum error for performance counter

- We note that the importance of local memory performance counters is irregular. Similar to VALUInsts, LDSInsts is quite a decisive component. LDSInsts is a reflection of LDS instructions executed on iGPU. A larger LDSInsts means that iGPU has to process more LDS instructions and hence consume more energy. Compared to LDSInsts, LDSBankConflict is less important. Among all the profiling files, LDSBankConflict is either equal to zero or near to zero, indicating that this hardware event is quite rare and statistically put little influence on power consumption.
- Lats, most global memory performance counters are less important than general performance counters. As shown in Fig. 6, models without global memory performance counters are of less error than those without ALU performance counters. Particularly, models without MemUnitBusy, MemUnitStalled, WriteUnitStalled and WriteSize are precise, indicating that these performance counters can affect model accuracy less. CacheHit is a special case and its role is similar to a general performance counter. This is reasonable because higher cache hit rate means fewer visits to the global memory and leads to lower energy dissipation in consequence.

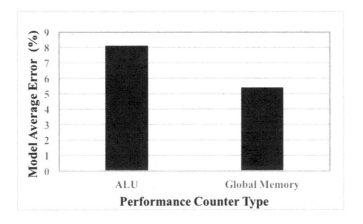

Fig. 6. Average error of performance counter

4.3 Multi-simplification Model

Based on the above performance counter importance analysis, we know that some performance counters have little impact on the model accuracy. For example, the accuracy of model without SFetchInsts is similar to that of the full model, indicating the possibility to simplify our model by removing SFetchInsts. Therefore, we study the relationship between the number of performance counters and model accuracy. Specifically, as shown in Fig. 7, we start with the full model and gradually remove the performance counters that are of less importance. We stop the simplification until there are only 6 performance counters left in the list, indicating 9 simplified models are built for further analysis. These simplified models are equipped with 14, 12, ..., 6 performance counters respectively.

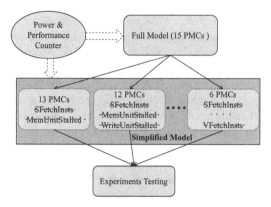

Fig. 7. Multi-simplification model building

Results and Analysis. Since now we remove more than one performance counter from the power model, its stability would be possibly weakened significantly. Normally, a power model with weak stability is not suitable for DVFS even if it is still of high accuracy. Therefore, apart from the median absolute error, we choose the root-mean-square deviation (RMSE) to evaluate the model accuracy and stability. RMSE can reflect the sample standard deviation of the differences between predicted values and observed values.

Figures 8 and 9 show the experimental results. If there are 11 or fewer performance counters, the median absolute errors are above 10% and these simplified models produce much misprediction. Furthermore, higher RMSE suggests the poor stability of these simplified models. When the number of performance counters is 12 or more, the median absolute errors are 4.75%, 3.86% and 2.21% respectively and RMSEs are below 0.02, which means that the three simplified models are still relatively precise and robust. The three reduced performance counters are SFetchInsts, WriteUnitStalled and WriteSize.

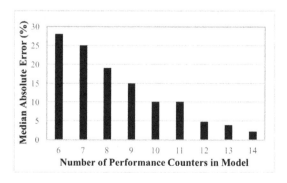

Fig. 8. Median absolute error of model with varied number of performance counters

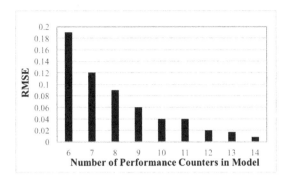

Fig. 9. Root mean square error of model with varied number of performance counters

5   Conclusion

In this paper, we proposed a statistical model for analysing the power consumption of OpenCL kernels on iGPU in APU. The median absolute error of our model is 2.12%. We then also evaluated the tightness of the relationship between power consumption and performance counters. Thereafter, we explored the possibility of simplifying the statistical model and concluded that the model accuracy is still acceptable if we only take 12 performance counters into consideration.

References

1. Nvidia Corparation, What is CUDA (2016). http://www.nvidia.com/object/what_is_cuda_new.html
2. Stone, J.E., Gohara, N., Shi, G.: OpenCL: a parallel programming standard for heterogeneous computing systems. Comput. Sci. Eng. **12**(3), 66–72 (2010)
3. Nvidia Corparation. Geforce GTX 280 (2016). http://www.nvidia.com/object/product_geforce_gtx_280_us.html
4. Intel Corparation. Intel Core i7–920 Processor (2016). http://ark.intel.com/product.aspx?id=37147
5. Branover, A., Foley, D., Steinman, M.: AMD fusion APU: Llano. IEEE Micro. **32**(2), 28–37 (2012)
6. AMD, Amd CodeXL (2016). http://developer.amd.com/tools-and-sdks/opencl-zone/codexl/
7. Zhang, Y., Owens, J.D.: A quantitative performance analysis model for GPU architectures. In: Proceedings of International Symposium on High-Performance Computer Architecture, pp. 382–393 (2011)
8. Luo, C., Suda, J.: A performance and energy consumption analytical model for GPU. In: 2011 IEEE Ninth International Conference on Dependable, Autonomic and Secure Computing, pp. 658–665 (2011)
9. Baghsorkhi, S.S., Delahaye, M., Patel, S.J., Gropp, W.D., Hwu, W.W.: An adaptive performance modeling tool for GPU architectures. ACM SIGPLAN Not. **45**(5), 105 (2010)

10. Hong, S., Kim, H.: An analytical model for a GPU architecture with memory-level and thread-level parallelism awareness. ACM SIGGARCH Comput. Archit. News **37**(3), 152 (2009)
11. Wu, G., Greathouse, J.L., Lyashevsky, A., Jaysasena, N., Chiou, D.: GPGPU performance and power estimation using machine learning. In: 2015 IEEE 21st International Symposium on High Performance Computer Architecture (HPCA), pp. 564–576 (2015)
12. Diop, T., Jerger, N.E., Anderson, J.: Power modeling for heterogeneous processors. In: Proceedings of Workshop on General Purpose Processing Using GPUs, pp. 90–98 (2014)
13. Zhang, Y., Hu, Y., Li, B., Peng, L.: Performance and power analysis of ATI GPU: a statistical approach. In: Proceedings of 6th IEEE International Conference on Networking, Architecture, and Storage, NAS 2011, pp. 149–158 (2011)
14. Karami, A., Khunjush, F., Anderson, S.A.: A statistic performance analyzer framework for OpenCL kernels on Nvidia GPUs. J. Supercomput. **71**(8), 2900–2921 (2014)

Application of OFDM-CDMA in Multi-user Underwater Acoustic Communication Based on Time Reversal Mirror

Yonggang Wang[✉], Jingwei Yin, Zhengrong Pan, and Pengyu Du

Harbin Engineering University, Harbin 150001, China
wongyonggang@163.com

Abstract. In the paper, a novel method, which exploits both the Orthogonal Frequency Division Multiplexing (OFDM)-Code Division Multiple Access (CDMA) technique and the time reversal mirror technique, is investigated to realize multi-user underwater acoustic communication. By taking advantages of these two techniques, the proposed method has the capability of effectively suppressing multi-path interference, reducing the error rate, enhancing the processing gain and improving user capacity. Simulations verify the correctness and effectiveness of the proposed method.

Keywords: OFDM-CDMA · Time reversal mirror · Multi-users · Underwater acoustic communication

1 Introduction

Multi-user communication in the underwater acoustic network has become a hot research topic in recent years [1]. Spread Spectrum (SS) technology, which is regarded to be the basis of the third generation of mobile telecommunication Code Division Multiple Access (CDMA) technology, has been widely used in underwater acoustic communication [2, 3], due to its distinct advantages of anti-fading and anti-interference abilities, low transmit power and high reliability. In addition, the well-known Orthogonal Frequency Division Multiplexing (OFDM) [4–8] technology has the capability of resisting frequency selective fading, enhancing spectrum efficiency and improving fast transmission speed. By exploiting both CDMA and OFDM techniques in the underwater acoustic communication, the proposed method enables to improve anti-interference and anti-fading abilities, and enhance the spectrum efficiency. Compared to the mere CDMA technique, the OFDM-CDMA technology has low error rate, and accommodates more communication users.

Inter-symbol interference caused by the channel multi-path extension is one of the main obstacles to the underwater acoustic communication. The time reversal mirror technology, which has the capability of spatial focusing and time compression, can adaptively match the acoustic channel without prior knowledge of the environment [9] and reduce the impact on the channel fading [1]. It can also improve the processing gain in space division multiple access, have great potential applications in multi-user underwater acoustic communication.

© Springer Nature Singapore Pte Ltd. 2017
B. Zou et al. (Eds.): ICPCSEE 2017, Part II, CCIS 728, pp. 325–334, 2017.
DOI: 10.1007/978-981-10-6388-6_27

According to the distinct advantages of both OFDM-CDMA and time reversal mirror techniques, a novel method is proposed to achieve multi-user underwater acoustic communication by effectively exploiting both techniques. Simulation results verify the correctness and effectiveness of the proposed method.

2 OFDM–CDMA System

OFDM and CDMA techniques are the core techniques in the communication technology, which have many advantages. On the one hand, the proposed method takes advantage of code 'division multiple access to achieve the resources sharing across multiple users; on the other hand, it makes use of the serial-to-parallel conversion of OFDM to convert the data points to each subcarrier to realize the high-speed transmission. Reconstruction of communication system based on OFDM-CDMA can further improve the anti-interference and anti-fading performance. The flow chart of the proposed method is shown in Fig. 1.

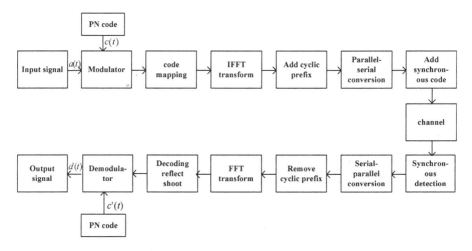

Fig. 1. Diagram of the OFDM-CDMA system

An input binary bit information $a(t)$ is transferred to a spread spectrum signal $a(t) \cdot c(t)$ by timing a pseudo random sequence $c(t)$. Then the spread spectrum signal is used as new input information for code mapping and BPSK modulated. The mapped sequence information then goes through IFFT transformation so as to change the spectrum expression of the data to the time domain. The sending signal $s(t)$ is:

$$s(t) = \sum_{m=-\infty}^{+\infty} \sum_{n=0}^{N-1} c_n a_m \exp(j2\pi f_n(t - mT))g(t - mT) \tag{1}$$

where c_n is the No.n chip of $c(t)$, a_m is the No.m input data of $a(t)$, T is symbol cycle for OFDM.

$$g(t) = \begin{cases} 1 & (0 \leq t \leq T) \\ 0 & (else) \end{cases} \tag{2}$$

For the anti-multipath effect of spread spectrum technology and the decoding characteristics of the spread spectrum code, the system can still ensure stable communication performance without cyclic prefix, improved the effective communication rate. Balance Gold code is selected as the synchronization code to facilitate the detection of received signal. On the receiving end, the system conducts the inverse transform process corresponding to the transmitting end, completes the demodulation by one step of FFT transform. It avoids high requirements for carrier synchronization compared to spread spectrum system.

The OFDM-CDMA technique can be regarded as a special kind of direct sequence CDMA technology, the emission data on the transmitting end is original data multiplied with a spreading sequence, the new spreading sequence is obtained from the original one by performing IFFT transform. The OFDM-CDMA technique also can be regarded as a kind of OFDM technology, but the OFDM data before IFFT carrier modulation is obtained by multiplying the original data with the spreading sequence [10]. The OFDM-CDMA technology combines the advantages of both OFDM and CDMA, which can preserve the spectrum resources, and improve the anti-interference ability.

3 OFDM-CDMA Underwater Acoustic Communication System Based on Virtual Time Reversal Mirror

Time reversal mirror (TRM) technology can automatically matches the underwater acoustic channel without a priori knowledge of any environment, with the "best" realization of the space and time filter. It uses the focusing effect, recombines multi-path signal, suppresses the inter-symbol interference and improves the SNR [11–13]. Time reversal mirror technology includes active time reversal mirror technology, passive time reversal mirror technology and virtual time reversal mirror technology. This system uses the virtual time reversal mirror technology. Compared to the active time reversal mirror technology, virtual time reversal mirror technology has higher communication speed and low complexity. Compared with the passive time reversal mirror, the virtual one uses the synchronization code $p(t)$ as the detection signal, and makes $p(t)$ go through underwater acoustic channel $h(t)$. Received detection signal is inverted to be $p(-t) \otimes h(-t)$. The preprocessed signals have eliminated the channel multipath effect, and preserve the information of the detection signal. By convoluting with the probe signal, the original information can be recovered. In multi-user communication, the inversion of $p(-t) \otimes h(-t)$ is difficult to match the superimposition of multi-user information code, the focusing effect is not ideal even in high SNR case. Therefore, the virtual time reversal mirror technology is the best choice in multi-user communication. On the receiving end, it does the copy-related processing of the received detection signal to estimate the channel impulse response. Performing the convolution of the received signal with the time reversal of the estimated channel, the reverse processing can be realized.

Multi-user communication can be divided into uplink communication and downlink communication, the uplink communication is that multiple sub-nodes send messages to the master node; the downlink communication is that the master node transmits messages to multiple sub-nodes [1]. In this paper, we only take the uplink communication mode, put sub-nodes in different horizontal distance and different depth, which can guarantee weak correlation between them with the channel impulse response of the master node. The communication principle diagram is shown in Figs. 2 and 3.

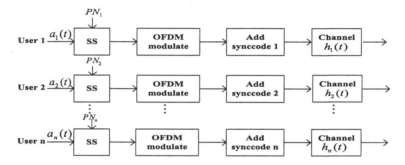

Fig. 2. Flow chart of multi-user communication transmitter

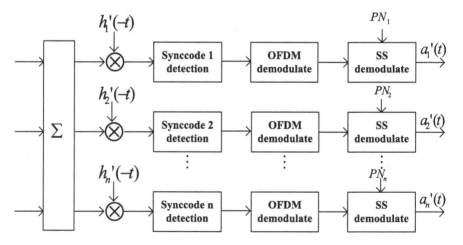

Fig. 3. Flow chart of multi-user communication receiver

The No.n user's input data information is extended by pseudo-random code spread spectrum and OFDM signal modulated using IFFT transformation. After the synchronizations, sn(t) is made as the sending signal, which will go through the channel hn(t).

And r(t) is received as superimposed signal. The estimated channel h'n(t) as time reversal channel is obtained by convoluting with the impulse response. The time reversal channel is inversed and made convolution with r(t). The new sequence information will be used for follows, which only focuses on the No.n user's information and inhibits interference of other unexpected users due to the weak cross-correlation of channel. And also the new sequence information can achieve separation with other users and reduce the multiple access interference in multi-user communication.

$$r(t) = \sum_{k=1}^{n} s_k(t) \otimes h_k(t) \tag{3}$$

Assuming that h'(t) is No.i user's estimated channel. make $r(t)$ convolution with h'(t) to recover the information symbols transmitted by the No.i user at the receiving end.

$$s_{i\_out}(t) = r(t) \otimes h'_i(-t) = [\sum_{k=1}^{n} s_k(t) \otimes h_k(t)] \otimes h'_i(-t)$$

$$\approx s_i(t) + [\sum_{k=1,k\neq i}^{n} s_k(t) \otimes h_k(t)] \otimes h'_i(-t) \approx s_i(t) + \varepsilon_i \tag{4}$$

When $h'_i(t)$ is estimated accurately, $h'_i(-t) \otimes h_i(t)$ can be approximated to the autocorrelation function of $h_i(t)$. The correlation peak can be regarded as the "single peak" that can be approximated to δ function. The main peak amplitude is significantly higher than the side-lobe, which can inhibit the underwater acoustic channel's inter-symbol interference from multi-path expansion, achieve focusing effect, and realize channel equalization. At the same time, for the weak correlation between channels, the normalized amplitude of $h_k(t) \otimes h'_i(-t)$ is less than 1 significantly, the recovered signal is approximated to the information symbols transmitted by No.i user, and ε_i can be regarded as multiple access interference of other users, it can be processed as noise.

The time reversal processing strengthens the signal of the expected user. After that, the sync-byte detection is made and the different gold codes are used. These codes have excellent auto-correlation and weak cross-correlation, which can make accurate detection of synchronous correlation, further weaken the unexpected users' interference and benefit the following demodulation and de-spreading processing.

4   Simulations

In order to verify the effectiveness of the OFDM-CDMA communication system based on time reversal mirror in multi-user underwater communication, this paper studies with computer simulation. The parameters shown in Table 1. Among them, the noise is band-limited white noise, the spread spectrum code is the order 7 balance Gold code, the synchronization code is the order 10 balance Gold code. Balance Gold code, with extremely excellent relevant characteristics, extend transmitted information in spread spectrum, compress information in the de-spreading, and make the interference power expansion, improving the anti-interference ability of the system.

Table 1. Parameters of OFDM-CDMA modulation

FFT modulation points	4800
The sampling frequency	48 kHz
The starting frequency of OFDM signal	4 kHz
The ending frequency of OFDM signal	8 kHz
The OFDM signal length	100 ms
The mapped modulation	BPSK
The balance gold signal length	20 ms

The underwater acoustic channel is obtained from the channel simulation software, constructing two, four, six different number of weakly related channels respectively to do multi-user communication simulation study. Table 2 shows BER(Bit-Error-Rate) statistics of different users when using time reversal technology and without it.

Table 2. BER statistics of different user

Num	User n	BER (%)			
		SNR = 5 dB		SNR = 10 dB	
		No TRM	TRM	No TRM	TRM
2	User 1	2.49	0	0	0
	User 2	0	0	0	0
4	User 1	1.25	0	0	0
	User 2	0	0	0	0
	User 3	23.93	0	23.69	0
	User 4	3.49	0	3.42	0
6	User 1	3.24	0	2.49	0
	User 2	0	0	0	0
	User 3	22.69	0	14.71	0
	User 4	7.48	0	4.74	0
	User 5	2.99	0	1.51	0
	User 6	10.72	0	9.23	0

Note: 1. The SNR in the table is measured after through a band-pass filter; 2. The BER in the table is counted from multiple communication simulation.

Table 2 indicates that OFDM-CDMA system can realize stable communication in the condition of a less number of users without using time reversal mirror. But in a large number of users, this system is not stable, the BER of individual user is high, and the number of user with high BER also increases. OFDM-CDMA system using time reversal mirror technology, can not only greatly improve the quality of communication, but also maintain excellent and stability communication performance under the more number of users. Compared to the former, OFDM-CDMA communication system based on the time reversal technology can accommodate more users and even give more excellent quality.

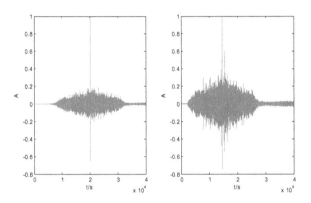

Fig. 4. Signal processing diagram at the receiving end

Figure 4 is the received data of the No.6 user after the master node has received superimposed signal. The left shows the synchronous detection with the processing of time reversal mirror technology, while the right is the synchronous detection without processing of the time reversal mirror. What can be seen is that time reversal makes both the anti-side-lobe and interference from other users better weakened. It is benefit to separate the desired user and reduce interference from other users.

To further verify the performance of OFDM-CDMA communication system based on time reversal mirror, we construct the impulse response function of 12 communication nodes, location distribution of each node is shown in Table 3. Figure 5 is correlation function diagram for the channel impulse response of the No.7 sub-node with the channel impulse response of all nodes. The auto-correlation peak of the No.7 user is obvious while the cross-correlation peak with other channel is much smaller, which verifies it meet the prerequisite for the use of time-reversal mirror at the multi-user receiving decoder. Table 3 is bit error rate (Bit-error-rate, BER) of 12 users using time reversal mirror under the SNR = 5 dB and 10 dB. In the low SNR, 12 users' BER is similar and the multi-user communication has stable performance. In higher SNR, each user node can nearly realize error-free transmission.

Table 3. BER statistics of each user's computer simulation

User n	Depth/m	Horizontal distance/m	BER (%)	
			$SNR = 5$ dB	$SNR = 10$ dB
User 1	4	500	2.00	0
User 2	4	900	1.75	0
User 3	4	1300	2.24	0
User 4	4	1700	1.75	0
User 5	8	700	2.74	0.25
User 6	8	1100	1.75	0
User 7	8	1500	1.25	0
User 8	8	1900	2.00	0
User 9	12	500	2.00	0
User 10	12	900	1.25	0
User 11	12	1300	1.25	0
User 12	12	1700	2.00	0

Note: 1. The SNR in the table is measured after through a band-pass filter;
2. The BER in the table is counted from multiple communication
simulation.

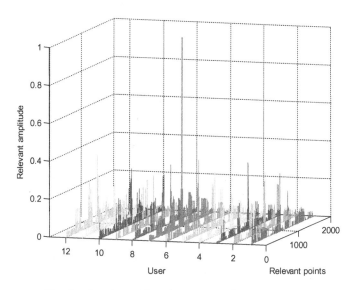

Fig. 5. The channel impulse response function diagram

5   Conclusions

OFDM-CDMA takes advantages of OFDM and CDMA technique, which not only has
high spectrum utilization, but also has low error rate and strong anti-interference and
anti-fading abilities. Compared to the mere CDMA technique, the OFDM-CDMA

technology accommodates more communication users and has great application prospect in underwater network communication. The time reversal mirror technology, which has the capability of spatial focusing and time compression, can adaptively match the acoustic channel without prior knowledge of the environment and reduce the impact on the channel fading. It can also improve the processing gain in space division multiple access, have great potential applications in multi-user underwater acoustic communication.

In this paper, the virtual time reversal mirror technology combined with OFDM-CDMA technology is applied in the multi-user underwater acoustic communication. We establish a multi-user communication model, verify the feasibility of OFDM-CDMA technology based on time reversal mirror used in multi-user underwater acoustic communication system through computer simulation. We also compare the performance of the system using time reversal mirror technology and the system without using time reversal mirror technology in the different SNR and different number of communication users, which verify the superiority of time reversal mirror technology for weakly correlated channel. The application of OFDM-CDMA technology in the multi-user underwater acoustic communication has better tolerance and higher stability. It also has a certain practical value in multi-node group communication of underwater sensor network, which lake test and sea trial can further improve the system.

Acknowledgements. This paper was supported by the National Natural Science Foundation of China (Nos. 61471137, 51179034).

References

1. Yin, J.: Underwater Acoustic Communication Principle and Signal Processing Technology, vol. 59. National Defence Industry Press, Beijing (2011)
2. Yu, Y., Zhou, F., Qiao, G.: M-ary code shift keying spread spectrum underwater acoustic communication. Acta Physica Sinica **61**(23) (2012)
3. Zhang, G., Hovem, J.M., Dong, H.: Experimental demonstration of spread spectrum communication over long range multipath channels. Appl. Acoust. **73**, 872–876 (2012)
4. Guo, D., Meng, X., Jiang, B., Gao, X., You, X.: Cholesky-decompositon-based co-channel interference suppression method in MIMO-OFDM systems. J. Southeast Univ. (Nat. Sci. Edn.) **45**(1) (2015)
5. Huang, Y., Hu, A., Cheng, Y., Xie, S.: Accurate time delay estimation scheme based on training sequence in OFDM systems. J. Southeast Univ. (Nat. Sci. Edn.) **44**(1) (2014)
6. Ai, B., Wang, J.T., Zhong, Z.D.: Broadband Wireless OFDM Synchronization Technology, pp. 13–16. Posts and Telecom Press, Beijing (2011)
7. Yin, J., Wang, C., Pan, Z.: Research on OFDM underwater acoustic communication based on MDAPSK. J. Huazhong Univ. Sci. Technol. (Nat. Sci. Edn.) **41**(3), 20–24 (2013)
8. Lv, T.-J., Tu, X.-F.: ICI management for efficient radio resource allocation in multi-cell OFDMA networks. China Commun. **8**(7), 84–94 (2011)
9. Yin, J., Wang, Y., Wang, L., et al.: Multiuser underwater acoustic communication using single-element virtual time reversal mirror. Chin. Sci. Bull. **54**(8), 1302–1310 (2009)

10. Lu, X.: Based on OFDM and spread spectrum communication technology research and implementation of algorithm. Chengdu Electronic Technology University (2012)
11. Shimura, T., Watanabe, Y., Ochi, H., Song, H.C.: Long-range time reversal communication in deep water: experimental results. J. Acoust. Soc. Am. **132**(1), EL49–EL53 (2012)
12. Sadler, J., Shapoori, K., Malyarenko, E., Severin, F., Maev, R.G.: Locating an acoustic point source scattered by a skull phantom via time reversal matched filtering. J. Acoust. Soc. Am. **128**(4), 1812–1822 (2010)
13. Song, A., Badiey, M., Newhall, A.E., et al.: Passive time reversal acoustic communications through shallow-water internal waves. IEEE J. Ocean. Eng. **35**(4), 756–765 (2010)

Hypergraph-Based Data Reduced Scheduling Policy for Data-Intensive Workflow in Clouds

Zhigang Hu, Jia Li, Meiguang Zheng[(✉)], Xinxin Zhang, Hui Kang,
Yong Tao, and Jiao Yang

School of Software, Central South University, Changsha 410075, China
{zghu,jiali1108,zhengmeiguang}@csu.edu.cn

Abstract. Data-intensive computing is expected to be the next-generation IT computing paradigm. Data-intensive workflows in clouds are becoming more and more popular. How to schedule data-intensive workflow efficiently has become the key issue. In this paper, first, we build a directed hypergraph model for data-intensive workflow, since Hypergraphs can more accurately model communication volume and better represent asymmetric problems, and the cut metric of hypergraphs is well suited for minimizing the total volume of communication. Second, we propose a concept data supportive ability to help the presentation of data-intensive workflow application and provide the merge operation details considering the data supportive ability. Third, we present an optimized hypergraph multi-level partitioning algorithm. Finally we bring a data reduced scheduling policy HEFT-P for data-intensive workflow. Through simulation, we compare HEFT-P with three typical workflow scheduling policies. The results indicate that HEFT-P could obtain reduced data scheduling and reduce the makespan of executing data-intensive workflows.

Keywords: Data-intensive workflow · Directed hypergraph · Data reduced scheduling · Cloud computing

1 Introduction

Data-intensive computing has emerged as a new and promising paradigm, which aims at gaining a better understanding of the problem under study and a refined search space for solutions by analyzing large volumes of data [1]. The on-demand access, scalability and availability the clouding computing offers, makes it ideal for storage, management, and analysis of big data [2]. Large-scale network applications based on cloud computing are distributed, heterogeneous, and data-intensive, such as scientific workflow systems, known as Data-Intensive Applications [3]. Tasks in these applications tend to acquire, transport and process large amounts of data. Unreasonable data selection or task scheduling policies would lead to excessive communication volume, which increases the execution time cost and economy cost of using cloud resources. And it could seriously affect execution efficiency. Therefore, how to optimize the data-intensive workflow execution, simplify the representation of data-intensive workflow application, reduce its communication volume and complexity to obtain better

© Springer Nature Singapore Pte Ltd. 2017
B. Zou et al. (Eds.): ICPCSEE 2017, Part II, CCIS 728, pp. 335–349, 2017.
DOI: 10.1007/978-981-10-6388-6_28

knowledge abstraction, have become a valuable issue in data-intensive application research.

In this paper, our purpose is to schedule data-intensive workflow in cloud environment, reduce the makespan by decreasing repeated data transmission and communication between different computing or storage nodes. The key three contributions in this paper are as follows:

1. New concept of support ability, which will be used to allocate tasks that need the same support ability to the same computing node as much as possible and enhance the localization;
2. A directed hypergraph model of data-intensive workflow considering data supportive ability is established, which could express the dependencies between tasks and the traffic in the application accurately;
3. Minimize the communication volume between tasks by partitioning the directed hypergraph model.

The remainder of this paper is organized as follows. Section 2 reviews related work. Section 3 presents the directed hypergraph model of data-intensive workflows. Section 4 derives a merge algorithm MAH for the hypergraph model and optimizes the multi-level partitioning algorithm. Section 5 describes the proposed schedule policy HEFT-P and Sect. 6 presents the experiment details and the simulation results. Section 7 concludes the paper and highlights future directions.

2 Related Work

The objective function of task scheduling problem is to map tasks onto processors and order their executions so that task-precedence requirements are satisfied and a minimum overall completion time is obtained. Because it is an NP-complete problem, a large number of heuristic algorithms are proposed, which can be further classified into three groups. The first one is list scheduling heuristics. It maintains a list of all tasks of a given graph according to their priorities. Some of the examples are the Modified Critical Path (MCP), Dynamic Level Scheduling (DLS), Earliest Time First (ETF) algorithms. For static model, the Heterogeneous Earliest Finish Time (HEFT) algorithm and the Critical Path on a Processor (CPOP) algorithm proposed in [4], and get a better performance and running time results. But these two algorithms cannot get better results while the number of processors is excessive and some simple cases. An energy-aware scheduling policy for data-intensive workflow is proposed in [5], compared with HEFT and CPOP, lower energy consumption is obtained under the condition of considerable scheduling length. In [6], a hybrid genetic algorithm combining genetic algorithm with HEFT algorithm is proposed, and a lower schedule time is obtained. The dynamic scheduling algorithm (ASA) is proposed for the dynamic computing environment in [7], which improves the elasticity of dynamic environment and the tasks, and obtains the better scheduling effect; the second is clustering heuristics, an algorithm in this group maps the tasks in a given graph to an unlimited number of clusters. Such as Dominant Sequence Clustering (DSC) [8]; the third is task duplication heuristics, the idea behind it is to schedule a task graph by mapping some of its tasks redundantly, which reduces the interprocess communication overhead. But the algorithms in this group have much higher complexity values than the algorithms in the other group.

The accurate problem model is the basis of the problem study, and the task scheduling model is the basis of task schedule and resource allocation. Currently, the task scheduling models representing dependencies mainly are Task Priority Graph (TPG) and Directed Acyclic Graph (DAG). But with the emergence of data-intensive applications in cloud environments, the relationship between tasks becomes more and more complex, and the traditional DAG graph cannot represent the real relationship between tasks accurately. This paper proposes the use of directed hypergraphs to represent data-intensive workflows. Directed hypergraphs can better represent asymmetric dependencies [9, 10] and the cut metric is well suited for minimizing the total volume of communication [11]. In addition to the concept of general graph, it also has the ability to partition and merge hyperedges which general graph don't have. The partitioning methods mainly include multi-level algorithms [12, 13], spectral methods [14], and various local optimization algorithms and so on. In order to meet the needs of partition, based on hMETIS, PaToH and other serial partitioning software packages, a parallel partitioning software Zoltan toolkit presented in [15] achieves good speedup on several large problems. Based on the advantages of hypergraph, a recursive hypergraph bipartitioning framework were proposed in [16] and it can reduce the total volume and total message count in a single phase, and compared with other models, it achieves better results. In [17], an optimizing method based on directed hypergraph and resource constraint is presented, transform the process-structure-optimizing problem into the directed hypergraph cutting problem. The paper [18] model the relationships holding between learning activities and the related competence by adopting directed hypergraph, and organize a large-scale repositories of learning objects.

This paper presents the concept of data support ability, and abstracts the data functions on the basis of the characteristics of data-intensive application; Based on the attributes of directed hypergraphs, the data support ability is coupled with the task to construct a directed hypergraph model considering the data support ability, so as to better study and optimize the data-intensive workflow scheduling problem; The task partition is realized by data reduction and partitioning in the directed hypergraph model, so that the smaller makespan can be obtained under the constraint of load balancing.

3   Workflow Hypergraph Model of Data-Intensive

This paper mainly aims at optimizing the workflow and shortening the task execution time by data reduction and partition of directed hypergraphs. Among them, the directed hypergraph model of data-intensive workflow is the basis of analysis and research. According to the correspondence between the attributes of the directed hypergraph and the traditional DAG graph, the data-intensive workflow is modeled, and the scheduling problem of data-intensive workflow is transformed into the partition and distribution problem of directed hypergraph model.

Compared to a directed edge can only connect two vertices of the general directed graph, there is a general hypergraph is a generalization of the general graph, can be connected to a number of hypernodes, and can be a good expression of multiple relationships. Extensive research proves that the hypergraph model can build communication model more accurately, especially in computer science, workflow management, etc. At the same time, hypergraph partitioning, as the basic problem of hypergraph theory, is the effective measure to minimize communication in scientific computing [12, 13].

3.1 Model Definition

In this paper, we construct the workflow model based on the correspondence between directed hypergraph attributes and data-intensive workflow attributes and the relationship between data and the task. Therefore, the input datasets and tasks in the definition correspond to two different types of nodes in the hypergraph; the dependencies between tasks and data support for tasks correspond to two different types of edges.

The directed hypergraph model of data-intensive workflow is $\overrightarrow{DAH} = (V, E)$, where V is the collection of all vertices, which is divided into two types of node sets. $V = DV \cup TV$, where DV represents the set of dataset vertices, and TV represents the set of task vertices, $DV \cap TV = \varnothing$. Where E is collections of all hyperedges, and it is also divided into two types of hyperedge sets. $E = DE \cup TE$, DE is the set of support for the data to the task, and TE is a workflow hyperedge set representing the dependencies between tasks, $DE \cap TE = \varnothing$. What's more, w represents the weight of the vertices.

For the task vertices set TV, $TV = \{tv_1, tv_2, \ldots, tv_n\}$, tv_i represents the subtasks in the data-intensive applications, $tv_i.pred$ and $tv_i.succ$ is the predecessor and successor vertex set of tv_i respectively. The task vertex without predecessor named entry vertex, tv_{entry}, and the one without successor named exit vertex, tv_{exit}. Based on this, we have the following definition of the workflow hyperedge.

Definition 1: Workflow Hyperedges *TE*. $\forall te_m \in TE$, from the current task tv_i points to its successor task tv_j, that is $\exists tv_i, tv_j \in TV$, $tv_i \in T(te_m)$, $tv_j \in H(te_m)$. $T(te_m)$ and $H(te_m)$ is the tail and head vertex sets of the hyperedge te_m respectively.

$\forall tv_i, tv_j \in TV$, if $\exists te_i \in TE$, then $tv_i = T(te_i), tv_j = H(te_i)$, named tv_i and tv_j as operational adjacent subtasks, written as $\wedge(tv_i, tv_j)$.

Data selection is the preparation phase of the task execution, in order to select the appropriate data efficiently for the task, to get a better analyzation of the function realized by data and obtain the data knowledge abstraction, the following definition defines data support ability in this paper.

Definition 2: Data Support Ability *a_k*. In a data-intensive application, we use the DWS service to extract and analyze the functions that can be implemented by the input data, and record the result as data support ability a_k, and assign the identification to its corresponding input data. Where k is the id of the data support ability, that is, the type of function that the data can achieve. In the application, all types of data support abilities constitute the ability type set *Ability* together, $Ability = \{a_1, a_2, \ldots, a_p\}$, where p is the total number of types of abilities that the *Ability* has.

For the dataset vertices set DV, $DV = \{dv_1, dv_2, \ldots, dv_m\}$, dv_j is the dataset vertex in data-intensive application. Based on this, we have the following definition of data support hyperedges.

Definition 3: Data Support Hyperedges *DE*. $\forall de_m \in DE$, from the dv_j points to task tv_i which takes dv_j as an the input data, that is, $\exists dv_j \in DV, tv_i \in TV$, $dv_j \in T(de_m), tv_i \in H(de_m)$, where, $T(de_m) \subseteq DV, H(de_m) \subseteq TV$.

Each dv_j has certain data support abilities, they are able to map the data support abilities in *Ability* to constitute the Data's Supportive Ability Sets (DAS) for each DV. $DAS_j = \{das_{j1}, das_{j2}, \ldots, das_{jp}\}$, $das_{jk} = 1$ means dv_j has the data support ability a_k, 0 means dv_j does not have a_k. And then all DAS_j of dv_j and *Ability* constitute the Dataset Supportive Ability Matrix (DM) of $m * p$ together, where m is the number of the DV.

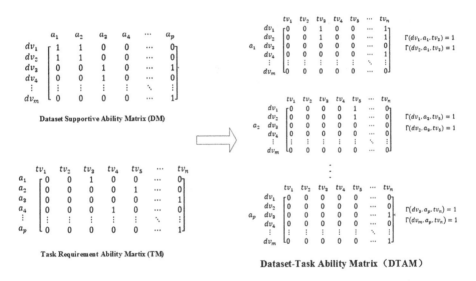

Fig. 1. Data set-task capability matrix DTAM

And tv_i are able to map the data support abilities they require in *Ability* to constitute the Task Request Ability (RA) for each TV. $RA_i = \{ra_{i1}, ra_{i2}, \ldots, ras_{ip}\}$, $ra_{ik} = 1$ means tv_i need data support ability a_k, 0 means don't need a_k. Then all RA_i of tv_i and *Ability* constitute a Task Requirement Ability Matrix (TM) of $p * n$ together, where n is number of tasks. And then we can construct a 3 dimensions Dataset-Task Ability Matrix (DTAM) of $m * p * n$ by integrating DM and TM, so that the relationship between the TV and the DV is clearly expressed, showed in Fig. 1. If $\Gamma(dv_j, a_k, tv_i) = 1, 1 \leq j \leq m$, $1 \leq k \leq p, 1 \leq i \leq n$, indicates that the dv_j can provide the support ability a_k required for the execution of tv_i.

After introducing the definition of data support ability, the vertices and the hyper-edges of the directed hypergraph for data-intensive workflow model $\overrightarrow{DAH} = (V, E)$, are re-expressed as follows.

Task vertex is represented as a quaternion $tv_i = (i, RA, InData, OutData) \in TV$, where i is the id of the task vertex, $RA \in Ability$. $InData$ and $OutData$ is the input and output data of the task respectively. Support data set of tv_i denoted: $tv_i.SDSet$.

Data vertex is represented as a quaternion $dv_j = (j, DAS, \Lambda, S_i) \in DV$, where j is the id of dataset vertex, $DAS \subset Ability$, S_i is the storage node where the dataset vertex is located.

$\forall de_n \in DE$, if $tv_i \in H(de_n)$, $dv_j \in T(de_n)$, then $\{tv_i.RA \in dv_j.DAS | RA \in Ability, DAS \subset Ability\}$.

In Fig. 2, for example, the DAS of dv_1 and dv_2 are: $dv_1.DAS = \{a_1, a_2\}$, $dv_2.DAS = \{a_1, a_2\}$, the RA of tv_3 and tv_5 are: $tv_3.RA = \{a_1\}$, $tv_5.RA = \{a_2\}$. From the matrix DTAM we can see: $\Gamma(dv_1, a_1, tv_3) = 1$, $\Gamma(dv_2, a_1, tv_3) = 1$, $\Gamma(dv_1, a_2, tv_5) = 1$, $\Gamma(dv_2, a_2, tv_5) = 1$. We can get the data support sets: $tv_3.SDSet = \{dv_1, dv_2\}$, $tv_5.SDSet = \{dv_1, dv_2\}$. Figure 2(a) is a traditional representation of data-intensive work-flow. This paper considers the data requirement of tasks in data-intensive workflows, and combines the DAG with DTAM matrix to obtain the directed hypergraph model considering support ability, showed in Fig. 2(b). Through the combination and mapping operation, we transform the scheduling problem of data-intensive workflow into data reduction and partition problem of directed hypergraph.

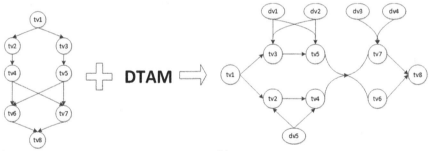

(a)An example of traditional DAG graph for data-intensive workflows

(b) A directed hypergraph model of data-intensive workflow considering data support ability

Fig. 2. DAG and DTAM matrix changed to be directed hypergraph model considering support ability by combining and mapping

3.2   Problem Formulation

The makespan of data-intensive workflow is determined by the Critical Path (CP) in the directed hypergraph model considering support ability. Makespan is calculated using the method in [4], as follows,

$$makespan = \max_{tv_i \in V}\{\{tlevel(tv_i) + blevel(tv_i)\}\} \tag{1}$$

Among them, $tlevel(tv_i)$ is the longest path from the tv_{entry} to tv_i, which does not include tv_i's weight. $blevel(tv_i)$ is the longest path from tv_i to the tv_{exit}. They are calculated as,

$$blevel(tv_i) = tv_i.w + \max_{tv_j \in succ(tv_i)}\{c_{i,j} + blevel(tv_j)\} \tag{2}$$

$$tlevel(tv_i) = \max_{tv_j \in pred(tv_i)}\{c_{i,j} + tlevel(tv_j)\} \tag{3}$$

$c_{i,j}$ is the communication volume of the task tv_i and its successor. By the above formula, we found we have no way to change the calculation of the critical tasks on the critical path, but we can reduce the communication between the tasks through the merging or the assignment of critical tasks to the same computing nodes to enhance the localization, and thus reduce the makespan.

In the operation of the hypergraph model of the data-intensive workflow, the hypergraph partitioning theory will be used as the main guidance. Given a directed hypergraph, $\overrightarrow{DAH} = (V,E)$ and an integer k, the TV is divided into k subsets of approximate weights, $Partition = \{p_1, p_2, \ldots, p_k\}$, where p_l is, $p_l.w = \sum_{tv_i \in p_l} tv_i.w$. They need to meet the following conditions:

1. $p_i \cap p_j = \varnothing, (i,j = 1,2,\ldots k)$;
2. $\cup_{l=1}^{k} p_l = TV$;
3. $thread = (1+\varepsilon)\frac{\sum_{tv_i \in TV} tv_i.w}{k}, p_l.w < thread$.

where ε is the load balancing threshold predefined. If the vertices of hyperedge te_i belong to at least two different partitions p_i and p_j, we consider the te_i has been cut. We try to minimize the connectivity, that is, the number of edges to be cut,

$$cuts\left(\overrightarrow{DAH}, P\right) = \sum_{i=0}^{|TE|-1}\left(\lambda_i\left(\overrightarrow{DAH}, P\right) - 1\right) \tag{4}$$

$\lambda_i\left(\overrightarrow{DAH}, P\right)$ is the number of partitions hyperedge i connected. $cuts\left(\overrightarrow{DAH}, P\right)$ is important due to it can reflect the communication overhead in parallel computing accurately. Minimized connectivity corresponds to the target of minimizing the communication volume between computing nodes.

4 Data Reduced Hypergraph Partition Algorithm

In this section, we will optimize the structure of the directed hypergraph model through the data reduction based on data support ability and the partition algorithm, thus reduce the scheduling time of the task. Data reduction were implemented by merging tasks, thereby reducing the repeated transmission of the same data. The hypergraph partitioning is an NP-hard problem [19], and the heuristic method can be used to solve this problem effectively. In this paper, the multi-level algorithm is used to partition the hypergraph model of data-intensive workflow, and the partition algorithm is optimized to obtain better performance. The paper will take different strategies according to the size of the task to partition the model and get better result.

4.1 Merge Operation Based on the Data Support Ability

In a data-intensive workflow, the amount of input and output data for a subtask is often very large, but there are some adjacent tasks in which there is a common support data, and if they are assigned to the same virtual machine, the time consumption due to waiting for data transmission can be reduced. In this paper, the support data of the task is obtained according to the DTAM matrix, and reduce the data according to the following merge lemmas to reduce the repeated transmission of the same supportive data, reducing the execution time of the task and the complexity of the algorithm. The merge algorithm for hypergraph (MAH) is illustrated in Algorithm 1.

Definition 4: Merge Operation. $\forall \wedge (tv_i, tv_j)$, if $tv_i.SDSet \cap tv_j.SDSet \neq \varnothing$, then tv_i and tv_j can be merged. And if the tasks are merged into tv_c, then,

$$tv_c.InData = tv_i.InData \cup tv_j.InData - tv_i.OutData \cap tv_j.InData,$$
$$tv_c.OutData = tv_i.OutData \cup tv_j.OutData - tv_i.OutData \cap$$
$$tv_j.InData, \quad tv_c.SDSet = tv_i.SDSet \cap tv_j.SDSet$$

Algorithm 1. MAH algorithm

```
Input: DTAM, DGH⃗ = (V, E)
Output: DGH⃗⁰ = (V⁰, E⁰)
/* Calculate the support data set for tasks          */
for tvᵢ ∈ TV and tvᵢ.RA ∈ Ability do
    aᵢ ← tvᵢ.RA, tvᵢ.SDSet ← {};
    for dvⱼ ∈ TV do
        if Γ(dvⱼ, aᵢ, tvᵢ) = 1 then
            tvᵢ.SDSet ← tvᵢ.SDSet ∪ dvⱼ;
    end for
end for
/* Determine the adjacent tasks need to be merged     */
for tvᵢ ∈ TV do
    if ∧(tvᵢ, tvⱼ) then
        if tvᵢ.SDSet∩tvⱼ.SDSet ≠ ∅ and tvᵢ.w< thread then
            Combine(tvᵢ, tvⱼ);
end for
```

After the merge operation, the model will be partitioned. The partition algorithms are divided into local and global categories. The Kernighan-Lin algorithm and the Fiduccia-Mattheyses algorithm are the effective partitioning methods of the local search strategy, but their main drawback is random initial partitioning of the vertex sets, which may affect the final partition quality. The global strategy relies on the properties of the entire graph and does not depend on any initial partition. Therefore, we use the multi-level algorithm in the global strategy to partition the merged hypergraph model, which can obtain the better partition results efficiently for the hypergraph model.

4.2 Multi-level Algorithm Based on Data Support Ability

The multi-level partition algorithm can get high quality partition results in a short time for the hypergraph model. In this section, we will use the multi-level algorithm to partition the large scale hypergraph model which has been merged. The multi-level algorithm is divided into three phases: coarsening, coarse partitioning and refinement. In coarsening, the vertices were matched and merged by multiple iterations until the termination condition is satisfied, and the multi-level coarse hypergraph is constructed, which reduces the number of hyperedges and the number of vertices in the original hypergraph effectively; in the coarse partitioning phase, we partition the smallest hypergraph obtained in the coarsening phase. In the phase of the refinement, the coarse partitioning was projected to a finer hypergraph and was optimized on each level, the problem of migrating the vertex group in the original hypergraph can be solved effectively.

Algorithm 2. Coarse algorithm based on support data

Input: $\overrightarrow{DGH_i} = (V_i, E_i)$
Output: $\overrightarrow{DGH_{i+1}} = (V_{i+1}, E_{i+1})$
if $r > 10\%$ **then**
   Sort all vertices in descending order of weight;
   Set all vertices, unmatched;
   Initialize $ip[tv] \leftarrow 0$, $tv \in TV$;
   for $te_i \in TE$ and $tv_j \in te_i$ **do**
      for $te_i \in TE$ and $tv_j \in te_i$ **do**
       $ip[tv_j] \leftarrow ip[tv_j] + 1$;
      end for
   end for
   /* there are multiple maximum common edges   */
   if $MaxIp(ip[tv_j], tv_j).lengh > 1$ **then**
      for $tv_p \in Max(ip[tv_i], tv_i)$ **do**
         Select tv_i whose edge with larger weight and the vertex with smaller id, $w \leftarrow tv_p$;
      end for
    match tv_i and w

Hypergraph Coarsening Based on Task Merging. After the merge processing of vertices, the directed hypergraph will be coarsed. A sequence of successively smaller hypergraphs $\overrightarrow{DGH_i} = (V_i, E_i)$ is constructed from the original hypergraph. $\overrightarrow{DGH} = (V, E)$, such that $|TV_i| < |TV_{i-1}| < \ldots < |TV_0|$, where $|TV|$ is the vertex number of TV. The TV in $\overrightarrow{DGH_i}$ are matching and merging to construct the next smaller coarse $\overrightarrow{DGH_{i+1}}$ by Inner Product Matching (IPM) [20]. The two vertices matching in $\overrightarrow{DGH_i}$ level will be merged into one vertex in the $\overrightarrow{DGH_{i+1}}$ level, and the weight of the vertex is the sum of the merged vertices weights.

However, hypergraph coarsening is the most time-consuming phase in multi-level algorithms [21], we will optimize the matching algorithm for the coarsening phase. In order to achieve better coarsening effect, this paper takes into account the communication and order between vertices. First, all vertices are sorted in ascending order of weight, and then the vertices are matched in order. The purpose of this practice is to make the smaller vertices have the priority to match, so as to avoid the combined vertex's weight is too large. We use greedy matching heuristic algorithm, which sort all hyperedges according to their common hyperedges in the decreasing order, and then merge the hyperedges with the largest number of common hyperedges. In the case of the number of inner product are same, the current vertex will first select the one whose edge with larger weight and vertex with smaller *id*. Until the reduction rate r satisfies the termination condition, the coarsening phase terminates, and it usually terminates when the value of r is less than 10%. The full algorithm is summarized in Algorithm 2.

Coarse Partitioning and Refinement. When the hypergraph is small enough, that is, the reduction rate is less than the threshold, the coarse partition phase is entered. The purpose of this phase is to obtain the initial partitions of a hypergraph by partitioning the minimum coarse hypergraph obtained from the previous phase. We run the Greedy Hypergraph Growing algorithm (GHG) to compute a different partitioning into k partitions, $Partition = \{p_1, p_2, \ldots, p_k\}$. Finally, we evaluate the cut metric (1) on each processor to pick the globally best partitioning.

However, partitioning the coarse graph into k partitions does not represent the high-quality partition of the original graph, but the repetitive local optimization operation in the optimization phase can solve this problem well.

The main purpose of the refinement phase is to reduce the connectivity $cuts\left(\overrightarrow{DAH}, P\right)$ between partitions under the conditions of load balancing. We select the optimal partition result in the coarse partitioning phase, and the refinement phase will map it back to the original hypergraph by multiple iterations and use the local optimization algorithm to improve its quality in each iteration. The most successful refinement methods are variations of Kernighan–Lin (KL) and Fiduccia–Mattheyses (FM). Compared to greedy algorithms, KL/FM generally produce better partition quality.

5   Scheduling Algorithm

In this paper, the algorithm HEFT-P is proposed based on the HEFT algorithm. The partition p_l obtained in the above section is used as a unit to schedule the tasks, in order to achieve the efficient scheduling of the data-intensive workflow. The priority of the task $rank_u(tv_i)$ is calculated as (3), and the traditional algorithm will schedule the individual tasks directly according to the priorities, but the proposed HEFT-P algorithm will be scheduled in units of tasks. Let tv_i be the task which has the highest priority in p_l, to satisfy the dependency of the task execution. The algorithm sets the earliest start time of p_l as the start time of tv_i, and as soon as possible to schedule the task which is already prepared. Thus the waiting time for the tasks is reduced. That is, let the maximum $rank_u(tv_i)$ of the tasks contained in p_l be the $rank_l$ of p_l,

$$rank_l = \max_{tv_i \in p_l}\{rank_u(tv_i)\} \tag{5}$$

p_l will be scheduled according to priority $rank_i$ in this paper.

Algorithm 3. HEFT-P scheduling algorithm

Input: *Partition*
Output: schedule result
Calculate the *rank_u* of all tasks and the *rank_l* of all
Partitions;
Sort the *Partitions* in the scheduling list by nonincreasing
order the *rank_l*;
while there is are unscheduled p_l in the list **do**
      Select the first p_l from the scheduling list;
      for each computing node in the computing node set
      ($v_m \in Q$) **do**
            Calculate the Earliest Finish time $EFT(p_l, v_m)$ of
            p_l on v_m;
      end for
      Assign p_l to the computing node v_m that minimizes
      EFT of p_l;
end while

After the partitions are scheduled, each p_l is assigned to the appropriate computing node. The number of partitions is the same as the number of computing nodes, and the time complexity of the algorithm is $O(k^2)$. After the partition is scheduled, we will schedule the tasks according to the order of $rank_u$ on the computing node which they are assigned to.

6 Performance Evaluation

6.1 Experimental Setup

Experiments were based on MATLAB R2014a platform, using the simulated Cyber-Shake and Montage standard scientific workflow for simulation experiments, and the standard scientific workflow data were obtained from [22]. In order to analyze and evaluate the performance of the HEFT-P algorithm proposed in this paper, we compare it with the classical algorithms HEFT, CPOP and MCP to verify its performance.

In this paper, the scale of the workflow is controlled by changing the number of tasks in the data-intensive scientific workflow; the communication to computation ratio (CCR) is the ratio of the average communication overhead to the average computation cost. The type of the workflow is changed by adjusting the CCR. We have selected Average Schedule Length (ASL) of 500 runs as the performance metric. The range bars for ASL of all algorithms shows a 95% of the confidence interval for corresponding ASL.

6.2 Experimental Analysis

In Fig. 3 the abscissa represents the number of computing nodes, and the number of it increases from 2 to 64; the ordinate represents the ASL of Cybershake workflows scheduled by different algorithms. When the number of computing nodes is small, the ASL of policy in this paper is equivalent to the ASL of HEFT and CPOP algorithm. However, with the increase in the number of computing nodes, our policy can always produce the lowest ASL. As shown in Fig. 3(a), when the number of tasks is less than the number of computing nodes, the ASL of the various scheduling algorithms is approximate. Because of the relatively abundant resources available at this time, various scheduling algorithms can perform scheduling decisions based on simple heuristic methods. At this point, the total amount of available resources become the main factor affecting ASL rather than the algorithm itself. In the experiment, it is found that when the load balancing threshold parameter ε is set to 20%, the results of this paper are the

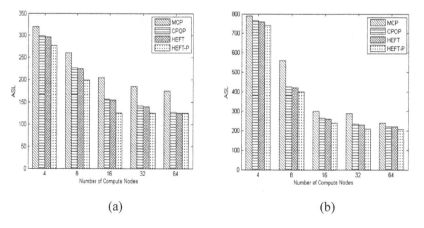

(a)　　　　　　　　　　　　　　(b)

Fig. 3. ASL of (a) 30 (b) 100 vertices CyberShake workflows

best. It can be seen from Fig. 3(b) that with the increase of the number of tasks, the policy can still get better scheduling results.

In Fig. 4, the ordinate represents the ASL of Montage workflows scheduled by different algorithms. The results of the Fig. 4 show that with the increase of the number of computing nodes, the ASL of the four algorithms has different degrees of decline, the policy of this paper can always produce the lowest ASL. In this paper, we partition the scientific workflow with the hypergraph partitioning algorithm. Compared with other scheduling algorithms, this paper can significantly reduce the communication volume between the computing nodes and obtain the lowest ADL. It is also found that the scheduling effect of CyberShake workflow is better than that of the Montage workflow. Because both types of workflows have different characteristics, CyberShake is a data-intensive workflow and has a higher CCR value as compared to the Montage workflow. This shows that this policy in this paper is more suitable for data-intensive workflow.

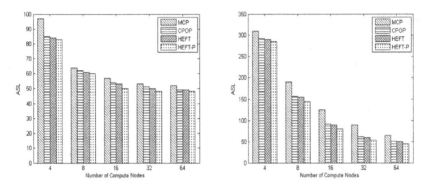

Fig. 4. ASL of (a) 25 (b) 100 vertices Montage workflows

The above experimental results show that in the cloud computing environment, as the size of data-intensive applications increases, the dependencies between tasks are more complex, the data transmission of the task and the computing node is more frequent, and the communication overhead of the task and its corresponding scheduling length are also significantly increased. Compared with HEFT, CPOP and MCP algorithm, this policy can reduce the communication volume between the tasks and get better scheduling effect.

7   Conclusion

In this paper, we first analyze the challenges of data-intensive workflow in clouds. And then we introduce the concept of data supportive ability by combining the hypergraph theory and the characteristics of task scheduling of data-intensive applications in clouds. Based on this, the directed hypergraph model of data-intensive applications is constructed, and a HEFT-P scheduling algorithm for data-intensive tasks based on directed hypergraphs is proposed. Experimental results show that the proposed policy is effective. Compared to HEFT, CPOP and MCP algorithms, especially in the case of large number of tasks, the proposed policy could get a lower scheduling time. In the future, we will adapt our policy to other real data-intensive applications. In addition, our work considers only merge operation. We plan to explore how to extend this work to decomposition operation for better data expression and data placement.

Acknowledgements. The authors warmly thank the reviewers for their insightful comments which helped to improve this work. This work was supported in part by National Natural Science Foundation of China (NSFC), project 61602525 and project 61572525.

References

1. Chen, C.L.P., Zhang, C.Y.: Data-intensive applications, challenges, techniques and technologies: a survey on Big Data. Inf. Sci. **275**, 314–347 (2014)
2. Armbrust, M., Fox, A., Griffith, R.: A view of cloud computing. Commun. ACM **53**(4), 50–58 (2010)
3. Gong, X., Jin, C.Q., Wang, X.L.: Data-intensive science and engineer: requirements and challenges. J. Comput. Sci. **35**(8), 1563–1578 (2012)
4. Topcuoglu, H., Hariri, S., Wu, M.: Performance-effective and low-complexity task scheduling for heterogeneous computing. IEEE Trans. Parallel Distrib. Syst. **13**(3), 260–274 (2002)
5. Xiao, P., Hu, Z.G., Qu, X.L.: Energy-aware scheduling policy for data-intensive workflow. J. Commun. **36**(1), 17-2015017 (2015)
6. Ahmad, S.G., Liew, C.S., Munir, E.U.: A hybrid genetic algorithm for optimization of scheduling workflow applications in heterogeneous computing systems. Journal of Parallel and Distributed Computing **87**, 80–90 (2016)
7. Hu, M., Luo, J., Wang, Y.: Adaptive scheduling of task graphs with dynamic resilience. IEEE Trans. Comput. **66**(1), 17–23 (2017)
8. Shin, K.S., Cha, M.J., Jang, M.S.: Task scheduling algorithm using minimized duplications in homogeneous systems. J. Parallel Distrib. Comput. **68**(8), 1146–1156 (2008)
9. Catalyurek, U.V., Boman, E.G., Devine, K.D.: Hypergraph-based dynamic load balancing for adaptive scientific computations. In: IEEE International Parallel and Distributed Processing Symposium, pp. 1–11. IEEE (2012)
10. Zhou, D., Huang, J., Schölkopf, B.: Learning with hypergraphs: clustering, classification, and embedding. In: Advances in Neural Information Processing Systems, pp. 1601–1608 (2010)
11. Zhao, H., Liu, X.: Hypergraph-based task-bundle scheduling towards efficiency and fairness in heterogeneous distributed systems. In: Parallel & Distributed Processing, pp. 1–12. IEEE (2010)
12. Çatalyürek, Ü., Aykanat, C.: PaToH (partitioning tool for hypergraphs). In: Padua, D. (ed.) Encyclopedia of Parallel Computing, pp. 1479–1487. Springer, New York (2011)
13. Çatalyürek, Ü.V., Deveci, M., Kaya, K.: Multithreaded clustering for multi-level hypergraph partitioning. In: 2012 IEEE 26th International Parallel & Distributed Processing Symposium (IPDPS), pp. 848–859. IEEE (2012)
14. Biswal, P., Lee, J.R., Rao, S.: Eigenvalue bounds, spectral partitioning, and metrical deformations via flows. J. ACM **57**(3), 13 (2010)

15. Devine, K.D., Boman, E.G., Heaphy, R.T.: Parallel hypergraph partitioning for scientific computing. In: Proceedings of 20th IEEE International Parallel & Distributed Processing Symposium, p. 10. IEEE (2010)
16. Selvitopi, O., Acer, S., Aykanat, C.: A recursive hypergraph bipartitioning framework for reducing bandwidth and latency costs simultaneously. IEEE Trans. Parallel Distrib. Syst. **28** (2), 345–358 (2016)
17. Sun Xuedong, F., Xu Xiaofei, S., Wang Gang, T.: Directed hypergraph based and re-source constrained enterprise process structure optimization. J. Softw. **17**(1), 59–67 (2006)
18. Laura, L., Nanni, U., Temperini, M.: The organization of large-scale repositories of learning objects with directed hypergraphs. In: Cao, Y., Väljataga, T., Tang, J., Leung, H., Laanpere, M. (eds.) ICWL 2014. LNCS, vol. 8699, pp. 23–33. Springer, Cham (2014). doi:10.1007/978-3-319-13296-9_3
19. Lengauer, T.: Combinatorial Algorithms for Integrated Circuit Layout. Springer Science & Business Media, Heidelberg (2012)
20. Çatalyürek, Ü.V., Aykanat, C., Uçar, B.: On two-dimensional sparse matrix partitioning: models, methods, and a recipe. SIAM J. Sci. Comput. **32**(2), 656–683 (2010)
21. Ümit, V.Ç., Mehmet, D., Kamer, K.: Multithreaded clustering for multi-level hypergraph partitioning. In: 6th IEEE International Parallel and Distributed Processing Symposium, IPDPS, Shanghai, China, pp. 848–859 (2012)
22. Pegasus Team: Workflow Generator. https://confluence.pegasus.isi.edu/display/pegasus/WorkflowGenerator

Software System Rejuvenation Modeling Based on Sequential Inspection Periods and State Multi-control Limits

Weichao Dang[1,2(✉)] and Jianchao Zeng[2,3]

[1] College of Electrical and Information Engineering,
Lanzhou University of Technology,
Lanzhou 730050, People's Republic of China
[2] Division of Industrial and System Engineering,
Taiyuan University of Science and Technology,
Taiyuan 030024, People's Republic of China
dangweichao@tyust.edu.cn
[3] School of Computer Science and Control Engineering,
North University of China, Taiyuan 030051, People's Republic of China
zjc@nuc.edu.cn

Abstract. This paper addresses the issue of software rejuvenation modeling. Rejuvenation strategies with sequential inspection periods and state multi-control limits are proposed here because the inspection-based approach involves the sampling of longer fixed periods of the state of system, which increases the probability of soft failure. The degradation process of the software system interferes with inspection and rejuvenation is modeled as a Markov chain. The steady-state probability density function of the system is thus derived, and a numerical solution of the function is provided. Expressions for mean unavailability time are derived during the inspection period when soft failure occurs. Finally, the steady-state availability of the system is modeled, and the solution to it is obtained using a genetic algorithm. The effectiveness of the model was verified by numerical experiments. Compared with rejuvenation strategies with fixed inspection periods, those with sequential inspection periods yielded greater steady-state availability of the software system.

1 Introduction

With the advances in Internet technology, the reliability of computer systems has emerged as a major concern for researchers. The failure of computer systems is usually caused by hardware or software failure, the latter being the more prominent source [1,2]. Huang et al. [3] proposed the concept of software aging, and pointed out that the performance of software that runs for a long time tends to deteriorate and its failure rate increases, eventually causing the system to crash due to software faults, memory leaks, unreleased file locks, data corruption, storage fragments, and cumulative errors. Software aging has since been reported in UNIX workstations, OLTP DBMS servers, the Apache Web server, Sun JVM,

© Springer Nature Singapore Pte Ltd. 2017
B. Zou et al. (Eds.): ICPCSEE 2017, Part II, CCIS 728, pp. 350–364, 2017.
DOI: 10.1007/978-981-10-6388-6_29

spacecraft flight systems [4–8], Web application systems [9], and cloud computing infrastructures [10]. A well-known example of software aging is in the US Patriot missile defense system, which failed after 100 h of continuous operation due to the accumulation of rounding-off errors, with the system incapable of tracking and intercepting incoming Scud missiles [11].

As a significant cause of software aging, aging-related software faults are non-deterministic, and difficult to eliminate in the software development and test phase. Therefore, errors triggered by specific user inputs occur in the operational phase [10,12]. As a proactive and preventative solution to software aging, software rejuvenation has been proposed by Huang et al. [3]. In software rejuvenation, to prevent serious failures, the running software is paused at appropriate times and its internal state is eliminated to restore the system to its initial state or a relatively healthy inter-mediate state.

In comparison with passive repair techniques, proactive and preventive techniques significantly reduce system unavailability time. The proactive and preventive techniques, however, lose effectiveness because they lack in correct scheduling and maintenance strategies. Therefore, a significant issue in software rejuvenation is when and how to trigger it.

There exist time-based and inspection-based rejuvenation approaches to deter-mine the timing for software rejuvenation [13]. The time-based approach determines the optimal rejuvenation timing by analyzing the relationship between different states and assumption failure distribution of the software system by adopting Markov models [3], semi-Markov models [14–16], stochastic Petri nets [17], and other mathematical tools to model the state transition process of software deterioration, in the process establishing functions between cost and time, and between cost and workload to provide the times for rejuvenation strategies and rejuvenation intervals. As a black-box approach, the time-based approach does not focus on degradation mechanisms of the system [18]. However, it is difficult to raise the optimization level of the time-based approach when applied to increasingly complex software systems with a wide variety of operating environments. The inspection-based approach determines the optimal rejuvenation schedule through statistical analysis of system data concerning resource usage (e.g., free physical memory and used swap space) and performance (e.g., throughput and response delay), using time-series analysis and machine learning to obtain the rate of resource consumption or performance degradation trends in order to predict system short-term failures according to actual load and resource consumption, and implement appropriate rejuvenation strategies. Compared with the time-based rejuvenation approach, the inspection-based approach has higher flexibility in real-time decision making, and is suitable for a wider variety of operation scenarios. By collecting operating system resource usage data at regular intervals from a networked UNIX workstation by using an SNMP-based distributed monitoring tool, Garg et al. [19] found that 33% of reported outages were due to resource exhaustion, and that software aging was a non-negligible source of failures in computer systems. Jia et al. [20] studied the effect of various performance indicators (e.g., response time, physical memory, and swap space) on software aging and predicted the effect of a single parameter on software aging using principal component analysis. Avritzer et al. [21] established a response time-based simulation model of software system performance degradation, and proposed corresponding software rejuvenation algorithms according to three methods of response-time measurement.

2 System Description

2.1 Characteristics of the Software System

Fixed and variable sampling periods are adopted to detect performance indicators that reflect the system state. Statistical analyses of performance indicators can be used to determine the health of the system at different times. When the state of the system degrades so seriously that it can no longer satisfy user requests, a failure occurs; this is called a soft failure [26]. When a soft failure occurs, the degradation level of the system is defined as D_f. The characteristics of the software system are then described as follows:

(a) The health of the software system can be expressed as a random variable X_t, associated with the occupancy of computer resources, the value of which can be obtained by running a special inspection program. The subscript t represents the running time of the software system.
(b) The software system running in an initial healthy state is represented by $X_0 = 0$. The performance of the software system gradually degrades, and its performance after running for a long time is denoted by $0 < X_t < D_f$. When $X_t \geq D_f$, system failure occurs. The failure can only be detected by an inspection.
(c) The inspection operation requires CPU time and other resources, resulting in performance degradation; therefore, the system cannot be inspected frequently. The running duration of the inspection program is denoted by T_i.

2.2 Rejuvenation Strategies

In this paper, rejuvenation strategies with sequential inspection periods and state multi-control limits are proposed. The specific rejuvenation strategies are as follows:

(a) According to the results of inspection, decisions on maintenance are made: no maintenance is conducted if $0 < X_t < D_r$; rejuvenation is pre-activated if $D_r \leq X_t < D_f$; when $X_t \geq D_f$, system recovery is reactivated, where the rejuvenation threshold is denoted by D_r.
(b) The subsequent inspection timing is determined according to system state on the basis of the sequential inspection mechanism. The interval $[0, D_r)$ is divided into N equal parts $[0, D_r) = \bigcup_{l=0}^{N-1} [\xi_l, \xi_{l+1})$. When $X_t \in$

$[\xi_l, \xi_{l+1}), l = 0, \ldots, N-1$, the subsequent inspection occurs after $N-l$ time units; and when $X_t \geq D_r$, the subsequent inspection occurs after N time units (see Fig. 1).

(c) The system restores to the initial healthy state after reactive recovery and pro-active-preventive rejuvenation. The recovery duration and rejuvenation duration are represented by T_f and T_r respectively.

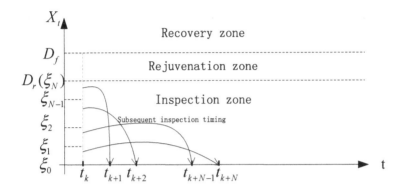

Fig. 1. Rejuvenation strategies with sequential inspection periods and multi-control limits

2.3   Degradation Processes of Software System

The software degradation process can be denoted by a non-decreasing continuous random process $\{X_t, t > 0\}$. When we suppose that the inspection program has no effect on degradation state, the given degradation state can be obtained [21]. The evolution of the system exhibits periodic characteristics if the restoring-as-good-as-new strategies described in Sect. 2.2 are adopted.

Following inspection and rejuvenation at periodic discrete maintenance timings $t_k = k\delta t$, the discrete-time stochastic process $X_k = X_{t_k}$ can be observed. The maintenance period δt is generally imposed in light of both system characteristics and the performance environment. Hereafter, this maintenance period δt is assigned the value 1 (in arbitrary time units). The interference of the evolution of the software system with inspection and rejuvenation behavior can be denoted by the stochastic process $\{X_k\}$, which is an embedded Markov chain in continuous state space. $X_k = 0$ implies that the system is in a new state at this point, called the regenerated point, and $X_k > 0$ means that the system is in the degradation state at another point, called the semi-regenerated point. The period between two successive inspection timings is called the semi-renewal cycle, represented by T. Figure 2 shows the realization of a representative sample path of the maintained system state for a four-threshold $(\xi_1, \xi_2, \xi_3, \xi_4, N = 4)$, featuring an inspection beginning at T_{n+3}, a rejuvenation at T_n, and a recovery at T_{n+2}. The degradation process of the software system can be approximated by a semi-renewal process.

The elementary increments in deterioration occurring between two successive maintenance times t_k and t_{k+1} are assumed to be positive, exchangeable, and stationary. System state degradation increment at $\Delta_{(k,k+1)}X$ two successive times t_k and $t_k + 1$ is assumed to follow pdf f. The quantity of cumulative deterioration during i maintenance periods is then a random quantity: $\Delta_{(k,k+i)}X = \sum_{l=1}^{i} \Delta_{(k+l-1,k+l)}X$. The associated deterioration pdf is the ith convolution of f, $f^{(i)}$.

A semi-renewal cycle consists of an inspection phase, a rejuvenation phase, a failure recovery phase, and an operating phase. According to the above-mentioned strategies for sequential inspection periods and state multi-control limits, the system state at the semi-renewal points might be in different zones, where system maintenance decisions and subsequent operating periods are different.

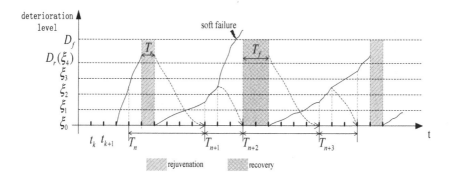

Fig. 2. A representative sample path realization of the maintained system state

3 Software System Availability Model Based on Above-Mentioned Rejuvenation Strategies

The different scenarios are analyzed at the semi-renewal points, also called maintenance decision points, and hence a steady-state probability density function of the software degradation processes is obtained in Sect. 3.1. The steady-state probability and the average length of time in different phases of the semi-renewal cycle are derived in Sect. 3.2. Since a system's soft failure can only be observed following inspection, the length of the mean unavailability time between a soft failure and the time of inspection needs to be determined. The details of this are described in Sect. 3.3.

3.1 The Steady-State Probability Density Function of the software System Degradation State

As described in Sect. 2.3, the system's state at the semi-renewal point is only related to the previous point; therefore, $\{X_k\}$ is an embedded Markov chain in continuous state space. According to the steady-state probability density formula of the irreducible, ergodic Markov chain $\pi(x) = \int \pi(y)K(y,x)dy$, the steady-state probability density function of the degradation state can be derived. $K(y,x)$, also called the transition probability kernel [27] [Chap. 3, p. 65], is conditional probability density from the previous semi-renewal point to the given one. The state at the beginning of the semi-renewal cycle is denoted by y and that at the end by x. The steady-state probability density function of the degradation state is represented by $\pi(x)$. The state at the be-ginning of the semi-renewal cycle consists of two scenarios. In one scenario, if $y \geq D_r$, rejuvenation or recovery needs to be carried out, due to which the system is restored to its initial healthy state. In the other scenario, if $0 < y < D_r$, no maintenance is needed. The two scenarios are analyzed as follows:

(a) In the scenario where $y \geq D_r$, at the beginning of the semi-renewal cycle, the system is restored to the initial healthy state after rejuvenation and recovery. The end of the semi-renewal cycle, i.e., the subsequent inspection timing, occurs after N time units. The probability of $y \geq D_r$, therefore, is expressed as $P(y \geq D_r) = \int_{D_r}^{\infty} \pi(y)dy = 1 - \int_0^{D_r} \pi(y)dy$. The conditional probability density of state x at the end of the semi-renewal cycle is the Nth convolution of $f(x)$, which is represented as $f^{(N)}(x)$.

(b) In the other scenario, where $0 < y < D_r$, at the beginning of the semi-renewal cycle, which is further expressed as $y = X_k \in [\xi_l, \xi_{l+1})$, $l = 0, \ldots, N - 1$, the sub-sequent inspection time occurs after $N - l$ time units. The conditional probability density of state x at the end of the semi-renewal cycle is the $N - l$th convolution of $f(x - y)$, which is represented as $f^{(N-l)}(x - y)$.

(c) The transition probability kernel of the system can be expressed as

$$K(y, x) = \sum_{l=0}^{N-1} I_{\{y \in [\xi_l, \xi_{l+1})\}} f^{(N-l)}(x - y) + I_{\{y \geq \xi_N\}} f^{(N)}(x) \qquad (1)$$

where $I_A(x)$ is an indicator function defined as $I_A(x) = 1$ for $x \in A$ and $I_A(x) = 0$ for $x \notin A$.

The steady-state probability density function of the degradation state can be de-rived as follows:

$$\pi(x) = \int_0^{\infty} \pi(y)K(y, x)dy$$

$$= \int_0^{\infty} \pi(y)(\sum_{l=0}^{N-1} I_{\{y \in [\xi_l, \xi_{l+1})\}} f^{(N-l)}(x - y) + I_{\{y \geq \xi_N\}} f^{(N)}(x))dy$$

$$= \sum_{l=0}^{N-1} \int_{\xi_l}^{\xi_{l+1}} \pi(y)f^{(N-l)}(x - y)dy + \int_{D_r}^{\infty} \pi(y)dy] f^{(N)}(x) \qquad (2)$$

$$= \sum_{l=0}^{N-1} \int_{\xi_l}^{\xi_{l+1}} \pi(y)f^{(N-l)}(x - y)dy + [1 - \int_0^{D_r} \pi(y)dy] f^{(N)}(x)$$

3.2 Steady State Availability of the Software System

With the steady-state probability density function of the degradation state obtained at the semi-renewal point, the mean time of different phases within the semi-renewal cycle can be further deduced as follows:

(a) The mean operating time in the operating phase is denoted by $E(D_o(T))$. As described above, if $y = X_k \in [\xi_l, \xi_{l+1}), l = 0, \dots, N - 1$, the subsequent inspection occurs after $N - l$ time units. The operating time is expressed as $N - l$. When we define $i = N - l$, the mean operating time is expressed as $\sum_{l=0}^{N-1} (N - l) \int_{\xi_l}^{\xi_{l+1}} \pi(y)\mathrm{d}y = \sum_{i=1}^{N} i \int_{\xi_{N-i}}^{\xi_{N-i+1}} \pi(y)\mathrm{d}y$. If $y = X_k \geq D_r$, i.e., $y \in [\xi_N, \infty)$, the subsequent inspection is after N time units, which means that the operating time is N. The mean operating time is expressed as $N \int_{\xi_N}^{\infty} \pi(y)\mathrm{d}y$. The expected length of the operating phase is therefore expressed as:

$$E(D_o(T)) = \sum_{i=1}^{N} i\left(\int_{\xi_{N-i}}^{\xi_{N-i+1}} \pi(y)\mathrm{d}y\right) + N \int_{\xi_N}^{\infty} \pi(y)\mathrm{d}y \tag{3}$$

If $y = X_k \geq D_f$, soft failure occurs before the next inspection. That is to say, the system is unavailable between failure and subsequent inspection. The mean duration of the downtime is expressed as $E(D_u(T))$, the solution to mitigate which is proposed in Sect. 3.3.

(b) The mean rejuvenation time in the rejuvenation phase is expressed as:

$$E(T_r) \int_{D_r}^{D_f} \pi(y)\mathrm{d}y \tag{4}$$

(c) The mean recovery time in the failure recovery phase is expressed as:

$$E(T_f) \int_{D_f}^{\infty} \pi(y)\mathrm{d}y \tag{5}$$

The mean inspection time in the inspection phase is denoted by $E(T_i)$, and the effect of inspection time on system unavailability is denoted by impact factor a.

According to the theory of semi-renewal processes, the steady state availability of the system is approximated by its availability in a semi-renewal cycle:

$$A_\infty \approx A(T) = \frac{E(D_o(T)) + (1 - a)E(T_i) - E(D_u(T))}{E(D_o(T)) + E(T_i) + E(T_r) \int_{D_r}^{D_f} \pi(y)\mathrm{d}y + E(T_f) \int_{D_f}^{\infty} \pi(y)\mathrm{d}y} \tag{6}$$

Obviously, the steady state availability of the system is affected by the maximum inspection period N and the rejuvenation threshold D_r. In the rejuvenation decision-making process, if the maximum inspection period N is too small, the inspection program runs frequently, and system unavailability time increases. With a short N, both the probability of the system state exceeding the rejuvenation threshold and the system rejuvenation time increase. Correspondingly, both the probability of the system state exceeding the failure threshold and the system recovery time decrease. On the contrary, if the maximum inspection period is too long, both the probability of failure occurrence and recovery time increase. If the rejuvenation threshold is set too low, rejuvenation decision becomes frequent, and the unavailable time of the system increases, whereas the probability of system failure increases as well as unavailable time if the rejuvenation threshold is set too high.

It has been found that the maximum inspection period and rejuvenation threshold affect the probability of rejuvenation and recovery as well as the availability of the system. The best strategy, therefore, is to choose the optimized maximum inspection period and rejuvenation threshold to maximize the system's steady state availability.

3.3 Mean Unavailable Time in the Operating Phase

Due to discrete time inspections, the exact failure time and $D_u(T)$ are unknown. The unavailability time is approached by an upper-bound $D_u^+(T)$ running from t_k, just prior to failure, until the next inspection operation, the expectation of which is given by $E(D_u^+(T)) = \sum_{k=1}^{N} kP(D_u^+(T) = k)$, where $P(D_u^+(T) = k)$ represents the probability that the unavailable time is time units. Let $H(k|z, l')$ be the probability that the unavailability time is equal to $k\delta t$ on a Markov renewal cycle, conditioned on deterioration value after maintenance $z = y - y_0$. If the inspection occurs after time units, the system state deteriorates from $z = y - y_0$ to D_f at $l' - k$, which can be expressed by the following equation:

$$
\begin{aligned}
P(D_u^+(T) = k) &= E(I_{\{D_u^+(T)=k\}}) \\
&= E(E(I_{\{D_u^+(T)=k\}}|z < \xi_N)) \\
&= E(P(D_u^+(T) = k|z < \xi_N)) \qquad (7) \\
&= E(\sum_{l'=1}^{N} H(k|z = y - y_0, l'; l' \geq k)) \\
&= E(\sum_{l'=k}^{N} H(k|z = y - y_0, l')(I_{\{y \in [\xi_{l'}, \xi_{l'+1}]\}} + I_{\{y \geq \xi_N\}}))
\end{aligned}
$$

where $I_A(x)$ is an indicator function defined as $I_A(x) = 1$ for $x \in A$ and $I_A(x) = 0$ for $x \notin A$. If y is located in the inspection zone, $H(k|z = y - y_0, l') = H(k|y, l') = f^{(l'-k)}(D_f - y)$, where $y_0 = 0$; if y is located in the rejuvenation zone and the failure zone, $H(k|z = y - y_0, l') = H(k|0, l') = f^{(l'-k)}(D_f)$ where $y_0 = y$. When $l' = N - l$, the following equation can be obtained:

$$E(D_u^+(T)) = \sum_{k=1}^{N} kP(D_u^+(T) = k)$$

$$= \sum_{k=1}^{N} kE(\sum_{l'=k}^{N} H(k|z = y - y_0, l')(I_{\{y \in [\xi_{l'}, \xi_{l'+1}]\}} + I_{\{y \geq \xi_N\}}))$$

$$= \sum_{k=1}^{N} k(\sum_{l=0}^{N-k} \int_{\xi_l}^{\xi_{l+1}} \pi(y)H(k|y, N-l)dy + H(k|0, N)\int_{D_r}^{\infty} \pi(y)dy)$$

$$= \sum_{k=1}^{N} k(\sum_{l=0}^{N-k} \int_{\xi_l}^{\xi_{l+1}} \pi(y)f^{(N-l-k)}(D_f - y)dy + \qquad (8)$$

$$f^{(N-k)}(D_f)\int_{D_r}^{\infty} \pi(y)dy$$

3.4   Numerical Solution to the Steady-State Probability Density Function

Equation (2), which expresses steady-state probability density, is an implicit integral equation, the analytical solution to which is difficult to obtain. Numerical methods are used to obtain the approximate solution. Equation (2) can be rewritten as:

$$\pi(x) = f^{(N)}(x) - f^{(N)}(x)\int_{0}^{D_r} \pi(y)dy + \sum_{l=0}^{N-1} \int_{\xi_l}^{\xi_{l+1}} \pi(y)f^{(N-l)}(x-y)dy \quad (9)$$

According to the numerical solutions to the integral equation and the quadrature rule:

$$\int_{a}^{b} y(x)dx = \sum_{j=1}^{N} y(x_j)\delta x \qquad (10)$$

When we apply Eq. (10), the numerical expression of Eq. (9) becomes:

$$\pi(x) = f^{(N)}(x) - f^{(N)}(x)h\sum_{j=1}^{D_r/h} \pi(jh) + h\sum_{l=0}^{N-1} \sum_{j=\xi_l/h}^{\xi_{l+1}/h} \pi(jh)f^{(N-l)}(x-jh) \quad (11)$$

When we define $p = x/h, q = D_r/h$, the approximate equation at the quadrature point can be expressed as:

$$\pi(ph) = f^{(N)}(ph) - f^{(N)}(ph)h\sum_{j=1}^{q} \pi(jh) + h\sum_{l=0}^{N-1} \sum_{j=\xi_l/h}^{\xi_{l+1}/h} \pi(jh)f^{(N-l)}(ph-jh)$$

$$(12)$$

Following transposition, the following equation can be obtained:

$$\pi(ph) + hf^{(N)}(ph)\sum_{j=1}^{q} \pi(jh) - h\sum_{l=0}^{N-1} \sum_{j=\xi_l/h}^{\xi_{l+1}/h} \pi(jh)f^{(N-l)}(ph-jh) = f^{(N)}(ph),$$

$$p \in [1, \ldots, i_{max}] \qquad (13)$$

In Eq. (13), where $D_{\max}$ is the censored number replacing ∞, $i_{\max} = \lfloor D_{\max}/h \rfloor$ is the maximum number of numerical integrations. Then, the matrix notation of Eq. (13) becomes

$$\pi(\mathbf{I} + \mathbf{K^1} - \mathbf{K^2}) = \mathbf{f^{(N)}} \tag{14}$$

Solving Eq. (14), we can derive

$$\pi = (\mathbf{I} + \mathbf{K^1} - \mathbf{K^2})^{-1}\mathbf{f^{(N)}} \tag{15}$$

π in Eq. (15) is the vector to be solved. $\pi = (\pi(h), \pi(2h), \dots, \pi(i_{\max}h))'$ and $\mathbf{f^{(N)}} = (f(h), f(2h), \dots, f(i_{\max}h))'$, where superscript $'$ represents the vector transpose. $\mathbf{I}$ in Eq. (15) represents the $i_{\max} \times i_{\max}$-dimensional identity matrix. $\mathbf{K^1}$ and $\mathbf{K^2}$ in Eq. (15) represent the matrices $\{K_{pj}^1\}_{i_{\max} \times i_{\max}}$ and $\{K_{pj}^2\}_{i_{\max} \times i_{\max}}$, respectively, where

$$K_{pj}^1 = \begin{cases} h * f^{(N)}(ph) & p = 1, \dots, i_{\max} \\ 0 & \text{else} \end{cases},$$

$$K_{pj}^2 = \begin{cases} h * f^{(N-l)}(ph - jh) & p = 1, \dots, i_{\max}, j = 1, \dots, \max(p, q), l = 0, N-1 \\ 0 & \text{else} \end{cases}.$$

3.5 Solution to Availability.Model

The above-mentioned analyses show that the maximum inspection periods and rejuvenation thresholds need to be adjusted in order to maximize the steady-state availability of the system, hence influencing the probability of rejuvenation and recovery. The rejuvenation problem under discussion, therefore, can be modeled as Eq. (16), which describes a single-objective optimization problem with constraints, namely:

$$\begin{aligned} \max \quad & A(N, D_r) \\ s.t. \quad & N \in \mathbb{N}^+ \\ & 0 < D_r < D_f \end{aligned} \tag{16}$$

As shown in Eq. (16), the model is characterized by variables with constraints, is non-linear, and represents single-objective optimization. The genetic algorithm can be applied to handle any target in the form of linear/non-linear constraints, discrete/continuous, and mixing the search space. The approximate optimal solution to the model can be solved through the joint optimization of the decision variables.

4 Numerical Experiments

The degradation process of the software system can be modeled as a Markov process. It can also be modeled as a gamma process with independent, non-negative increments with a gamma distribution with an identical scale parameter [18,23,28]. We chose gamma distribution as the experimental distribution of the software system degradation processes. If we suppose that the software system degradation increment $\triangle_{(k,k+1)}X$ accords with $\Gamma(\alpha, \beta)$ within unit time and the pdf of $\triangle_{(k,k+1)}X$ is $f(x|\alpha,\beta) = \frac{\beta^\alpha}{\Gamma(\alpha)}x^{\alpha-1}e^{-\beta x}, x > 0$, where $\Gamma(\alpha) = \int_0^\infty x^{\alpha-1}e^{-x}dx, \alpha > 0$, the degradation increment accords with within time units. Gamma distribution shape parameters $\alpha = 2$ $\alpha = 4$ and scale parameter $\beta = 1$ were selected in this study. The differences in the values of α/β reflected different degradation rates of the system. The failure threshold of the system was denoted by $D_f = 20$, the mean running time of inspection program was denoted $E(T_i) = 0.02$, the impact factor by $a = 0.8$, and the mean rejuvenation and recovery times by $E(T_r) = 0.14$ and $E(T_f) = 1$ respectively.

4.1 Verification of the Optimization Model

We first calculated system availability when no rejuvenation strategies were adopted. If $\alpha = 2$, the mean degradation rate could be derived as $\alpha/\beta = 2$ and the mean time to failure (MTTF, also called life) as 10, which showed that the system's steady-state availability was $10/11 = 0.909091$; If $\alpha = 4$, the mean degradation rate could be de-rived as $\alpha/\beta = 4$ and the MTTF as 5, which showed that the steady-state availability is $5/6 = 0.833333$.

The effect of each decision variable on the optimization objectives was analyzed to show that for each decision variable, there existed an optimal value that optimized the objective. In this way, the correctness of the maintenance optimization model was verified. Figure 3(a) shows the trends of steady-state availability with varying maxi-mum inspection periods (N) on the condition of rejuvenation threshold $D_r = 13.1$. Figure 3(b) shows the probability trends of three maintenance decisions with varying maximum inspection periods (N). Obviously, the steady-state availability of the system had a maximum value with gradual increase in the inspection period. Due to the effects of the maximum inspection period on availability, the system was inspected so frequently that lower system availability resulted if the inspection period was too short. With the increase in N, the probability of rejuvenation and recovery increased. Trade-offs between the probability of rejuvenation and recovery, and that of the inspection, was obtained at a special time, when the steady-state availability of the system was maximal. As N continued to increase, the probability of recovery increased accordingly; meanwhile, the steady-state availability decreased. Figure 4(a) shows the trends of steady-state availability with varying rejuvenation thresholds in the maxi-mum inspection period $N = 5$. Figure 4(b) shows the probability trends of occurrence of the three maintenance decisions with varying rejuvenation thresholds D_r. The smaller the value of D_r, the higher the probability of rejuvenation and the lower the probability of recovery, and the greater the availability of the system; As D_r increased, the probability of rejuvenation decreased, whereas the probability of recovery increased, and steady-state availability decreased drastically.

The probability density function (see Eq. 15) was solved by applying the numerical solutions proposed in Sect. 3.4. $D_{\max} = 10D_f$ represents censored number replacing ∞ for gamma distribution pdf. If the interval $[0, 10D_f]$ was equally divided into 2,000 parts, $h = 0.1$. The parameters of the genetic algorithm were configured as follows: the number of individuals was 30, the maximum

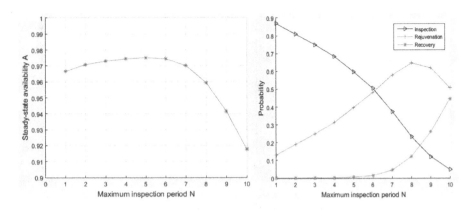

Fig. 3. Effects of maximum inspection period on availability(L) and maintenance decisions(R)

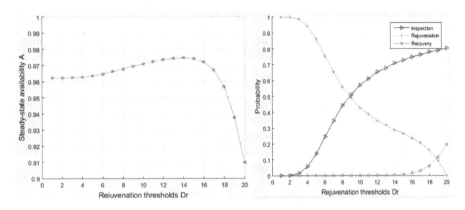

Fig. 4. Effects of rejuvenation thresholds on availability(L) and maintenance decision(R)

number of generations was 30, the generation gap was 0.8, the mutation probability was 0.2, and the crossover probability was 0.8. The upper bound of the maximum inspection period $T_{max} = 10$. Figures 5(a) and (b) show examples of the optimization result of GA when $\alpha = 2, \beta = 1$ and $\alpha = 4, \beta = 1$. If $\alpha = 2$, the approximate optimal solutions obtained were $T^* = 5$ and $D_r^* = 12.357179$. Accordingly, the maximum availability was 0.976940; If $\alpha = 4$, the approximate optimal solutions obtained were $T^* = 2$ and $D_r^* = 12.393670$. Accordingly, the maximum availability was 0.951690. Obviously, the steady-state availability of the system greatly increased by adopting the above-mentioned sequential inspection and rejuvenation strategies. It was found that if the degradation rate of the system increased, the maximum inspection period significantly decreased to increase the probability of rejuvenation and reduce that of recovery. Accordingly, greater availability was obtained.

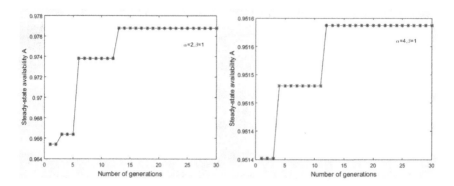

Fig. 5. An example of the optimization result of GA

4.2   Comparison of Optimization Results Between Fixed and Sequential Inspection Periods

Table 1 shows the optimization results of the maximum steady-state availability of the system with different degradation rates under the conditions of not adopting any rejuvenation strategies (P_0), adopting rejuvenation strategies with fixed inspection periods (P_1), and adopting rejuvenation strategies with sequential inspection periods (P_2), respectively. As we can see, the steady-state availability of the system with the same degradation rates could be increased by adopting strategies P_1 and P_2. Obviously, when the rejuvenation strategies of P_2 were adopted, higher availability was obtained, and the increase in availability was more significant with gradual increase in degradation rate.

Table 1. Effects of fixed and sequential inspection periods on steady-state availability

α, β	$A|P_0$	$A|P_1$	↑	$A|P_2$	↑
0.5, 1	0.975610	0.992728	1.75%	0.993481	1.83%
1, 1	0.952381	0.985972	3.53%	0.987299	3.67%
2, 1	0.909091	0.934188	2.76%	0.974940	7.24%
4, 1	0.833333	0.937318	12.48%	0.950690	14.08%

5   Conclusions

The following has been discussed in this study:

(a) Rejuvenation strategies with sequential inspection periods and state multi-control limits were proposed according to performance degradation characteristics of the software system. The state evolution process, interfered with by inspection and rejuvenation behavior, was modeled as a Markov chain; hence, the analytical expressions of steady-state availability were derived.

(b) The steady-state probability density function of the system's state under varying conditions related to inspection period and rejuvenation threshold was derived, and a numerical solution to the function was proposed.

(c) The occurrence of software system failure was detected through the introduction of an inspection mechanism, and the mean unavailability time was derived in the operating phase when soft failure occurred, which makes the optimization model more realistic and more practical.

(d) The effectiveness of the proposed optimization model was verified by numerical experiments. Compared with rejuvenation strategies with fixed inspection periods, those with sequential inspection periods enhanced the steady-state availability of the software system.

Acknowledgments. This research was supported in part by the Chinese National Natural Science Foundation under Grant No. 61573250, the Key Research and Development Program of Shanxi Province under Grant No. 201703D121042-1, the Key Science and Technology Program of Shanxi Province under Grant No. 20130321006-01, the Youth Foundation of Shanxi Province under Grant No. 201601D021065 and the PhD Research Startup Foundation of TYUST under Grant No. 20152021.

References

1. Gray, J., Siewiorek, D.P.: High-availability computer systems. Computer **24**(9), 39–48 (1991)
2. Sullivan, M., Chillarege, R.: Software defects and their impact on system availability-a study of field failures in operating systems. In: Proceedings of the Twenty-First International Symposium on Fault-Tolerant Computing, FTCS-21 (1991)
3. Huang, Y., Kintala, C., Kolettis, N., Fulton, N.D.: Software rejuvenation: analysis, module and applications. In: Proceedings of 25 International Symposium, Fault-Tolerant Computing, FTCS-25 (1995)
4. Garg, S., Puliafito, A., Telek, M., Trivedi, T.: Analysis of preventive maintenance in transactions based software systems. IEEE Trans. Comput. **47**(1), 96–107 (1998)
5. Cassidy, K.J., Gross, K.C., Malekpour, A.: Advanced pattern recognition for detection of complex software aging phenomena in online transaction processing servers. In: Proceedings of International Conference on Dependable Systems and Networks, DSN 2002 (2002)
6. Grottke, M., Li, L., Vaidyanathan, K., Trivedi, K.S.: Analysis of software aging in a web server. Discuss. Pap. **55**(3), 411–420 (2005)
7. Cotroneo, D., Orlando, S., Pietrantuono, R., Russo, S.: A measurement-based ageing analysis of the JVM. Softw. Testing Verif.Reliab. **23**(3), 199–239 (2013)
8. Tai, A.T., Alkalai, L., Chau, S.N.: On-board preventive maintenance: a design-oriented analytic study for long-life applications. Perform. Eval. **35**(3–4), 215–232 (1999)
9. Matias, R., Andrzejak, A., Machida, F., Elias, D.: A systematic differential analysis for fast and robust detection of software aging. In: 2014 IEEE 33rd International Symposium on Reliable Distributed Systems (2014)
10. Araujo, J., Matos, R., Alves, V., et al.: Software aging in the eucalyptus cloud computing infrastructure: characterization and rejuvenation. ACM J. Emerg. Technol. Comput. Syst. **636**(8), 1557–1564 (2014)

11. Marshall, E.: Fatal error: how patriot overlooked a scud. Science **255**(5050), 1347 (1992)

12. Cotroneo, D., Pietrantuono, R., Russo, S., Trivedi, K.: How do bugs surface? A comprehensive study on the characteristics of software bugs manifestation. J. Syst. Softw. **113**, 27–43 (2016)

13. Cotroneo, D., Natella, R., Pietrantuono, R., et al.: A survey of software aging and rejuvenation studies. ACM J. Emerg. Technol. Comput. Syst. **10**(1), 104 (2014)

14. Bao, Y., Sun, X., Trivedi, K.S.: A workload-based analysis of software aging, and rejuvenation. IEEE Trans. Reliab. **54**(3), 541–548 (2005)

15. Zhao, T.H., Yong, Q.I., Shen, J.Y., et al.: Application server multi-state aging model and optimal rejuvenation strategy research. J. Syst. Simul. **19**(8), 1705–1709 (2007)

16. Rinsaka, K., Dohi, T.: Toward high assurance software systems with adaptive fault management. Softw. Qual. J. **24**(1), 1–21 (2016)

17. Wang, D., Xie, W., Trivedi, K.S.: Performability analysis of clustered systems with rejuvenation under varying workload. Perform. Eval. **64**(3), 247–265 (2007)

18. Bobbio, A., Sereno, M., Anglano, C.: Fine grained software degradation models for optimal rejuvenation policies. Perform. Eval. **46**(1), 45–62 (2001)

19. Garg, S., Van Moorsel, A., Vaidyanathan, K., et al.: A methodology for detection and estimation of software aging. In: Proceedings of Ninth International Symposium on Software Reliability Engineering (1998)

20. Jia, Y.F., et al.: On the relationship between software aging and related parameters. In: International Conference on Quality Software (2008)

21. Avritzer, A., Bondi, A., Grottke, M., et al.: Performance assurance via software rejuvenation: monitoring, statistics and algorithms. In: International Conference on Dependable Systems and Networks, DSN 2006 (2006)

22. Meng, H., Liu, J., Hei, X.: Modeling and optimizing periodically inspected software rejuvenation policy based on geometric sequences. Reliab. Eng. Syst. Saf. **133**(133), 184–191 (2015)

23. Grall, A., Dieulle, L., Berenguer, C., Roussignol, M.: Continuous time predictive maintenance scheduling for a deteriorating system. IEEE Trans. Reliab. **51**(2), 150–155 (2001)

24. Zhao, T.H., et al.: Application server rejuvenation policy research based on aging accumulative damage model. J. Syst. Simul. **18**, 226–229 (2006)

25. Zhang, J.H., et al.: Approach of virtual machine failure recovery based on hidden Markov model. J. Softw. **25**(11), 2702–2714 (2014)

26. Avritzer, A., Weyuker, E.J.: Monitoring smoothly degrading systems for increased dependability. Empirical Softw. Eng. **2**(2), 59–77 (1997)

27. Meyn, S., Tweedie, R.L.: Markov Chains and Stochastic Stability. Springer, Heidelberg (1993). p. xxviii+594

28. Noortwijk, J.M.V.: A survey of the application of gamma processes in maintenance. Reliab. Eng. Syst. Saf. **94**(1), 2–21 (2009)

Research on Power Quality Disturbance Signal Classification Based on Random Matrix Theory

Keyan Liu[1], Dongli Jia[1], Kaiyuan He[1], Tingting Zhao[2],
and Fengzhan Zhao[2(✉)]

[1] China Electric Power Research Institute, Haidian, Beijing, China
[2] College of Information and Electrical Engineering,
China Agricultural University, Haidian Beijing, China
zhaofz@cau.edu.cn

Abstract. In this paper, a method of power quality disturbance classification based on random matrix theory (RMT) is proposed. The method utilizes the power quality disturbance signal to construct a random matrix. By analyzing the mean spectral radius (MSR) variation of the random matrix, the type and time of occurrence of power quality disturbance are classified. In this paper, the random matrix theory is used to analyze the voltage sag, swell and interrupt perturbation signals to classify the occurrence time, duration of the disturbance signal and the depth of voltage sag or swell. Examples show that the method has strong anti-noise ability.

Keywords: Power quality disturbance · Random matrix theory · Mean spectral radius

1 Introduction

With the extensive application of distributed power supply and the diversified development of power load and the improvement of the demand level of power supply, the problem of power quality disturbance has caused widespread concern. Quickly identify the type of power quality disturbance and determine the disturbance time, are of great significance to the corresponding compensation measures.

In recent years, the power quality disturbance signal classification method is endless, the traditional methods are Fourier transform [1, 2], wavelet transform [3, 4], and S transform [5, 6], the new method are improved wavelet transform [7], symmetrical components [8], ensemble empirical mode decomposition [9], sparse signal decomposition [10] and so on. These methods extract the characteristics of the disturbance signal, and then identify the type of disturbance, the time of occurrence and other information. The above methods have their own advantages, but the ubiquitous problem is that the anti-noise ability is poor, generally need to be de-noised before the application.

Random matrix as a big data analysis method, its application in the power industry is increasing. In [11], a smart grid architecture based on random matrix theory is proposed, and its application in transient analysis, stability analysis, critical node evaluation and

© Springer Nature Singapore Pte Ltd. 2017
B. Zou et al. (Eds.): ICPCSEE 2017, Part II, CCIS 728, pp. 365–376, 2017.
DOI: 10.1007/978-981-10-6388-6_30

fault detection is discussed. This provides a new method for power system analysis in big data scenarios. In [12], they use the random matrix to analyze the correlation between the running state data of the distribution network and the various influencing factors data. This is a method to quickly excavate the intrinsic relationship of massive data. Reference [13] proposed a statistical indicator system based on linear eigenvalue statistics of large random matrices. In [14], a method of multi-source heterogeneous data analysis based on random matrix theory and time series analysis is proposed for power equipment condition monitoring, voltage stability analysis and low-frequency oscillation detection. Reference [15, 16] proposed methods of static and transient stability analysis based on the random matrix theory. The relationship between the mean spectral radius and the historical data, simulation data is studied respectively.

The power quality disturbance signal is often doped with a certain noise signal, but the random matrix spectrum analysis is not affected by the noise. In this paper, the method of power quality disturbance classification based on random matrix theory is proposed. We study the correspondence between the change of the mean spectral radius (MSR) and the occurrence time, the duration and the voltage sag or swell depth of the disturbance signal based on the random matrix spectrum analysis of the power quality disturbance signal. The method of disturbance localization and classification is proposed.

2 Power Quality Disturbance Signal

The definitions of the variables are shown in Table 1:

Table 1. Variable description.

Variable symbol	Variable meaning
A	Baseband voltage amplitude
f_0	Fundamental frequency (50 Hz)
w_0	Fundamental frequency spectrum $w_0 = 2\pi f_0$
N_1	Number of sampling points per week
N_2	Total sampling points
T	Sampling interval $T = 1/(f_0 N_1)$
n	Sampling point
ϕ	Initial phase angle of the baseband

2.1 Standard Baseband Voltage Signal

The standard baseband voltage signal refers to the fundamental frequency sinusoidal signal without noise and any other disturbances, hereinafter referred to as 'baseband voltage signal', the mathematical model is:

$$x(nT) = A \sin (w_0 nT + \varphi), \quad 0 \leq n \leq N_2 - 1 \tag{1}$$

2.2 Baseband Voltage Sag/Swell Signal

Voltage sag, refers to the short time voltage fluctuation phenomenon with the voltage root mean square value reduced to 0.1–0.9 times the baseband voltage amplitude between the duration of 0.5 cycles to 1 min under power frequency conditions. In most cases, voltage sag is associated with a system failure. A typical failover time is 3–30 cycles. The mathematical model of the voltage sag is:

$$x(nT) = A \sin (w_0 nT + \varphi), \quad 0 \le n \le n_1, n_2 \le n \le N_2 - 1$$
$$x(nT) = kA \sin (w_0 nT + \varphi), \quad n_1 \le n \le n_2$$

$$(2)$$

Where, n_1 and n_2 are respectively the start and end time of the voltage sag, $n_2 - n_1$ is the disturbance duration, and k is the voltage sag amplitude. The typical range is $0.1 \le k \le 0.9$.

The voltage swell is defined similarly to the voltage sag, but its voltage root mean square are increased to 1.1–1.8 times the baseband voltage amplitude. The signal model is similar to Eq. (2), except that where k is the voltage swell amplitude, which is equal to the ratio of the amplitude of the voltage during the voltage transient disturbance to the amplitude of the baseband voltage before the disturbance. The value ranges from $1.1 \le k \le 1.8$. n_1 and n_2 are the starting and ending times of the voltage swell.

2.3 Baseband Voltage Interrupt Signal

The voltage interruption means that the voltage root mean square value drops below 0.1 times the baseband voltage and the duration does not exceed 1 min. The mathematical model of the signal is similar to that of Eq. (2). The difference is that k is the amplitude of the voltage interrupt signal, and its typical range is $0 \le k \le 0.1$. n_1 and n_2 correspond to the start and end of the voltage interruption.

3 Random Matrix Theory

3.1 Fundamentals of Random Matrix

When the dimension of the random matrix is infinite or reaches a certain scale, the empirical spectral distribution of its eigenvalues converges to its limit spectrum distribution [17]. This theory makes it possible to apply random matrix theory to practical engineering problems.

Suppose that $X = \{X_{ij}\}_{N \times M}$ is a random matrix with non-Hermitian features, each element is a random variable that satisfies the independent identically distributed, and its expectation and variance satisfy $E(X_{ij}) = 0$, $\sigma(X_{ij}) = 1$. When N and M are infinite, and the ratio $c = N/M \le 1$ is fixed, the empirical spectral distribution of the eigenvalues λ of X converges to the probability density function as shown in (3), this is Ring Law [18].

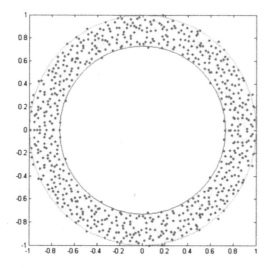

Fig. 1. Ring Law (Color figure online)

$$f(\lambda) = \begin{cases} \frac{1}{\pi cL}|\lambda|^{-1}, & (1-c)^{\frac{1}{2}} \le |\lambda| \le 1 \\ 0, & \text{other} \end{cases} \tag{3}$$

The eigenvalues of the random matrix is drawn in the complex plane (red dot), as shown in Fig. 1, it is obvious that evenly distributed between the two rings. Where, the outer ring radius is 1, the inner ring radius is $\sqrt{1-c}$, its size is only related to the ratio c.

The linear eigenvalue statistics can reflect the eigenvalues distribution of a random matrix and can comprehensively reflect the characteristics of the matrix. MSR is the simplest and most widely used as a linear eigenvalue statistic, which is defined as follows:

$$k_{MSR} = \frac{1}{N}\sum_{i=1}^{N}|\lambda_i| \tag{4}$$

Where $|\lambda_i|$ is the distance between i-th eigenvalue of the random matrix to the origin on the complex plane.

The physical mean of the MSR: the mean of the distance between all the eigenvalues of the random matrix to the origin on the complex plane, which can reflect the characteristics of the information contained in the matrix. So the MSR can be used as an indicator to reflect the statistical characteristics of the random matrix.

3.2 Power Quality Disturbance Random Matrix Construction and Classification Method

In this paper, the power quality disturbance signal model is used to generate the power quality disturbance signal sample. The sampling time is a = 10 power cycle period, and the sampling point N_1 = 128 per cycle. The total sampling points are N_2 = a × N_1 =

1280. Copy the disturbance signal to N rows and add white noise to each row, SNR = 16.8, thus constructing the data source of the power quality disturbance signal. The disturbance signal is extracted by the sliding window of scale $N \times M$ to form the random matrix X, and the matrix X is normalized according to the following formula:

$$\tilde{x}_{ij} = (x_{ij} - \bar{x}_i) \times (1/\sigma(x_i)) \quad 1 \leq i \leq N; 1 \leq j \leq M \tag{5}$$

The processed matrix becomes a random matrix with expectation 0 and variance 1. The matrix is decomposed and its MSR is calculated according to (4). Thus, as the sliding window moves, the MSR variation of the signal can be obtained.

The specific steps of classify the power quality disturbance using the random matrix theory are as follows:

Step 1: Construct the data source. Copy the disturbance signal to N rows and add white noise to each row.

Step 2: Construct a random matrix. From the first sample point in the data source, the disturbance matrix is extracted by the sliding window of $N \times M$ to form a random matrix X.

Step 3: Random matrix normalization. The random matrix is normalized by (5) to obtain a normalized matrix.

Step 4: Calculate the MSR. The characteristic spectrum of the matrix is analyzed and the MSR is calculated by (4).

Step 5: Slide the window in step size N_1, construct the random matrix of the next sampling cycle, and perform step 3, until the sampling time is $(aN_1 + 1)$, perform step 6.

Step 6: Analyze the change of the MSR with the window sliding to determine the disturbance information.

4 Power Quality Disturbance Classification Based on RMT

4.1 Voltage Sag/Swell Disturbance Positioning

The moment of the voltage sag/swell corresponds to the abrupt point of the disturbance signal. Figure 2(a) is the baseband voltage signal, Fig. 2(b) and (c) are the voltage sag signal and voltage swell signal. A voltage sag/swell disturbance occurred in the fourth cycle to the seventh cycle.

The random matrix is composed of the baseband voltage signal, the voltage sag signal and the voltage swell signal respectively. The MSR of each random matrix is calculated respectively. As the sliding window moves, the MSR variation of each random matrix is shown in Fig. 3.

It can be seen from the figure that the MSR of the random matrix during the voltage sag/swell period is significantly different from that of no disturbance signal random matrix. When the voltage sag occurs, the MSR of the random matrix increases, the average spectral radius of the random matrix decreases when the voltage swell. Therefore, according to the change of the MSR of the random matrix, it is possible to

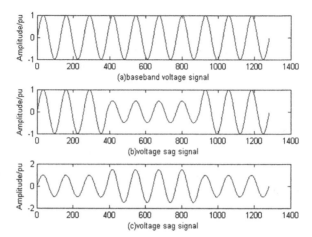

Fig. 2. Voltage sag/swell signal

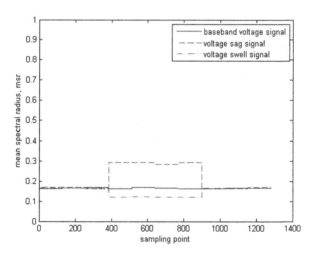

Fig. 3. MSR of different matrices

judge not only the time of disturbance generation but also the two different disturbance signals of voltage sag and voltage swell.

4.2   Voltage Sag/Swell Amplitude Recognition

The amplitude of the voltage sag/swell reflects the severity of the disturbance, and the effective identification of the amplitude can guide the electric workers to take different remedial measures.

Figure 4 is the voltage sag signals for the voltage sag amplitude were 70%, 50%, 30%. Figures 5, 6, 7 and 8 shows the distribution of the eigenvalues of random matrices at different voltage sag amplitude. As can be seen from the figures, as the voltage sag amplitude k becomes smaller, the eigenvalues of the matrix tend to be distributed to the outer circle, that is, the MSR tends to 1.

The timing chart in Fig. 9 shows the MSR of the signal that is obtained according to the timing variation of the different voltage sag amplitude signals shown in Fig. 4.

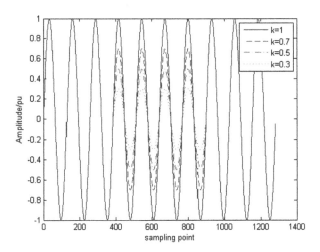

Fig. 4. Different voltage sag amplitude signal

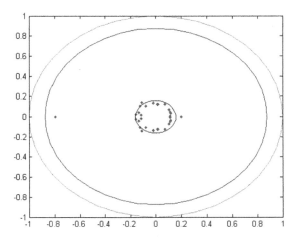

Fig. 5. The distribution of eigenvalues when k = 1

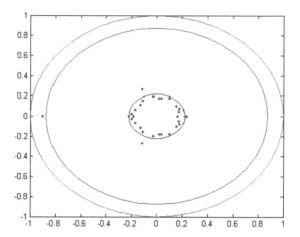

Fig. 6. The distribution of eigenvalues when k = 0.7

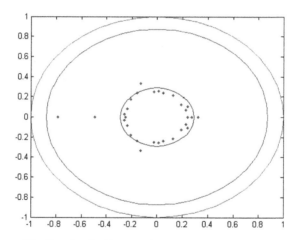

Fig. 7. The distribution of eigenvalues when k = 0.5

As can be seen from Fig. 9, the larger the amplitude of the voltage drop, the greater the change in the MSR. Therefore, we can determine the amplitude of the voltage sag based on the size of the MSR of the random matrix.

4.3   Voltage Interrupt Disturbance Classification

The method of classifying the voltage interrupt disturbance is the same as that of the voltage sag disturbance (Fig. 10). The voltage interrupt disturbance signal corresponds to a special voltage sag disturbance signal (the voltage sag amplitude k is between 0 and 0.1), so the result of its MSR is similar to the voltage sag as shown in Fig. 11.

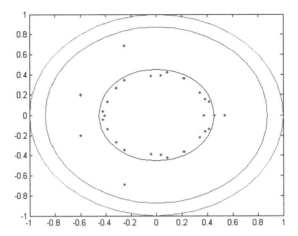

Fig. 8. The distribution of eigenvalues when k = 0.3

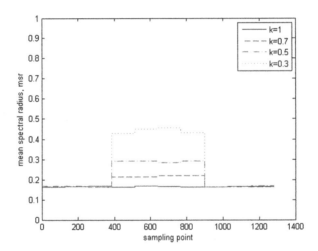

Fig. 9. MSR of different voltage sag amplitude signals

As can be seen from the figure, the MSR during voltage interruption is close to one. This shows that the MSR of the random matrix approaches 1 when the voltage amplitude drops to the limit.

Accordingly, we can distinguish the voltage sag disturbance and voltage interrupt disturbance according to the value of the random matrix MSR.

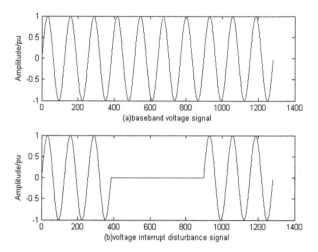

Fig. 10. Voltage interrupt signal

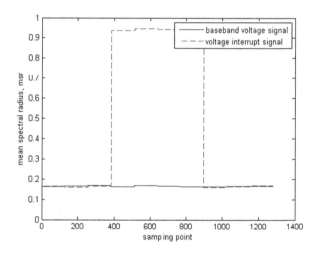

Fig. 11. MSR of the voltage interrupt signal

5 Conclusion

In this paper, a method based on random matrix theory is proposed to classify the power quality disturbance. The following conclusions are drawn:

(1) The power quality disturbance classification method based on random matrix spectrum analysis proposed in this paper can effectively identify the voltage sag disturbance, voltage swell disturbance, voltage interrupt disturbance.

(2) According to the change of the MSR of the random matrix, the occurrence time and duration of the power quality disturbance can be determined.

(3) Although the method proposed in this paper can qualitatively analyze the amplitude of the voltage sag/swell, but does not get the exact relationship between the MSR change value and the voltage sag/swell amplitude, which needs further study.

Acknowledgement. The authors gratefully acknowledge the key technology project of state grid corporation of China (EPRIPDJK[2015]1495).

References

1. Singh, U., Singh, S.N.: Application of fractional Fourier transform for classification of power quality disturbances. IET Sci. Meas. Technol. **11**, 67–76 (2017)
2. Zhao, F., Yang, R.: Voltage sag disturbance detection based on short time fourier transform. Proc. CSEE **27**, 28–34 (2007)
3. Santoso, S., Powers, E.J., Grady, W.M., Hofmann, P.: Power quality assessment via wavelet transform analysis. IEEE Trans. Power Delivery **11**, 924–930 (1996)
4. Thirumala, K., Umarikar, A.C., Jain, T.: Estimation of single-phase and three-phase power-quality indices using empirical wavelet transform. IEEE Trans. Power Delivery **30**, 445–454 (2015)
5. Li, J., Teng, Z., Tang, Q., Song, J.: Detection and classification of power quality disturbances using double resolution S-Transform and DAG-SVMs. IEEE Trans. Instrum. Meas. **65**, 2302–2312 (2016)
6. Zhao, F., Yang, R.: Power-quality disturbance recognition using S-Transform. IEEE Trans. Power Delivery **22**, 944–950 (2007)
7. Dalai, S., Dey, D., Chatterjee, B., Chakravorti, S., Bhattacharya, K.: Cross-spectrum analysis-based scheme for multiple power quality disturbance sensing device. IEEE Sens. J. **15**, 3989–3997 (2015)
8. Kumar, R., Singh, B., Shahani, D.T.: Symmetrical components-based modified technique for power-quality disturbances detection and classification. IEEE Trans. Ind. Appl. **52**, 3443–3450 (2016)
9. Liu, Z., Cui, Y., Li, W.: A classification method for complex power quality disturbances using EEMD and rank wavelet SVM. IEEE Trans. Smart Grid **6**, 1678–1685 (2015)
10. Manikandan, M.S., Samantaray, S.R., Kamwa, I.: Detection and classification of power quality disturbances using sparse signal decomposition on hybrid dictionaries. IEEE Trans. Instrum. Meas. **64**, 27–38 (2015)
11. He, X., et al.: A big data architecture design for smart grids based on random matrix theory. IEEE Trans. Smart Grid **8**, 674–686 (2017)
12. Xu, X., He, X., Ai, Q., Qiu, R.C.: A correlation analysis method for power systems based on random matrix theory. IEEE Trans. Smart Grid **PP**, 1–10 (2016)
13. He, X., et al.: Designing for situation awareness of future power grids: an indicator system based on linear eigenvalue statistics of large random matrices. IEEE Access **4**, 3557–3568 (2016)
14. Han, B., Luo, L., Sheng, G., Li, G., Jiang, X.: Framework of random matrix theory for power system data mining in a non-Gaussian environment. IEEE Access **4**, 9969–9977 (2016)

15. Wu, Q., Zhang, D., et al.: A method for power system steady stability situation assessment based on random matrix theory. Proc. CSEE **36**, 5414–5420 (2016)
16. Liu, W., Zhang, D., et al.: Power system transient stability analysis based on random matrix theory. Proc. CSEE **36**, 4854–4863 (2016)
17. Bai, Z., Silverstein, J.W.: Spectral Analysis of Large Dimensional Random Matrices, 2nd edn. Springer, New York (2010). doi:10.1007/978-1-4419-0661-8
18. Guionnet, A.: The single ring theorem. ArXiv e-prints (2009). http://arxiv.org/pdf/0909.2214.pdf

DCC: Distributed Cache Consistency

Shenling Liu$^{(\boxtimes)}$, Chunyuan Zhang, and Yujiao Chen

College of Computer, National University of Defence Technology,
Changsha 410073, Hunan, China
liushenling@nudt.edu.cn

Abstract. Replicating caches across distributed nodes to convert remote data transferred for local data reading, are widely used to reduce latency for applications and enhance storage availability in a distributed software system. DCC (Distributed Cache Consistency) is an mixed geo-replicated protocol to allow consistency of distributed caches by properly the relaxing read-write constraint. DCC reduces the number of required message round-trips in synchronization by taking advantage of a combination of Fast Paxos and Lease. At the same time, a lease protocol and a local synchronization group is used to reduce the message transmission and collision probability. A simple calculation model is set up to predict the performance of the model and analyze the potential application fields of DCC. The results, based on actual system observation parameters, show that DCC meets different consistency requirements for a distributed cache in scalable applications by adjustment of configuration parameters.

Keywords: DCC · Eventually consistency · Combination · Fast Paxos · Lease

1 Introduction

Distributed architecture is a general scheme to solve the load balancing problem and guarantee reliability in large scale systems, and has become the foundation of scaleable application system. In this design, service provided by local agents improves the response speed by limiting data transmission delay. In such a framework, each distributed node is highly autonomous, and replicated data that is accessed frequently by local nodes is usually stored in caches. Since the internet is dynamic in nature, maintaining remote caching node consistency is a challenging problem due to unstable network connections. Various innovative strategies have been proposed by researchers to solve inconsistency problems among different nodes, and can be broadly categorized into time-based mechanisms and quorum-based schemes. Time-based mechanisms highly depend on controlling the time of the node to cache, and the data in the cache can be updated only if the data ownership has not expired. In contrast, "majority set" schemes depend on polling, using a vote of the majority to decide on a consistent result.

Lease [1] is a fault-tolerant mechanism that provides efficient consistent access to cached data in distributed systems that was originally proposed by Gray. Lease is the foundation of time-based mechanisms, and grants the holder the right to full control over the corresponding data within the lease term, and other nodes require permission

© Springer Nature Singapore Pte Ltd. 2017
B. Zou et al. (Eds.): ICPCSEE 2017, Part II, CCIS 728, pp. 377–387, 2017.
DOI: 10.1007/978-981-10-6388-6_31

from the lease holder before modification of the represented data. The work in [2] proposed adaptive leases, where different leases are used for different types of data objects to avoid network delay. Kishor Kumar et al designed EDCM [3], CR [4], CBCC [5], SDCI [6], and an IR-based algorithm named DCCC [7], relying on a flexible TTL definition for specific application areas to reduce network traffic and read-write collisions.

The work in [8] presented Paxos, a decentralized synchronous protocol using master control as determined by a voting majority for data consistency in each round. This algorithm offers high availability and high fault tolerance but requires more rounds of message passing. Lamport optimized Paxos and proposed Fast Paxos [9, 10], a strategy that avoids use of a master by distinguishing between classic and fast ballots, allowing a reduction to only one message round to the master without collisions. MDCC, Multi-Ring Paxos, HT-Paxos [11–13] are other modifications of the Paxos protocol that focus on fault tolerance and latency, but also promote throughput and scalability by reducing the average number of message rounds. Paxos and its variants provide excellent solutions to ensure the consistency of distributed systems.

Overall, time-based mechanisms can effectively reduce the number of message passing and avoid network fluctuation, and quorum-based schemes provide higher reliability due to the decentered architecture. With the extensive application of distributed architecture in various application fields, a general protocol to guarantee consistency of multiple versions of cache is needed for system construction. To address this requirement, this paper presents a combination protocol called DCC (Distributed Cache Consistency), which supports both traditional sequentially-consistent and weakly consistent relationships, providing a universal mechanism that adapts different synchronization requirements. In summary, the key contributions of DCC are as follows.

(1) The cache synchronization of a group can be defined according to the local principle of the program. The cache data in a group can be internalized by synchronization in real - time, allowing data synchronization between groups using an adapted lease approach.
(2) The lease protocol can be optimized to reduce total data transmission and fast Paxos can be used to process data collision.
(3) The simulated results suggest that DCC is efficient and fault tolerant, that can be adapted to different distributed computational environments.

In Sect. 2, we present the overall architecture of DCC. Section 3 elaborates the optimistic lease protocol, and the simulation results of the model are presented in Sect. 4.

2 Architecture Overview

Similar to other distributed storage designs such as DBS3 [14] and MDCC [11], DCC also uses a library-centric approach. As shown in Fig. 1, DCC is a centralized scheme that contains three components such as the Synchronization Node Group, the Centric Node, and the Cache Node.

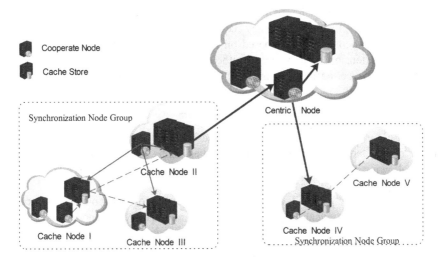

Fig. 1. DCC architecture.

The Synchronization Node Group consists of a set of cache nodes and maintains sequential consistency in internal nodes. Therefore, every cache store contains a full replica of each other cache in the group. The data storage of cache in this system is unbalanced, and related application data is stored in the local cache, reducing the communication delay.

The Centric Node is defined as a center coordinate and is also a cache store node, but it is responsible for managing the data lease. As a result, the first copy of data is created by the Centric Node and the storage locations of all data are recorded in it. When there is a cache miss for some node, the node always contacts the Centric Node and the coordinator node retrieves the data from the correct node.

As illustrated in Fig. 1, the distributed cache store and cooperate nodes are modules of the Cache Node and are logically independent. The cache stores are components for storage and for data coordination, and data access services are provided by cooperate nodes. Hence, cache stores and cooperate nodes can be deployed independently, allowing DCC to support multiple cache storage models and have higher extensibility. The cooperate nodes provide an interface for storage services, and cooperate nodes and cache stores have a corresponding relationship, so the corresponding module should be deployed in the same cluster to avoid network delay. The detailed protocol for each phase is discussed in the next section.

3 Protocol

In this section, we describe our new optimistic consistency protocol for cache synchronization across geographical cache nodes. In our hypothetical environment, network latency between long distance cache nodes tends to be higher than that between nodes in close range. The goal of DCC is to provide a fault-tolerant, efficient protocol with reduced latency from fewer message rounds. At the same time, DCC meets the practical needs of different consistency requirements by parameter adjustment.

The core protocol of DCC is based on Lease [1] and Fast-Paxos [10]. Our combination and extension of these two protocols is takes advantages of their separate advantages, as described in detail in the following sections of this chapter.

3.1   Optimization of the Lease Protocol

Lease is a time-based mechanism that provides efficient consistent access to cached data in distributed systems, Lease distinguishes cache nodes between clients and master, allowing direct mapping. In our scenario, the clients are the Synchronization Node Groups and the Centric Node is the master.

The classic Lease protocol, as applied for consistency of read-your-writes for data copy, includes three typical processes: (1) Lease grant. The master node sends a data duplicate to the client when it requests data and simultaneously generates a lease. The client that receives the data transcription and lease is called the data holder. (2) Maintaining the lease term. The master node blocks data updating when the lease is not expired, reading requests from other clients will receive a data copy with no lease, correlating data transcription and data on the master node. Some work [15] optimizes this process, and the master node sends a message before it is time for data updating to inform the data holder that the lease term has expired. This strategy can avoid data blocking and reduce false sharing. (3) Lease expiration management. When the lease expires and there are no data modification requests from other clients, the master node will extend the lease term. Otherwise, the master node can perform the data update and grant the lease to a new client.

The lease protocol can significantly reduce network message data, but there is an apparent single node failure problem, so if there are frequent cache updates, synchronization efficiency will be affected. To address these questions, optimized Lease, with a flow similar to that of basic Lease, may be used. Optimized Lease includes three main aspects of improvement.

The first contribution of DCC is the improvement of master node selection. Different from the classical Lease protocol, the main node is replaced by several candidate master nodes, and the running master node is not assigned but instead is elected by the synchronization node groups. In DCC, we use classic Paxos to select the master node. Only the candidate master nodes can be elected as a master node, and each synchronization group gets one ballot to participate in the vote. Other candidate nodes will grant a lease to the master node after election, and then check the master node's working state at the end of the lease to decide between renewing the lease or electing a new master. This not only solves the single-point failure problem of master nodes, but also effectively reduces the number of message rounds.

Secondly, DCC effectively reduces the data modification blocking problem by optimizing the data modification strategy. As in the classic Lease protocol, data read and write requests from non-data-holding nodes will generate cache miss and long distance data transmission. Therefore, we propose the concept of a synchronization group according to the principle of program locality. The group consists of a compatible number of nodes that are geographically located lose together (data transmission delay is small), that can read and write the same data, allowing sequenced synchronization across nodes. Another important adjustment is relaxing the read-your-writes consistency. Unlike the classic lease agreement that data updates are conducted by the master node, any internal cache node can update the data when a data lease is held by a synchronization node group. As a result, nodes in a synchronization group can access the data directly, reducing the number of communication messages between the client node and the master node and increasing the efficiency of data access. At the same time, the data modification request of nodes from other groups must be sent to the lease holding node to be executed through the master node.

Finally, we define the data write-back process to ensure the eventual data consistency. On the basis of the lease information, a new description field version is added to all data records. Any updates to records creates new versions. This can be represented in the form vread $\rightarrow$ vwrite, where vread is the record version read by cooperate nodes, and vwrite is the new version of the record. This provides a simple method for monitoring data changes. The records that were modified during the lease term will be sent to the master node at the end of the lease term, and then the master node and its candidate nodes will update the data in their cache store.

3.2 Data Synchronization Method

In the optimized protocol described in the previous section, there are two contributions to the overall consistency of the model. The first is the real-time synchronization of data between a synchronous node group during the lease term, and the second is the synchronization between the data synchronization group and the master node when the lease expires.

In the synchronous node group, the Fast-Paxos algorithm is used to achieve sequential consistency. At the same time, to simplify data management within the group and facilitate communication with master nodes, we define a proxy node, generated by node voting within the group followed by processing messages with the master node. The synchronization of the data in the group is performed as follows.

(a) When a cache miss occurs in the synchronization node group, the proxy node requests data from the master node and obtains a lease.
(b) The proxy node informs other cache nodes within the group to update their data via the classical Paxos algorithm. The proxy node can be directly used as the master node in the Paxos algorithm (other nodes do not have this data record, so there will be no conflict), reducing the number of votes to achieve rapid synchronization.
(c) During the period of the Lease, as shown in Fig. 1, the Fast-Paxos algorithm is used for data synchronization.

All versions start as an implicitly fast ballot number, unless a master has changed the ballot number through a Phase1a message [10] (First step of Paxos algorithm). As a result, any node in the group can propose an option directly to the cache nodes, which in turn can only accept the first proposed option. When a Paxos collision occurs, a new classic ballot must be started with Phase 1a (First step of Paxos algorithm) to reach an agreement.

The above steps guarantee data consistency between cache nodes within the synchronization group, however, synchronization between the group and the master node is needed to ensure consistency of the entire model. This process is called data write-back. Because the data across the nodes within the group is consistent, in the data write-back step, only the proxy nodes of each group, the master node, and the candidate nodes are involved in the synchronization process.

When the lease of the data record expires, the proxy node first determines whether the data has changed by comparing it to the recorded version, and then writes the modified data back to the master node. The Fast-Paxos algorithm is also used for the data writing-back step. Together with this process, the proxy node is the proposer, and the master node and the candidate nodes are acceptors. The proxy node submits the modified data record directly to all acceptors, which is always successful because the proxy node has a higher version of the data more recently.

Periodic synchronization of DCC guarantees the eventual consistency of the data across cache nodes. However, the above synchronization algorithm implies that other cache nodes without a data lease are not able to modify the relevant data. To achieve this demand within this framework only requires setting the lease term to 0. Better node group classification can reduce the cross-group data update frequency, in this paper, we divide nodes into groups according to geographical location of node based on the Internet of things device management features.

4 Experiment

The detailed design of the DCC model is proposed in Sect. 3. Here, we use a simple calculation model to simulate the analysis, and then we use client cache access data of QUANWFI, a distributed content sharing platform based on the Internet of things, to assess the performance of the model and analyze the potential of DCC applications.

4.1 A Simple Model

For the simulation we used a system composed of a single master node and a set of N groups, where each group consists of S cache nodes. The related parameters are listed in Table 1. All cache nodes have the same message transmission delay, D, to the master node, and the communication delay within the group is D'. At the same time, we use P to represent the message processing time.

In our algorithm, we the nodes that are most likely to access and modify data are distributed in a particular synchronization group. The read and write operation frequencies for clients in the group obey a Poisson distribution of R and W, and the read and write operation frequencies of other cache nodes, those not in the group, obey a Poisson distribution of R' and W'.

To verify the validity of the DCC, we analyzed master node load and the data read and write delay, which represent the system stability, reliability, availability, and efficiency.

Table 1. Defined parameters.

Symbol	Description
N	Number of synchronization node group
S	Number of cache nodes in the group
d	Message transport delay between cache nodes and master node
d'	Message transport delay between cache nodes in the group
p	Time to process a message
R	Rate of reads for cache nodes within the group where data is shared
R'	Rate of reads for cache nodes not in the group where data is shared
W	Rate of writes for cache nodes within the group where data is shared
W'	Rate of writes for cache nodes not in the group where data is shared
T_s	Lease term

If a cache handles the desired M read operation within a lease term and then counts the read request caused by the lease request, the cost of a lease request is spread to $1 + M$ read operations. According to the above definition, M can be calculated as:

$$M = RT_s + (N - 1)R'T_s \tag{1}$$

Hence, the processing speed for the read request of the master node can be calculated using the following expression:

$$L_R = \frac{2(RS + (N-1)SR')}{1 + RT_s + (N-1)R'T_s} \tag{2}$$

Where the numerator is the processing rate of the read request for the master node per unit time. If we define $\eta = R + (N - 1) * R'$, then Formula 2 can be simplified to:

$$L_R = \frac{2S\eta}{1 + \eta T_s} \tag{3}$$

The additional load of the master node comes from the write operation of the client node. There are two ways to handle data write operations in DCC. The data write request from the client nodes in the specified synchronization group, which holds the lease, does not require notification of the master node, but the nodes in the other groups must obtain authorization before the data is written. The master node load for data writing is $L_w = (N - 1) * S * W'$. Therefore, the total load of the master node can be written as:

$$L_m = \frac{2S\eta}{1 + \eta T_s} + (N - 1)SW' \tag{4}$$

As illustrated in the above formula, the main factors that determine the load of the master node are T_s and W'. A sufficiently long lease term will be able to reduce the load of the master nodes. At the same time, the minimum load is related to the data write probability, and this is independent of how the lease is set.

Another important performance parameter of this system is the average delay of data read and write. The use of the Lease management and Paxos coordination algorithms require additional overhead for data reading and writing. We can calculate a total increase delay, defined as t_d, during a lease. In the DCC framework, when the client needs to request data from the master node, the message is sent through the proxy node within the group. In this model, a message needs to be processed three times with two passes, and each group only needs to request the data once. Unlike a read request, a write request by a node other than the lease holder will generate an additional message processing event:

$$t_a = 2N(3p + d + d') + 2(3p + d + d')(N - 1)SW'T_s + 2r_c(p + d')ST_s \qquad (5)$$

Where r_c represents the collision of the Fast-Paxos algorithm, and each collision needs another two additional message processing cycles. The r_c can be determined from the following formula, as the probability of writing data also follows the Poisson distribution.

$$r_c = P\{\varepsilon > 1\} = \sum_{k=2}^{\infty} \frac{\lambda^k}{k!} e^{-\lambda} (\lambda = SW) \qquad (6)$$

The number of expected data readings in the lease period is:

$$f = 1 + ST_s(R + W) + (N - 1)ST_s(R' + W') \qquad (7)$$

Through formula 5 and formula 6, we can represent the average increased delay of data read and write as:

$$t_d = t_a/f = \frac{2N(3p + d + d') + 2(3p + d + d')(N - 1)SW'T_s + 2r_c(p + d')T_sS}{1 + ST_s(R + W) + (N - 1)ST_s(R' + W')} \qquad (8)$$

Similarly, we define three variables such as $\alpha = 3p + d + d'$, $\beta = p + d'$, and $\delta = R + W + (N - 1)(R' + W')$, so Eq. 7 can be simplified as:

$$t_d = \frac{2N\alpha}{1 + ST_s\delta} + \frac{2\alpha(N - 1)W' + 2r_c\beta}{1 + \delta} \qquad (9)$$

It can be seen that Eqs. 8 and 4 are similar in form, so an appropriate lease length can effectively reduce the average delay.

4.2   Analysis Based on the Simple Model

We use the formulas in the previous section to predict the performance of the DCC in the common IOT cache mechanism. In this section, We use actual run data of QUANWIFI [16], a distributed router management cloud platform, for model performance simulation. The system is deployed in the Ali cloud's Qingdao data center and the Shanghai data center, it manages a total of 10,000 wireless routers, the distributed nodes collect running data of the routers in real time and send the control instruction to realize the remote devices management. The system performance parameters were measured, as shown in Table 2.

Table 2. Parameters for QUANWIFI

Description	Symbol	Observed values
Number of groups	N	5
Message propagation time between nodes and master	d	65.6 ms
Message propagation time between nodes in the group	d'	≈1 ms
Message processing time	p	≈15 ms
Rate of reads for nodes within specified group	R	0.864/s
Rate of reads for nodes not in specified group	R'	0.3/s
Rate of writes for nodes within specified group	W	0.1/s
Rate of writes for nodes not in specified group	W'	0.035/s

Figure 2 shows the change curve of the relative load of the server with different numbers of nodes and lease term for consistency using Eq. 4 in Sect. 4.1.

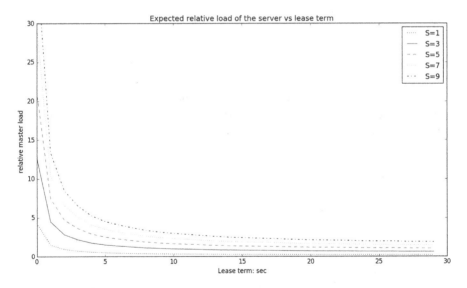

Fig. 2. Load of relative master nodes vs. lease term

As illustrated in Fig. 2, the overhead of the master node for consistency was rapidly reduced with the extension of the lease. Especially for a short lease term, the curve of overhead declined obviously, meaning that the early increase in the lease can obtain the most benefits. As shown in Fig. 2, the load for consistency was reduced to 10% for a lease of about 10 s. Considering that the master node's overhead for consistency is only a fraction of its total load, the performance improvement of the extended lease will be small. Therefore, we recommend a short lease of no more than 10 s in DCC.

Equation 8 defines the relationship of the average delay and lease duration for each read and write operation to guarantee consistency. The calculated results according to the set parameters are shown in Fig. 3. Consistent with the load curve of the master node, the average delay was also significantly reduced when the lease is shorter.

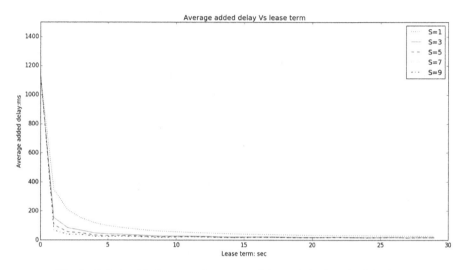

Fig. 3. Relationship between the average delay and the length of the lease

Therefore, a shorter lease (such as 10 s, can show the most benefits of DCC. Because the information transmission delay is low, collision caused by the Fast-paxos algorithm did not affect the overall system effectiveness.

5 Conclusion

We addressed a significant challenge of distributed system design to guarantee data application and sharing efficiency, particularly for a geographically distributed system. This challenge is significant, as each system requires a data synchronization mechanism. Thus, the DCC strategy will greatly facilitate service design.

DCC, presented in this paper, is a complete consistency framework based on the Lease protocol and the Paxos algorithm. DCC can provide data consistency for distributed systems in different applications and allows distributed system design to be freed from data consistency maintenance. In this way, developers can focus on the implementation of the application without needing to address the problem of consistency.

We established a general simple analysis model, analyzed the performance and execution efficiency of the DCC framework in detail, and performed performance prediction using relevant data from a real system. In future work, we will use an actual application system to build a DCC prototype system to further explore the use of DCC for distributed application systems.

References

1. Gray, C.G., Cheriton, D.R.: Leases: an efficient fault-tolerant mechanism for file cache consistency. In: Proceedings of Symposium on Operating Systems Principles (1989)
2. Duvvuri, V., Shenoy, P., Tewari, R.: Adaptive leases: a strong consistency mechanism for the World Wide Web. In: INFOCOM 2000. Nineteenth Joint Conference of the IEEE Computer and Communications Societies. Proceedings, vol. 2, pp. 834–843. IEEE. IEEE Xplore (2000)
3. Kishor Kumar, R., Sujatha, S.G.: EDCM: Efficient Data Caching Method for maintaining cache consistency in wireless mobile networks. Int. J. Eng. Sci. Comput. 7, 693–698 (2014)
4. Bibiana, J.J.: Cache replication-an efficient cache consistency scheme for the server based, networks. Int. J. Eng. Sci. Res. Technol. 3(1) (2014)
5. Ganesarathinam, S., Kanniappan, V.: Cluster based cache consistency using agent technique in mobile environment. J. Comput. Sci. 10(4), 614–622 (2014)
6. Lingamaiah, G.: SDCI: Scalable Distributed Cache Indexing for cache consistency for mobile environments. IJRET Org 02(4), 428–435 (2013)
7. Ros, A., Jimborean, A.A.: Hybrid static-dynamic classification for dual-consistency cache coherence. IEEE Trans. Parallel Distrib. Syst. 27(11), 3101–3115 (2016)
8. Gafni, E., Lamport, L.: Disk Paxos. Distrib. Comput. 16(1), 1–20 (2003)
9. Lamport, L.: Generalized consensus and Paxos. Microsoft Research 7(7), IV-809–IV-812 (2005)
10. Lamport, L.: Fast Paxos. Distrib. Comput. 19(2), 79–103 (2006)
11. Franklin, M.J., Kraska, T., Madden, S., et al.: MDCC: Multi-Data Center Consistency. Comput. Sci. 13(4), 257–259 (2012)
12. Kumar, V., Agarwal, A.: HT-Paxos: high throughput state-machine replication protocol for large clustered data centers. Sci. World J. 2015, 704049 (2015)
13. Marandi, P.J., Primi, M., Pedone, F.: Multi-ring Paxos. In: DSN, pp. 1–12 (2012)
14. Brantner, M., et al.: Building a database on S3. In: Proceedings of SIGMOD (2008)
15. Lee, B.H., Lim, S.H., Kim, J.H., et al.: Lease-based consistency schemes in the web environment. Future Gener. Comput. Syst. 25(1), 8–19 (2009)
16. Liu, S., Zhang, C., Fan, H.: An open and extensible platform for online-to-offline service. J. Balk. Tribol. Assoc. 4, 5284–5296 (2016)

Harmonic Pollution Level Assessment in Distribution System Using Extended Cloud Similarity Measurement Method

Tianlei Zang[1,2(✉)], Yan Wang[1], Zhengyou He[1], and Qingquan Qian[1]

[1] School of Electrical Engineering, Southwest Jiaotong University,
Chengdu 610031, Sichuan Province, China
zangtianlei@126.com
[2] Key Laboratory of Electric Power Big Data of Guizhou Province,
Guizhou Institute of Technology, Guiyang 550003, Guizhou Province, China

Abstract. The purpose of harmonic pollution level assessment is to judge whether the harmonic pollution of a bus or a region is beyond the limit. Considering the fuzziness of the index criteria values caused by harmonic variation, this paper proposes an assessment method for harmonic pollution level based on the extended cloud similarity measurement. Firstly, according to the main characteristics of harmonic pollution, a set of evaluation index of harmonic pollution level is formed, including the total harmonic distortion rate of voltage, the fifth harmonic current RMS value, the seventh harmonic current RMS value, the eleventh harmonic current RMS value, and thirteenth harmonic current RMS value. Secondly, considering the group decision-making behavior of the harmonic pollution level assessment, the group eigenvalue method is utilized to integrate the weight of multiple operators and to calculate the comprehensive weight of the assessment index. On this basis, the harmonic pollution are divided into four levels, and the index criteria values are expressed as triangular fuzzy numbers. Besides, the fuzzy numbers are unified by the extended cloud. Finally, the extended cloud similarity and the Technique for Order Preference by Similarity to Ideal Solution (TOPSIS) is utilized to get the evaluation results. The monitoring data of seven 10 kV buses in Guangzhou is adopted to verify the validity and rationality of the evaluation method.

Keywords: Harmonic · Pollution level assessment · Extended cloud · Group eigenvalue · TOPSIS

1 Introduction

With the development of electrical power system, a large quantity of nonlinear loads have been connected to the power grid, and harmonics have become the most prominent problem in power quality [1–5]. Therefore, it is significant to study the method of evaluating the harmonic pollution level.

Presently, research on the assessment of the harmonic pollution level is scarce. However, there has been varieties of research focusing on power quality evaluation. In [6], the probability statistics and vector algebra method is employed to obtain the

© Springer Nature Singapore Pte Ltd. 2017
B. Zou et al. (Eds.): ICPCSEE 2017, Part II, CCIS 728, pp. 388–400, 2017.
DOI: 10.1007/978-981-10-6388-6_32

quantitative index for the evaluation of power quality. Considering the fuzziness of some indexes of power quality, such as voltage deviation and voltage fluctuation, a fuzzy mathematics method is applied to evaluate the power quality in [7]. The authors of [8] evaluated the power quality by combining the analytic hierarchy process and fuzzy mathematics. In [9], the matter-element analysis method is adopted to evaluate the power quality. The authors of [10] introduced a power quality assessment method based on extension theory and analytic hierarchy process. The matter-element theory is utilized to analyze the power quality and three kinds of matter-element models are constructed. In the application of artificial intelligence algorithms, an evaluation model of power quality is established based on the radial basis function neural network (RBFNN) in [11]. A projection pursuit method and Shepard interpolation genetic algorithm is used to evaluated the power quality in [12] and [13] respectively.

In current research, only the total harmonic distortion rate of voltage (THD_v) or the harmonic ratio of voltage (HRV) is utilized as the index of the harmonic pollution level [14, 15]. In view of this, this paper established an index set of harmonic pollution level evaluation. Besides, considering the fuzziness of the index criteria values caused by the harmonic variation, we present a method evaluating the harmonic pollution level based on the group eigenvalue, extended cloud and TOPSIS.

2   Establishment of Index Set for the Harmonic Pollution Level Assessment

The harmonic pollution level should be comprehensively evaluated through a series of indexes. Figure 1 shows the 95% probability statistic results of harmonic currents of a 10 kV bus. Correspondingly, the 95% probability statistic results of harmonic voltage ratios are as shown in Fig. 2. I_h is the h-th harmonic current. HRV_h is the h-th harmonic ratio of voltage. The data were collected from the harmonic analysis report of a 10 kV bus in Guangzhou power grid, China.

As shown in Figs. 1 and 2, the 95% probability values of the 5th harmonic current and the 7th harmonic current are greater than the limits of Chinese national standard.

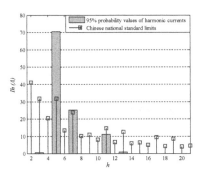

Fig. 1. The 95% probability statistical results of harmonic currents

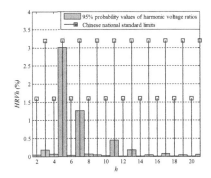

Fig. 2. The 95% probability statistical results of harmonic voltage ratios

However, the 95% probability values of harmonic voltage ratios do not exceeds the limits of Chinese national standard. Therefore, the harmonic current should be taken into account when the level of harmonic pollution is evaluated. It can be seen from Fig. 1, the values of even harmonic currents are small. With the increase of the harmonic order, the content of harmonic current will decrease. Therefore, the THD_v, the RMS values of the 5th harmonic current, the 7th harmonic current, the 11th harmonic current and the 13th harmonic current are selected as the evaluation index of the harmonic pollution level.

According to Chinese national standard GB/T 14549-1993, four levels of harmonic pollution are divided and their criteria values for all the indexes are set. Specifically, the Chinese national standard limits of GB/T 14549-1993 are the second level, indicating that the level of harmonic pollution is mild. Level 1 indicates that harmonic level is qualified. Level 3 and level 4 indicate that the level of harmonic pollution is beyond the Chinese national standard. The distance between the harmonic current index levels is 20% of the Chinese national standard limits. The level difference of the THD_v is 2%. The harmonic pollution level assessment indexes and its level reference values of the 10 kV distribution network are shown in Table 1. The four reference levels of harmonic pollution as well as positive and negative ideal solutions, are denoted as L_1, L_2, L_3, L_4, F_+ and F_-, respectively. The intervals between the four level and positive & negative ideal solutions represent as qualified, mild pollution, moderate pollution, serious pollution and severe pollution, respectively. According to statistics, the maximum value of harmonic current of 10 kV bus in Guangzhou power grid can reach 80A. In order to adapt to the vast majority of conditions, the negative ideal solutions of the five indexes in this paper are all set to 100.

Table 1. Evaluation indexes and levels division

Reference levels and ideal solutions	Total harmonic distortion rate of voltage THD_v (%)	RMS value of the 5th harmonic current I_5 (A)	RMS value of the 7th harmonic current I_7 (A)	RMS value of the 11th harmonic current I_{11} (A)	RMS value of the 13th harmonic current I_{13} (A)
F_+	0	0	0	0	0
L_1	2	16	12	7.44	6.32
L_2	4	20	15	9.30	7.90
L_3	6	24	18	11.16	9.48
L_4	8	28	21	13.02	11.06
F_-	100	100	100	100	100

3   Comprehensive Assessment Model of Harmonic Pollution Level

3.1   The Basic Principle of the Harmonic Pollution Level Assessment

The assessment of the harmonic pollution level can be transformed into a multiple criteria decision making problem for various schemes. Suppose a set of schemes

$X_1, X_2, \cdots, X_m$, and the indexes of the scheme are denoted as $y_1, y_2, \cdots, y_n$. The decision matrix of the above problems is constructed as follows:

$$X = \begin{bmatrix} x_{11} & x_{12} & \cdots & x_{1n} \\ x_{21} & x_{22} & \cdots & x_{2n} \\ \vdots & \vdots & \ddots & \vdots \\ x_{m1} & x_{m2} & \cdots & x_{mn} \end{bmatrix}. \tag{1}$$

Where x_{ij} is the value of the j-th index y_j in scheme X_i.

According to Table 1, the four levels and positive & negative ideal solutions can be regarded as 6 schemes, each of which is corresponding to 5 indexes, namely, THD_v, I_5 I_7, I_{11}, and I_{13}. Table 1 can be expressed as the decision matrix of the harmonic pollution level:

$$X = \begin{bmatrix} 0 & 0 & 0 & 0 & 0 \\ 2 & 16 & 12 & 7.44 & 6.32 \\ 4 & 20 & 15 & 9.30 & 7.90 \\ 6 & 24 & 18 & 11.16 & 9.48 \\ 8 & 28 & 21 & 13.02 & 11.06 \\ 100 & 100 & 100 & 100 & 100 \end{bmatrix}. \tag{2}$$

For a period of time, the assessment of harmonic pollution level is affected by the change of harmonic parameters. Due to the harmonic voltage and current analysis and report data usually include the maximum value, the average value and the 95% probability value, the attribute value x_{ij} can be described as triangular fuzzy number.

Definition 1: $x_{ij} = \langle a, b, c \rangle$ is triangular fuzzy number, where $0 < a \leq b \leq c$, its membership function is

$$\mu(x) = \begin{cases} 0 & x \leq a \\ \frac{x-a}{b-a} & a < x \leq b \\ \frac{x-c}{b-a} & b < x \leq c \\ 0 & x > c \end{cases}. \tag{3}$$

According to the numerical value, a, b and c represent the mean value, the 95% probability value and the maximum value of the THD_v or the RMS value of each harmonic current, respectively.

3.2   Extended Cloud and Extended Cloud Similarity Measurement

In order to uniformly calculate fuzzy numbers, the triangular fuzzy number can be converted into the triangular cloud. The specific conversion method is as follows.

Definition 2: The triangle cloud is composed of two semi-clouds with half rising and half declining, and these two semi-clouds have the same expectations. However, the amplitudes of increase and decrease and the distribution of cloud droplets on the semi-cloud are different. Thus, the triangular cloud can be denoted as $C_s(Ex, En_1, En_2, He_1, He_2)$.

According to the definition of triangular fuzzy number, the triangular fuzzy number $x_{ij} = \langle a, b, c \rangle$ can be converted into the triangular cloud:

$$
\begin{cases}
Ex = b \\
En_1 = \frac{b-a}{3} \\
En_2 = \frac{c-b}{3} \\
He_1 = He_2 = k
\end{cases}
\tag{4}
$$

Where, Ex is the expectation of the triangular cloud, En_1 and En_2 are the entropy of the triangle cloud, He_1 and He_2 are the hyper entropy of the triangle cloud, and k is a constant.

Definition 3: The extended cloud can be represented through six characteristic values, denoted as $C_g(Ex_1, Ex_2, En_1, En_2, He_1, He_2)$, where (Ex_1, En_1, He_1) represents the upper semi-cloud, while (Ex_2, En_2, He_2) represents the bottom semi-cloud. When $Ex_1 = Ex_2, En_1 \neq En_2, He_1 \neq He_2$, the extended cloud will degrade into the triangular cloud.

For two extended cloud models $C(Ex_1, Ex_2, En_1, En_2, He_1, He_2)$ and $C'(Ex'_1, Ex'_2, En'_1, En'_2, He'_1, He'_2)$, assuming that $\vec{v}_1 = (Ex_1, En_1, He_1)$ and $\vec{v}_2 = (Ex_2, En_2, He_2)$ are the upper semi-cloud and the bottom semi-cloud of the extended cloud $C(Ex_1, Ex_2, En_1, En_2, He_1, He_2)$. Correspondingly, $\vec{v}'_1 = (Ex'_1, En'_1, He'_1)$ and $\vec{v}'_2 = (Ex'_2, En'_2, He'_2)$ are the upper semi-cloud and the bottom semi-cloud of the extended cloud $C'(Ex'_1, Ex'_2, En'_1, En'_2, He'_1, He'_2)$.

Definition 4: For the two clouds i and j, the similarity between i and j can be expressed as

$$
sim(i, j) = cos\left(\vec{v}_i, \vec{v}_j\right) = \frac{\vec{v}_i \cdot \vec{v}_j}{\|\vec{v}_i\| \cdot \|\vec{v}_j\|}.
\tag{5}
$$

Where $\|\vec{v}_i\| = \sqrt{(Ex_1)^2 + (En_1)^2 + (He_1)^2}$, $\|\vec{v}_j\| = \sqrt{(Ex_2)^2 + (En_2)^2 + (He_2)^2}$.

According to similarity of the cloud model, the similarity between the extended clouds $C(Ex_1, Ex_2, En_1, En_2, He_1, He_2)$ and $C'(Ex'_1, Ex'_2, En'_1, En'_2, He'_1, He'_2)$ can be calculated as follows:

$$
\begin{aligned}
sim(C, C') &= \alpha \frac{\vec{v}_1 \cdot \vec{v}'_1}{\|\vec{v}_1\| \cdot \|\vec{v}'_1\|} + (1 - \alpha) \frac{\vec{v}_2 \cdot \vec{v}'_2}{\|\vec{v}_2\| \cdot \|\vec{v}'_2\|} \\
&= \alpha \cos\left(\vec{v}_1 \cdot \vec{v}'_1\right) + (1 - \alpha) \cos\left(\vec{v}_2 \cdot \vec{v}'_2\right).
\end{aligned}
\tag{6}
$$

Where α is the scale coefficient, in this paper, $\alpha = 0.5$.

3.3 Weight Determination of the Harmonic Pollution Level Evaluation Index Based on the Group Eigenvalue

The actual harmonic pollution level should be evaluated by multiple power grid operators. Thus, the group eigenvalue method [16] is adopted to determine the index weight of the harmonic pollution level in this paper. It is assumed that there are n buses whose harmonic pollution levels are to be evaluated, m evaluation indexes, and o power grid operators participating in the evaluation, which are denoted as $M = \{1, 2, \cdots, m\}$, $N = \{1, 2, \cdots, n\}$, and $O = \{1, 2, \cdots, o\}$, respectively.

The index set of the harmonic pollution level evaluation is assumed to be $I = \{I_1, I_2, \cdots, I_m\}$. The set of the evaluation experts of the harmonic pollution level is $E = \{E_1, E_2, \cdots, E_o\}$. Each expert independently endow the weights for all the assessment indexes. The weight vector endowed by expert E_i is $W_i = [w_{i1}, w_{i2}, \cdots, w_{im}]$, where $w_{ij} \in [0, 1], i \in E, j \in M$, and $\sum_{j=1}^{m} w_{ij} = 1$.

The ideal evaluation expert is denoted as E^*, and the weight endowed by E^* represents the view of the whole expert group.

Thus, assuming that G which makes f maximum is the weight W^* of each index given by the ideal evaluation expert E^*.

$$f = \sum_{i=1}^{o} (G^T W_i)^2. \tag{7}$$

Where, $G = [g_1, g_2, \cdots, g_n]^T \in E^n$.

When $\|G\|_2 = 1$, $\max\limits_{\|G\|_2 = 1} \sum_{i=1}^{o} (G^T W_i)^2 = \sum_{i=1}^{o} (W^{*^T} W_i)^2$.

Since $w_{ij} > 0$, the matrix $M = W^T W$ composed of the comprehensive weight evaluation matrix W is the nonnegative irreducible matrix.

According to Perron-Frobenius theorem [16], M has the maximum positive eigenvalue $\lambda_{\max}$; $\lambda_{\max}$ is corresponding to all positive eigenvectors; and $\lambda_{\max}$ increases with the elements of matrix M.

Moreover,

$$\max\limits_{\|G\|_2 = 1} \sum_{i=1}^{o} (G^T W_i)^2 = \sum_{i=1}^{o} (W^{*^T} W_i)^2 = \lambda_{\max}. \tag{8}$$

Where, W^* is the positive eigenvectors, $\|W^*\|_2 = 1$, denoted as the comprehensive weight integrating the opinions of all experts.

The comprehensive weight W^* can be obtained by the numerical power method as follows.

Step 1: iteration count $k = 0$, $y_0 = [1/n, 1/n, \cdots, 1/n]^T \in E^n$;
Step 2: let $y_1 = M y_0, z_1 = y_1 / \|y_1\|_2$;
Step 3: update $k = k + 1, y_{k+1} = M z_k, z_{k+1} = y_{k+1} / \|y_{k+1}\|_2$;

Step 4: calculate $\varepsilon_z = \max\limits_{j \in N}|z_{k+1,j} - z_{k,j}|$, if it meets the accuracy requirements, the comprehensive weight of the harmonic pollution level assessment is z_{k+1}; otherwise, go to Step 3.

3.4   Evaluation Process of Harmonic Pollution Level Based on the Extended Cloud Similarity Measurement

According to the above description, the evaluation model of harmonic pollution level based on the extended cloud similarity measurement is shown as follows. First of all, the evaluated values of all the harmonic pollution indexes are calculated. Then, the harmonic pollution level of the evaluated bus is further evaluated.

Step 1: Construct the reference levels and positive and negative ideal solutions. The four reference levels and positive & negative ideal solutions mentioned in Sect. 2 are constructed. In accordance with the fuzzy characteristic of each index, the fuzzy criteria values of the indexes are given. Meantime, the comprehensive weight is calculated through the group eigenvalue method.

Step 2: Normalize the criteria values. In this paper, the evaluation indexes of harmonic pollution level are cost indexes, which can be normalized as

$$n_{ij}^k = \frac{x_{ij}^k}{\max\left[x_{ij}^k\right]} \tag{9}$$

Where, $\max\left[x_{ij}^k\right]$ are the maximum of each fuzzy number under a criterion.

Step 3: Convert all the triangular fuzzy criteria values of indexes into the extended clouds according to Eq. (4).

Step 4: Calculate the positive and negative ideal solutions under each criterion. The hyper entropy of the positive ideal solution is set to 0, and the hyper entropy of the negative ideal solution is set to 0.02. The positive and negative ideal solutions are I^+ and I^-, respectively.

$$\begin{aligned} I^+ &= \{max[Ex_{1i}], max[Ex_{2i}], min[En_{1i}], min[En_{2i}], 0, 0\} \\ I^- &= \{min[Ex_{1i}], min[Ex_{2i}], max[En_{1i}], max[En_{2i}], 0.02, 0.02\} \end{aligned} \tag{10}$$

Step 5: According to Eq. (6), the similarity between each reference level and the positive & negative ideal solutions are calculated and denoted as S_i^+ and S_i^-.

Step 6: Substitute the weight of each index, calculate the distances between the reference level and the positive & negative ideal solutions d_i^+ and d_i^-. And the relatively close degree between the reference level and the positive ideal solution R_i.

$$d_i^+ = \sum_{j=1}^{5} W_j S_{ij}^+ \; ; d_i^- = \sum_{j=1}^{5} W_j S_{ij}^-. \tag{11}$$

$$R_i = d_i^- / (d_i^+ + d_i^-). \tag{12}$$

Step 7: Give the evaluation value interval of qualified, mild pollution, moderate pollution, serious pollution and severe pollution.

Step 8: Input the index criteria values of the evaluated bus, and repeat Step 2–Step 7. Finally, divide the harmonic pollution level of each bus according to the obtained evaluation values.

4 Example Analysis

The statistics of A phase monitoring data during 24 h from seven 10 kV buses (Bus 1– Bus 7) in Guangzhou power grid of China are collected, the relevant data is shown in Table 2, of which 3 values in parentheses are the mean value, the 95% probability value and the maximum value of the $THDv$ or the RMS value of harmonic current, respectively.

Table 2. The index fuzzy criteria values of each bus

Bus	THD_v (%)	I_5 (A)	I_7 (A)	I_{11} (A)	I_{13} (A)
Bus 1	(7.93, 8.13, 8.14)	(13.75, 25.24, 30)	(0.06, 2.32, 30)	(0, 0, 0)	(0, 0, 0)
Bus 2	(0.91, 1.42, 1.51)	(20.08, 37.57, 40)	(3.40, 6.65, 8)	(0.02, 0.62, 8)	(0.32, 2.89, 8)
Bus 3	(7.83, 8.07, 8.08)	(2.75, 19.42, 40)	(0, 0, 0)	(0, 0, 0)	(0, 0, 0)
Bus 4	(7.19, 7.39, 7.40)	(11.38, 24.55, 30)	(8.81, 23.80, 30)	(0, 0, 0)	(0, 0, 0)
Bus 5	(3.25, 6.66, 8.15)	(5.61, 6.55, 7.44)	(2.13, 2.51, 2.88)	(0.69, 0.89, 0.96)	(0.35, 0.57, 0.72)
Bus 6	(7.03, 7.20, 7.27)	(0, 0, 0)	(0, 0, 0)	(0, 0, 0)	(0, 0, 0)
Bus 7	(1.59, 1.84, 1.85)	(6.35, 10.35, 12)	(1.16, 5.07, 6)	(1.53, 5.83, 6)	(0, 0, 0)

Assume that five experts E_1–E_5 participated in evaluating the harmonic pollution level. The five experts endowed the weight of each index, and the vectors are

$$W = \begin{array}{c} E_1 \\ E_2 \\ E_3 \\ E_4 \\ E_5 \end{array} \begin{bmatrix} 0.50 & 0.25 & 0.10 & 0.10 & 0.05 \\ 0.50 & 0.25 & 0.15 & 0.05 & 0.05 \\ 0.55 & 0.2 & 0.15 & 0.05 & 0.05 \\ 0.55 & 0.2 & 0.10 & 0.10 & 0.05 \\ 0.60 & 0.2 & 0.10 & 0.05 & 0.05 \end{bmatrix}$$

Based on the group eigenvalue method, the comprehensive weight of harmonic pollution level assessment is obtained.

$$M = W^T W = \begin{bmatrix} 1.4650 & 0.5900 & 0.3225 & 0.1875 & 0.1350 \\ 0.5900 & 0.2450 & 0.1325 & 0.0775 & 0.0550 \\ 0.3225 & 0.1325 & 0.0750 & 0.0400 & 0.0300 \\ 0.1875 & 0.0775 & 0.0400 & 0.0275 & 0.0175 \\ 0.1350 & 0.0550 & 0.0300 & 0.0175 & 0.0125 \end{bmatrix}$$

The power method is employed to calculate the comprehensive weight when the iterative accuracy is set to $\varepsilon = 0.0001$. The iterative calculation process as shown in Table 3.

Table 3. The iterative calculation process for evaluating the comprehensive weight

k	y_k	z_k
0	[0.2, 0.2, 0.2, 0.2, 0.2]	–
1	[0.5400, 0.2200, 0.1200, 0.0700, 0.0500]	[0.5400, 0.2200, 0.1200, 0.0700, 0.0500]
2	[0.9795, 0.3966, 0.2166, 0.1259, 0.0905]	[0.5414, 0.2192, 0.1197, 0.0696, 0.0500]
3	[0.9810, 0.3972, 0.2169, 0.1261, 0.0906]	[0.5415, 0.2192, 0.1197, 0.0696, 0.0500]
4	[0.9810, 0.3972, 0.2169, 0.1261, 0.0906]	[0.5415, 0.2192, 0.1197, 0.0696, 0.0500]

As can be seen from Table 3, the comprehensive weight is

$$W^* = \begin{bmatrix} 0.5415 & 0.2192 & 0.1197 & 0.0696 & 0.0500 \end{bmatrix}$$

Prior to calculating the evaluation value of each reference level of the harmonic pollution, the index criteria values of different levels in Table 1 are rewritten as the triangular fuzzy numbers as shown in Table 4.

Table 4. The fuzzy index criteria values of different reference levels and ideal solutions

Level	THD_v (%)	I_5 (A)	I_7 (A)	I_{11} (A)	I_{13} (A)
F_+	(0, 0, 0)	(0, 0, 0)	(0, 0, 0)	(0, 0, 0)	(0, 0, 0)
X_1	(2, 2, 2)	(16, 16, 16)	(12, 12, 12)	(7.44, 7.44, 7.44)	(6.32, 6.32, 6.32)
X_2	(4, 4, 4)	(20, 20, 20)	(15, 15, 15)	(9.3, 9.3, 9.3)	(7.9, 7.9, 7.9)
X_3	(6, 6, 6)	(24, 24, 24)	(18, 18, 18)	(11.16, 11.16, 11.16)	(9.48, 9.48, 9.48)
X_4	(8, 8, 8)	(28, 28, 28)	(21, 21, 21)	(13.02, 13.02, 13.02)	(11.06, 11.06, 11.06)
F_-	(100, 100, 100)	(100, 100, 100)	(100, 100, 100)	(100, 100, 100)	(100, 100, 100)

After Step 2–5, the similarity between each reference level and positive and negative ideal solution is obtained:

$$S_{ref}^+ = \begin{bmatrix} 0.0200 & 0.0200 & 0.0200 & 0.0200 & 0.0200 \\ 0.7211 & 0.9946 & 0.9895 & 0.9707 & 0.9592 \\ 0.9032 & 0.9968 & 0.9937 & 0.9817 & 0.9741 \\ 0.9548 & 0.9980 & 0.9959 & 0.9876 & 0.9824 \\ 0.9748 & 0.9987 & 0.9972 & 0.9912 & 0.9874 \\ 1.0000 & 1.0000 & 1.0000 & 1.0000 & 1.0000 \end{bmatrix}$$

$$S_{ref}^- = \begin{bmatrix} 1.0000 & 1.0000 & 1.0000 & 1.0000 & 1.0000 \\ 0.7071 & 0.1240 & 0.1644 & 0.2596 & 0.3017 \\ 0.4472 & 0.0995 & 0.1322 & 0.2102 & 0.2454 \\ 0.3162 & 0.0830 & 0.1104 & 0.1764 & 0.2064 \\ 0.2425 & 0.0712 & 0.0948 & 0.1518 & 0.1779 \\ 0.0200 & 0.0200 & 0.0200 & 0.0200 & 0.0200 \end{bmatrix}$$

Substitute the comprehensive weight W^* to obtain the distance between the reference level and the positive & negative ideal solutions:

$$d_{ref}^+ = [0.0200 \quad 0.8425 \quad 0.9436 \quad 0.9729 \quad 0.9845 \quad 1.0000]$$
$$d_{ref}^- = [1.0000 \quad 0.4629 \quad 0.3067 \quad 0.2252 \quad 0.1778 \quad 0.0200]$$

Obtain the relatively close degree between each reference level and the positive & negative ideal solutions:

$$R_{ref} = [0.9804 \quad 0.3546 \quad 0.2453 \quad 0.1880 \quad 0.1529 \quad 0.0196]$$

According to the evaluation value vector, the evaluation value intervals of the qualified, mild pollution, moderate pollution, serious pollution and severe pollution are as shown in Table 5.

Table 5. The corresponding relationship between the evaluation value interval and the harmonic level

Level no.	Evaluation value interval	Harmonic level
1	[0.3546, 0.9804]	Qualified
2	[0.2453, 0.3546)	Mild pollution
3	[0.1880, 0.2453)	Moderate pollution
4	[0.1529, 0.1880)	Serious pollution
5	[0, 0.1529)	Severe pollution

According to the evaluation index criteria values of the buses to be evaluated in Table 2, the index fuzzy criteria values of the positive and negative ideal solution are added to form the decision matrix. Similarly, after Step 2–5, the similarity between the harmonic level of each bus and positive & negative ideal solution is obtained:

$$S_{bus}^+ = \begin{bmatrix} 0.0200 & 0.0200 & 0.0200 & 0.0200 & 0.0200 \\ 0.3911 & 0.6853 & 0.0770 & 0.0200 & 0.0200 \\ 0.0781 & 0.8052 & 0.2607 & 0.0359 & 0.0993 \\ 0.3879 & 0.4286 & 0.0200 & 0.0200 & 0.0200 \\ 0.3612 & 0.6464 & 0.5972 & 0.0200 & 0.0200 \\ 0.2573 & 0.3097 & 0.1350 & 0.0594 & 0.0430 \\ 0.3539 & 0.0200 & 0.0200 & 0.0200 & 0.0200 \\ 0.1053 & 0.3987 & 0.1714 & 0.1976 & 0.0200 \\ 1.0000 & 1.0000 & 1.0000 & 1.0000 & 1.0000 \end{bmatrix}$$

$$S_{bus}^- = \begin{bmatrix} 0.9992 & 0.9816 & 0.9743 & 0.9981 & 0.9990 \\ 0.9273 & 0.7258 & 0.9968 & 0.9981 & 0.9990 \\ 0.9976 & 0.5889 & 0.9505 & 0.9998 & 0.9948 \\ 0.9287 & 0.8574 & 0.9743 & 0.9981 & 0.9990 \\ 0.9389 & 0.7522 & 0.7772 & 0.9981 & 0.9990 \\ 0.9680 & 0.9425 & 0.9693 & 0.9975 & 0.9988 \\ 0.9415 & 0.9816 & 0.9743 & 0.9981 & 0.9990 \\ 0.9956 & 0.9137 & 0.9623 & 0.9775 & 0.9990 \\ 0.0200 & 0.0196 & 0.0195 & 0.0200 & 0.0200 \end{bmatrix}$$

Substitute the comprehensive weight to obtain the distance between the harmonic level of the buses to be evaluated and the positive & negative ideal solutions:

$$d_{bus}^+ = [0.0200 \quad 0.8073 \quad 0.7111 \quad 0.7409 \quad 0.8626 \quad 0.8675 \quad 0.5337 \quad 0.7664 \quad 1.0000]$$
$$d_{bus}^- = [0.6915 \quad 0.3251 \quad 0.5545 \quad 0.3180 \quad 0.2741 \quad 0.3111 \quad 0.3280 \quad 0.4934 \quad 0.0138]$$

Obtain the relatively close degrees of harmonic pollution level of the buses and the positive and negative ideal solutions, namely, the evaluation values of harmonic pollution level of the buses is

$$R_{bus} = [0.2871 \quad 0.4381 \quad 0.3003 \quad 0.2412 \quad 0.2640 \quad 0.3806 \quad 0.3917]$$

The evaluation values and the harmonic pollution levels of the buses are shown in Fig. 3.

The evaluation results indicate that the harmonic levels of bus 2, bus 6 and bus 7 are among the qualified range. Bus 1, bus 3 and bus 6 are mild pollution. And bus 4 has reaches moderate pollution. By referring to IEC standard IEC61000-4-7, Chinese national standard GB/T 14549-1993 and other related standards, it can be seen that the proposed method evaluate the harmonic pollution level with fuzzy index criteria values is reasonable.

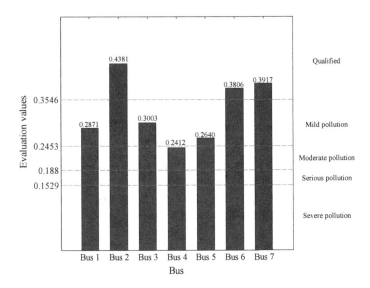

Fig. 3. Evaluation values and harmonic pollution levels of each bus

5   Conclusion

Considering the group decision making behavior and fuzziness of evaluation index criteria values in harmonic pollution level assessment, this paper proposed a method evaluating the harmonic pollution level based on the extended cloud similarity measurement.

(1) The total harmonic distortion rate of voltage and the RMS values of harmonic current are adopted as the index of harmonic pollution level to form the evaluation index set which can characterize the harmonic pollution level of the bus in the power distribution system.
(2) The group eigenvalue method can integrate the weight information given by multiple experts and get the comprehensive weight of the expert group.
(3) Triangular fuzzy number can describe the fuzziness of index criteria. The extended cloud model can uniformly calculate different triangular fuzzy numbers.
(4) On the basis of the reference levels together with the positive and negative ideal solution, TOPSIS can give a reasonable evaluation result of the harmonic pollution level.

The proposed method can also be utilized to solve the power system decision problems such as service restoration scheme evaluation of power distribution network, black start scheme evaluation and power quality evaluation.

Acknowledgment. The authors acknowledge the support by Science and Technology Support Program of Sichuan Province (No. 2016RZ0079), Open Project of Key Laboratory of electric Power Big Data of Guizhou Province and Guizhou Fengneng Science and Technology Development Co., Ltd.

References

1. Boby, M., Pramanick, S., Kaarthik, R.S., Arun Rahul, S., Gopakumar, K., Umanand, L.: Fifth- and seventh-order harmonic elimination with multilevel dodecagonal voltage space vector structure for IM drive using a single DC source for the full speed range. IEEE Trans. Power Electron. **32**(1), 60–68 (2017)
2. Xavier, L.S., Cupertino, A.F., de Resende, J.T., Mendes, V.F., Pereira, H.A.: Adaptive current control strategy for harmonic compensation in single-phase solar inverters. Electr. Power Syst. Res. **142**(1), 84–95 (2017)
3. Sun, Z.M., He, Z.Y., Zang, T.L., Liu, Yilu: Multi-interharmonic spectrum separation and measurement under asynchronous sampling condition. IEEE Trans. Instrum. Meas. **65**(8), 1902–1912 (2016)
4. Zang, T.L., He, Z.Y., Fu, L., Wang, Y., Qian, Q.Q.: Adaptive method for harmonic contribution assessment based on hierarchical K-means clustering and Bayesian partial least squares regression. IET Gener. Transm. Distrib. **10**(13), 3220–3227 (2016)
5. Zang, T.L., He, Z.Y., Fu, L., Chen, J., Qian, Q.Q.: Harmonic source localization approach based on fast kernel entropy optimization ICA and minimum conditional entropy. Entropy **18**(6), 214 (2016)
6. Jiang, H., Peng, J.C., Ou, Y.P., Li, Z.Y.: Power quality unitary quantification and evaluation based on probability and vector algebra. J. Hunan Univ. (Nat. Sci.) **30**(1), 66–70 (2003)
7. Jia, Q.Q., Song, J.H., Lan, H., Yang, Y.H., Yin, C.X.: Quality of electricity commodity and its fuzzy evaluation. Power Syst. Technol. **24**(6), 46–49 (2000)
8. Zhao, X., Zhao, C.Y., Jia, X.F., Li, J.Y.: Fuzzy synthetic evaluation of power quality based on changeable weight. Power Syst. Technol. **29**(6), 11–16 (2005)
9. Huang, J., Zhou, L., Li, Q.H., Zhang, F., Liu, H.Y.: Evaluation of power quality based on the method of matter-element. J. Chongqing Univ. (Nat. Sci. Ed.) **30**(6), 25–29 (2007)
10. Ding, L., Jia, X.F., Zhao, C.Y., Li, G.Y.: Synthetic evaluation of power quality based on extenics. Electr. Power Autom. Equip. **27**(12), 44–47, 52 (2007)
11. Liu, Y.Y., Li, G.D., Gu, Q., Xu, Y.H.: The comprehensive evaluation of power quality based on the RBF neural network. Electrotech. Appl. **26**(1), 45–48 (2007)
12. Zhou, L., Li, Q.H., Zhang, F.: Application of genetic projection pursuit interpolation model on power quality synthetic evaluation. Power Syst. Technol. **31**(7), 32–35 (2007)
13. Li, Q.H., Zhou, L., Liu, H.Y., Zhang, F.: Evaluation of power quality by accelerating genetic algorithm and Shepard interpolation. High Volt. Eng. **33**(7), 139–143 (2007)
14. Cavallini, A., Cacciari, M., Loggini, M., Montanari, G.C.: Evaluation of harmonic levels in electrical networks by statistical indexes. IEEE Trans. Ind. Appl. **34**(4), 1116–1126 (1994)
15. Lai, L.L., Chan, W.L., Tse, C.T., So, A.T.P.: Real-time frequency and harmonic evaluation using artificial neural networks. IEEE Trans. Power Delivery **14**(1), 52–59 (1999)
16. Zang, T.L., He, Z.Y., Qian, Q.Q.: Distribution network service restoration multiple attribute group decision-making using entropy weight and group eigenvalue. In: Proceedings of the 3rd International Conference on Intelligent System Design and Engineering Applications, Hong Kong, China, pp. 602–606 (2013)

Fusion of Multimodal Color Medical Images Using Quaternion Principal Component Analysis

Qamar Nawaz$^{(\boxtimes)}$, Xiao Bin, Li Weisheng, and Isma Hamid

Chongqing University of Posts and Telecommunications, Chongqing, China
mqamarnawaz@hotmail.com

Abstract. Multimodal medical image fusion is used to merge functional and structural information of the same body organ. Most of the multimodal image fusion algorithms are designed to fuse grayscale images that are produced by different imaging modalities. It is likely that problems of colour distortion and information loss will occur in fused image when source images are fused by using algorithms that are not originally designed to fuse colour images. These problems can be avoided by representing and processing source images as quaternion numbers. Quaternion representation of a colour pixel encodes information of its colour channels on the imaginary parts of a quaternion number and provides the advantage to processing colour information holistically as a vector field. In this paper, we proposed an image fusion algorithm based on Quaternion Principal Component Analysis (QPCA), to fuse multimodal colour medical images. Quaternion principal components are calculated by decomposing quaternion covariance matrix using Quaternion Eigenvalue Decomposition (QEVD). Fusion rule is designed, based on the fusion weights that are extracted from the highly influential principal component. To test the performance of the proposed algorithm, experiments have been performed on six image-sets of multimodal colour images of the brain. Experimental results are compared objectively with existing image fusion algorithms. Comparison results show that the proposed algorithm performed better than existing algorithms in fusing colour medical images.

Keywords: QCPA based image fusion · Multimodal image fusion · Colour image fusion · Medical image fusion

1 Introduction

Medical imaging is an essential and cost-effective clinical tool that is used to obtain inside images of a human body by using different medical imaging modalities. Due to operational limitations, an imaging modality is able to obtain either structural or functional image of a body organ [1]. For better diagnostics, it is sometimes necessary to get structural and functional images of the body organ into a single composite image, to get more understanding of the problem related to the organ under study. The process of combining multiple images into a single composite image is called image fusion [2]. The fused image provides more information specific to an application by reducing the ambiguity and minimising the redundancy [3]. To get all relevant information from

© Springer Nature Singapore Pte Ltd. 2017
B. Zou et al. (Eds.): ICPCSEE 2017, Part II, CCIS 728, pp. 401–413, 2017.
DOI: 10.1007/978-981-10-6388-6_33

source images into fused image, without introducing noise or unexpected features, is the basic and essential requirement of image fusion process [4, 5].

In recent years, many image fusion algorithms have been proposed in different fields of applications. Most of the existing algorithms have been designed to fuse grayscale images. The algorithm designed to fuse grayscale images can also be used to fuse colour images through a straight forwarded implementation i.e. split each colour image into three colour channels, separately apply fusion algorithm on each set of corresponding colour channels to obtain fused colour channels, and obtain final fused image by merging all fused channels. The main problem with this implementation is its inability to process colour information holistically that leads to loss of colour information.

In past few years, quaternion numbers [6] are being used extensively in colour image processing. Quaternion representation of a colour image encodes its colour channels on imaginary parts of a quaternion number that provides the advantage to processing colour information holistically as vector field [7–9]. In this paper, we proposed an algorithm for multi-modal colour image fusion that is based on Quaternion Principal Component Analysis (QPCA) [10]. Up to best of our knowledge, this is the first ever attempt to use QPCA to fuse multimodal colour medical images. To test the performance of the proposed algorithm, we performed experiments on six different image-sets. Each image-set contains two perfectly registered multimodal brain images. We compared the performance of the proposed algorithm with three existing multimodal image fusion algorithms using objective quality assessment method. Comparative analysis of quality measures shows the superiority of the proposed algorithm over existing algorithms in the same domain.

The rest of the paper is organised as follows. Section 2 gives an overview of quaternion numbers. Section 2 elaborates representation of the color image as quaternions. Section 3 describes Quaternion Principal Component Analysis. Section 4 presents proposed methods. Section 5 includes experiments and discussion on the results, and Sect. 6 concludes the paper.

2 Quaternions and Color Image Representation

Quaternion numbers, introduced by William Rowan Hamilton in 1894, are a generalisation of complex numbers [6]. Unlike a complex number, one real and three imaginary numbers constitute a quaternion number [11, 12], as represented by Eq. 1.

$$q = w + xi + yj + zk \tag{1}$$

where w, x, y & z are real parts, whereas i, j & k are imaginary parts, having the following properties.

$$i^2 = j^2 = k^2 = ijk - 1 \tag{2}$$

$$ij = k, \quad jk = i, \quad ki = j \tag{3}$$

$$ji = -k, \quad kj = -i, \quad ik = -j \tag{4}$$

Equations 3 and 4 show that the quaternion multiplication operation is not commutative which is a major difference between the computations of quaternions and complex numbers. Conjugate and modulus of quaternion numbers can be taken by using Eqs. 5 and 6, respectively.

$$\bar{q} = w - xi - yj - zk \qquad (5)$$

$$|q| = \sqrt{w^2 + x^2 + y^2 + z^2} \qquad (6)$$

A quaternion number having $w = 0$ is called a pure quaternion and a quaternion number having unit modulus is called a unit quaternion. A unit quaternion can be represented as:

$$q = \frac{i + j + k}{\sqrt{3}} \qquad (7)$$

A quaternion number can also be represented as the sum of its vector and scalar parts as:

$$q = S(q) + V(q) \qquad (8)$$

where $S(q)$ represents scalar part i.e. $S(q) = w$ and $V(q)$ represents vector part i.e. $V(q) = xi + yj + zk$. For more detail about quaternion algebra, please refer to [6].

An RGB colour image is composed of three channels i.e. red, green and blue. Each pixel in colour image represents values of all three channels. As, a pure quaternion has three imaginary parts, so it is possible to represent a colour pixel as a pure quaternion by mapping red, green and blue components over imaginary parts of the pure quaternion [13]. The whole colour image can be represented as an array of pure quaternion numbers. In recent years, quaternion representation of colour images is actively used in image processing [3, 5–8]. Equation 9 shows the quaternion representation of a colour pixel $f(x, y)$.

$$f(x, y) = r(x, y)i + g(x, y)j + b(x, y)k \qquad (9)$$

where red, green and blue component are represented by $r(x, y)i, g(x, y)j$ and $b(x, y)k$, respectively. Let, two colour images are represented as quaternion matrices e.g. $a, b \in \mathbb{H}^{M \times N}$, then cross-correlation of these two quaternion matrices can be defined as [14]:

$$C(m, n) = \sum_{x=1}^{M} \sum_{y=1}^{N} a(x, y)(b(x - m, y - n))' \qquad (10)$$

where (x, y) is the row and column index of images a and b. The shift operation on b is implemented cyclically using modulo arithmetic [15]. When $a = b$, the autocorrelation of these two images is computed. To obtain cross-variance, the mean value of each image is subtracted first. Representation of an RGB colour image as quaternion numbers takes the advantage in processing colour information as a whole unit and therefore, higher accuracy can be achieved [16].

3   Quaternion Principal Component Analysis

Principal Component Analysis (PCA) [17] is a powerful and widely used technique in signal and image processing. PCA is based on Eigenvalue Decomposition (EVD) of real numbered covariance matrix of input signals. The EVD is a method of factorization of the covariance matrix. After factorization, the covariance matrix is represented in the form of eigenvalues and corresponding eigenvectors [18]. In [10], the authors proposed Quaternion Principal Component Analysis (QPCA) for colour images which is based on Quaternion Eigenvalue Decomposition (QEVD) of quaternion covariance matrix. EVD of quaternion matrices is a complex task as compared to EVD of real matrices. The QEVD algorithm states that every quaternion matrix can be decomposed into three matrices.

Let, $Q \in \mathbb{H}^{N \times N}$ is quaternion matrix that can be decomposed as $Q = UDU^\triangleleft$, where $\triangleleft$ represents conjugate transpose, $U \in \mathbb{H}^{N \times N}$ is a unitary matrix, and $D \in \mathbb{C}^{N \times N}$ is a diagonal matrix. The unitary matrix U contains eigenvectors and the diagonal matrix D contains corresponding eigenvalues on its diagonal. The QPCA algorithm to represent colour images, proposed in [10], is as follows.

Let, $I \in \mathbb{H}^{M \times N}$ is quaternion representation of an RGB colour image. The covariance matrix C_i of ith column of I is calculated as:

$$C_i = (c_i - \mu(c_i))(c_i - \mu(c_i))^\triangleleft \tag{11}$$

After calculating covariance matrices of all columns, mean covariance matrix is calculated using Eq. 12.

$$C = \frac{1}{N} \sum_{i=1}^{N} C_i \tag{12}$$

The mean covariance matrix now can be decomposed by using QEVD, and can be written as:

$$C = UDU^\triangleleft \tag{13}$$

As the mean covariance matrix is quaternionic Hermitian, therefore, the diagonal values of the matrix D are real values [19]. By using Eq. 14, all columns of the original covariance matrix are moved from its original basis to uncorrelated basis after diagonalization of the covariance matrix. Repeating this process on all columns regenerates an image.

$$z_i = U^T c_i \tag{14}$$

4   Image Fusion Using QPCA

Let, X and Y are two pre-registered RGB source images of size $m \times m$. Pixels of each image is converted into pure quaternion numbers by using Eqs. 15 and 16.

$$a_{i,j} = (X_{R_{i,j}})i + (X_{G_{i,j}})j + (X_{B_{i,j}})k \tag{15}$$

$$b_{i,j} = (Y_{R_{i,j}})i + (Y_{G_{i,j}})j + (Y_{B_{i,j}})k \tag{16}$$

Repeating this process on each pixel leads to quaternion matrices A and B of input images X and Y, respectively. To calculate covariance matrix, we first converted each quaternion matrix into column matrices of size $m \times 1$ and formed a new matrix Z of size $m \times 2$. Then each element of covariance matrix can be calculated as:

$$c_{x,x} = \frac{1}{m}(Z_x - \mu(Z_x))^\triangleleft(Z_x - \mu(Z_x)) \tag{17}$$

$$c_{x,y} = \frac{1}{m}(Z_x - \mu(Z_x))^\triangleleft(Z_y - \mu(Z_y)) \tag{18}$$

$$c_{y,x} = \frac{1}{m}(Z_y - \mu(Z_y))^\triangleleft(Z_x - \mu(Z_x)) \tag{19}$$

$$c_{y,y} = \frac{1}{m}(Z_y - \mu(Z_y))^\triangleleft(Z_y - \mu(Z_y)) \tag{20}$$

The quaternionic Hermitian covariance matrix C is formed as:

$$C = \begin{bmatrix} c_{x,x} & c_{x,y} \\ c_{y,x} & c_{y,y} \end{bmatrix} \tag{21}$$

Decomposition of a covariance matrix using QEVD gives two matrices V and D of size 2×2. Matrix V is a full matrix that contains eigenvectors, and the matrix D is a diagonal matrix that contains corresponding eigenvalues. Both matrices are sorted in a way that first vector in the matrix V is the corresponding vector of a highest eigenvalue in the matrix D. Finally, fused image is obtained by using following fusion rule.

$$F = A \times \mu\left(\frac{V(1,1)}{\sum V(1)}, \frac{V(2,1)}{\sum V(1)}\right) + B \times \mu\left(\frac{V(1,2)}{\sum V(2)}, \frac{V(2,2)}{\sum V(2)}\right) \tag{22}$$

In the above fusion rule, the order of multiplication is important because quaternion multiplication [6] is non-commutative. Figure 1 shows the block diagram of proposed algorithm.

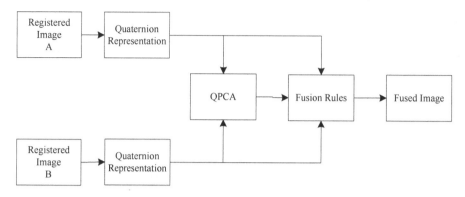

Fig. 1. Block diagram of QPCA based image fusion algorithms

5 Experiments and Results

We used Quaternion and Octonion toolbox for Matlab [20] to implement our proposed algorithm. To perform experiments, we selected six image-sets from a publically available database of brain image [21]. Each image-set consists of two images such that image-sets 1, 2 and 3 contain pair of colour images and image-sets 4, 5 and 6 contain one colour and one grey image. For comparison, we used three existing image fusion algorithms that are based on PCA, Gradient Pyramid, and Discrete Wavelet Transform. Image fusion toolbox [22] provides the implementation of these existing algorithms, but we modified the implementation of PCA based fusion algorithm as: applied PCA based image fusion on R, G and B individually and then merged R, G and B fused results to get the final fused image. For each image-set, fused results are obtained by applying proposed algorithm along with existing image fusion algorithms. Visual results of the experiments are shown in Figs. 2, 3, 4, 5, 6 and 7 for image-sets 1, 2, 3, 4, 5 and 6, respectively.

Subjective quality assessment is a method to visually assess the quality of a fused image. Visual inspection is only one aspect of the comparison and only an expert of medical field can decide whether a fused image represents useful information or not. Even, the opinion of a group of experts could be different about the same fused image. By performing a subjective quality assessment of the fused results, it seems that our proposed algorithm generated acceptable results as compared to existing algorithms. Objective quality assessment [23] is another way to mathematically measure the quality of a fused image using standardised image quality assessment matrices. It also overcomes the shortcoming of artificial factors and takes advantage of statistical properties of the image to establish a steady standard of performance evaluation.

To measure the quality of fused images, we used four full reference image quality matrices, that includes MS-SSIM [24], PSNR [25], RMSE [23] and MI [26]. Full-reference matrices require a reference image in order to assess the quality of fused image. In multimodal image fusion domain, reference image is not usually available therefore, we used a three steps method [27] to measure the quality of fused image: (a) quality is measured using source image 1 as a reference image, (b) quality is measured using source image 2 as reference image, and (c) final quality value is calculated by taking average of quality values produced in steps a and b. Tables 1, 2, 3, 4, 5 and 6 show quality assessment results of fused images in image-sets 1, 2, 3, 4, 5 and 6, respectively.

As per the quality assessment results, in fusing images in image-set 1, the proposed algorithm performed best in two quality matrices (MS-SSIM and RMSE) whereas PCA performed best in remaining two quality matrices (PSNR and MI) followed by the proposed method. The visual result of PCA based image fusion method does not fully incorporate visual information of both source images. Therefore, to assess the fusion quality, we have to incorporate both subjective and objective results. Combination of subjective and objective quality assessment proves the superiority of the proposed algorithm in fusing images in image-set 1. Similarly, the proposed algorithm performed best, in two quality matrices (PSNR, RMSE) for image-set 2, in three quality matrices (MS-SSIM, PSNR, RMSE) for image-set 3, in three quality matrices (MS-SSIM, RMSE, MI) for image-set 4, in all four quality matrices (MS-SSIM, PSNR, RMSE, MI) for image-set 5, and in three quality matrices (PSNR, RMSE, MI) for image-set 6. By combining visual and objective quality assessment, it can be safely stated that the proposed algorithm performed best in fusing all six image-sets.

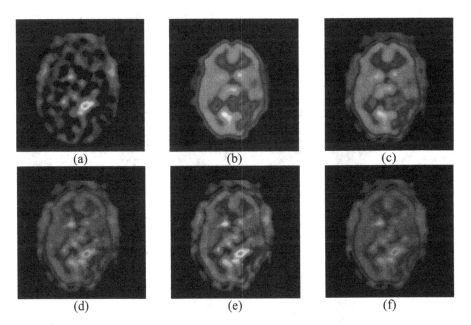

Fig. 2. Fusion results of image-set 1 (a) SPECT-T1 (b) SPECT-TC (c) PCA (d) Gradient Pyramid (e) DWT-DBSS (2,2) (f) Proposed

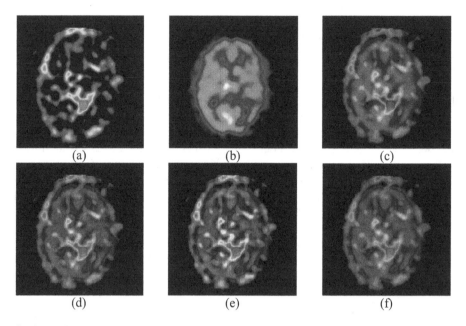

Fig. 3. Fusion results of image-set 2 (a) SPECT-T1 (b) SPECT-TC (c) PCA (d) Gradient Pyramid (e) DWT-DBSS (2,2) (f) Proposed

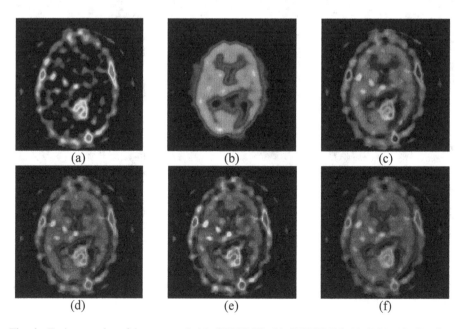

Fig. 4. Fusion results of image-set 3 (a) SPECT-T1 (b) SPECT-TC (c) PCA (d) Gradient Pyramid (e) DWT-DBSS (2,2) (f) Proposed

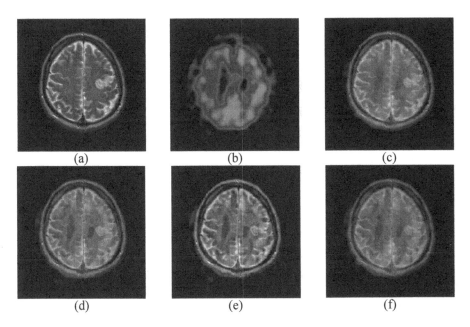

Fig. 5. Fused results of image set 4 (a) MR-T2 (b) SPECT-TC (c) PCA (d) Gradient Pyramid (e) DWT-DBSS (2,2) (f) Proposed

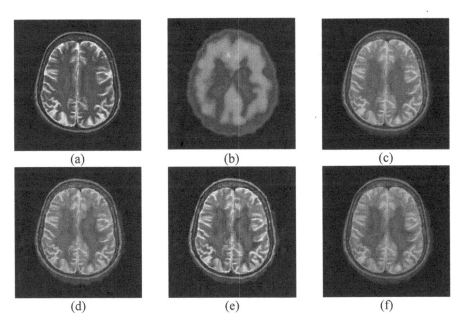

Fig. 6. Fused results of image set 5 (a) MR-T2 (b) SPECT-TC (c) PCA (d) Gradient Pyramid (e) DWT-DBSS (2,2) (f) Proposed

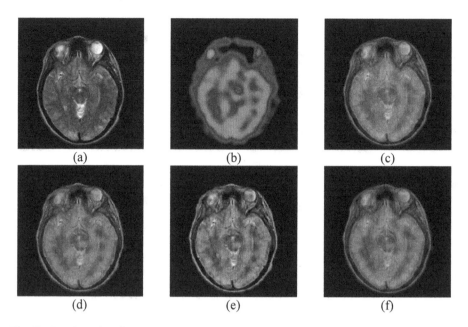

(a)   (b)   (c)

(d)   (e)   (f)

Fig. 7. Fused results of image set 6 (a) MR-T2 (b) SPECT-TC (c) PCA (d) Gradient Pyramid (e) DWT-DBSS (2,2) (f) Proposed

Table 1. Quality assessment values of fused images obtained by fusing image-set 1

Algorithms	Quality matrices			
	MS-SSIM	PSNR	RMSE	MI
PCA	0.646	**20.323**	0.113	**1.058**
Gradient pyramid	0.683	19.183	0.109	0.800
DWT-DBSS (2,2)	0.668	18.916	0.113	0.708
Proposed	**0.684**	19.384	**0.107**	0.880

Table 2. Quality assessment values of fused images obtained by fusing image-set 2

Algorithms	Quality matrices			
	MS-SSIM	PSNR	RMSE	MI
PCA	**0.694**	18.558	0.119	**1.033**
Gradient pyramid	0.686	18.550	0.118	0.792
DWT-DBSS (2,2)	0.676	18.146	0.124	0.724
Proposed	0.687	**18.755**	**0.115**	0.835

Table 3. Quality assessment values of fused images obtained by fusing image-set 3

Algorithms	Quality matrices			
	MS-SSIM	PSNR	RMSE	MI
PCA	0.630	17.394	0.135	**1.025**
Gradient pyramid	0.640	17.895	0.127	0.740
DWT-DBSS (2,2)	0.628	17.542	0.133	0.650
Proposed	**0.644**	**18.102**	**0.124**	0.889

Table 4. Quality assessment values of fused images obtained by fusing image-set 4

Algorithms	Quality Matrices			
	MS-SSIM	PSNR	RMSE	MI
PCA	0.805	**22.404**	**0.077**	1.152
Gradient pyramid	0.780	21.589	0.083	0.976
DWT-DBSS (2,2)	0.760	20.728	0.093	0.898
QPCA	**0.809**	22.251	**0.077**	**1.181**

Table 5. Quality assessment values of fused images obtained by fusing image-set 5

Algorithms	Quality matrices			
	MS-SSIM	PSNR	RMSE	MI
PCA	**0.842**	23.412	0.068	1.248
Gradient pyramid	0.822	22.416	0.076	1.132
DWT-DBSS (2,2)	0.801	21.489	0.088	1.077
QPCA	**0.842**	**23.441**	**0.067**	**1.257**

Table 6. Quality assessment values of fused images obtained by fusing image-set 6

Algorithms	Quality matrices			
	MS-SSIM	PSNR	RMSE	MI
PCA	**0.816**	23.014	0.071	1.115
Gradient pyramid	0.800	22.400	0.076	1.031
DWT-DBSS (2,2)	0.780	21.526	0.085	0.958
QPCA	0.815	**23.078**	**0.070**	**1.118**

6 Conclusion

In this paper, we proposed multimodal medical image fusion algorithm that is based on the quaternion principal component analysis. To test the performance of the proposed algorithm, we performed experiments to fuse two colour/colour and colour/grey images, contained in six image-sets. Fused results are compared with existing image fusion algorithms by using objective quality assessment method. Four image quality assessment matrices are used to get quality values of fused images. The aggregate result of quality values shows that the proposed algorithm performed better than existing multimodal image fusion algorithms. The proposed algorithm is simple and easy to implement hence, can be used as an alternative to fusing multi-modal colour medical images.

Acknowledgements. This work was partially supported by Natural Science Foundation of China (No. U1401252, 61472055, 61572092), Program for New Century Excellent Talents in University of China (NCET-11-1085) and Chongqing outstanding Youth Fund (cstc2014jcyjjq40001).

References

1. Pietrzyk, U., Herholz, K., Fink, G., Jacobs, A., Mielke, R.: Image registration: validation for PET, SPECT, MRI and CT brain studies. J. Nucl. Med. **35**, 2011–2018 (1994)
2. Tu, T.-M., Su, S.-C., Shyu, H.-C., Huang, P.S.: A new look at IHS-like image fusion methods. Inf. Fusion **2**, 177–186 (2001)
3. Ardeshir Goshtasby, A., Nikolov, S.: Image fusion: advances in the state of the art. Inf. Fusion **8**, 114–118 (2007)
4. Nikolov, S., Hill, P., Bull, D., Canagarajah, N.: Wavelets for image fusion. In: Petrosian, A. A., Meyer, F.G. (eds.) Wavelets in Signal and Image Analysis. Computational Imaging and Vision, vol. 19, pp. 213–241. Springer, Dordrecht (2001)
5. Rockinger, O., Fechner, T.: Pixel-level image fusion: the case of image sequences. In: Aerospace/Defense Sensing and Controls, pp. 378–388. International Society for Optics and Photonics (1998)
6. Kantor, I.L., Solodovnikov, A.S.: Hypercomplex Numbers: An Elementary Introduction to Algebras. Springer, New York (1989)
7. Ell, T.A., Sangwine, S.J.: Hypercomplex fourier transforms of color images. IEEE Trans. Image Process. **16**, 22–35 (2007)
8. Chen, B.J., Shu, H.Z., Zhang, H., Chen, G., Toumoulin, C., Dillenseger, J.-L., Luo, L.M.: Quaternion Zernike moments and their invariants for color image analysis and object recognition. Signal Process. **92**, 308–318 (2012)
9. Chen, B., Shu, H., Coatrieux, G., Chen, G., Sun, X., Coatrieux, J.L.: Color image analysis by Quaternion-type moments. J. Math. Imaging Vis. **51**, 124–144 (2015)
10. Le Bihan, N., Sangwine, S.J.: Quaternion principal component analysis of color images. In: 2003 Proceedings of International Conference on Image Processing, ICIP 2003, vol. I, p. 809. IEEE (2003)
11. Wang, X., Li, W., Yang, H., Wang, P., Li, Y.: Quaternion polar complex exponential transform for invariant color image description. Appl. Math. Comput. **256**, 951–967 (2015)
12. Guo, L., Dai, M., Zhu, M.: Quaternion moment and its invariants for color object classification. Inf. Sci. **273**, 132–143 (2014)
13. Khalil, M.I.: Applying quaternion Fourier transforms for enhancing color images. Int. J. Image Graph. Signal Process. IJIGSP **4**, 9 (2012)
14. Xu, Y., Yu, L., Xu, H., Zhang, H., Nguyen, T.: Vector sparse representation of color image using quaternion matrix analysis. IEEE Trans. Image Process. **24**, 1315–1329 (2015)
15. Sangwine, S.J., Ell, T.A.: Hypercomplex auto- and cross-correlation of color images. In: 1999 Proceedings of International Conference on Image Processing, ICIP 1999, vol. 4, pp. 319–322 (1999)

16. Dubey, V.R.: Quaternion fourier transform for colour images. Int. J. Comput. Sci. Inf. Technol. **5**, 4411–4416 (2014)
17. Smith, L.I.: A tutorial on principal components analysis. In: Elementary Linear Algebra. Wiley, New York (2002)
18. Zhou, B.B., Brent, R.P.: Jacobi-like algorithms for eigenvalue decomposition of a real normal matrix using real arithmetic. In: IPPS, p. 593. IEEE (1996)
19. Zhang, F.: Quaternions and matrices of quaternions. Linear Algebra Appl. **251**, 21–57 (1997)
20. Sangwine, S., Le Bihan, N.: Quaternion and octonion toolbox for Matlab. http://qtfm. sourceforge.net/
21. The Whole Brain Atlas. http://www.med.harvard.edu/AANLIB/home.html
22. Rockinger, O.: Image Fusion Toolbox. http://www.metapix.de/indexp.htm
23. Raut, G.N., Paikrao, P.L., Chaudhari, D.S.: A study of quality assessment techniques for fused images. Int. J. Innov. Technol. Explor. Eng. **2**, 290–294 (2013)
24. Wang, Z., Simoncelli, E.P., Bovik, A.C.: Multiscale structural similarity for image quality assessment. In: 2004 Conference Record of the Thirty-Seventh Asilomar Conference on Signals, Systems and Computers, pp. 1398–1402. IEEE (2003)
25. Hore, A., Ziou, D.: Image quality metrics: PSNR vs. SSIM. In: 2010 Proceedings of the 20th International Conference on Pattern Recognition (ICPR), pp. 2366–2369. IEEE, Istanbul (2010)
26. Qu, G., Zhang, D., Yan, P.: Information measure for performance of image fusion. Electron. Lett. **38**, 313–315 (2002)
27. Wang, Z., Bovik, A.C., Sheikh, H.R., Simoncelli, E.P.: Image quality assessment: from error visibility to structural similarity. IEEE Trans. Image Process. **13**, 600–612 (2004)

Research on Adaptive Mobile Collaborative Learning System

Ling Luo[(⊠)], You Yang, and Yan Wei

Computer and Information Science College, Chongqing Normal University,
Chongqing 401331, China
luoling@cqnu.edu.cn

Abstract. Adaptive learning system supports more personalized learning and improves learning outcome. Mobile learning gives learners new learning mode that learners can learn in anywhere and at any time. People can easily achieve success by collaborative learning. So this paper proposes a common model of adaptive mobile collaborative learning system (AMCLS). The AMCLS has six parts which are Learner Model, Collaborative Learning Model, Domain Model, Adaptive Recommend Model based on Context-Aware, Evaluation Model and Presentation Model. The AMCLS can provide the appropriate learning contents and learning paths to learners according to their characteristics, learning context-aware, devices, time and location. Besides, it also provides better collaborative learning mode to help learners to accomplish their collaborative learning more efficiently.

Keywords: Adaptive learning system · Mobile learning · Collaborative learning system

1 Introduction

The core of E-Learning is the construction of Online learning system, but currently, there are some problems of these Online learning system: (1) the learning resource of these systems provided to all learners are the same. That mean these systems cannot provide personal and intelligent service to learner according to his personality and his real demand, (2) learner easily feels exhausted and alone because these systems haven't better interactive function to support the communication between teachers and students, as well as between students and students, (3) these systems don't support mobile learning so that these system can't meet the demand the learner who wants to learn in anywhere, at any time. For these reasons, the utilization of these Online learning system is lower and even most systems are free.

Adaptive learning is thought as "the learning process provides each learner a unique learning experience according to learner's personalities so that the learner can get higher learning achievements, learning satisfaction, learning effectiveness and so on." [1, 2]. Mobile learning is a fashion and newer learning mode, which is called as "learners can study through content and social interactions by using several different kinds of mobile electronic devices" [3]. That means learners can learn by the use of mobile devices in anywhere and at any time [4]. Collaborative learning is that two or more learners come

© Springer Nature Singapore Pte Ltd. 2017
B. Zou et al. (Eds.): ICPCSEE 2017, Part II, CCIS 728, pp. 414–423, 2017.
DOI: 10.1007/978-981-10-6388-6_34

together to be groups to study [5]. Collaborative learning is not same as individual learning, learners engaged in collaborative learning capitalize on one another resources and skills (asking one another for information, evaluating one another ideas, monitoring one another work, etc.) [6]. Therefor the research is meaningful that combine adaptive learning and mobile learning, as well as collaborative learning to construct adaptive mobile collaborative learning system (AMCLS) to realize learner's personalized, collaborative and mobile learning more effectively.

2 Related Work

The development of Adaptive Mobile Collaborative Learning System is based on the construction of system mode. Actually, there were many Adaptive Learning System (ALS) mode developed for various purposes of education. Adaptive Hypermedia Architecture (AHA) is a Web-based adaptive hypermedia system, which can support on-line courses with different adaptive features, such as conditional explanations and links [7]. The AHA consists out of three models: user model, domain model, adaptive model. The user model represents the users' knowledge and preferences. The domain model defines the aspects of the application which can be adapted or which are otherwise required for the operation of the adaptive system. The adaptive model contains all rules which are concerned with the relationships which exist between the representation of the users (the user model) and the representation of the application (the domain model) [8]. Henze and Nejdl introduced a logical characterization for the definition of adaptive educational hypermedia systems (AEHS) as a quadruple (DOCS, UM, OBS, AC): DOCS (Document Space) describes documents and knowledge topics; UM (User Model) stores, describes, and infers individual user's information, knowledge, preferences; OBS (Observations) observes individual user's knowledge state and interactions with the system for updating UM; AC (Adaptation Component) contains rules for the describing the adaptive functionality of the system.

The research of adaptive mobile learning system (AMLS) began in 2009. Researchers did their AMLS mostly with the views of location-based [9], content adaptive and device adaptive [10], Ontology-based, furthermore, only small number of developed system were just apply in English course learning. Actually, there are some problems needed to do further research, such as how to construct and present mobile learning content according to the characteristics of a variety of different device, how to acquire learner' learning context-aware efficiently, how to improve mobile user's experience, etc.

Collaborative Networked Learning is stated as "Collaborative Networked Learning (CNL) is that learning which occurs via electronic dialogue between self-directed co-learners and learners and experts. Learners share a common purpose, depend upon each other and are accountable to each other for their success. CNL occurs in interactive groups in which participants actively communicate and negotiation meaning with one another within a contextual framework which may be facilitated by an online coach, mentor or group leader." [11]. Due to the numerous technological resources available to students and tutors, collaborative learning is usually supported by computers in what is known as computer supported collaborative learning (CSCL). One of the main challenges when designing a learning experience with collaboration is that of structuring the overall learning process as to trigger productive argumentation among students.

3 Adaptive Mobile Collaborative Learning System Model

According to the characters of ALS, M-learning and CNL, this paper proposes a structure of Adaptive Mobile Collaborative Learning System (AMCLS) is shown in Fig. 1. The model contains 6 parts, namely Learner Model, Collaborative Learning Model, Domain Model, Adaptive Recommend Model Based on Context-Aware, Evaluation Model and Presentation Model.

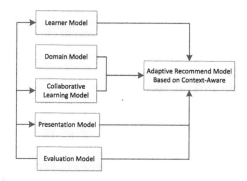

Fig. 1. Structure of adaptive mobile collaborative learning system

3.1 Learner Model

The Learner model which is the most important component of AMCLS as it stores all the information about the learner determines the accuracy and reliability of the penalization implementation. The Learner Model (LM) mainly describes learner's characteristics, such as learning style, cognitive level, interesting, automotive and collaborative learning abilities, and etc. In addition, a learner behavior diagnostic model should be instructed which can modify LM dynamically by detecting individual learning status, collaborative learning status, and knowledge mastering situation etc. The LM structure is shown in Fig. 2.

Among those factors, learning styles has been recognized by researchers as being important factors [12]. Learning styles help learners understand their own strengths for more efficient learning. Keefe indicated that learning style is both a characteristic which indicates how a student learns and likes to learn, as well as an instructional strategy informing the cognition, context and content of learning. Previous studies have reported that

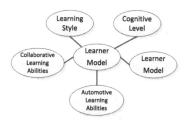

Fig. 2. LM structure

students' learning performance could be improved if proper learning style dimensions could be taken into consideration when developing adaptive learning systems [13, 14].

This study combines the explicit access with implicit access to estimate learner's learning style. Firstly, using learning style questionnaire for get learning style explicitly, meanwhile initializing learning style model, and then amend and improve the learning style model constantly by digging Netware learning behaviors with Bayesian Network (BN) so that the estimate learning style is most similar learner's real learning style.

(1) **Explicit access method**

There are several different learning style models including Kolb, Honey and Mumford, and Felder and Silverman [16]. Among these various learning styles, the Felder–Silverman Learning Style Model (FSLSM) developed by Felder and Soloman [17] have been recognized by many researchers as being a highly suitable model for developing adaptive learning systems [15]. The FSLSM contains the Index of Learning Style (ILS) questionnaire for evaluating learning styles. There are four dimensions in FSLSM and each learner is characterized by a specific preference for each of these dimensions: sensitive/intuitive dimension (how information is processed), active/reflective dimension (how information is presented), visual/verbal dimension (how information is input), and sequential/global dimension (how information is understood), [16] (see Table 1). Therefore, this study utilizes the FSLSM to investigate learning style explicitly.

Table 1. Felder-Silverman learning dimension and learner characteristics

Learning dimension	Learner characteristics	
Processing	Active: Retain and understand information best by doing something active with it such as discussing it, applying it, or explaining it to others	Reflective: Prefer observation rather than active experimentation. Tend to think about information quietly first
Perception	Sensor: Like learning factors, often like solving problems by well-established methods and dislike complications and surprises. Patient with details and good at memorizing facts and doing hands-on work. More practical and careful than intuitors	Intuitive: Prefer discovering possibilities and relationships. Like innovation and dislike repetition. Better at grasping new concepts and comfortable with abstractions and mathematical formulations. Tend to work faster and more innovative than sensors
Input	Visual: Remember best what they sees from visual representations such as graphs, chart, pictures and diagrams	Verbal: More comfortable with verbal information such as written texts or lectures

In order to detect both the preference and the degree of preference of learners for each dimension, the Index of Learning Styles (ILS) has been developed by Felder and Soloman [17]. ILS is a 44 item questionnaire aimed at identifying the learning styles according to FSLSM.

(2) **Implicit access method**

After obtain the basic learning style by FSLSM, we will use Bayesian networks (BNs) to represent and detect students' learning styles according to their behaviors in the adaptive learning system. A BN is a compact, expressive representation of uncertain relationships among parameters in a domain. A BN is a directed a cyclic graph where nodes represent random variables and arcs represent probabilistic correlation between variables [18]. Figure 3 is a BN model which describes the learning style.

We model the four dimensions of Felder's framework according to Fig. 1. Each dimension with a variable in the BN will be modeled. The values these variables can take are sensory/intuitive, active/reflective, visual/verbal, and sequential/global respectively.

Bayes' theorem is the mathematical model underlying BN. Bayes' theorem is shown in Eq. (1), which relates conditional and marginal probabilities. Bayes' theorem yields the conditional probability distribution of a random variable A. Assuming we know information about another variable B in terms of the conditional probability distribution of B given A, and the marginal probability distribution of A alone. Equation (1) reads: the probability of A given B equals the probability of B given A times the probability of A, divided by the probability of B.

$$P(A/B) = \frac{P(B/A)P(A)}{P(B)} \tag{1}$$

Bayesian network can be noted by B = <G, θ> which represents the joint probability distribution on a random variable set X = {X_1, X_2, ..., X_n}, where G denotes a directional graph with no cycles with its nodes to be random variables (e.g. Xi as random variable and $\prod_i$ as respective parent variable). Bayesian network is structured

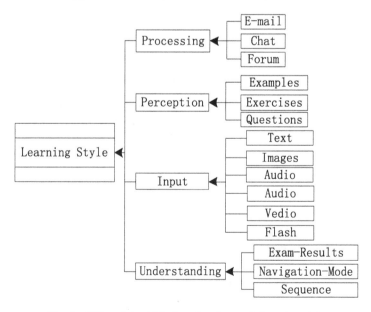

Fig. 3. BN model which describes learner's learning style

according to Markov independence equations. Set of variables which define the conditional probability distribution $P(X_i|\prod_i)$ for each random variable X_i is denoted by θ assuming that the variable has a parent variable $\prod_i$. Joint probability distribution of the variable X is defined by network structure which identifies the conditional independence of X and the set P which is the local probability distribution for each variable. Absence of an edge between two nodes indicates the conditional independence. Having the network structure the joint probability distribution can be written as follows:

$$P(x_1, x_2, \ldots, x_n) = \prod_{i=1}^{n} P(x_i \Pi_i) \qquad (2)$$

where, for each variable xi, $\prod_i\{x_1, x_2, \ldots, x_{n-1}\}$ is a set of variables of which xi is conditionally dependent.

Therefore we construct a BN model by building a directed a cyclic graph that encodes assertions of conditional independence. We can induce the joint probability distribution of K by utilizing the chain rule of probability as follows by giving a domain K = $\{K_1, K_2, \ldots, K_n\}$ and an ordering on the variables $(K_1, \ldots, K_n)$,

$$P(k_1, \ldots, k_{n/\varepsilon}) = \prod_{i=1}^{n} P(k_i/v_1, \ldots, k_{i-1,\varepsilon}) \qquad (3)$$

where, for every K_i, there will be some subset $\Pi_i \subseteq \{K_1, \ldots, K_n\}$ such that K_i and $\{K_1, \ldots, K_n\}$ are conditionally independent given Π_i. That is,

$$P(k_i/k_1, \ldots, k_{i-1,\varepsilon}) = P(v_i/\Pi_{i,\varepsilon}) \qquad (4)$$

3.2 Collaborative Learning Model

Collaborative Learning Model (CLM) can recommend some learners with similar learning characteristics to learners according to social network analysis method (SNA) and the learner model, and on this basis, constructing effective collaborative learning commodity. CLM also contains some collaborative learning methods and algorithm to help students collaborate more effectively with their peers, maximizing individual student and group learning. Figure 4 is the collaborative learning model.

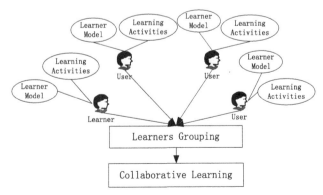

Fig. 4. Collaborative learning model

3.3   Domain Model

We mainly combines Ontology and Topic Maps technology, as well as some classic knowledge model such as Dublin Core, IEEE LOM, CELTS-3, to Constructing domain model according to ultra-brief characteristic of mobile learning material. There are two steps to accomplish the construction of Domain Model:

(1) Building ontology library based on domain knowledge. Domain ontology library can provide normalized annotated words and reference standard which is used to mark documents base on ontology,
(2) Accomplish Semantic annotation of documents with Topic Maps standards and construct document logic view, then save these information into Topic Maps annotation library with XTM format which is convenient to build link relationship between concept semantic and document resource on indexing.

3.4   Adaptive Recommend Model Based on Context-Aware

The function of adaptive recommend model based on Context (ARMBC) is recommend appropriate learning content and learning path to learner on the basis of LM and context-aware. The context is stated as the time, the location, the circumstances and the device. The time contains two parameters which are real date-time and learning progress. The location indicates the learner's geographic location. The location-awareness of the AMCLS can be employed to sense the current geographic location of the mobile learner who processes the mobile device being used to conduct mobile learning. So the location-based learning contents can be implemented to enhance the contextual interaction for learners in a mobile learning environment. Location-based learning contents are those learning objects that are tied with particular locations. When a mobile learner is physically at or near a particular location, the learner could be assigned to conduct location-based learning activities. The circumstances contain some information such as noise, seat, climate, space, light, etc. The device contains the display performance, storage performance, bandwidth, operating system, etc. The Adaptive Recommend Model Based on Context-Aware is showed in Fig. 5.

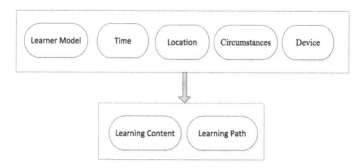

Fig. 5. Adaptive recommend model based on context-aware

According to Fig. 4, Adaptive Recommend Model Based on Context-Aware Adaptation Mechanism is represented in Eq. (5),

$$R(LC(i), LP(i)) = S(LM(i), T(i), L(i), C(i), D(i)) \tag{5}$$

where S is the mathematical representation, R(LC(i), LP(i)) is the output of the ARMBC adaptation mechanism representing the appropriate learning contents and appropriate Learning path. LM(i) is the Learning Model, T(i) is the input representing the time constrains, L(i) is the input representing the location constrains, C(i) is the circumstances, D(i) is the mobile device constrains.

3.5 Presentation Model

The presentation model contains a transformation engine which can transform the learning resource learner request into proper format and then sent them to learner's mobile terminal according to characteristics of heterogeneous devices and the circumstance of the network. That mean this process can shield the difference of terminal equipment and the complexity of the he environment and reduce the external interference factors on the pan in the study. In order to accomplish the target, we considered 5 factors: (1) adaption of media size and file size: the differences of terminal screens will induce the difference of the media size and file size appropriate to the devices. For example, the proper image size of hand-phone is 25 * 25 pixels, while the proper image size of PAD might be 120 * 120 pixels. (2) Layout adaption of learning resources. Layout mainly is decided by two elements: screen size and distinguishability. The layout mode of the same resources will be different with several different size of screen. (3) Network adaption: the request come from terminal will be sent to server through network and then the terminal will receive some data come from server at good network condition. On the contrary, the engine will search for these data automatically saved into the database in online station when the terminal send request at the worse network connection status.

3.6 Evaluation Model

We will construct evaluation model with two factors: knowledge acquisition and ability level, on this basis build adaptive exercises library. There are four steps to accomplish learner's evaluation: (1) the evaluation model will predict the situation of learner's knowledge according to the learners' learning track, prediction learners' knowledge, (2) selecting some exercises with appropriate difficulty to compose paper according to predicted results, and then execute the test, (3) at the end of test, the evaluation model will combine the data come from learner's answer and the whole learning progress to analyze the learner's knowledge defects by design data mining algorithm (K-Means, Apriori, Naive Bayes, etc.), (4) design learning assistant agent to guide student to learn these knowledge that the learner hasn't grasp.

4 Conclusion

Adaptive learning provides a personal learning to learner. Mobile learning give learner's new learning mode that the learner can learn in anywhere and at any time. Collaborative learning, unlike individual learning, people engaged in collaborative learning capitalize on one another resources and skills (asking one another for information, evaluating one another ideas, monitoring one another work, etc.). People can easily achieve success by collaborative learning. This paper proposed a common model of adaptive mobile collaborative learning system (AMBCLS). The AMBCLS can provide the appropriate learning content and learning path according to learner's characteristics, time, location, context-aware and device. Meanwhile, the system also provide better collaborative learning environment and many efficient collaborative learning mode and method to help learners to accomplish their autonomic and collaborative study more efficiently. But we don't verify the model of AMBCLS through actual teaching experiment, we will execute some experiments to prove the effectiveness of the model, at the same time to find the shortage of the model, and then improve the model constantly.

Acknowledgment. This work was supported by Chongqing Municipal Education Commission (Nos. KJ1400512, 152017 and yjg152001).

References

1. Monova-Zheleva, M.: Adaptive learning in Web-based educational environments. Cybern. Inf. Technol. **5**(1), 44–55 (2005)
2. Olfman, L., Mandviwalla, M.: Conceptual versus procedural software training for graphical user interfaces: a longitudinal field experiment. MIS Q. **18**(4), 405–426 (1994)
3. Crompton, H.: A historical overview of mobile learning: toward learner-centered education. In: Berge, Z.L., Muilenburg, L.Y. (eds.) Handbook of Mobile Learning, pp. 3–14. Routledge, Florence (2013)
4. Crescente, M.L., Lee, D.: Critical issues of m-learning: design models, adoption processes, and future trends. J. Chin. Inst. Ind. Eng. **28**(2), 111–123 (2011)
5. Dillenbourg, P.: Collaborative Learning: Cognitive and Computational Approaches. Advances in Learning and Instruction Series. Elsevier Science Inc., New York (1999)
6. Chiu, M.M.: Group problem solving processes: social interactions and individual actions. J. Theory Soc. Behav. **30**(1), 27–50, 600–631 (2000)
7. De Bra, P., Aerts, A., Berden, B., de Lange, B., Rousseau, B., Santic, T., Stash, N.: AHA! The adaptive hypermedia architecture. Paper Presented at the Fourteenth Conference on Hypertext and Hypermedia, Nottingham, United Kingdom (2003)
8. De Bra, P., Aroyo, L., Cristea, A.: Adaptive web-based educational hypermedia. In: Levene, M., Poulovassilis, A. (eds.) Web Dynamics, pp. 387–410. Springer, Heidelberg (2004). doi:10.1007/978-3-662-10874-1_16
9. Tan, Q., Zhang, X., Kinshuk, R.M.G.: The 5R adaptation framework for location-based mobile learning systems. In: 10th World Conference on Mobile and Contextual Learning, pp. 18–21 (2011)
10. Ghadirli, H.M., Rastgarpour, M.: An adaptive and intelligent tutor by expert systems for mobile devices. arXiv preprint arXiv:1304.4619 (2013)

11. De Laat, M., Lally, V., Lipponen, L., et al.: Investigating patterns of interaction in networked learning and computer-supported collaborative learning: a role for Social Network Analysis. Inter. J. Comput.-Support. Collab. Learn. **2**(1), 87–103 (2007)
12. Filippidis, S.K., Tsoukalas, L.A.: On the use of adaptive instructional images based on the sequential-global dimension of the Felder-Silverman learning style theory. Interact. Learn. Environ. **17**(2), 135–150 (2009)
13. Graf, S., Liu, T.C., Kinshuk: Analysis of learners' navigational behaviour and their learning styles in an online course. J. Comput. Assist. Learn. **26**(2), 116–131 (2010)
14. Harris, K., Reid, D.: The influence of virtual reality play on children's motivation. Can. J. Occup. Ther. **72**(1), 21–30 (2005)
15. Yang, T.C., Hwang, G.J., Yang, S.J.H.: Development of an adaptive learning system with multiple perspectives based on students' learning styles and cognitive styles. J. Edu. Technol. Soc. **16**(4), 185 (2013)
16. Felder, R.M., Silverman, L.K.: Learning styles and teaching styles in engineering education. Eng. Educ. **78**(7), 674–681 (1988)
17. Felder, R.M., Soloman, B.A.: Index of learning styles questionnaires. Online version (1997). http://www.engr.ncsu.edu/learningstyles/ilsweb.html. Accessed 6 May 2006
18. Jensen, F.V.: An Introduction to Bayesian Networks. Springer, New York (1996)

Plagiarism Detection in Homework Based on Image Hashing

Ying Chen[✉], Liping Gan, Shiqing Zhang, Wenping Guo,
Yuelong Chuang, and Xiaoming Zhao

Institute of Intelligent Information Processing,
Taizhou University, Taizhou 317000, China
Ychen222@foxmail.com

Abstract. The problem of high similarity in homework has troubled teachers with time. Previous plagiarism detection systems are mainly realized by string matching which has a limitation, i.e., image homework cannot be detected. To this issue, we propose a new method of plagiarism detection in homework. First, we get fingerprint features of image homework by converting text homework into images. Then, we use image hashing algorithm and hamming distance to calculate the similarity of these features. Finally, we perform the empirical study on course of Computer Network Experiment, the test shows that our method not only reliably keeps the detection speedily, but also consistently ensures precision and false positive rate.

Keywords: Plagiarism detection · Image fingerprinting · Image hashing · Hamming distance

1 Introduction

Learning is an indispensable part of human life, especially for students. However, many students are not very clear about the meaning of learning. They do not spend much time on study, and their homework is often perfunctory, or is directly plagiarism. Plagiarism repeatedly, the cause is not only because of lazy working of students, but also because of there is no effective method to make plagiarism detection in homework.

For a long time, plagiarism detection in homework is mainly relying on the teacher's manual recognition. However, limited by the teacher's energy and time, it's difficult to detect plagiarism. Plagiarism is not only contributed to the lazy inertia of students, but also undermines the teaching order. It is hard for the teacher to master the students' grasp of the knowledge point from the homework. In addition, when the teacher cannot identify whether the homework is copied or not, he can only make a score according to the quality of the homework, which may lead to opportunistic students get better results to the students who complete the work seriously and independently. It's easy to hurt students' learning enthusiasm of known the truth. To curb the plagiarism from the source, it is necessary for students to recognize the harmfulness of copying homework. But more important, it is necessary to do the similarity comparison in homework, especially a system of plagiarism detection in homework to reduce the workload on teachers.

B. Zou et al. (Eds.): ICPCSEE 2017, Part II, CCIS 728, pp. 424–432, 2017.
DOI: 10.1007/978-981-10-6388-6_35

The previously homework plagiarism detection systems [1–3] mainly use the string matching algorithm and word frequency statistics through the text information to carry out comparative testing result. This approach has high accuracy. However, the time complexity and the amount of computation are relatively high when in the detection of homework with large number of text and homework in the form of images are cannot be detected.

By converting homework into images and calculating the similarity of the images to determine whether the homework is a plagiarism overcomes the limitation. Images corresponding to the homework are easy to obtain and retaining the original visual information of homework, thus, the method has a high feasibility.

2 Proposed Method

2.1 Perceptual Image Hashing

The realization of image similarity comparison is mainly using "perceptual image hashing" [4] currently. Perceptual image hashing is the process of examining the contents of an image and then constructing a value that uniquely identifies an image based on these contents. Given an input image, a hash function is applied and an "image hash" is computed based on the image's visual appearance. Images that are "similar" should have hashes that are "similar" as well. Using image hashing algorithms makes performing near-duplicate image detection substantially easier. Since the perceptual image hash function maps an image to a short binary string based on an image's visual appearance, it's not only reduces the storage cost of data, but also shorten the retrieval time in a large number of image data retrieval.

Perceptual image hashing should be collision-resistant and robust [5]. The ability of collision-resistant is that it will not be mapped to the same digest when it is perceived different images. And robust is proposed by using the invariance of the image digest after geometric deformations or format changes.

So many algorithms are introduced with perceptual image hashing development. Image hash is developed as a result of feature extraction and the coding of intermediate result [6–8]. The basic steps of the perceptual image hashing are shown in Fig. 1.

Fig. 1. Flow chart of the perceptual hash sequence generation

The image is composed of pixels, contains a lot of detail information. So before the feature extraction it is need for certain processing of the image, such as unified image specifications or converting the image to grayscale to reduce the complexity of subsequent operations.

The extraction of image features is a key step in the in the whole process, the ability of the image express will directly determine the final effect. In perceptual image hashing, both global and local features are used in forming the hash sequence. The global features are based on Zernike moments representing luminance and chrominance characteristics of the image as a whole. The local features include position and texture information of salient regions in the image.

Researchers have been improving on this step, many feature extraction methods with ability of collision-resistant and robustness were proposed in a number of different domain, such as the perception hash based on quaternion discrete Fourier transform and log-polar transform [9, 10], perceptual hash based on image feature points [11], and so on.

However, the feature values thus extracted are often have a certain redundancy, and in order to facilitate storage and operation, it is necessary to quantize and encode the feature values to obtain the as short as possible hash sequences for ease of the subsequent application, such as the dHash and Hamming distance.

2.2   dHash and Hamming Distance

According to the characteristics of students' homework, we can use the "difference hash", or simply dHash algorithm to compute the image fingerprints. Simply put, the dHash algorithm looks at the difference between adjacent pixel values [12]. Then, based on these differences, a hash value is created. This algorithm compares the brightness difference of adjacent pixels of the preprocessed images, which is realized on the image gradient and has better image expression ability.

The implementation process is as follows:

```
input: images in dataset of images
output: difference
for row in range (size):  #size indicates the size of the
image after zooming
for col in range(size):
pixel_left = image.getpixel((col, row))
pixel_right = image.getpixel((col + 1, row))
difference.append(pixel_left > pixel_right)
```

Students rarely copy a whole homework in a plagiarism, but do some simple small changes more or less. Thus it cannot only examine whether the image features is the same, but also need to calculate the similarity of the homework. The copied part has basically the same appearance of the original one [13], thus features extracted in the forged region will be quite similar to the original ones. Therefore, matching between the hashed image and the test image helps to detect the copy-move attack.

The similarity of the image is obtained by calculating the hamming distance of the hash sequence.

$$\text{HammingD}\,(A, B) = \sum_{i=1}^{n} A_i \oplus B_i \tag{1}$$

Here, n is sequence length, A and B are two hash sequences.

Basically, a piece of homework can be converted into a number of images. Each image can be generated an image fingerprint by image Hash algorithm, the comparison between the two homework need to compare all the generated fingerprints. We propose a method as follows:

We convert each page of homework A into an image, totally n images. And each image is generated an image fingerprint through the perceptual image hash algorithm as F_1, F_2 ... F_n. So as the homework B, which generate m image fingerprints, G_1, G_2 ... G_m.

Let $D_{ij} = D(F_i, G_j)$ means the distance of F_i and G_j with the hamming, here $1 \leq i \leq n, 1 \leq j \leq m$. Thus the average hamming distance obtained by homework A and B is:

$$\overline{D(A, B)} = \frac{1}{n} \sum_{i=1}^{n} \min(D_{i1}, D_{i2}, D_{i3}, \ldots, D_{im}) \tag{2}$$

Here, the value of $\overline{D(A, B)}$ may be a decimal and in the range of [0, 64].

The smaller the distance of hamming, the more similar the two images, which means there is a suspected plagiarism.

3   Experiment Verification

There are many indicators for measuring image similarity performance, where Recall, Precision, and FCR (False Positive Rate) are more important in students' homework detection.

Recall is the proportion of all relevant images that were retrieved, i.e., the proportion of similar images in the dataset that were actually retrieved. The reality is that most current algorithms fail to find all similar images, and many of the retrieved images contain dissimilar images (false positives). Precision refers to the proportion of retrieved images that are relevant, i.e., the proportion of all retrieved images that the user was expecting. FPR represents the probability that the dissimilar homework is judged to be similar.

They are calculated as follows:

Recall = (The number of similar images detected)/(The number of all relevant images in the dataset)

Precision = (The number of similar images detected)/(The number of images detected)

FPR = (The number of dissimilar images detected)/(The number of images detected)

3.1   Experimental Preparations

It is arranged 6–8 experiments per semester in the course of "Computer Network Experiment". The experimental content is relatively constant. After each experiment, students will need to submit an experimental report in PDF format. The experimental report requires experimental steps, experimental process and the results. Particularly, the processes and results need screenshots or photos as the support material. As it is something cumbersome to fill the experimental report, and the experimental screenshots or photos are likeness, a part of the students choose to copy someone's report to cope with the teacher. Especially, some students splice multiple experimental reports as one piece of homework in order to avoid plagiarism being found.

We collected a total of 1000 experimental reports as a sample set, in which 100 experimental reports have been verified as plagiarism, and 50 experimental reports are considered to have plagiarism suspects.

Before the start of the experiment, we need to set two thresholds. Threshold T_1 is used to determine whether the homework is plagiarized, threshold T_2 is to determine whether the homework is suspected of copying. In this experiment, based on years of teaching experience, we set the threshold value of 5 to determine the plagiarism of homework, which means the hamming distance of less than 5 (including 5) will be identified as plagiarism. And we set the suspected plagiarism threshold value of 10, which means the hamming distance of less than 10 (including 10) and greater than 5 will be considered plagiarism suspects.

3.2   Experimentation

Firstly, we need to preprocess the sample images. We convert all the images into the grayscale of the same size, and generate the hash sequence through the difference hash function. Secondly, we need to calculate the hamming distance of the sample image and the hash value in the dataset, obtain the similarity between the sample image and all the relevant images in the dataset.

If an image is calculated that the hamming distance is greater than the similarity threshold T_2 in the dataset, the hash sequence of the image will be added to the dataset. Repeat the steps until the sample images all detected. If the similarity is greater than the plagiarism threshold T_1 but less than T_2, it will output the sample image together with the similar image(s) in the dataset, and manually judged by the teacher to ensure that the false positive rate is as low as possible. If the similarity is less than the plagiarism threshold T_1, it is directly judged as plagiarism homework to reduce the workload of the teacher.

The experimental flow is shown in Fig. 2.

We choose homework of "experimental II: Switch configuration" as an example. There are two copies of the homework without plagiarism, marked as H_1 and H_2, respectively.

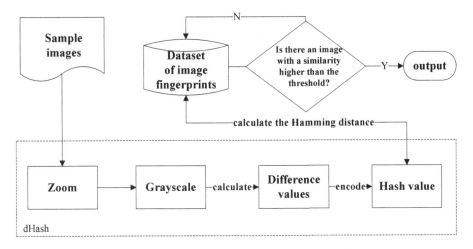

Fig. 2. Flow chart of homework plagiarism detection

First, according to the number of pages, H_1 and H_2 were converted to 7 and 8 images, respectively. Then, use the difference hash algorithm to get the hash value for each image. Finally, calculate the hamming distances. The result is shown in Table 1.

The rightmost column and the last row in the Table 1 are the minimum of the column or row, indicating the maximum similarity between the pages in H_1 and the pages in H_2.

From the Table 1, it can be seen that the hamming distance is 1 on the home page of the two documents, it is absolutely very similar. This is because the experimental report in the course of "Computer Network Experiment" has a fixed format, so the documents' home pages have same structure and a similar content. Thus we need to ignore the hamming distance of home pages in practice. The other hamming distance in the Table is greater than 11, indicating that the remaining images have a low similarity.

Table 1. The hamming distances between H_1 and H_2

H_1	H_2								
	1	2	3	4	5	6	7	8	Distance
1	1	23	27	34	30	34	27	24	1
2	22	21	23	32	32	26	25	22	21
3	32	23	23	22	32	28	27	26	22
4	24	27	21	22	22	26	21	24	21
5	20	17	25	24	24	24	21	20	17
6	24	29	29	26	24	22	17	22	17
7	28	21	23	22	22	14	19	16	14
Distance	1	17	21	22	22	14	17	16	18

Similar to the detection process of documents H_1 and H_2, the images generated from the 1000 reports in the sample set are compared to our dataset of image finger-prints. If the fingerprint of the homework does not match the fingerprint of any images in the dataset, we add the fingerprint of the homework to the dataset, so that create a constantly updated dataset of image fingerprints that can be used in subsequent homework detection.

The final experimental result of the 1000 reports plagiarism detection is shown in Table 2.

Table 2. Detection results of the 3 indicators

Classification approach	Indicators		
	Recall (%)	Precision (%)	FPR (%)
dHash	58.3	83.6	3.6

3.3 Experiment Results Analysis

In the result of Table 1, it can be seen that images with the same structure and slightly different content obtained the same image fingerprint by using the image hashing algorithm, indicating that it has a certain degree of robustness. And the other images of different structure and different content calculated relatively large hamming distance value, indicating that the algorithm has a collision-resistant ability.

In the result of Table 2, it is obviously that image hashing has a relatively high Precision and low FPR, but the Recall is not very satisfactory. This is due to we adjust the threshold T_2 for taking into account the risk of miscarriage of justice and the workload of teachers, resulting in many splicing reports do not to be found, and difficult to find all the similar images.

Besides, in order to reduce the computational complexity, we did not make further preprocessing on the images, resulting in some of the copied homework get a low similarity because of change the layout of the pages.

Considering the speed and validity, the dHash algorithm can achieve ideal detection effect under the condition of guaranteeing the low FPR, so it is suitable for the pla-giarism detection of student homework.

4 Conclusions

We propose a new method which uses the image hashing for the detection of copy-move plagiarism in homework of images. As a fingerprint can identify a person, image hash value can only identify an image. By comparing homework images' similarity, it can effectively solve the limitation of traditional means in which images cannot be detected for plagiarism. Our method is based on the image's visual appearance, which extracts image features as fingerprint and unidirectional mapping to a short hash sequence, and compares the hash value for plagiarism detection. The experimental results show that the accuracy of our method is ideal, and the calculation complexity is low, the storage capacity is small, thus it has a good application value. Our process is not very perfect, but it is worked. By using this algorithm we managed to reduce the number of copied reports by over 90%. Slowly but surely, the outbreak slowed - however, it never fully stopped.

Because our method takes into account the complexity of the calculation and discards a lot of details, it affects the detection results. And the dHash is realized on the image gradient, a small change in layout of the homework's image may have a great change in hash value, which easily leads to the final hash sequence changes, thus may lead to a real plagiarized homework cannot be detected. We will continue to optimize the application of image hashing in homework detection and improve the expression ability of image features in order to obtain better detection results in the future.

Acknowledgment. This research was supported by Zhejiang Provincial Natural Science Foundation of China under Grant No. LY14F020036.

References

1. Krizkova, S., Tomaskova, H., Gavalec, M.: Preference comparison for plagiarism detection systems. In: IEEE International Conference on Fuzzy Systems, pp. 1760–1767 (2016). doi:10.1109/FUZZ-IEEE.2016.7737903
2. Xiaoping, Z., Xiaoxuan, M., Honghong, S.: Research on a VSM-based e-homework anti-plagiarism system. In: International Conference on Information Management, Innovation Management and Industrial Engineering (ICIII), pp. 102–105 (2012). doi:10.1109/ICIII. 2012.6339788
3. Rosales, F., Garcia, A.F., Rodriguez, S., Pedraza, J.L., Mendez, R., Nieto, M.: Detection of plagiarism in programming assignments. IEEE Trans. Educ. **51**(2), 174–183 (2008). doi:10. 1109/TE.2007.906778
4. Hadmi, A., Ouahman, A.A., Said, B.A.E., Puech, W.: Perceptual image hashing. INTECH Open Access Publisher (2012). doi:10.5772/37435
5. Weng, L., Preneel, B.: A secure perceptual hash algorithm for image content authentication. In: International Conference on Communications, pp. 108–121 (2011). doi:10.1007/978-3-642-24712-5_9
6. Davarzani, R., Mozaffari, S., Yaghmaie, K.: Perceptual image hashing using center-symmetric local binary patterns. Multimed. Tools Appl. **75**(8), 4639–4667 (2016). doi:10.1007/s11042-015-2496-6
7. Liu, Y., Xin, G.J., Xiao, Y.: Robust image hashing using radon transform and invariant features. Radioengineering **25**(3), 556–564 (2016). doi:10.13164/re.2016.0556
8. Qin, C., Chen, X., Ye, D., Wang, J., Sun, X.: A novel image hashing scheme with perceptual robustness using block truncation coding. Inf. Sci. **361**, 84–99 (2016). doi:10.1016/j.ins. 2016.04.036
9. Ouyang, J., Coatrieux, G., Shu, H.: Robust hashing for image authentication using quaternion discrete Fourier transform and log-polar transform. Digital Signal Process. **41**, 98–109 (2015). doi:10.1016/j.dsp.2015.03.006
10. Qin, C., Chang, C.C., Tsou, P.L.: Robust image hashing using non-uniform sampling in discrete Fourier domain. Digital Signal Process. **23**(2), 578–585 (2013). doi:10.1016/j.dsp. 2012.11.002

11. Monga, V., Evans, B.L.: Robust perceptual image hashing using feature points. In: International Conference on Image Processing, pp. 677–680 (2004). doi:10.1109/ICIP.2004. 1418845
12. Ahmed, F., Siyal, M.Y.: A novel hashing scheme for image authentication. In: Proceedings of the 2006 Innovations in Information Technology, pp. 1–5 (2006). doi:10.1109/ INNOVATIONS.2006.301971
13. Abraham, A.K., Haroon, R.P.: An improved hashing method for the detection of image forgery. IOSR J. Comput. Eng. 6(5), 13–19 (2014). doi:10.9790/0661-16531319

A Multi-objective Genetic Algorithm Based on Individual Density Distance

Lianshuan Shi[(⊠)] and Huahui Wang

School of Information Technology Engineering,
Tianjin University of Technology and Education, Tianjin, China
shilianshuan@sina.com

Abstract. The uniform and extension distribution of the optimal solution are very important criterion for the quality evaluation of the multi-objective programming problem. A genetic algorithm based on agent and individual density is used to solve the multi-objective optimization problem. In the selection process, each agent is selected according to the individual density distance in its neighborhood, and the crossover operator adopts the simulated binary crossover method. The self-learning behavior only applies to the individuals with the highest energy in current population. A few classical multi-objective function optimization examples were used tested and two evaluation indexes U-measure and S-measure are used to test the performance of the algorithm. The experimental results show that the algorithm can obtain uniformity and widespread distribution Pareto solutions.

Keywords: Individual density distance · Multi-objective optimization · Multi-agent · Self-learning · S-measure · U-measure

1 Introduction

Multi-objective optimization problem (MOP) exists widely in engineering. Many optimization problems in real life are composed of multiple goals, and these goals are often conflicting, the solution won't be one, but is made up of multiple solutions. Multi-objective programming research has attracted the attention of many scholars. A number of different kinds of multi-objective optimization algorithm have been proposed.

The traditional mathematical programming method dependent on the mathematical properties of optimization functions, which search ability for the target space is low, and the amount of calculation is large.

Genetic algorithm, as all other random-search oriented optimization algorithms, doesn't require any information about structure of the function to be optimized. Also, GA doesn't require information about initial point and successfully find optimal solution. It have shown great advantage when GA is used to solve the multi-objective optimization problem. However, when the dimension of the function increases, the search space increases and the coupling between the variables increases, which leads to the increase of the probability that the genetic algorithm falls into the local optimum [1].

© Springer Nature Singapore Pte Ltd. 2017
B. Zou et al. (Eds.): ICPCSEE 2017, Part II, CCIS 728, pp. 433–441, 2017.
DOI: 10.1007/978-981-10-6388-6_36

Zhong *et al.* [2] proposed an effective evolutionary algorithm, namely Multi-Agent Genetic Algorithm (MAGA), which combine the ability of the agent to perceive and react with the environment and the search method of genetic algorithm. MAGA has much higher convergence rate to solve the complex function optimization problem. In Agent-based generate algorithm, each individual will be defined as an agent, in its neighborhood each agent competes and cooperate with the other individuals to improve itself energy, while taking advantage of their own knowledge to learn to adapt to the environment better. So that Agent-based GA can achieve much higher convergence rate and better optimization effect.

Individual density distance is an important factor for populations' diversity. In this paper, the individual density distance is used in multi-agent genetic algorithm to solve multi-objective optimization problems. The purpose is that, in the case of higher convergence rate for this algorithm, uniform and extension distribution Pareto solutions can be obtained.

The rest of this paper is organized as follows: In Sect. 2, the Mathematical Model for Multi-objective Optimization Problem and a few basic concepts are introduced. Section 3 describes the improved multi-agent genetic algorithm for numerical optimization problems. Section 4 gives U-Measure and S-Measure, an extensive measure of non-dominated solutions for multi-objective programming. Section 5 gives approach steps. The experimental studies are shown in Sect. 6. Finally, conclusions are presented in Sect. 7.

2   The Mathematical Model for Multi-objective Optimization Problem (MOP)

A multi-objective optimization problem can describe as follows, the set of decision variables with the number of n, the set of the objective function with the number of k and the set of constraints with the number of m. The mathematical model for the multi-objective optimization problem can be given as follow:

$$\min \mathbf{y} = f(\mathbf{x}) = (f_1(\mathbf{x}), f_2(\mathbf{x}), \ldots, f_k(\mathbf{x}))$$
$$s.t.: g(\mathbf{x}) = (g_1(\mathbf{x}), g_2(\mathbf{x}), \ldots, g_m(\mathbf{x})) \geq 0$$

where, $\mathbf{x}$ represents the decision variable vector, $\mathbf{y}$ represents the objective function vector, Ω represents the decision variable space, the constraint of $g(\mathbf{x}) \geq 0$ defines the feasible region $\Omega = \{\mathbf{x}|g(\mathbf{x}) \geq 0\}$.

Definition 1: If $\mathbf{x}* \in \Omega$ and $f_i(\mathbf{x}^*) \leq f_i(\mathbf{x})$, $i = 1, 2, \cdots, k$ for each $\mathbf{x} \in \Omega$, then $\mathbf{x}^*$ is called an optimal solution of the MO (multi-objective optimization).

Definition 2: Suppose $x_1, x_2 \in \Omega$, if for any $i(1 \leq i \leq m)$, $f_i(x_1) \leq f_i(x_2)$ and at least exists one i, $f_i(x_1) \leq f_i(x_2)$, then called x_1 dominate x_2.

If one solution $\mathbf{x}'$ can not be dominated by any $\mathbf{x}(\mathbf{x} \in \Omega)$, $\mathbf{x}'$ is called the Pareto optimal solution.

Because the objectives of MOP are often competing and conflict with each other, in many cases, it is impossible to achieve a solution in which all objectives obtain their optimum value. In general, the Pareto optimal solution set is considered for multi-objective optimization problem.

An agent is a physical or abstract individual that acts on itself and the environment and responds to changes of the environment. The evolutionary idea of multi-agent system is that each chromosome in the traditional genetic algorithm is replaced by an agent, and the evolution mechanism is applied to agents. Agent can exchange information with other agents and can learn by himself in his neighbor environment in the evolutionary process. By the competition and collaborative behavior between the agents, agents will become more adaptation to the environment. An agent is a candidate solution for a multi-objective optimization problem, whose energy is expressed by the inverse of the mean of its objective function.

Trust-degree [1] of Agent a to b at time t is denoted as trust (a, b, t), trust (a, b, t) ϵ [−1, 1]; When t = 0, trust (a, b, 0) = 0. In the evolutionary process, the trust-degrees of an agent to other individuals in the neighborhood are determined based on the success of cooperation with other individuals. If the cooperation is successful, then the individual's trust-degree is increased a value δ; if the cooperation fails, then reduce a value γ. In order to encourage better cooperation between agents, so that to speed up the evolution, let $\gamma > \delta$. When cooperation between a and b succeed, trust (a, b, t + 1) = trust (a, b, t) + δ; when a and b fail to cooperate, trust (a, b, t + 1) = trust (a, b, t) − γ, where the value γ and δ can be set according to the specific problem, generally set $\gamma = 0.1$, $\delta = 0.05$.

3 Establishment and Updating of Neighborhood

In the multi-agent evolutionary system, the perceived ability and behavior of the agent are only in the limited and local environment. How to define the neighborhood of an agent is a very critical. If the neighborhood is too large, the convergence is slow. If the neighborhood is too small, the agent only can exchange information with small number of individuals that reduce the ability to find excellent individuals. In the existing multi-agent evolution system, most of neighborhood is selected based on the adjacent location of the agent, that is, select multiple agents in adjacent position of an agent as its neighborhood.

In the square lattice structure [3], each agent will compete or collaborate with other agents in the neighborhood in order to gain better resources for it to evolve better. Each agent will only perceive a local environment around himself, so competition and collaboration will only occur in the individual neighborhood. Through the neighborhood information will gradually spread to the entire body of the grid.

4 Algorithm Design

In the process of evolution, following strategies are used: In the selection process, agent is selected according to not only agent's energy, but also its density in its neighborhood. Only the agent with higher energy and bigger density distance [4] can get the chance of genetic. These not only ensure the higher convergence speed, but also maintain the diversity of the population. Neighborhood competition operator and crossover operator perform the competition and cooperation behavior of agents, the self-learning operator perform the evolution of agents by using themselves knowledge.

4.1   Competition Operator

In competition process, failure agent will not be able to survive in the environment. Each agent in the population will compete with the agent with the biggest energy in its neighborhood. If the energy of the agent is less than the maximum value of agent's energy in its neighborhood, the agent will not be able to survive, and its position will be replaced by the new agent. Through competition operator, the individuals with low energy will be removed, and the energy of population will increase.

Let $A_{i,j}$ is an agent in the neighbor, if meet expression (1), then the agent is a winner, otherwise, the agent is a loser, will not be suitable for survival in the neighbor. The energy of each agent is taken as the inverse of the mean of its objective functions.

$$Energy(A_{i,j}) > Energy(Max_{i,j}) \tag{1}$$

If the energy of $A_{i,j}$ is greater than the energy of the agent with the biggest energy, then it can survive in the neighbor, and take part in crossover and mutation operator; Otherwise, it cannot survive, its position will be replaced by $Max_{i,j}$. There are two replacing strategies. The new agent $New_{i,j} = (e_1, e_2, \cdots, e_n)$ can be generated as (2).

$$e_i = \begin{cases} m_i & U(0,1) \geq 0.5 \\ m_i + U(-1,1) \times (m_i - a_i) & \text{other} \end{cases}, \quad i = 1, 2, \cdots, n \tag{2}$$

where $U(*, *)$ is a random function.

The density distance of agent can be calculated by formula $d_i = \sum_{l=1}^{m} (f_l(x_{i+1}) - f_l(x_{i-1}))$. If the deference of $Energy(A_{i,j})$ and $Energy(Max_{i,j})$ is less than ε, where ε is a small value, then one of $A_{i,j}$ and $Max_{i,j}$ is selected randomly, which energy will be decreased. At the same time, the density of agents with higher energy is lower, so that don't fall into the local convergence.

4.2   Crossover Operator

In the crossover process, the agent will perform collaborative operation with agents in its neighbors with probabilities $P_c + trust \times 0.1$. The higher the trust-degree of an agent to another agent, another agent will think the individual is very easy to cooperate with, and will produce higher energy offspring after crossover.

In the multi-objective evolutionary algorithm with real number coding, arithmetic crossover is a more commonly used operator, and its operation speed is very fast, and the performance is very stable, so here we use the simulated binary crossover operator (SBX). Assume that two individuals A and B in the neighborhood generate two sub-individuals A', B' according to (3).

$$\begin{cases} a'_i = ((1+\beta)a_i + (1-\beta)b_i)/2 \\ b'_i = ((1-\beta)a_i + (1+\beta)b_i)2 \end{cases}, \quad i \in (1, 2, \cdots, n) \tag{3}$$

where, β is a random variable which must be regenerated as follow formula in each dimension.

$$\beta = \begin{cases} \mu, \mu \le 0.5 \\ 2(1-\mu), \mu > 0.5 \end{cases}, \quad \mu \in U(0, 1) \tag{4}$$

The crossover operator makes full use of the information of the parent to generate the offspring.

4.3   Self-learning Operator

The agent has knowledge related to the problem, so it can use its own knowledge to learn to improve the energy. Recent studies have shown that for the problem of function optimization, when genetic algorithm is combined with local search strategy, better solutions can be obtained. So the local search for agents, not only can find a better solution, but also shows the self-learning behavior of the agent. In order to carry out local search faster, the self-learning behavior only applies to the individuals with the highest energy on the current generation. Assuming that the Pareto optimal solution of the generation is $Best_{ij} = (b_1, b_2, \cdots, b_n)$, then a small random perturbation is performed to find a better solution. The formula for generating a new solution is as follows [5]:

$$e_i = \begin{cases} b_i, U(0, 1) < 1/n \\ b_i + G(0, 1/t), other \end{cases}, \quad i = 1, 2, \cdots, n \tag{5}$$

where $G(0, 1/t)$ represents the Gaussian distribution of the random number generator, t is the number of evolutionary generations. If the new solution dominates the original solution, then replace the original solution as the optimal solution, otherwise continue to find.

After the evolution is complete, a new generation of descendants is generated. In the parent group, there are also excellent agents with good genes. In order to avoid the loss of the excellent agents in the parent, the new and old groups are merged into the candidate sets of replication, and according to the Pareto domination relationship to classify individuals of the candidate set into different levels. According to levels of individual from lower to higher level, some individuals will be copy into the next generation of groups. If all individuals of some level were copied, the number of entire group are more than the size of the population, the individuals of this level will be selected according to the density of agents from higher to lower to copy into the next generation, until the required size.

4.4   The Measure Function

In the multi-objective optimization problem, a good algorithm evaluation criterion [6] is: more non-inferior solution, the Pareto interface were uniform and wide distribution range. In order to evaluate the performance of the proposed algorithm in solving the multi-objective optimization problem, two evaluation indexes are used to judge the performance of the algorithm, namely U-measure and S-measure [7].

U-measure is used to evaluate the uniformity of a set of non-inferior solutions for function optimization problems. The smaller the U-measure, the better is the distribution of the Pareto solution. S-measure is used to evaluate extension of non-inferior solution set. According to the definition of S-Measure, the smaller the S-measure is, the smaller the distance between the Pareto solution and each of its reference solutions, so that the distribution of the Pareto solutions are wider in objective space.

5   Algorithm Description

Let $Best^t$ is the best individual at the t-th generation, P_c is the crossover probability that used to crossover operator in the neighborhood.

Step 1: Initialize the population, agent lattice, size of population, number of generation t = 0;

Step 2: Perform the competition operator for each agent;

Step 3: Perform crossover operator;

Step 4: Find $Best^t$ in population, then perform the self-learning operator on $Best^t$ to find better agent;

Step 5: Combine the parent and the offspring population, perform non-dominated sorting and density distance calculating, then select the first N agents and put into next generation population;

Step 6: If the termination condition is satisfied, output the result and terminate the algorithm; otherwise, go to step 2.

6   Experiments

6.1   Test Functions

In this paper, four test functions are used to test the proposed algorithm performance. The test function contains two objective functions. The feasible fields of the objective spaces are different, the dimensions are different. The performance and convergence speed of an algorithm can be tested more comprehensively. Where the SCH is a convex function and Pareto is continuous; the number of decision variable of ZDT1 is n = 30, the Pareto front is convex and continuous. The dimension of ZDT2 is also 30, the Pareto frontier is non-convex and continuous, the Pareto frontier of ZDT3 is convex and discontinuous. The following table is a detailed description of the test function:

(1) SCH: The convex Pareto optimal solution set

$$\min \begin{cases} f_1(x) = x^2 \\ f_2(x) = (x-2)^2 \end{cases}, \quad x_i \in [-10^3, 10^3], \quad i = 1, \cdots, n$$

(2) ZDT1: The convex Pareto optimal solution set

$$\min \begin{cases} f_1(x) = x_1 \\ f_2(x) = g(1 - \sqrt{x_1/g}) \end{cases}, \quad g = 1 + 9(\sum_{i=2}^{n} x_i/(n-1)), \quad x_i \in [0, 1], \quad i = 1, \cdots, 30$$

(3) ZDT2: The non-convex Pareto optimal solution set

$$\min \begin{cases} f_1(x) = x_1 \\ f_2(x) = g(1 - (x_1/g)^2) \end{cases}, \quad g = 1 + 9(\sum_{i=2}^{n} x_i/(n-1)) \, x_i \in [0, 1], \quad i = 1, \cdots, 30$$

(4) ZDT3: The convex and non-continuous Pareto optimal solution set.

The test function ZDT3 represents the discreteness features: its Pareto-optimal front consists of several non-contiguous convex parts.

$$\min \begin{cases} f_1(x) = x_1 \\ f_2(x) = g[1 - \sqrt{x_1/g} - x_1 \sin(10\pi x_1)/g] \end{cases}, \quad x_i \in [0, 1], \quad i = 1, \cdots, 30$$

$$g = 1 + 9(\sum_{i=2}^{n} x_i/(n-1))$$

The introduction of the sine function in $f_2(x)$ causes discontinuity in the Pareto optimal front. However, there is no discontinuity in the objective space.

6.2   Experiment Results

In order to verify the advantages and disadvantages of the algorithm performance, the perform results are compared with NSGA-II. In the experiment, the population size is 100 for current algorithm and NSGA-II, the crossover probability is 0.8, the mutation probability is 0.05, and the neighborhood size of this algorithm is set to 8, the termination criterion also is 100 iterations, and each algorithm runs 50 times.

The following is an experimental simulation of the two algorithms to optimize the test function. From the above simulation, we can see that when solving the SCH problem and the ZDT3 problem, the performance of this algorithm is not very different from that of NSGA-II, and it can converge to the optimal region of the problem. When solving ZDT1 and ZDT2, the algorithm shows better performance, and the distribution and convergence of the Pareto solution are better than NAGA-II. Furthermore, in order to analyze the performing, the non-inferior solutions in the objective space are analyzed statistically, and the U-measure and S-measure are used as statistical analysis.

According to the U-measure definition, the smaller the U-measure value, the better the uniformity of distribution of the corresponding Pareto solution. Let U1 and U2 are respectively the U-measure of the Pareto solution for current algorithm and the NSGA-II. Then, when U1 < U2, the Pareto solutions of current algorithm is more evenly distributed in the objective space than NSGA-II. The U-measure of the Pareto solution obtained from the two optimization algorithms for the test function are shown in Table 1, where Function is the test function and Algorithm is the corresponding algorithm, the U-measure of the obtained Pareto solution is expressed by U (F, A).

Table 1. Comparison of experiment results of U-measure between current algorithm and NAGA II

U(F,A) Function Algorithm	SCH	ZDT1	ZDT2	ZDT3
Current algorithm	0.2016	0.3358	0.6205	0.1531
NSGA-II	0.2168	0.3765	0.6537	0.1520

It shows us in the Table 1, the distribution of solutions which obtained by current algorithm are more uniformity than NSGA-II for the test functions SCH, ZDT1 and ZDT2, specially for ZDT1 and ZDT2, advantages of uniformity are obvious. For ZDT3, the uniformity of distribution of the pareto solutions obtained by the current algorithm is are slightly worse than NSGA-II, but the difference is very small. This expresses that the algorithm can obtain better in obtaining uniform distribution Pareto solution.

According to the definition of s-measure, the smaller the s-measure value, the better the widespread of distribution of the corresponding Pareto solution. Let S1 and S2 are respectively the s-measure of the Pareto solution for current algorithm and the NSGA-II. Then, when S1 < S2, the Pareto solutions of current algorithm is more widely distributed in the objective space than NSGA-II. The S-measure of the Pareto solution obtained from the two optimization algorithms for the test function are shown in Table 2, where Function is the test function and Algorithm is the corresponding algorithm, the S-measure of the obtained Pareto solution is expressed by S (F, A).

The result show in the Table 2, the S-measure of the Pareto solution obtained for the test functions by current algorithm are smaller than NSGA-II. Only for ZDT3, the S-measure of the pareto solutions obtained by the current algorithm is greater than NSGA-II. This expresses that the algorithm can obtain more widespread distribution Pareto solution.

Table 2. Comparison of experiment results of S-measure between current algorithm and NAGA II

S(F,A) Function / Algorithm	SCH	ZDT1	ZDT2	ZDT3
Current algorithm	8.3628e-005	1.2658e-004	0.0069	1.3537
NSGA-II	0.2500	1.9347e-004	0.0105	1.2568

7   Conclusions

Genetic algorithm based on agent has obtained a lot of good results when solving multi-objective optimization. It exploited the known characteristics of some benchmark functions to achieve outstanding results. In this paper, we propose a multi-objective evolutionary algorithm based on multi-agent density. Each agent competes and collaborates with other individuals in the neighborhood to increase its own energy. Agent can also use their own knowledge to self-learning to increase energy. In the process of competitive operation, if the number of individuals with high energy in the neighborhood is too large, the density of individual will be used to determine which agent will be selected. Four classic benchmark functions are used test the effects of the algorithm. The experimental results show that the proposed algorithm is better than NSGA-II in solving the optimization function problem of two objectives. Result show the given algorithm can obtained uniformity and widespread distribution Pareto solution.

Acknowledgment. This work was supported by Tianjin Research Program Application Foundation and Advanced Technology (14JCYBC15400).

References

1. Pan, X.Y., Liu, F., Jiao, L.C.: Multi-objective social evolutionary algorithm based on multi-agent. J. Softw. **20**(7), 1703–1713 (2009)
2. Zhong, W., Liu, J., Xue, M., Jiao, L.: A multiagent genetic algorithm for global numerical optimization. IEEE Trans. Syst. Man Cybern. Part B: Cybern. **34**(2), 1128–1141 (2004)
3. Zhong, W., Liu, J., Liu, F., Jiao, L.: Combinatorial optimization using multi-agent evolutionary algorithm. Chin. J. Comput. **27**(10), 1341–1353 (2004)
4. Lei, D.M., Wu, Z.M.: Crowding-measure based multi-objective evolutionary algorithm. Chin. J. Comput. **28**(8), 1320–1326 (2005)
5. Pan, X.Y., Jiao, L.C.: Social cooperation based multi-agent evolutionary algorithm. J. Xidian Univ. **36**(2), 274–280 (2009)
6. Zheng, J.: Multi-objective genetic algorithm and its application. Postdoctoral research report of Chinese Academy of Sciences, p. 12 (2004)
7. Ma, G.J.: Multi-objective genetic algorithm based on new model. Excellent master's degree thesis, Xidian University (2009)

An Improved Binary Wolf Pack Algorithm Based on Adaptive Step Length and Improved Update Strategy for 0-1 Knapsack Problems

Liting Guo[(⊠)] and Sanyang Liu

School of Mathematics and Statistics, Xidian University,
Xi'an 710126, Shaanxi, China
guo.liting@yahoo.com

Abstract. Binary wolf pack algorithm (BWPA) is a kind of intelligence algorithm which can solve combination optimization problems in discrete spaces. Based on BWPA, an improved binary wolf pack algorithm (AIBWPA) can be proposed by adopting adaptive step length and improved update strategy of wolf pack. AIBWPA is applied to 10 classic 0-1 knapsack problems and compared with BWPA, DPSO, which proves that AIBWPA has higher optimization accuracy and better computational robustness. AIBWPA makes the parameters simple, protects the population diversity and enhances the global convergence.

Keywords: Binary wolf pack algorithm · 0-1 knapsack problem · Adaptive step length · Update strategy

1 Introduction

The knapsack problem is a typical combinatorial optimization problem which is a classical NP-hard problem of computer science [1]. It offers many practical applications in many areas. Besides accurate algorithms (branch and bound method, dynamic programming method) and approximation algorithms (greedy algorithm and ant colony algorithm) [2], many swarm intelligence algorithms are applied well in 0-1 knapsack problem. For instance, artificial fish swarm algorithm is applied in combinatorial optimization problems [3] and 0-1 knapsack problems [4]; firefly algorithm [5] and its improved algorithm [6] are applied in 0-1 knapsack problems. To some extent, swarm intelligence algorithms improve shortages of existing traditional algorithms in 0-1 knapsack problems. There are many researches about improvements of algorithms to solve 0-1 knapsack problems, such as adaptive genetic annealing algorithm which improves convergence speed and strengthens the capability of the species out of local optimum [7].

Wolf pack algorithm (WPA) [8] is a heuristic swarm intelligent method that simulates wolves hunting behaviors and mode of preys distribution, has good computational robust and global searching ability. WPA has been applied in multiple complex function optimization problems successfully. Based on WPA, redesigned by binary coding, a binary wolf pack algorithm (BWPA) [9] is referred. With studies of 10 examples, BWPA is found that it has good stability, astringent and global searching ability which has better calculating capabilities than DPS, GGA and QGA. However,

© Springer Nature Singapore Pte Ltd. 2017
B. Zou et al. (Eds.): ICPCSEE 2017, Part II, CCIS 728, pp. 442–452, 2017.
DOI: 10.1007/978-981-10-6388-6_37

BWPA has shortages such as too many parameters, fixed search step length and incomplete pack update strategy. Thus, improved binary wolf pack algorithm based on adaptive step length and improved update strategy of wolf pack is referred.

2   0-1 Knapsack Problems

0-1 knapsack problems can be described as follows: given n items and a knapsack, the volume of each item is w_i, and the value of each item is $p_i(i = 1, 2, \cdots, n)$, the volume of the knapsack is C. How to choose m items of n items and put into the knapsack can make the sum of value of the m items maximized and the sum of volume of them doesn't exceed C. If $x_i = 1$ means the i item was put into the knapsack, $x_i = 0$ means the i item wasn't, then the 0-1 knapsack problem can be described as follows:

$$\begin{cases} \max f(x) = \sum_{i=1}^{m} p_i x_i \\ s.t. \sum_{i=1}^{m} w_i x_i \leq C, x_i = 0, 1 \end{cases} \tag{1}$$

0-1 knapsack problems belong to discrete optimization problems with constraint, the model of which has to be transformed when it will be solved by BWPA. Here, punish coefficient is imported to punish unfeasible solutions. Finally, the transformed non-constrained optimization problems can be described as follows:

$$\max F(x) = f(x) - \lambda \cdot \max(0, (\sum_{i=1}^{m} w_i x_i - C)) \tag{2}$$

where λ is punish coefficient, is a big enough number which can ensure that the worst value of feasible solution is better than the best value of unfeasible solution.

3   Improved Binary Wolf Pack Algorithm

3.1   Adaptive Step Length

BWPA consists of scouting, summoning and beleaguering three intelligent behaviors mainly. BWPA updates the group position and finds the optimal solution by scouting step length $step_a$, summoning step length $step_b$ and beleaguering step length $step_c$ which are all integer and present the degree of searching accuracy of artificial wolves in solution space. Based on experience, there exists relation in code space whose length is m as follows:

$$\lceil m/10 \rceil \geq step_a \geq step_b \geq step_c = 1 \tag{3}$$

In fact, the value of step length is set fixedly as follows:

$$step_a/8 = step_b/4 = step_c = 1 \tag{4}$$

The above setting method of step length limits the accuracy of the algorithm. When the search scope is small and step length is too large, it will add unnecessary searching and time; when the search scope is large and step length is too small, it will make searching imprecisely so that the best solution is found hardly.

Thus, the adaptive step length method is introduced so that step length can adjust consciously according to the distance between artificial wolves and the leader. Step length can be calculated as follows:

$$step = \lceil rand \cdot norm(x(i, :) - X_{lead}) \rceil \tag{5}$$

where $\lceil * \rceil$ means ceil, $rand$ means random number from 0 to 1, $x(i, :)$ means the position of present artificial wolf.

Scouting Behavior with Adaptive Step Length. $N - 1$ best artificial wolves except the leader in solution space are regarded as searching wolves. The searching wolf i perceives the prey odor concentration of current position, which means calculating the objective function value Y_i of the searching wolf i. If Y_i is higher than Y_{lead}, $Y_{lead} = Y_i$, and Y_{lead} represents the prey odor concentration of leader wolf's position; if Y_i is lower than Y_{lead}, the searching wolf i makes a step forward in h directions respectively, which means the position X_i of the searching wolf i is executed kinematic operator $\Theta(X_i, M, step)$ by h times, where $M = \{1, 2, \cdots, m\}$, $step$ means adaptive scouting step length and is calculated by function (5). Record the prey odor concentration Y_{ip} of every direction and return to its origin position, where $p \in H$, $H = \{1, 2, \cdots h\}$. Make a step forward in the direction p^* and update X_i, repeat scouting behaviors until the prey concentration Y_i of some searching wolf i is higher than Y_{lead} or the times T of scouting maximize the biggest times T_{max}.

p^* is stated as follows:

$$\begin{cases} p^* \in \max\{Y_{ip}, p \in H\} \\ Y_{ip^*} > Y_{i0} \end{cases} \tag{6}$$

where h means a random number in $[h_{min}, h_{max}]$.

Summoning Behavior with Adaptive Step Length. The leader wolf convenes $N - 1$ fierce wolves to run towards it by summoning step length $step$. The position X_i of fierce wolf i is stated as follows:

$$X_i' = \Theta(X_i, M, step) \tag{7}$$

where $step$ is calculated by function (5), M is stated as follows:

$$M(k) = \begin{cases} j, k = k + 1, j = j + 1, x_{dj} \neq x_{ij} \\ null, k = k, j = j + 1, x_{dj} = x_{ij} \end{cases} \tag{8}$$

where $j = 1, 2, \cdots, m$; k starts from 1; $null$ means no data. M is the set of different code positions between fierce wolf position X_i and the leader wolf position X_{lead}. Function (8) means the values of same code positions aren't inverted, but the values of different code positions are inverted. If M is empty set, execute random kinematic operator $\Theta(X_i, M^*, 1)$ once, and $M^* = \{1, 2, \cdots, m\}$.

On the way, if the prey odor concentration Y_i of some fierce wolf i is higher than Y_{lead}, $Y_{lead} = Y_i$, and this fierce wolf replaces the leader wolf and summons again; if Y_i is lower than Y_{lead}, this fierce wolf keeps summoning until the distance between it and the leader wolf is lower than

$$d_{near} = \lceil m/\omega \rceil \tag{9}$$

where d_{near} means determined distance, ω means distance determined factor, $\lceil * \rceil$ means ceil.

Beleaguering Behavior with Adaptive Step Length. Fierce wolves unite searching wolves and beleaguer the prey to catch it. The position of the leader wolf is regarded as the position of the prey, and the position X_i of artificial wolf i is stated as follows:

$$X_i' = \Theta(X_i^*, M^*, step) \tag{10}$$

where $M^* = \{1, 2, \cdots, m\}$, $X_i^* = \Theta(X_i, M, step)$, M is stated as function (8), $step$ is calculated as function (5).

3.2 Improved Update Strategy of Wolf Pack

In 0-1 knapsack problems, because the solution space consists of 0 and 1, the solutions are relatively limited. In the process of iteration, there will be several same best values and schemes appearing frequently. "Survivor" as the update strategy of BWPA means that removing the R artificial wolves with worst values and adding R new randomly.

For example, the pack with N members, after one iteration, generates $N - R$ same best values and schemes, showed as follows (Fig. 1):

$$
\begin{aligned}
x_1 &= (1,1,0,0,1,0,1,0,1,0) \\
x_2 &= (1,1,0,0,1,0,1,0,1,0) \\
x_3 &= (1,1,0,0,1,0,1,0,1,0) \quad \text{best schemes} \\
x_4 &= (1,1,0,0,1,0,1,0,1,0) \\
x_5 &= (1,1,0,0,1,0,1,0,1,0) \\
x_6 &= (1,0,0,0,1,0,1,0,1,0) \\
x_7 &= (1,0,0,0,1,0,1,0,1,0) \\
x_8 &= (1,1,0,0,1,0,1,0,0,0) \\
x_9 &= (1,1,0,0,1,0,1,0,0,1) \\
x_{10} &= (1,1,0,0,1,0,0,0,1,0,1)
\end{aligned}
$$

Fig. 1. The pack after one iteration

When the R artificial wolves with worst values are removed, the left artificial wolves are showed as follows (Fig. 2):

Removing all schemes except the best one, there is only one solution left, which destroys the population diversity.

$$x_1 = (1,1,0,0,1,0,1,0,1,0)$$
$$x_2 = (1,1,0,0,1,0,1,0,1,0)$$
$$x_3 = (1,1,0,0,1,0,1,0,1,0)$$
$$x_4 = (1,1,0,0,1,0,1,0,1,0)$$
$$x_5 = (1,1,0,0,1,0,1,0,1,0)$$

Fig. 2. The left artificial wolves

Hence, an improved update strategy is put forward which ranks the objective function values of the pack after one iteration from big to small. If several same best values and schemes appear, retain only one and remove others as follows (Fig. 3):

$$x_1 = (1,1,0,0,1,0,1,0,1,0)$$
$$x_6 = (1,0,0,0,1,0,1,0,1,0)$$
$$x_7 = (1,0,0,0,1,0,1,0,1,0)$$
$$x_8 = (1,1,0,0,1,0,1,0,0,0,0)$$
$$x_9 = (1,1,0,0,1,0,1,0,0,0,1)$$
$$x_{10} = (1,1,0,0,1,0,0,0,1,0,1)$$

Fig. 3. The left artificial wolves with improved update strategy

Updating the left pack with an improved update strategy can protect the population diversity, avoid local optimum and searching with low speed.

3.3 Flow of AIBWPA Algorithm

Detailed procedure of AIBWPA algorithm as follows:

step1 Initialization. Initialize the positions of artificial wolves X_i, the number of pack N, the biggest iteration times K_{max}, the rate factor of searching wolves α, the biggest times of summoning T_{max}, distance determined factor ω, the rate factor of updating β.

step2 Choose the artificial wolf with the best objective function value as the leader wolf, record its position X_{lead} and objective function value Y_{lead}. The last $N - 1$ wolves are regarded as searching wolves which execute scouting behaviors according to formula (3) and (4). Until scouting times of every searching wolf reach T_{max}. Turn to *step3*.

step3 Choose $N - 1$ wolves except the leader as fierce wolves which run to the prey according to formula (5). If fierce wolf i perceives the odor concentration of prey Y_i is higher than Y_{lead}, $Y_{lead} = Y_i$, fierce wolf i replaces the leader and executes summoning behavior. If Y_i is lower than Y_{lead}, keep running until $d_{is} \leq d_{near}$, turn to *step4*.

step4 Update positions of wolves which execute summoning behavior according to formula (6).

step5 Update the leader wolf position according to "Survivor" rule. Update the wolf pack according to improved update strategy.

step6 If solution meets optimization accuracy or iterations times reach K_{max}, output the position of leader wolf which is the best solution; if not, turn to *step2*.

4   Experimental Result and Analysis

4.1   10 Classical 0-1 Knapsack Problems

To show the advantages of improved binary wolf pack algorithm compared with BWPA, test algorithms' abilities of optimization performance with 10 classical 0-1 knapsack problems supplied by reference 9, as shown in Table 1:

Table 1. 10 classical 0-1 knapsack problems.

Num	Dim	Parameters (w, p, C)	Optimum
k1	10	$w = (95, 4, 60, 32, 23, 72, 80, 62, 65, 46)$, $p = (55, 10, 47, 5, 4, 50, 8, 61, 85, 87)$, $C = 269$	295
k2	15	$w = (56.357531, 80.874050, 47.987304, 89.596240, 74.66048, 85.894345, 51.353496, 1.498459, 36.445204, 16.589862, 44.56923, 0.4669, 37.788018, 57.118442, 60.716575)$, $p = (0.125126, 19.330424, 58.500931, 35.029145, 82.284005, 17.410810, 71.050142, 30.399487, 9.140294, 14.731285, 98.852504, 11.908322, 0.891140, 53.166295, 60.176397)$, $C = 375$	481.07
k3	20	$w = (92,4,43,83,84,68,92,82,6,44,32,18,56,83,25,96,70,48,14,58)$, $p = (44,46,90,72,91,40,75,35,8,54,78,40,77,15,61,17,75,29,75,63)$, $C = 878$	1024
k4	23	$w = (983,982,981,980,979,978,488,976,972,486,486,972,972,485,485,969,966,483,964, 963,961,958,957)$, $p = (981,980,979,978,977,976,487,974,970,485,485,970,970,484,484, 976,974,974,482,962,961,959,958,857)$, $C = 10000$	9767
k5	50	$w = (80,82,85,70,72,70,66,50,55,25,50,55,40,48,50,32,22,60,30,32,40,38,35,32,25,28,30, 22,50,30,45,30,60,50,20,65,20,25,30,10,20,25,15,10,10,10,4,4,2,1)$, $p = (220,208,198,192, 180,180,165,162,160,158,155,130,125,122,120,118,115,110,105,101,100,100,98,96,95,90, 88,82,80,77,75,73,72,70,69,66,65,63,60,58,56,50,30,20,15,10,8,5,3,1)$, $C = 1000$	3103
k6	50	$w = (80,82,85,70,72,70,66,50,55,25,50,55,40,48,50,32,22,60,30,32,40,38,35,32,25,28,30, 22,25,30,45,30,60,50,20,65,20,25,30,10,20,25,15,10,10,10,4,4,2,1)$, $p = (220,208,198,192, 180,180,165,162,160,158,155,130,125,122,120,118,115,110,105,101,100,100,98,96,95, 90,88,82,80,77,75,73,72,70,69,66,65,63,60,58,56,50,30,20,15,10,8,5,3,1)$, $C = 1000$	3119
k7	50	$w = (428,754,699,587,789,912,819,347,511,287,541,784,676,198,572,914,988,4,355,569, 144,272,531,556,741,489,321,84,194,483,205,607,399,747,118,651,806,9,607,121,371,999, 494,743,967,718,397,589,193,369)$, $p = (72,490,651,833,883,489,359,337,267,441,70, 934,467,661,220,329,440,774,595,98,424,37,807,320,501,309,834,851,34,459,111,253,159, 858,793,145,651,856,400,285,405,95,391,19,96,283,152,473,448,231)$, $C = 11258$	16102
k8	60	$w = (135,133,130,11,128,123,20,75,9,66,105,43,18,5,37,90,22,85,9,80,70,17,60,35,57,35, 61,40,8,50,32,40,72,35,100,2,7,19,28,10,22,27,30,88,91,47,68,108,10,12,43,11,20,37,17,4, 3,21,10,67)$, $p = (350,310,300,295,290,287,283,280,272,270,265,251,230,220,215,212, 207,203,202,200,198,196,190,182,181,175,160,155,154,140,132,125,110,105,101,92,83, 77,75,73,72,70,69,66,60,58,45,40,38,36,33,31,27,23,20,19,10,9,4,1)$, $C = 2400$	8362
k9	80	$w = (40,27,5,21,51,16,42,18,52,28,57,34,44,43,52,55,53,42,47,56,57,44,16,2,12,9,40,23, 56,3,39,16,54,36,52,5,53,48,23,47,41,49,22,42,10,16,53,58,40,1,43,56,40,32,44,35,37,45,52, 56,40,2,23,49,50,26,11,35,32,34,58,6,52,26,31,23,4,52,53,19)$, $p = (199,194,193,191,189,178,174,169,164,164,161,158,157,154,152,152,149,142,131, 125,124,124,124,122,119,116,114,113,111,110,109,100,97,94,91,82,82,81,80,80,80,79,77, 76,74,72,71,70,69,68,65,65,61,56,55,54,54,53,47,47,46,41,36,34,32,32,30,29,29,26,25,23, 22,20,11,10,9,5,4,3,1)$, $C = 1173$	5183
k10	100	$w = (54,95,36,18,4,71,83,47,18,31,62,78,48,88,45,94,64,14,80,4,23,75,36,90,20,77,32,58, 6,14,86,84,59,71,21,30,22,96,49,81,48,37,28,6,84,19,55,88,38,51,52,79,55,70,53,64,99, 61,86,1,64,32,60,42,45,34,22,49,37,33,1,78,43,85,24,96,32,99,57,23,8,10,74,59,89,95,40,46, 65,6,89,84,83,6,19,45,59,26,13,8,26,5,9)$, $p = (297,295,293,292,291,289,284,284,283, 283,281,280,279,277,276,275,273,264,260,257,250,236,236,235,235,233,232,232,228,218,217, 214,211,208,205,204,203,201,196,194,193,193,192,191,190,187,187,184,184,184,181,179, 176,173,172,171,160,128,123,114,113,107,105,101,100,100,99,98,97,94,94,93,91,80, 74,73,72,63,63,62,61,60,56,53,52,50,48,46,40,40,35,28,22,22,18,15,12,11,6,5)$, $C = 3820$	15170

4.2   Simulation Experiment and Results Comparison of Improved Update Strategy and Adaptive Step Length Binary Wolf Pack Algorithm (TEST 1)

AIBWPA represents binary wolf pack algorithm with improved update strategy and adaptive step length, OBWPA represents the original binary wolf pack algorithm. The parameters settings are supplied by reference 9. The number of pack and the biggest iteration times are all 4 m and every problem runs 20 independent simulations. Evaluate algorithm in many aspects such as mean, best, worst, standard deviation (SD) and mean time, as shown in Table 2:

Table 2. Simulation experiment results comparison of 10 0-1 knapsack problems.

Num	Dim	Algorithm	Mean	Best	Worst	SD	Mean time
K1	10	AIBWPA	295	295	295	0	0.24
		OBWPA	294.5	295	288	1.57	0.12
		DPSO	293.3	295	287	2.49	0.01
K2	15	AIBWPA	481.07	481.07	481.07	0	0.66
		OBWPA	481.07	481.07	481.07	0	0.31
		DPSO	478.43	481.07	450.67	6.92	0.01
K3	20	AIBWPA	1024	1024	1024	0	1.35
		OBWPA	1020.78	1024	1013	4.12	0.65
		DPSO	1011	1024	986	11.75	0.02
K4	23	AIBWPA	9764.29	9765	9762	1.11	2.21
		OBWPA	9758.17	9760	9755	2.23	0.95
		DPSO	9761.18	9766	9756	2.64	0.02
K5	50	AIBWPA	3071.25	3084	3050	13.55	19.34
		OBWPA	3015.3	3045	2984	28.25	9.67
		DPSO	2981.7	3006	2954	18.85	0.14
K6	50	AIBWPA	3092.44	3112	3066	21.77	9.63
		OBWPA	3033.8	3077	2960	30.83	9.36
		DPSO	2990.6	3034	2932	31.00	0.14
K7	50	AIBWPA	15733.6	15971	15293	192.52	20.31
		OBWPA	15035.6	15399	14584	265.79	13.18
		DPSO	14732	15130	14385	278.65	0.14
K8	60	AIBWPA	8359	8362	8356	3.08	34.26
		OBWPA	8317	8356	8219	41.48	15.74
		DPSO	7776.89	7898	7616	87.18	0.23
K9	80	AIBWPA	4937.5	5066	4807	122.71	83.02
		OBWPA	4804.17	4945	4608	116.06	41.86
		DPSO	4514.6	4655	4331	78.09	0.51
K10	100	AIBWPA	15135.6	15170	15041	49.38	172.50
		OBWPA	15074.7	15170	14894	83.08	113.43
		DPSO	13498.45	13711	13295	97.97	1.09

From the Table 2, problems' dimensions increase from 10 to 100, AIBWPA has gradually increased abilities of optimization performance compared with AIBWPA and OBWPA. Mean and best can represent the solving accuracy and the ability of optimization performance. For k1–k3, AIBWPA succeed 100%; for k4–k10, the results of AIBWPA are closed to the theory best values. Worst and standard deviation represent the stability of algorithm and the capability of the species out of local optimum. The standard deviations of AIBWPA are smaller than those of other algorithms, which represents AIBWPA has stronger stability and robustness. AIBWPA costs time badly because of the increase of calculation about adaptive step length and population ranking. However, it improves the solving accuracy and stability effectively.

Figure 4 is given convergences curves of K5–K10 for showing that AIBWPA has stronger constringency effect more intuitively.

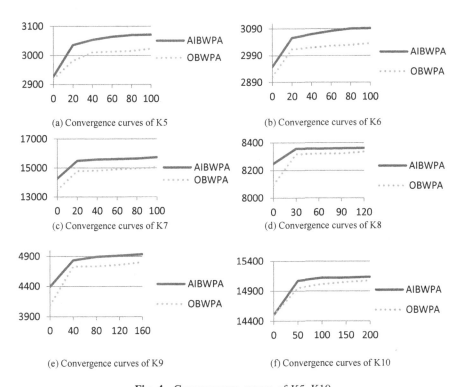

Fig. 4.   Convergences curves of K5–K10

4.3   Simulation Experiment and Results Comparison of Improved Update Strategy and Repair Strategy [9] (TEST 2)

IBWPA represents binary wolf pack algorithm with improved update strategy, BWPA represents binary wolf pack algorithm with repair strategy [9], OBWPA represents the original binary wolf pack algorithm. With above three algorithms, test 10 classical 0-1 knapsack problems 20 times respectively, regard mean, best, worst, standard deviations and mean time, comparatively analyze.

From the Table 3, for k1–k4, problems dimensions increase from 10 to 23, IBWPA succeed 100% whose solving accuracy is highest, standard deviations are smallest and robustness is highest. With the increase of problems dimensions, the mean time of IBWPA change from high to low relatively, which represents that IBWPA has obvious advantages in high dimension 0-1 knapsack problems. The improved update strategy has obvious effects on the promotion of optimization performance and stability. On the contrary, BWPA has unstable solving accuracy. As for k3–k6, k9–k10 problems, BWPA has worse effects than OBWPA, adds the complexity and time and lowers the stability of algorithm. Thus, this repair strategy doesn't work on the promotion of optimization performance.

Table 3. Results comparison of improved update strategy and repair strategy.

Num	Algorithm	Mean	Best	Worst	SD	Mean time
K1	IBWPA	295	295	295	0	0.43
	BWPA	295	295	295	0	0.12
	OBWPA	294.5	295	288	1.57	0.06
K2	IBWPA	481.07	481.07	481.07	0	0.33
	BWPA	481.07	481.07	481.07	0	0.18
	OBWPA	481.07	481.07	481.07	0	0.31
K3	IBWPA	1024	1024	1024	0	0.70
	BWPA	1019.4	1024	1013	4.40	0.67
	OBWPA	1020.78	1024	1013	4.12	0.65
K4	IBWPA	9767	9767	9767	0	8.24
	BWPA	9758.1	9761	9753	2.56	0.98
	OBWPA	9758.17	9760	9755	2.23	0.95
K5	IBWPA	3066.3	3084	3046	12.89	12.44
	BWPA	3007	3044	2970	27.07	8.86
	OBWPA	3015.3	3045	2984	28.25	9.67
K6	IBWPA	3076.5	3099	3044	16.06	10.50
	BWPA	3030.9	3070	2983	27.47	8.85
	OBWPA	3033.8	3077	2960	30.83	9.36
K7	IBWPA	15827.5	16021	15608	123.06	10.02
	BWPA	15255.3	15793	15033	255.62	9.23
	OBWPA	15035.6	15399	14584	265.79	13.18
K8	IBWPA	8355.1	8362	8356	9.60	16.79
	BWPA	8317.3	8356	8213	41.09	18.10
	OBWPA	8317	8356	8219	41.48	15.74
K9	IBWPA	4996	5062	4923	49.89	41.62
	BWPA	4766.2	4938	4541	117.20	51.10
	OBWPA	4804.17	4945	4608	116.06	41.86
K10	IBWPA	15123.5	15170	15077	44.15	93.74
	BWPA	14985.1	15104	14755	116.00	91.52
	OBWPA	15074.7	15170	14894	83.08	113.43

4.4 Simulation Experiment and Results Comparison of Adaptive Step Length and Fixed Step Length (TEST 3)

ABWPA represents binary wolf pack algorithm with adaptive step length, OBWPA represents the original binary wolf pack algorithm.

From the Table 4, for low dimension 0-1 knapsack problems, the mean and SD of ABWPA are lower than that of OBWPA, which represents that adaptive step length can improve the solving accuracy effectively and enhance the stability of algorithm.

Table 4. Results comparison of adaptive step length and fixed step length.

Num	Algorithm	Mean	Best	Worst	SD	Mean time
K1	ABWPA	295	295	295	0	0.09
	OBWPA	294.5	295	288	1.57	0.06
K2	ABWPA	481.07	481.07	481.07	0	0.28
	OBWPA	481.07	481.07	481.07	0	0.31
K3	ABWPA	1024	1024	1024	0	0.65
	OBWPA	1020.78	1024	1013	4.12	0.65
K4	ABWPA	9761.56	9766	9759	2.40	1.00
	OBWPA	9758.17	9760	9755	2.23	0.95
K5	ABWPA	2980.33	3035	2940	26.53	10.27
	OBWPA	3015.3	3045	2984	28.25	9.67
K6	ABWPA	2991.22	3033	2961	24.47	10.04
	OBWPA	3033.8	3077	2960	30.83	9.36
K7	ABWPA	15655.3	15955	15373	215.82	9.97
	OBWPA	15035.6	15399	14584	265.79	13.18
K8	ABWPA	8331.11	8356	8280	32.22	17.45
	OBWPA	8317	8356	8219	41.48	15.74
K9	ABWPA	4640.22	4770	4387	124.37	49.68
	OBWPA	4804.17	4945	4608	116.06	41.86
K10	ABWPA	14974.56	15126	14859	108.33	97.79
	OBWPA	15074.7	15170	14894	83.08	93.43

5 Conclusion

Based on binary wolf pack algorithm, I research adaptive step length's and improved update strategy's influence on this algorithm, and put forward a kind of binary wolf pack algorithm based on adaptive step length and improved update strategy. For low dimension problems, adaptive step length can improve the solution effect of algorithm. Improved update strategy which applies to discrete combinatory problems in limited solution space can protect the population diversity effectively and make searching more fully. Next, the improved update strategy can be applied repeatedly to the pack, remove more even all repeated solutions, protect the diversity of solution space and enhance the ability to search the best solution.

References

1. Mavrotas, G., Diakoulaki, D., Athanasios, K.: Selection among ranked projects under segmentation, policy and logical constraints. Eur. J. Oper. Res. **187**, 177–192 (2008)
2. Tian, F., Wang, Y.: The review of algorithms for 0-1 knapsack problems. Softw. Guide **8**(1), 59–61 (2009)
3. Li, X., Lu, F., Tian, G., et al.: Applications of artificial fish school algorithm in combinatorial optimization problems. J. Shandong University (Engineering Science) **34**(5), 64–67 (2004)
4. She, X., Zhu, M., Zhao, Y.: Improved artificial fish school algorithm to solve knapsack problem. Comput. Eng. Appl. **47**(21), 43–46 (2011)
5. Cheng, K., Ma, L.: Artificial glowworm swarm optimization algorithm for 0–1 knapsack problem. Appl. Res. Comput. **30**(4), 993–994, 998 (2013)
6. Guo, L., Shen, Q.: Improved glowworm algorithm to solve 0-1 knapsack problems. Softw. Guide **15**(1), 54–56 (2016)
7. Lv, X., Chen, S., Lin, J.: Adaptive genetic annealing algorithm of solving 0/1 knapsack. J. Chongqing Univ. Posts Telecommun. (Natural Science Edition) **25**(1), 138–142 (2013)
8. Wu, H., Zhang, F., Wu, L.: New swarm intelligence algorithm – wolf pack algorithm. Syst. Eng. Electron. **35**(11), 2430–2438 (2013)
9. Wu, H., Zhang, F., et al.: A binary wolf pack algorithm for solving 0-1 knapsack problem. Syst. Eng. Electron. **36**(8), 1660–1667 (2014)
10. Zhang, P., Shan, X., Gu, W.: A modified harmony search algorithm for 0-1 knapsack problems. Appl. Mech. Mater. **365–366**, 182–185 (2013)

Reform of Teaching Mode in Universities Based on Big Data

Bing Zhao$^{(\boxtimes)}$ and Li Fu

Department of Electronic Engineering, Heilongjiang University,
Harbin, Heilongjiang, China
zb0624@163.com

Abstract. With the rapid development of information technology, college teaching into the era of big data, the traditional teaching model has been unable to meet the needs of future innovative personnel training, big data technology brings challenges and opportunities to college education. This paper expounds the impact of big data on the traditional teaching mode, and proposes a new teaching model based on the big data mind set. It aims to improve the teaching effect, enhance the student interest and inject new vitality into the college teaching.

Keywords: Big data · Teaching mode · Teaching reform

1 Introduction

In 2016 Ministry of Education issued education informative thirteen five plan clearly pointed out that deepening the integration of information technology and education teaching development, the service education and teaching expand to the service for the whole process of education is one important tasks in the 'thirteen' period [1]. Actively explore the application of information technology in the educational model, focus on improving students' information literacy, innovation consciousness and innovation ability, develop digital learning habits, promote the comprehensive development of students, play the support role of informatization in high-quality talent future development.

Big data is the product of digital technology development to a certain stage, it refers to the analysis and processing of massive data [2], the era of big data is also known as post-information age. With the extensive application of information technology in the field of education, massive data will be produced every day, the structure of the data type has a variety of characteristics, including text, pictures, audio and video and other data, through the massive data processing and analysis to improve the teaching model of colleges and universities, that is, use big data technology to reform the teaching model [3,4]. College teaching in the era of big data is not limited to multimedia teaching, but also extends to teachers' research work and all aspects of student learning [5,6]. Big data processing technology and thinking ideas were introduced into classroom teaching, experimental teaching, practical activities, research project research and

© Springer Nature Singapore Pte Ltd. 2017
B. Zou et al. (Eds.): ICPCSEE 2017, Part II, CCIS 728, pp. 453–458, 2017.
DOI: 10.1007/978-981-10-6388-6_38

students daily life. How to effectively use the big data technology to carry out the reform of university model is one of the research hot spots in the field of education, and has the positive theory and practical significance.

At present, the research on teaching reform with big data is focused on the reform of teaching methods. However, the urgent problem of personnel training is the mutual adaptation between social needs and talent output, that is, the goal of personnel training needs to meet the social needs, and the big data technology is the effective technical means to dig the current needs and the potential demand of social needs. In addition, in accordance to the enterprise requirement of talents innovation and diversity, colleges and universities need to reform the existing personnel assessment mechanism and increase the process evaluation. The big data technology can help teachers collect various data in the process of student learning and establish more scientific evaluation system. This paper first analyzes the challenge of the big data technology to the traditional teaching mode, and carries on the questionnaire survey and the interview to the student about the current teaching method, finally, this paper studies the teaching mode reform of the university in the high data age from the following four aspects, adjusts the goal of personnel training, classroom teaching path, increases the process guidance and process evaluation, and establishes the database of teaching experience.

2   The Challenge of Traditional Teaching Model in Big Data Era

First, in the traditional teaching mode, the teaching method is relatively fixed, the teacher according to the syllabus and prescribed class hours arranges school teaching content and time, mainly to the teacher's teaching experience as the main prepared lesson content, students follow the teacher's arrangements for targeted learning. The traditional teaching mode is easy to master and apply the theoretical knowledge points, and has certain convenience. However, with the diversity of notebook computers, PAD, mobile phone terminal functions, the traditional knowledge acquisition approach has brought challenges. The method of students acquire knowledge is not limited to the classroom, but also through the network video, We Chat public, circle of friends and others acquire a large number of diversified information real time, the survey results of 200 students methods of knowledge acquisition is shown in Fig. 1.

The survey showed that more than half students learned about the knowledge through the Internet, 18.5% students obtain information by borrowing books, 17% students through WeChat or circle of friends to expand the relevant knowledge points, this acquisition can be seen as a network effect extension, only 8% students obtain information through the classroom completely. Therefore, in the context of the informative big data era, the traditional single teaching material, repeat what the books say, passive education model is difficult to fully mobilize the enthusiasm and initiative of students.

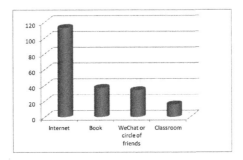

Fig. 1. The method of students acquire knowledge

Second the teaching effect of the traditional teaching mode is obvious, but the goal is single. Due to the increasing number of students in colleges and universities, the overall education resources are limited, in current, colleges and universities are faced with the problem of educational resources reducing distributed to individuals, the phenomenon of talent training tending to homogenize is serious increasingly. The interactive communication in traditional teaching corresponds to the corresponding teaching activities, students complete the same after-school exercises on time, the entire teaching process has a strong order, easy to carry out teaching, but limit the students innovation consciousness and innovation ability in a large extent. In addition, teachers in the classroom use their own practice cases to explain knowledge usually, to strengthen the students memory and master of knowledge point, but more limited to a single discipline, lack cross and integration between different disciplines.

About this question Whether the current class communication and teaching activities to meet your learning needs Extracting 80 students from every grade, this paper conducts a questionnaire survey, the survey result is shown in Fig. 2.

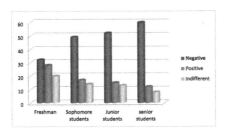

Fig. 2. Whether the current class communication and teaching activities to meet your learning needs

In the senior students, 75% students want to more communicate with teachers and further study, only 15% students express that the class content can meet their own learning needs, there are 8 students have not considered about this issue. In addition, the survey result shows that, with the depth of professional courses, high grade needs of interactive communication and class activities are higher than junior students significantly.

Third the evaluation standard of traditional teaching mode adopts a unified model standard. Teachers make a unified evaluation method according to the arrangement of teaching content in the teaching process, to assess student classroom work and the learning effect. However, in the era of informative big data, the traditional statistical assessment is difficult to reflect the individual learning effect, cannot meet the requirements of the times. The third question: whether the final exam mode can reflect own learning situation? 110 students in the same grade are investigated, the survey result is shown in Fig. 3.

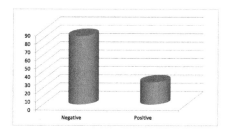

Fig. 3. Whether the final exam mode can reflect own learning situation

The result shows that more than half students want to change the assessment methods, of which 25% students expressed the idea of personalized network assessment.

3   The Reform of Teaching Mode in Big Data Era

3.1   Collect Social Needs Using Big Data Technology, Adjust the Personnel Training Goal

The first principle of personnel training is to meet the needs of economic and social development, that is, colleges and universities must understand the social needs. Therefore, colleges and universities should apply the big data technology to the reform of personnel training mode, make full use of the network, enterprise feedback and other means to collect the relevant data of social needs, extract the relevant training data from the university training status database, establish a wide range of personnel training goal combining the needs of the community with the talent and their own teaching resources, equipment, venues and other teaching resources, train multi-type compound talents, explore and attempt to train research and development talents, technical talents, management talents, service talents and other talents in the professional. Colleges and universities should rely on characteristics, play to strengths, break the convergence of the types of personnel training to meet the country, enterprises and students' needs, ease the employment difficulties of college graduates.

3.2   Classroom Learning Path

The teachers using classroom teaching need to face dozens or even hundreds students, teach the same content, but the students learn the starting point, learning habit and learning schedule are not the same. Through the analysis of big data, teachers can help every student adapt to their own learn methods and rhythm in a short time, every student can be improving and expanding based on the original knowledge. In the era of big data, classroom learning path is shown in Fig. 4.

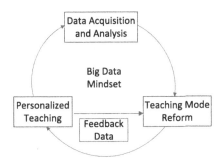

Fig. 4. Classroom learning path

3.3   Increase the Learning Process Analysis, Process Evaluation

Using the big data mindset, teachers collect data from all kinds of learning environment, store and analyze data, analysis results can provide guidance for teaching activities, ameliorate teaching mode, and achieve digital teaching. At the same time, collect student feedback data constantly, the teaching model is amended to achieve individualized teaching and personalized teaching. The teaching model data is detailed interpreted through the big data technology, the results not only focus on student learning behavior, but also concern about the subtle problems in teachers teaching activities, better display the characteristics of teaching and learning between teachers and students.

In traditional education, teachers cannot take into account the individual differences of all students. In the era of big data, teachers can digitize the student's entire learning process, record data information such as knowledge coverage in the students post-class exercises, the completion time and the error rate, use computer to analyze these data, master each student understanding of the content of teaching, evaluate the learning process and on this basis, propose Individual corresponding suggestions. For example, teacher can recommend different follow-up exercises based on the test results of each student. The analysis and use of a large number learning process data makes the development of students more personalized and justified. In addition, process evaluation using big data technology can also provide more scientific decision information for personnel training, improve the quality of personnel training. At the same time, the diversification of assessment methods can promote the learning method expansion, and can establish a more complete and reliable evaluation system.

3.4 Establish Database of Teaching Experience

In the traditional education model, the old teachers have more than ten years or even decades teaching practice experience, the quantity and merits of the experience determine the level and position in professional development, and new teachers have not these experiences. In the era of big data, the computer can retrieve the teaching problems recorded and the solution by the previous teachers, establish a database of teachers' teaching experience and analyze the data. When the data reaches a certain amount of information, the analysis results can recommend alternative solutions to the teachers who encounter similar problems. The solution of teaching problem is no longer limited to teachers own teaching accumulation, but on the analysis of large number teaching problems characterization and their solutions. It is foreseeable that the professional development of teachers will usher in a new transformation, and this change will help the reform of teaching mode.

4 Conclusion

The demand for talents in colleges and universities has changed greatly, the future talent should be innovative talent, be good at independent learning using information data. The existing teaching model cannot be very good to meet the high-quality training needs in information age, so the existing teaching model need to reform. In big data era, higher education has been digitized gradually, forming a massive amount of data, therefore, how to use big data technology to promote the deep integration of information technology and education teaching, to build information technology teaching model based on big data has a positive theoretical and practical significance.

Acknowledgements. This work is supported by 2015 Heilongjiang University New Century Education and Teaching Reform Project (NO. 2015B76). Many thanks to the anonymous reviewers, whose insightful comments made this a better paper.

References

1. Ministry of Education: Education Informative Thirteen Five Plan. http://www.moe.edu.cn. Accessed 7 June 2016
2. Zhu, S.: Construction of digitizing teaching model based on big data. Microcomput. Appl. **31**(5), 42–49 (2015)
3. Liang, W.: Big data era-classroom teaching will usher in real change. J. Beijing Inst. Educ. (Natural Science Edition) **8**(1), 14–16 (2013)
4. Zhao, J., Hu, Z.: Research on the informative teaching mode of university under the environment of big data. Inf. Sci. **34**(1), 92–103 (2016)
5. Yu, X.: Research on the establishment of personalized teaching mode based on the big datas application. Inf. Sci. **33**(11), 53–56 (2015)
6. Xu, Y., Sun, Q.: Discussion on innovative teaching of computer application basic course in big data age. China Comput. Commun. (13), 172–175 (2015)

The Construction and Application of MOOCs University Computer Foundation in Application-Oriented University

Ying San[1], Hui Gao[2](✉), Qilong Han[3], and Junyu Lin[4]

[1] Harbin Guangsha University, Harbin, China
249590520@qq.com
[2] Harbin Huade University, Harbin, China
44117252@qq.com
[3] Harbin Engineering University, Harbin, China
6837218@qq.com
[4] Institute of Information Engineering, CAS, Beijing, China
linjunyu@iie.ac.cn

Abstract. The paper took SPOC teaching in University Computer Foundation for example, discussed how to make online education meet the basic objectives and requirements of traditional teaching, meanwhile fulfilling the potential applications of MOOCs to meet the needs of all walks in computer application. Based on the sharing of resources in many application-oriented universities, from the perspective of teaching reform around innovation and entrepreneurship in the era of internet, the paper focused on the exploitation and significance of MOOCs. By discussing the means to develop and build the MOOCs University Computer Foundation, which was based on curriculum sharing, the paper demonstrated the work flow. It underlined that, in the educational circumstances nowadays, it was necessary to introduce and spread MOOCs. At the same time, it elaborated on the problems existing in the construction of MOOCs University Computer Foundation, and proposed some concrete methods on further development of MOOCs to meet the requirements of students in application-oriented universities.

Keywords: MOOCs · Innovative education · Individual development · Curriculum reform

1 Introduction

With the prevalence of "internet plus" concept, education resources will inevitably be opened up and constructed by means of internet. Under this background, universities are undoubtedly the main force in education resources open-up. MOOC (Massive Open Online Courses), as a new way of education, fits the times. Compared with traditional online courses, MOOC has a complete curriculum structure, especially SPOC (Small Private Online Course), which is applied in course teaching of universities. MOOC provides video materials, text materials, Q&A online service, interactive community and study monitoring, of which all serve the needs of independent study.

© Springer Nature Singapore Pte Ltd. 2017
B. Zou et al. (Eds.): ICPCSEE 2017, Part II, CCIS 728, pp. 459–466, 2017.
DOI: 10.1007/978-981-10-6388-6_39

The construction and application of MOOCs University Computer Foundation, which is based on curriculum resources sharing, accord with students' needs and the concept of state-advocated "internet plus education". It aims to improve teaching efficiency, develop students' ability of independent study and awareness of innovation, reduce teachers' workload and provide scientific and rational trainings that fit with their characteristics. University Computer Foundation is part of the public basic courses in computer, of which the video production could be finished by demonstrating and interpreting by MOOC teams. With practising, discussing, testing, online and offline, we compose the necessities of MOOCs.

2 Basic Steps of the Research on Curriculum Construction and Application of MOOCs University Computer Foundation Which is Based on Curriculum Sharing

2.1 Research on Application Value of MOOCs University Computer Foundation in University Teaching

2.1.1 Position of the Curriculum University Computer Foundation

University Computer Foundation is one of general curriculum offered to those first-year undergraduates, which is composed of computer basics, theory and application of operating system, network basics, office software application and other relevant knowledge and skills. The teaching contents are complex, which relate to not only computer theory but also practical application.

2.1.2 MOOCs Could Effectively Solve the Existing Problems in University Computer Foundation Courses

University Computer Foundation is usually offered during the first semester, three to four class hours per week (including courses in theory and practice), which aims at the freshmen of non-computer majors. Out of the complexity and wide range, it is difficult to reconcile the demands of courses neither in theory nor practice. Although we made many adjustments during teaching reform, it has not been solved yet. When MOOC is brought into teaching, students could follow the courses in theory and teacher's demonstration online independently, which makes it possible for flipped classroom to be carried out in left class hours. Teachers are supposed to monitor students' online study, and check learning effect during testing and practising offline, which resolves the problem of being short of class hours.

2.1.3 It Gives Full Play to Teachers' Personal Attributes in MOOCs

To meet students' needs and receive good effects, MOOC teams have to spend plenty of time and energy in curriculum design before those courses. During the courses, they need, according to the feedback on learning data, making adjustments repeatedly and constantly. So, teamwork is inevitably the foundation for well-designed MOOCs.

As public basic curriculum, University Computer Foundation courses are offered to more students than other courses, so more teachers are needed, which provides a base case for team cooperating, with each one playing a role according to his characteristic.

Teachers skilled at lecturing, who can always make the interpretations vivid and exhaustive in a logical way, in MOOC teams, play the role of lecturer in videos. Some others good at communicating, who have time to help students out with any queries enthusiastically, in MOOC teams take the role in providing Q&A online service and conducting group discussion. Some else, who are expert in computer and internet, can be offered the role of system engineer.

2.2 Research on the Way of Utilizing Available MOOCs Resources Scientifically and Effectively

In China, MOOC teams are mainly from project 985 universities. Curriculum resources are generally public basic courses, with Public English in the majority, which depends on the quantity of learners and the quality of educational resources. The construction of MOOCs aims at sharing and utilizing educational resources, which lay the base for the research.

2.2.1 Take C Language Program Design as an Introduction to Research on the Utilization of Available MOOCs Resources

Before the construction of MOOCs according to teaching practice, available platforms and resources should be applied in university teaching. As the base for this research is the cooperation between universities on sharing and construction of MOOC resources, we applied the MOOCs C Language Program Design from another project 985 university with permission in our curriculum teaching. These courses are designed comprehensively, including unit testing, FAQs, performance rating and a series of subsidiary contents, in which the system of achievement evaluation on subjective program-design questions is funded by The National Natural Science Fund. The system gives quite objective scores. Although some programs can not generate any result, but the system will give scores accordingly, as long as they reflect the correct programming ideas. Through the practice for one term, we got a quite positive feedback. Students said, although under monitoring, MOOCs gave them much more flexibility. Teachers felt that workload was reduced greatly, at the same time teaching efficiency was enhanced.

2.2.2 Make Full Use of Available Curriculum Resources of MOOC and SPOC to Build MOOC Resources in Accordance with Our Own Teaching

During the research on MOOC teaching in C Language Program Design, we found that MOOCs from other universities could not meet all the requirements of our own. For example, questions in test bank are difficult, rewards to students who give questions online is not enough. So, the application of available MOOC resources should conform to the teaching practice in MOOC-cooperating universities. It has brought us a new task that in order to promote MOOC teaching in application-oriented universities and institutions for vocational higher education, we should improve existing MOOC resources to meet the needs of teaching and learning.

2.3   MOOC Design of University Computer Foundation

2.3.1   Objectives of MOOC Teaching in University Computer Foundation

At beginning, MOOC resources are mainly from the SPOC for undergraduates of MOOC-cooperating universities. On one hand, MOOC teaching should help achieve the teaching objective of the course University Computer Foundation in application-oriented universities, on the other hand, we should fulfill its potential applications to meet the learning needs of all walks in computer application.

2.3.2   Curriculum Strategies of MOOCs University Computer Foundation

The curriculum strategies of MOOC teaching are composed of three parts, they are strategy of teaching resource construction, strategy of teaching activity and strategy of teaching process.

Basic strategy of teaching resource construction. Video resource is the most important one in basic teaching resources. First, course video should support mobile terminals, subtitled and not too long. At the same time, text of lecture should be available to meet the needs of different learners. Secondly, on MOOC platform, learners should be able to upload and download those materials they need, such as syllabus, process, test bank, lesson plan and presentation, key points guiding, school work, reference list, case bank, professional lectures and material bank. Thirdly, considering further development, we should chose the MOOC platform that meets certain criteria, which at least contain such factors as curriculum notice, introduction, teacher's introduction, videos, study resources, forum, self-testing bank, functions of job submission and score announcement.

Strategy of teaching activity, of which the most important is to conduct online activities, such as exercising, collaborating, correcting, discussing and Q&A, with offline activities as supplement. In SPOC teaching, the activities are mainly online study and flipped-classroom interaction, such as discussing, communicating, exercising and tutoring. Due to the large number of students, the main lecturer can not afford that much interaction, so assistants are needed by a ratio. In that way, team management and labor division are necessary, as well as evaluation and encouragement mechanism.

Strategy of teaching process design, which contains preparation, implementation and assessment. In preparation, we should finish curriculum design, make videos of lecture and chose a proper MOOC platform. When implementing, the SPOC teaching should match with the curriculum planning. Assessment consists of two parts, assessment of learners and of curriculum itself. The assessment of learners is composed of usual performance and exam performance, in which the former should take a large ratio than that in classroom teaching. The assessment of curriculum is based on many means. We could analyze studying data of learners collected by the platform, perform queries and in-depth surveys, or sort out reflections from internet and society, all of which help us to make necessary adjustments before a new round MOOC teaching.

3 Construction of MOOCs University Computer Foundation

3.1 Team Construction of MOOC Teaching and Researching

3.1.1 Composition

MOOC construction needs coordination and cooperation, while every member of team should do well in his specific jobs. At the beginning, from three project 985 universities of the same kind, whose MOOCs were constructed well, we invited seven key members, with four Readers therein. Thereafter four assistants and one lecturer took part in. As online courses, MOOCs need internet technical support, we invited two platform technicians. In addition, we had one executive staff in charge of the coordination and communication with the academic affairs office and other departments. Hereafter, the composition could meet the requirements of MOOC construction in expertise and technology.

3.1.2 Work Division

In course video, there should be only one lecturer, though two are supposed to back up according to the work load, and the main lecturer should take on one third of the video courses at least. Q&A online service and exercise correction are mainly performed by assistants, while other members are also responsible for answering questions, and together they collect and analyze the feedback. Teaching and practicing in flipped classroom is carried on by the teacher independently appointed in teaching program.

3.2 Platform

As the MOOCs C Language Program Design we applied in our SPOC teaching are from other universities, of which the platform was "iCourse", so we chose "iCourse" as the platform of our MOOCs University Computer Foundation. We chose it not only from the perspective of familiarity, but also its credibility. "iCourse" is a resource sharing platform of higher education curriculum, and interactive online-studying platform, sponsored by "the Education Quality and Teaching Reform Program in Higher Education", which was initiated during the 12th five-year plan by the Ministry of Education and the Treasury.

3.3 Course Arrangements

While designing MOOCs, we must take motivations into consideration, because heavy course load will reduce learning efficiency, and make students lose interest. It also works in SPOC teaching, although there are flipped classroom and strict monitoring mechanism. According to the performance and analysis of three main MOOC platforms, more students chose to take courses between four to ten weeks long. In view of this, MOOCs University Computer Foundation mentioned above does not apply to further promotion, (sixteen weeks every term, three to four hours per week, totaling forty-eight to sixty-four hours). Although the SPOC under construction is supposed to meet the learning needs of undergraduates mainly, we should also regulate the length of each video, as well as how many weeks every semester and courses' total hours. Each piece of video should be within 10 min, five to six pieces per week, and four to ten hours each course in total. The surplus hours could be used as flipped classroom, in which we could ask for feedback, discuss and practise. In this way, we crack the problem of insufficient hours for learning and practising, which is caused by the complexity of subject itself, and lay a good foundation for online education from SPOC to MOOC, to meet the requirements of more learners.

3.4   Making MOOCs Videos

3.4.1   Utilizing Macro Learning Resources

In recent years, Macro-lecture contests have been booming, institutions and universities required participants make a video less than 10 min on a knowledge point. The video should be short and pithy. Through these contests, some videos have been collected, most of which are concerned with University Computer Foundation. So, we decided to supplement the MOOCs University Computer Foundation with these macro learning resources. At the same time, we still followed the principle of one main lecturer in course video, unless the work load was heavy, other lecturers were needed. In that case, we insisted that one chapter should only be presented by one lecturer. As the main lecturer shouldered other arduous tasks of teaching and researching, another four participated in video lectures. So, in the course videos, six chapters were presented by five lecturers.

3.4.2   Adopting Different Methods to Record According to Course Contents

The contents of University Computer Foundation are impressively broad, which include academic disciplines such as computer basics, operational disciplines such as office automation applications, and practical disciplines for example computer network construction. So that, we adopted different methods to record according to the contents. Under the technical support of the Able-elec Technology (Shanghai) Co., Ltd, we went to teaching scenes to record the academic and practical contents, with the real scene of students interactions reserved. To get better effect, we made videos of operational disciplines in recording base, where teachers could remake the videos by Camtasia studio 6 and other screen recording applications, until they got satisfied.

3.5   Construction of Supporting Resources of MOOCs

MOOCs do not mean network teaching resources only, especially applied in higher education. MOOCs should provide online videos, as well meet the needs in flipped classroom. MOOC teaching in universities must conform to all the requirements in course teaching.

3.5.1   Construction of Questions Bank for Online Testing and Offline Computer-Based Examination

Previously, students took examinations of University Computer Foundation on the network examination platform of Wanwei. Other network examination platforms were also used in MOOCs-cooperating universities. But with a lack of labor and resources, contents of test bank, more or less, were not in accordance with teaching, through which teaching effects could not be presented. However, recently with the promotion of MOOCs cooperation and timely financial supports, a refined and applicative test bank will be built along with MOOC teaching resources.

3.5.2   Construction of Text Materials

For learners online, text and lecture notes are essential materials, so they will be provided on the platform in form of hyperlinks. Videos of MOOCs University Computer Foundation will be presented in text in form of QR codes, by scanning these codes, learners could watch and download course videos in mobile terminals.

3.5.3   Testing and School Work

Testing of MOOCs University Computer Foundation is composed of video-embedded tests and phased test. Video-embedded tests are mainly easy quizzes in class, which are put to attract students' attention and check learning effect, scores of which are not counted. Phased test and final exam are carried on offline in flipped classroom.

In the early stage of MOOCs construction, school work and testing are corrected by teachers and assistants, which, with the development, should be improved to new forms, including manual correction, system evaluation and peer assessment. Teachers are supposed to correct the tests that could reflect the teaching contents, which could help them to know the teaching effect. Objective items should be marked by computer, and subjective items that do not concern with major issues of principle would be left to students to discuss among themselves to meet the needs of collaborative learning.

3.6   Problems that Still Need to be Solved in MOOCs Construction and Application

3.6.1   Network and Technical Support

MOOCs rose with the prosperity of internet. Due to the costs of campus network construction, students can not use internet free of charge, especially learning online through mobile terminals. This is the main problem in the process of MOOCs construction and application. Therefore, network equipment and related technology support is the key to the effective application of MOOCs.

3.6.2   Formation of Learning Habits

During the application and construction of MOOCs, we found that, although we set up the SPOC learning monitoring mechanism and flipped-classroom online learning test, some students, who did not developed good learning habits, did not complete online learning on schedule, which made the whole MOOC teaching effect greatly reduced. So, it is the essential needs for achieving the expected teaching effect to develop students discipline and learning habits.

3.6.3   Adjustment of Teachers' Workload in MOOCs Construction

In the most intense phase of MOOCs construction, it was under the premise of not reducing workload that main members of the teaching team completed all the work, which was bound to damage the quality of MOOC teaching, and this was indeed the case in the following teaching. We had to do some repetitive work, again, designing, planning and recording. Although MOOCs will eventually reduce teachers' workload, but in the stage of construction, teaching team is supposed to pay a lot of hard work. This requires a reasonable reduction in basic workload at the first beginning.

4 Conclusion

Nowadays, internet applications and mobile terminals have been increasingly popular, the concept of innovative education and personalized learning leads the trend of thoughts, and the reform of higher education and teaching mode is extremely urgent. In such a situation, the introduction and popularization of MOOCs make the perfect combination of technology development and learning needs, and also bring new opportunities and challenges for the development of education in China. Under the background of "internet plus education", teachers and universities should keep up with the pace of the times, adjust in educational ideas, educational mechanism, educational content and educational form. We should seize opportunities and work together to deepen curriculum reform, provide high quality courses for students and lifelong learners, showing the social responsibility of higher education and teachers.

Acknowledgements. This work is supported, in part, by Key Program of Heilongjiang Education and Science during the 12th Five-Year Plan period: the Grant numbers are GJB1215026 and GJB1215025. Heilongjiang Association of Higher Education: the Grant number is 16Z071.

References

1. Kang, Y.: "Post MOOC Era" of online education: analysis of SPOC. Tsinghua J. Educ. (1) (2014)
2. Sun, H.: Discussion on reform of university computer basic course based on MQOC/SPOC. Inheritance and Innovation: Improve the Quality of Higher Education (2014)
3. Wang, Q.: MOOCs will change traditional teaching methods. China Education Info (19) (2013)
4. Xia, H.: MOOCs Provide New Opportunities for Teachers' Professional Development. Shanghai Education (28) (2013)
5. Li, Q.: Discussion on teaching reform of computer course in colleges and universities. China Comput. Commun. (09) (2014)
6. Zha, J., Li, X.: The significance and research review of MOOC platform construction in higher vocational colleges. J. Vocat. Technol. Coll. (05) (2015)
7. Deng, H., Li, M., Chi, Y., Tan, S.: Curriculum Knowledge System Construction in MOOC Era. Course Education Research, vol. 7, 3rd edn. (2013)
8. San, Y.: The necessity of computational thinking in computer basic courses for non-computer majors. Education Teaching Forum **7**(28) (2013)
9. Pappano, L.: 2012 The Year of the MOOC (12) (2012)
10. Gao, H., Qiu, Z., Wu, D., Gao, L.: Research and reflection on teaching of C programming language design. In: The International Conference of Young Computer Scientists, Engineers and Educators, 10 January 2015
11. Gao, H., Qiu, Z., Liu, Z., Huang, L., San, Y.: Research and practice on college students' innovation and entrepreneurship education. In: The International Conference of Young Computer Scientists, Engineers and Educators, 22 August 2016

Empirical Analysis of MOOCs Application in Sino-Foreign Cooperative Design Major Teaching

Tiejun Zhu[✉]

Anhui Polytechnic University, Wuhu 241000, Anhui, China
ztj@ahpu.edu.cn

Abstract. In China, MOOCs have been widely implemented and deeply applied in the teaching system of Sino-foreign cooperative design majors, and have led a sustained and profound educational reform and innovation. Taking aim at this development trend advocated by Chinese higher education department, the researchers specially selected 7 samples in all levels Chinese higher universities and colleges, to deeply analyze the effects of MOOCs in the teaching of Sino-foreign cooperative design major by using various methods such as tracking investigation, questionnaire and interview, statistical comparison and so on. The paper not only systematically elaborates MOOCs advantages through combining the features of Sino-foreign cooperative design major, but also reveals and highlights its potential drawbacks and shortcomings, and simultaneously puts forward the corresponding improvement measures. It aims to provide references of MOOCs weighing utilization for Sino-foreign cooperative design majors which have been carried out on MOOCs teaching or intent to implement.

Keywords: MOOCs application · Teaching reform · Sino-foreign cooperative design major · Empirical analysis

1 Introduction

Under the active advocating of Chinese education departments, all levels Chinese schools, especially universities and colleges are starting to carry out teaching reform based on MOOCs exploitation and application. Hereinto, the MOOCs teaching mode reform of Sino-foreign cooperative design major (hereinafter referred to as SFCDM)

This research was financially supported by Anhui Province key education and teaching projects of universities and colleges "School running characteristics exploration and innovation of international engineer institute of Anhui Polytechnic University" (2016jyxm0091); 2017 National Social Science Fund pre-research Project of Anhui Polytechnic University "Collaborative innovation assimilation and alienation research of intangible cultural heritage protection of the Yangtze River Delta" (2017yyrw01); Anhui provincial philosophy and social science planning project "Research on the internationalization strategy of higher education in Anhui Province" (AHSKY2016D29); "International compound students training mode of Local colleges and universities in global MOOCs era" (2014jyxm43).

© Springer Nature Singapore Pte Ltd. 2017
B. Zou et al. (Eds.): ICPCSEE 2017, Part II, CCIS 728, pp. 467–482, 2017.
DOI: 10.1007/978-981-10-6388-6_40

has the highest voice as well as the fastest action. Indeed, SFCDM, as another rising force and the development direction in the field of Chinese higher education, has significantly developed since the beginning of 1990s. Therefore, integration of the MOOCs teaching mode and the teaching system of SFCDM naturally walk in the forefront of Chinese higher education teaching reform.

Due to historical reasons, Chinese modern design is lack of abundant connotation, and design education has a late start so as to form a weak foundation. Face to Europe and the United States profound modern design history culture and advanced design education idea and system, many Chinese schools and design majors have a strong desire to cooperate with foreign universities so that SFCDMs have blossomed everywhere. According to the statistical data, by the end of August 2016, China existing approved SFCDMs by the Ministry of education have 63 and cover 21 provinces, municipalities and autonomous regions. Approved SFCDMs by provincial education department have 49 and cover 19 provinces, municipalities and autonomous regions. (A total of 112) Its division is given in Table 1. Among these SFCDMs, 52% have utilized MOOCs resource, and 15% have constructed their own design MOOCs platform.

Table 1. The number and distribution of SFCDMs in China (by the end of August 2016).

Division by region of the foreign cooperative partner territory		Division by region of Chinese universities and colleges		Division by education level	
Europe	41	Eastern China	76	Chinese-foreign cooperative organization	12
North America	23	Central China	18	Master education	3
Oceania	17	Western China	7	Undergraduate education	56
Asia	31	Northeast China	11	Junior college education	41

2   Background

There are a large number of colleges and universities in China. In terms of level and types, they can be divided into 985 and 211 colleges and universities, first-batch universities, second-batch universities, private colleges and universities, public institutions. In order to improve the representation of the research and the wide application, the researchers selected a representative university in each of the mentioned colleges and universities. And, in the survey of 7 colleges and universities, the researchers chose 7 SFCDMs which foreign partners were from different regions and countries. By using interview, questionnaire, statistics, comparison and evaluation methods, the researchers did a deep investigation and tracking records of MOOCs application state during SFCDMs teaching. Basic situation of selected samples is listed in Table 2. The main contents of this questionnaire(15 students are selected from the first and second grades of SFCDM in each sample university who have completed at least two MOOCs, and a total of 210 students from 7 colleges and universities) are as follows: The questionnaire included 44 questions: Of which 1 to 7 were basic information about students; 8 to 13 inquired as to students similar extent to MOOCs before they experienced the MOOC teaching model; 14 to 22 inquired students about their various of MOOC learning states and related reasons such as motives for MOOCs learning, the principles of elective courses, the feelings, problems and causes in the learning process, the reasons of dropping out of MOOCs, etc.; 23 to 30 investigated students' views about advantages of the MOOCs and their achievements such as acquired abilities and enhanced qualities get through by learning MOOCs. 31 to 39 were on the basis of students' self-assessment to collect all kinds of problems they met during the MOOCs learning period and their suggestions; 40 to 44 mainly surveyed students' satisfaction degree about MOOCs by comparing to other teaching methods and teaching resources after MOOCs introduction mechanism applied in SFCDMs' teaching. As for the results of the questionnaire, the paper will give specific analysis in the following content.

Table 2. Research samples' basic situation (by the end of August 2016).

Chinese universities and colleges	Type	Cooperative partner	SFCDM name	Enrollment scale/Year	Time length of MOOCs application/Year
Xi'an Jiaotong-Liverpool University	Chinese-foreign Cooperative Organization	The University of Liverpool (UK)	Architectural design	32 (Master)	1
Wuhan University of Technology	985 Innovation Platform for Superiority Subject	University of Wales (UK)	Arts and design	100 (Undergraduate)	3.5
Nanjing University of Science and Technology	211 Engineering University	Coventry University (UK)	Industrial design	30 (Undergraduate)	2
Anhui Polytechnic University	First-batch University	Bridgeport University (US)	Visual communication design	40 (Undergraduate)	1.5
Hunan City University	Second-batch University	Whitireia Community Polytechnic (New Zealand)	Visual communication design	100 (Undergraduate)	1
China Australia Business College of Shanxi	Private Institution	Woosong College	Figure image design	30 (Junior college)	2
Zhejiang Fashion Institute of Technology	Public Institution	Sugino Fashion College (Japan)	Costume design	90 (Junior college)	2.5

The research data collection processes and data analysis procedures are very strict. They are reflected in three aspects. Firstly, data collection adopted by means of the whole process of orientation, important node involvement and online follow-up. For example, the researchers applied interview and questionnaire methods to real-timely collect data and documents of MOOCs implementation crucial stages, and used online platform, social media software or APP to closely track to acquire efficient data.

Secondly, scientific analysis of comprehensive data. On the one hand, using the comparison method, trend analysis, modeling construction and so on to do the quantitative analysis, on the other hand, prediction, induction, deduction and research team discussion method were used to do qualitative analysis, to exemplify this, as analyzing the principles of student MOOCs selection, the researchers not only carried out questionnaire and discussed these survey results, but also they utilized prediction method to list all the possible principles first, then compared according to the real questionnaire results and in-depth probed into the detailed reason of deviation so as to precisely explore and understand the students' learning state and learning psychology. Thirdly, real-time adjustment, timely evaluation and feedback. For instance, in the process of data collection and analysis, once discovering the deviation from the actual situation by observation, the researchers timely adjusted research method and path. At the same time, the researchers introduced the evaluation and feedback mechanism including the research team self-assessment, initial assessment of data source objects, Formal assessment of data sources unit, submission and application department re-assessment, to ensure the scientific, effective and practical research data and conclusions.

This paper comprehensively analyzes the advantages and disadvantages of MOOCs application in the teaching system of SFCDMs. The research results can not only improve the function of SFCDMs' MOOCs teaching system, but can also help SFCDMs' students to benefit from the advantages and avoid the disadvantages during MOOCs learning processes so as to effectively improve students' learning efficiency and effectiveness of MOOCs. In addition, the research findings can also provide references of MOOCs weighing utilization for SFCDMs which have been carried out on MOOCs teaching or intent to implement.

3 Advantage Analysis

3.1 An Excellent Tool for the Effective Improvement of English Level

There is a big difference in English language proficiency requirements between SFCDMs and paralleled Non-SFCDMs. Because SFCDMs students need to study abroad, even non-English speaking countries also accept the TOEFL, IELTS scores, their English competence matters a lot. "In China, English comprehensive level of art and design students is relatively low. Quite a lot of MOOCs are based on English language teaching. Students can improve their English level through MOOCs learning". [1] Due to the reason, the current SFCDMs, which have already introduced MOOCs, all mainly recommend and select English MOOCs, and encourage students to choose English MOOCs in three global MOOCs platforms: Coursera, Udacity, edX. "Statistical data show that Coursera and Udemy MOOCs platforms only provide only 12% non-English courses". [2] Because the online MOOCs resources are enormous and extensive, students will choose courses according to their own learning interests, even hobbies. Naturally, the students will learn and enjoy English with pleasure and preferences. On the other hand, the majority of MOOCs set interactive activities such as Q&A quiz, learning feedback questionnaire, peer assessment and so on. These activities provide the opportunity to practice English and create an authentic English language environment. It brings a great advantage to SFCDMs students and effectively enhances their English level. According to the data of questionnaire, 87% of students reported that their English proficiency was significantly improved after learning MOOCs.

3.2 Self-assessment of Interest Points and Beneficial Supplement of Extensive Knowledge

"MOOCs provide a powerful framework for teaching and personalized learning experience." [3] Most MOOCs are concentrated essence. They are concise but rich in connotation. Students can carry out selective learning according to their own preferences. On the one hand, they can combine their interests to autonomously learn, self-assess learning ability, learning status and learning stimulation point, thereby eliminating the students' sense of distance and resistance to learning. On the other hand, design talents require acquiring diverse, connecting and crossing knowledge system. MOOCs introduction can make students contact and attain pluralistic knowledge on unlimited network, and provides a broad platform for the extension of students design thinking and burst of design inspiration. Figure 1 is design thinking development category of MOOC (Thinking, Fast and Slow) learning interface. This course was selected by a SFCDM student of Anhui Polytechnic University according to his learning interest. Anhui Polytechnic University recommended this MOOC to SFCDM students from MOOCs platform of University of Michigan in the United States, which was one of partner universities of Anhui Polytechnic University.

Fig. 1. Design thinking development category of MOOC (Thinking, Fast and slow) learning interface: selected by a SFCDM student of Anhui Polytechnic University according to his learning interest.

3.3   The Acquisition Platform of Design Creativity and Innovation Consciousness

"Sino-foreign cooperative education takes international talents cultivation as the goal, and the most important criteria for high level international talents is innovation." [4] Most MOOCs are not only from international top universities, but are also designed by famous teachers based on accumulating their extensive knowledge and rich teaching experience. Furthermore, the majority of design MOOCs also integrates some design masters' design understanding, design creativity and design innovation consciousness. Students can zero-distance contact, share, enjoy and learn their wonderful design idea and superb design practice, and at the same time, students can percept and feel their potential and outstanding design wisdom and extraordinary innovation vision field.

3.4   Viewfinder of Design Frontier and Fashion

SFCDMs should not only introduce quality international education resources, but also face international design frontier and follow the international fashion. In terms of most Chinese local universities and colleges, the difficulty to meet such requirements is self-evident. These schools are very hard to build up an international fashion design platform. On the other hand, for the SFCDMs students, they are also required to pay a much higher cost of learning. It is not realistic for many Chinese students. Nevertheless, the intervention of MOOCs just makes up for the defects and deficiencies in this area. MOOCs focus on the design frontier and latest fashion style like as a camera viewfinder. It grabs many dynamic pictures from the network to reflect the latest design trends in the world picture and presents in front of students. In the sampling survey questionnaire, there was an issue to inquire students why they are keen to select MOOCs to learn. 29% of the students chose the option "MOOCs resources can bring the latest fashion information, and have the opportunity to touch and experience the actual design trend".

3.5   Break the Limits of Time and Space

The biggest advantage of MOOCs is complete elimination of the physical limits of time and space, "any person in any place of the earth, as long as you can connect to the Internet, you will be able to participate in the MOOCs without any formal and rigorous requirements" [5]. Due to this reason, today's students, especially the students growing up in the Internet age, they can naturally gain a free and pleasant sensation, even the intimate feeling in MOOCs learning. MOOCs can not only let students link to any high quality courses all over the world, but also helps realizing circular learning. Learners can achieve full freedom of self control in learning time, learning content and learning progress. If learners do not understand a design technique or design concept or English sentences, they can repeatedly watch and listen and not be restricted in one moment or one place. This learning form greatly stimulates students' learning enthusiasm and participation. "The boom of the MOOCs actually and gradually creates a new virtual education business community which takes courses and schools as the core and crosses time, space and national boundaries" [6].

3.6   Effectively Making up for the Shortage of Teachers and Funds

Because of the particularity of SFCDM international design talents, it determines more requirements such as high quality teachers, hardware and software resources, small class teaching, and more design practice and so on. Therefore, SFCDMs education funding and personnel investment are much higher than the paralleled non-SFCDMs. Table 3 shows the statistical data of educational funds of sample university SFCDMs and paralleled non-SFCDMs. But the actual situation exists lots of difficulties. Owing to Chinese education status of the shortage of the overall educational resources, many SFCDMs' funding is difficult to meet the actual needs. Some of them even spend more than their income but not to respond to these requirements. (In China, tuition standard is approved by the local finance department, and some provinces and municipalities' finance charge standard is low. Therefore, under this situation, these schools' target of holding SFCDM is just to get social benefits. The economic benefit is no way to gain.) For another example, in the teaching field, SFCDMs prefer to invite more senior design talents or entrepreneurs with rich experience to some special lectures; however, even if the school can pay more lecture fees to them, they usually cannot guarantee the time. But "MOOCs can alleviate the problems of shortage of funds, venues and teaching resources." [7] MOOCs are not only fruitful in classic design course resources and famous lecturers, but also have many free courses. It alleviated SFCDM teacher shortage, fund shortage and other bottlenecks to a large extent.

Table 3. Statistical data of educational funds of sample university SFCDMs and paralleled non-SFCDMs (by the end of August 2016) (Unit: RMB).

Chinese universities and colleges	SFCDM name	TuitionYear RMB(domestic learning stage)	Teaching and personnel expenditure/each student/Year	Paralleled Non-SFCDM tuition/Year	Teaching and personnel expenditure of Paralleled Non-SFCDM/each student/Year
Xi'an Jiaotong-Liverpool University	Architectural design	90000	63950	N/A	N/A
Wuhan University of Technology	Arts and design	46800	35569	6750	4638
Nanjing University of Science and Technology	Industrial design	18000	16773	6800	4825
Anhui Polytechnic University	Visual communication design	10000	11901	7000	5143
Hunan City University	Visual communication design	27500	20912	8000	5669
China Australia Business College of Shanxi	Figure image design	20000	8265	8800	5926
Zhejiang Fashion Institute of Technology	Costume design	15000	9387	6000	4322

3.7 The Window of Understanding and Adapting to Foreign Countries' Studying and Living Style

According to the survey, the large majority of SFCDMs students did not have the long-term overseas learning and living experience, and only a few students have travelled abroad. However, the SFCDMs students need long-term study, live, even work abroad in the near future. Although these universities and colleges set up many short-term overseas travelling and learning programs, culture camp, training and internship projects, and also invited foreign experts and international students to teach, visit, exchange and cooperate in Chinese campus, it cannot fundamentally solve the problems of SFCDM students after they full access to unfamiliar foreign lands. MOOCs intervention timely opens a window to understand and adapt to foreign learning and life style in advance for SFCDMs students. MOOCs become a powerful supplement. Through choosing the MOOCs set up by the country which the learners will go to study, students can not only have a chance to feel and experience the target country's unique education concept and teaching features, but also get an opportunity to experience the world people thoughts, life situation, psychological and behavioral habits as well as his country's culture, history, economy, politics, society and customs, and effectively cultivate the intercultural communication ability, international cooperation consciousness, living and studying abroad adaption ability. "MOOCs application realized the integration of learning and life" [8].

3.8 Enhancing the Awareness of Autonomous Learning

"The design professional MOOCs learning can encourage learners by setting their own learning objectives to achieve autonomous learning." [9] MOOCs itself is an autonomous learning system. Its emphasis and dependence are all the independent learning awareness. Recommending, leading and encouraging the students to utilize the MOOCs resources to a large extent is to cultivate students' ability of autonomous learning. For SFCDMs students, they especially need self-improvement and autonomous learning ability. To reach high and profound design realm, it always requires the designer's self-cultivation and innovative learning. The spirit of "Never Stop" and "Independent Innovation" is one goal of SFCDMs talent cultivation. This is also in consistent with contemporary design international innovative talent standard. Figure 2 is proportional relation graph of ways and means to enhance the autonomous learning awareness and ability by samples' SFCDMs students' questionnaire. The figure shows MOOCs account for the highest proportion. Compared with other media ways, MOOCs even higher than mobile devices learning application resource (APP learning resource) and reached 27.5%. It is sufficient to practically prove the improvement effectiveness of MOOCs to enhance students' autonomous learning ability.

Proportional relationship of ways and means to enhance the autonomous learning awareness and ability

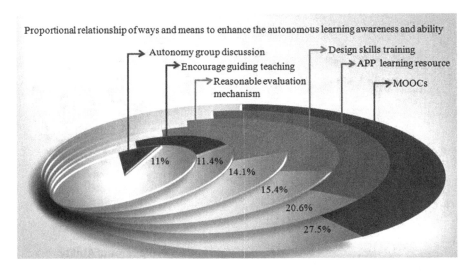

Autonomy group discussion

Encourage guiding teaching

Reasonable evaluation mechanism

Design skills training

APP learning resource

MOOCs

11%   11.4%
14.1%
15.4%
20.6%
27.5%

Fig. 2. Proportional relation graph of ways and means to enhance the autonomous learning awareness and ability.

3.9 Shaping of the International Visual Field and the Cultivation of the Multi Design Culture and Innovative Quality

The purpose of SFCDMs is to cultivate innovative international talents. Therefore, it is extremely important to shape SFCDMs students' international vision and understand multi design cultures. Comprehensive, diversified and extensive MOOCs resources can lead students into a broad international environment. Although this is a virtual network environment, MOOCs' systematical teaching structure composition, sharing resources, non-utilitarian and other fundamental attributes determine that participants can truly understand and learn the real international education and teaching, even the international society. Therefore, MOOCs help creating and expanding students' global vision.

In addition, students can communicate and talk with foreign teachers and their peers during MOOCs learning processes. "A Chronicle survey of MOOC professors last month found a median of 33,000 registrants for the courses that have been offered so far. One course, offered by Duke University via Coursera, saw 180,000 students sign up." [10] Chinese and foreign culture and art design thinking have been constantly collided and exchanged at MOOCs platform. It will give participants broad thinking and design creative space and inspire the infinite artistic inspiration. Except for design field, even the natural science and engineering MOOCs, as long as the students like, such MOOCs can also bring students knowledge diversity and promote conversion of thinking angle. Therefore, for SFCDMs students, MOOCs have already become the driving force of pluralistic design culture and innovation quality cultivation. Figure 3 shows the improved and developed design innovation qualities under the action of MOOCs which were listed by sampling SFCDMs students in the questionnaire.

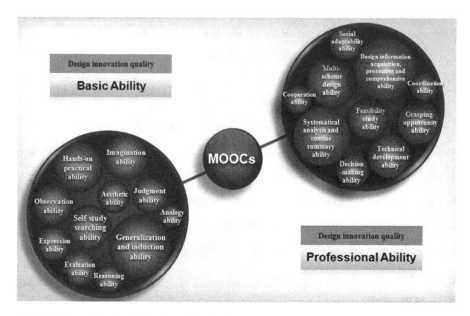

Fig. 3. The cultivation of students' design innovation quality under the action of MOOCs

3.10   In Favor of Promoting Teachers' Teaching Ability and Level

Because MOOCs promoted students' active learning, independent thinking, diverse knowledge and skills acquiring ability, and the students' comprehensive and professional knowledge quantity has reached a high level before the course starts, it brings a great pressure and challenge to professional course teachers. Furthermore, MOOCs require that "teachers have a better understanding on curriculum system and can accordingly adjust courses knowledge structure and content." [11] Therefore, MOOCs intervention undoubtedly promotes SFCDMs teachers' deep teaching introspection of practical teaching situation and teaching skills, and then push teachers to actively and deeply carry out the complement and integration of teaching contents, exploration and research, innovation and breakthrough of teaching methods. At the same time, MOOCs intervention makes teachers really realize the identity transformation from an instructor to the students' learning assistant and inspiring person of design creativity, so as to promote teachers' teaching level and a true understanding of the teaching behavior and teaching target. In addition, MOOCs have also led to the reform of higher education management concept and teaching mode.

3.11   Balancing Consideration

MOOCs intervention can balance students learning extent and learning level. Students are able to make up their deficiencies by using MOOCs resources. This is beneficial to the teachers to organize teaching activities and smoothly arrange teaching processes. On the basis of the MOOCs help, teachers can not only easily control the difficulty extent of English teaching and professional teaching, but can also accurately and reasonably arrange the key teaching points. On the other hand, MOOCs intervention can realize the balance of learning resources for the students whose family economic situation is not good. In China, normally, SFCDMs students need to pay more tuition than non-SFCDMs and other majors' Sino-foreign cooperative programs. It is difficult for poor students to acquire more extra learning resources. But MOOCs can change this situation and help these students to obtain abundant, free and excellent English and professional learning resources, and then help them to gain more domestic and foreign scholarships and grants so as to successfully complete their overseas study like other students.

4 Disadvantage Analysis

4.1 The Prominent Problem of the Higher Drop-Out Rate of MOOCs

High dropout rate is a long-term problem of MOOCs application and implementation process. In sampling universities and colleges, this situation also exists. Table 4 shows elected MOOCs number and drop-out rate situation of 7 samples. In this table, regardless of the type of schools, the difference of MOOCs dropout rate is not particularly high, even the first-class universities, the students' enrollment scores are much higher than other schools, but there is still a high MOOCs dropout rate. Therefore, we can see that MOOCs drop-out rate in SFCDMs does on depend on the students' scores.

Table 4. Elected MOOCs number and drop-out rate situation of samples (by the end of August 2016)

Chinese university	Cooperative major	Actual Students number in Chinese campus	Elected MOOCs number/Year	Uncompleted MOOCs number/Year	Drop-out rate/%
Xi'an Jiaotong-Liverpool University	Architectural design	32	227	102	44.93
Wuhan University of Technology	Arts and design	391	1681	811	48.25
Nanjing University of Science and Technology	Industrial design	89	329	173	52.58
Anhui Polytechnic University	Visual communication design	74	481	303	62.99
Hunan City University	Visual communication design	92	323	197	60.99
China Australia Business College of Shanxi	Figure image design	57	199	143	71.86
Zhejiang Fashion Institute of Technology	Costume design	165	511	357	69.86

Through random interviewing MOOCs drop-out reason of sampling SFCDMs students, the researchers summarized as follows: First, the self-existing problems of designing MOOCs, such as although designing MOOCs focus on the theoretical explanation and case analysis, actual practical operation and training are not enough, design interaction is not strong as well; Second, MOOCs learning flexibility is strong, but the majority of Chinese design majors' curriculum is full, and because of design majors' strong social application property, the spare time of design students is full as well. They tend to participant in various social activities, English language and design professional skills training, internship and even more part-time jobs, therefore, the learning time of SFCDMs students cannot be fully guaranteed; Third, the evaluation and examination mechanism of MOOCs are not strict. At the same time, most universities and colleges which introduced MOOCs do not have mandatory restrictive measures or compulsory assessment mechanism to promote students to complete MOOCs, thus, dropping out of MOOCs for students is without any cost; Finally, compared with non-SFCDMs, SFCDMs students have more opportunities to reach and meet foreign teaching ideas and methods, foreign teachers and even the real foreign teaching environment, so they are more likely to lose interests which are caused by these factors during MOOCs learning process, which result in dropout.

4.2   The Appearance of Excessive Dependence Status

According to the questionnaire, with continuous improvement of learning ability, knowledge level and MOOCs using proficiency degree, a small amount of students' MOOCs understanding ability and control ability were significantly stronger than the previous period, so their interests and confidence of selected MOOCs has been constantly rising. Moreover, free MOOCs learning time and space, high quality and richness of MOOCs resources, newly teaching activities, and even comfort and convenient individual learning environment greatly attracted these students, and then occurred the state of excessive dependence of MOOCs resources. In the questionnaire, 11% of the students selected MOOCs as their only favorite teaching mode and teaching resources, but did not choose the others. On the other hand, a part of students with weak self-control were easy to excessive indulge in MOOCs because of its virtual form, what is worse, some are even losing interest and are indifferent to the traditional classroom teaching.

4.3   Counteractive of Inappropriate Selection or Inappropriate Anticipation

Although the research results show MOOCs are helpful to improve students' learning initiative, learning enthusiasm and self-confidence, in many cases, lots of students can not accurately judge and decide which course is more suitable to them. The selected course may turn out to be either too hard or simple, which inevitably damage their interest. This phenomenon not only causes high drop-out rate, but is also harmful to students' learning confidence and learning enthusiasm, and then they are afraid of the difficult MOOCs or think MOOCs do not have much effect. MOOCs application obtains the opposite effect. Furthermore, because of the overall weak English language skills of SFCDMs students, most of them are difficult to understand and adapt to English MOOCs so as to produce nervous mood and enormous learning pressure. To a minor extent, their learning progress will be very slow; Seriously, some students' self-confidence and learning target will be destroyed. Therefore, the researchers advocate that MOOCs intervention mechanism in SFCDMs must match and construct psychological and professional guidance mechanism of MOOCs learning.

4.4   Lack of Practical Courses and Instant Interactive Activities

A large number of SFCDMs students complain that the MOOCs resources are lack of design practice courses, especially the design education practice training courses. At present, MOOCs seldom set up the simulated training experience courses which directly link to social practice. "A big obstacle to prevent the MOOCs from playing its transformation potential is failing to continuously provide a learning resource to participate in social practice." [12] In addition, another drawback of MOOCs is the absence of synchronous communication and real-time interactive sections. Owing to the MOOCs feature of online asynchronous teaching, although it brings the unlimited convenience in time and space, it simultaneously loses the advantageous section of the real-time communication and other synchronous activities. In terms of SFCDMs students, lively, vivid and direct communication with foreign teachers and students during MOOCs learning process can not only bring the deep and simulated international classroom experience, but can also carry out the actual combat of English language ability. At the same time, real-time discussion can also be more conducive to realizing the collision of creative thinking and burst of inspiration. In the questionnaire, in terms of existing issues of MOOCs, there are 22.9% students clearly pointed out the largest drawbacks of MOOCs is lack of design practice courses; 29.5% of the students put forward the degree of emotional communication is weak and even does not exist in many MOOCs; 13.8% of the students think that MOOCs lost real time communication, discussion, cooperation of fun which exist in the traditional classroom teaching.

To achieve this goal, SFCDMs need to mobilize social forces to create the MOOCs design practice platform. For universities and colleges, on the one hand, they should make full use of its extensive alumni resources and the cooperation objects such as local government, enterprises and foreign partners. Develop online and offline dual course resources with the frontier designer, project manager and related industry people. Expand the new MOOCs teaching form such as real-time image display, design project on-site teaching by using technology, online leading students to participate in the design practice or design project investigation or design exhibition or relevant design competition. On the other hand, these schools should release the incentive and innovative policy to timely adjust current introduced MOOCs structure and encourage teachers to set up some real-time online courses, or even the application of artificial intelligent technology may become the future development direction of the MOOCs.

4.5   The Standardization and Linearization of the MOOCs Make Students Feel Bored After Many Times of Contact

Because MOOCs aim at online public group, and its teaching system development is insufficient, although the MOOCs teaching process inserted newly activities such as peer evaluation, quiz, brainstorming, its core organized form is still stylized linear teaching process and structured knowledge analysis. It seems to be back to the traditional classroom teaching after experiencing some curvilinear motion. Such as Fig. 4, it is the standardization and linearization processes of MOOCs and the circular function diagram of MOOCs and offline learning environment. "MOOCs are generally hosted on a centralized platform, including video playback, discussion boards, standard ability test and course reading materials. The courses are still dominated by teachers and usually fit in with the linear, week-by-week learning mode." [2] Taking aim at different groups, different learning experiences, insights, perceptions, and different feedbacks, MOOCs cannot reach nonlinear path differentiation adjustment, which is contrary to the SFCDMs students' cultivation of individualization, flexibility and design perceptual ability.

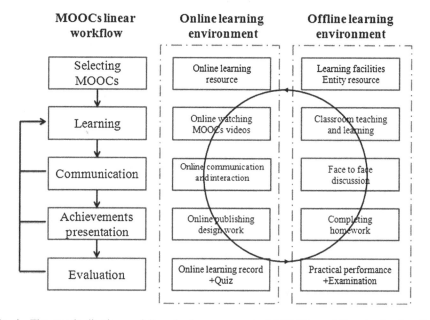

Fig. 4. The standardization and linearization processes of MOOCs and the circular function diagram of MOOCs and offline learning environment.

4.6   Adverse Factors for Teachers and Teaching Management

MOOCs played an important role in SFCDMs curriculum system construction and teachers' classroom teaching development, but the research found that although the

MOOCs mechanism promoted teachers self-improvement under the invisible pressure, it simultaneously affected some teachers self confidence and brought more psychological anxiety, which sometimes may lead to the out-of-control phenomenon in classroom teaching.

In teaching management aspect, because the majority of SFCDMs only orientated MOOCs into teaching assistant system, students MOOCs learning process was separate from teacher management, teachers cannot directly grasp students' learning dynamics, and only through indirect channels to know the details. Therefore, teachers cannot be synchronized to obtain the students' MOOCs actual learning status and real-time problems feedback so as to cause undesirable phenomena such as overlapped learning contents between MOOCs and classroom teaching, the dislocation of targeted counseling and so on. Taking aim at these issues, to explore the establishment of MOOCs learning communication mechanism between teachers and teachers, teachers and students as well as students and students, realize teachers' teaching skills and students' MOOCs learning dynamic real-time information sharing under the age of MOOCs is an effective path.

5 Conclusion

Through multi-region and multi-type sampling investigation, the researchers conclude that the intervention utilities of MOOCs teaching mode in SFCDMs have more advantages than disadvantages. MOOCs bring a substantive effect on promoting students' autonomous learning, adaptive learning of foreign teaching concepts and methods, international vision of design thinking, design of innovation ability and other aspects. At the same time, MOOCs intervention also has obvious benefits on improving teachers' teaching ability, classroom control ability, communication ability with students, overall balance ability, and adaptive adjusting teaching difficulty and progress. In addition, it plays a positive role in improving the teaching management mechanism, especially the teaching reform and innovation of Chinese SFCDMs.

But on the other hand, as for the drawbacks of MOOCs' application, we should pertinently avoid or weaken its undesirable consequences and influences, and according to the SFCDMs and students' actual, take effective countermeasures such as corresponding construction of MOOCs design practice complementary platform, teachers' regular appointment counseling mechanism, MOOCs tracking reminding and supervision assessment mechanism and so on, and then fundamental improve the MOOCs' defects in the application process, and form the more mature and advanced SFCDMs MOOCs teaching mode.

The research achievement not only improves MOOCs learning efficiency, SFCDMs overall teaching quality, and design talents cultivation quality, but also promotes extensive application and in-depth implementation of MOOCs education idea and teaching mode in Chinese universities and colleges.

References

1. Sadykova, G.: Mediating knowledge through peer-to-peer interaction in a multicultural online learning environment: A case study of international students in the US. Int. Rev. Res. Open Distance Learn. **15**(3), 24–49 (2014)
2. Wilson, L., Gruzd, A.: MOOCs–international information and education phenomenon. Bull. Assoc. Inf. Sci. Technol. **40**(5), 35–40 (2014)
3. Sonwalkar, N.: The first adaptive MOOC: A case study on pedagogy framework and scalable cloud architecture—Part I. MOOCs Forum **1**, 22–29 (2013). doi:10.1089/mooc.2013.0007
4. Lin, J., Liu, M.: On the quality construction of Chinese-foreign cooperation in school running. Educ. Res. **10**, 72–78 (2013)
5. Abeera, W., Miria, B.: Students' preferences and views about learning in a MOOC. Procedia – Soc. Behav. Sci. **152**, 318–323 (2014)
6. Li, W., Xiong, Q., Cai, Y.: On education sovereignty of higher education internationalization in the MOOC conditions. High. Educ. Res. **2**, 22–26 (2015)
7. Li, L.: The new idea of the development of confucius college – the application of the teaching mode of the MOOCs. J. Southwest Univ. Nationalities **12**, 224–229 (2014)
8. Li, F., Huang, M.: The opportunity and challenge of "MOOCs" to the university. China High. Educ. **7**, 22–26 (2014)
9. Milligan, C., Littlejohn, A.: Supporting professional learning in a massive open online course. Int. Rev. Res. Open Distrib. Learn. **15**(5), 197–213 (2014)
10. Kolowich, S.: Coursera takes a nuanced view of MOOC dropout rates. Chronicle of Higher Education, 8 April 2013
11. Deng, H., Li, M., Chi, Y., Tan, S.: The construction of course knowledge system in the period of "MOOCs". Course Educ. Res. **7**, 5–7 (2013)
12. Yang, D., Adamson, D., Rose, C.P., Sinha, T.: "Turn on, Tune in, Drop out": Anticipating student dropouts in massive open online courses. In: Proceedings of the 2013 NIPS Data-Driven Education Workshop, vol. 11, p. 14 (2013)

Crossing-Scene Pedestrian Identification Method Based on Twice FAS

Yun Chen[1], Xiaodong Cai[1(✉)], Yan Zeng[2], and Meng Wang[1]

[1] School of Information and Communication,
Guilin University of Electronic Technology, Guangxi Guilin 541004, China
caixiaodong@guet.edu.cn
[2] School of Mechanical and Electrical Engineering,
Guilin University of Electronic Technology, Guangxi Guilin 541004, China

Abstract. In the field of crossing-scene pedestrian identification, the recognition accuracy is low due to the large local variation of the samples. A method based on twice Feature-Aggregation-Separation (FAS) is proposed in this paper. Firstly, a novel network structure aggregating the same types and separating different types of features twice respectively is proposed. Secondly, a method of cross-input neighborhood differences is applied to deal with the features produced by the first aggregation-separation, and the results are taken as the input of the second aggregation-separation. Finally, the features produced by twice FAS are chosen for splicing, and the results are used for Softmax classifier. Compared with MCPB-TC [8] method based on features aggregation-separation, the proposed scheme can provide directional aggregation-separation of positive samples and negative samples. Compared with AIDLA [4] based on cross-input neighborhood differences, it offers better ability of discriminating inter-class and aggregating intra-class. It also outperforms those methods by the tests of CUHK01 and VIPeR data set.

Keywords: Feature-Aggregation-Separation · Crossing-scene · Cross-input

1 Background and Related Research

Crossing-scene pedestrian identification method is to match pedestrian features in different scenarios. It can be applied in many applications, such as cross-camera tracking and pedestrian retrieval. In the field of cross-scene pedestrian identification, there are different objects with high visual similarity and some other factors objects such as occlusion leading to partly invisible, pedestrian pose changes, lighting changes, complex background. It leads to pedestrian features with a high degree of complexity. These factors pose a great challenge. In this field, some methods were proposed.

There are methods applying local feature for matching, such as [1–3, 6]. In [1], input images are divided into blocks, and then features are extracted by using Convolution Neural Network (CNN). [2, 3, 6] differs from [1] in that each block feature is concatenated using a fully connected layer before performing a feature match. Its advantage is that the lack of pedestrian information caused by local occlusion is weakened, but these methods are prone to misjudgment when there are pedestrian parts of the region similar.

© Springer Nature Singapore Pte Ltd. 2017
B. Zou et al. (Eds.): ICPCSEE 2017, Part II, CCIS 728, pp. 483–491, 2017.
DOI: 10.1007/978-981-10-6388-6_41

There are other methods applying global feature for matching, such as [4]. A cross-input neighborhood differences method dealing with features obtained by CNN is proposed. It provides the ability to distinguish pedestrians, but when different individuals have a high degree of similarity or the same object has a great local change, this method does not perform well.

Based on [2, 5], the work in [8] obtained well-expressed features by aggregation-separation when features are divided. However, it is difficult to find the finest inter-class distance because the direction of aggregation-separation of different target features is ignored.

In order to solve the problem of inter-class separation in the case of large local change and high similarity among objects, a crossing-scene pedestrian identification method based on twice FAS is proposed with a novel network structure. The network adopts cross-input neighborhood differences method [4] in which the features are processed by twice FAS, the results are utilized as joint features for Softmax classification. Directional FAS for different target features and precise features expression are obtained by this method. It also provides promising test results on CUHK01 and VIPeR.

2   Crossing-Scene Pedestrian Identification Method Based on Twice FAS

2.1   Structure of the Network

The network of crossing-scene pedestrian identification method based on twice FAS consists of three parts. In the first part, $\langle I_i^o, I_i^+, I_i^- \rangle$ denotes a triple of three input images, where I_i^o and I_i^+ denote the same target while I_i^- denotes a different target with I_i^o and I_i^+. Firstly, a CNN is used to extract the features of I_i^o, I_i^+, and I_i^-, denoted as $f(I_i^o), f(I_i^+)$ and $f(I_i^-)$. Then, the resulting features are normalized after mapping to the high-dimensional feature-vector through the fully connected layer processing, denoted as $g(f(I_i^o))$, $g(f(I_i^+))$ and $g(f(I_i^-))$. Finally, triplet-loss is used to the first FAS of $g(f(I_i^o))$, $g(f(I_i^+))$ and $g(f(I_i^-))$.

By the characteristics of triplet-loss, Euclidean distance of $g(f(I_i^o))$ and $g(f(I_i^+))$ is reduced while Euclidean distance of $g(f(I_i^o))$ and $g(f(I_i^-))$ is increased. Due to the complexity of pedestrian samples, it is difficult to find the most appropriate feature expression meeting variability fully by one FAS. Therefore, it is essential to carry out a second FAS to serve the following points: (1) While distance of $g(f(I_i^o))$ and $g(f(I_i^-))$ is increased, distance of $g(f(I_i^+))$ and $g(f(I_i^-))$ is increased; (2) To prevent distance of $g(f(I_i^o))$ and $g(f(I_i^-))$ become too large leading to over-fitting, so it is necessary to keep distance between $g(f(I_i^o))$ and $g(f(I_i^-))$ in an appropriate direction.

Thirdly, the features obtained by twice FAS are taken as joint features for CNN. Using two fully connected layer, the results obtained are mapped to be high-dimensional feature vectors used for Softmax classification. The algorithm flowchart is shown in Fig. 1.

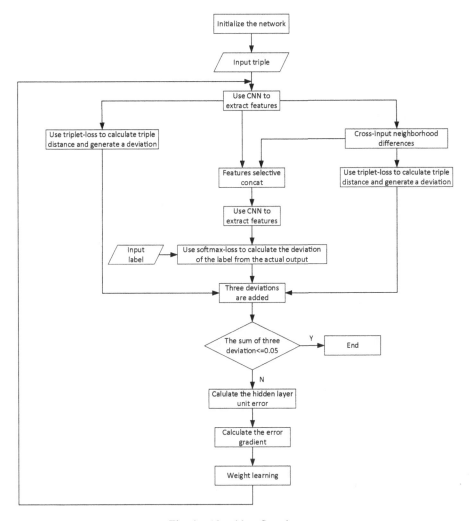

Fig. 1. Algorithm flowchart

2.2 Twice FAS Network

The proposed twice FAS network consists of three parts: the first and second FAS and a jointed features classification.

The First FAS. This part of the network settings are shown in Fig. 2: $\langle I_i^o, I_i^+, I_i^- \rangle$ are the i th triplet as the input of the network, where I_i^o, I_i^+, denotes the same type of samples while I_i^- denotes different samples with I_i^o, I_i^+. CNN is used to extract the features of I_i^o, I_i^+, and I_i^-. Parameters of this part of the network for each layer are shown in Tables 1 and 2.

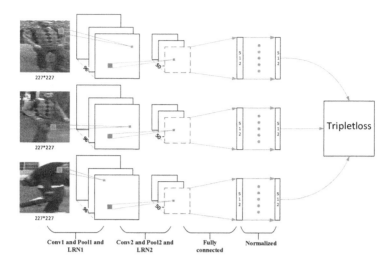

Fig. 2. Network of the first FAS.

Table 1. Convolution layer and pooling layer parameters.

Name	Type	Output Dim	Kernel size	Stride
Conv1	Convolution	56*56*96	7*7	4
Pool1	Max-pooling	27*27*96	4*4	2
Conv2	Convolution	23*23*256	5*5	1
Pool2	Max-pooling	11*11*256	3*3	2

Table 2. Lrn layer parameters.

Name	Size	Alpha	Beta
Lrn1	5	0.0001	0.75
Lrn2	5	0.0001	0.75

The results obtained through the LRN2 layer are represented as $f(I_i^o), f(I_i^+)$ and $f(I_i^-)$. Fully-connected layer is used for mapping $f(I_i^o), f(I_i^+)$ and $f(I_i^-)$ into three 512-dimensional features, respectively. The results are denoted as $F(Fc1)$:

$$F(Fc1) = \left(g\left(f\left(I_i^o\right)\right), g\left(f\left(I_i^+\right)\right), g\left(f\left(I_i^-\right)\right) \right) \tag{1}$$

The normalized layer is used to normalize three 512-dimensional feature vectors to [0,1] respectively for FAS used for triplet-loss.

The Second FAS. The network structure of the second FAS is shown in Fig. 3: cross-input neighborhood differences is used to deal with $f(I_i^o), f(I_i^+)$ and $f(I_i^-)$.

There are three parameters in LRN3 layer, the size is 5, alpha is 0.0001, beta is 0.75. The results obtained by the LRN3 layer are represented as $F(Lrn3)$:

$$F(Lrn3) = \left(\begin{cases} f(I_i^o) - f(I_i^+) \\ f(I_i^+) - f(I_i^o) \end{cases}, \begin{cases} f(I_i^o) - f(I_i^-) \\ f(I_i^-) - f(I_i^o) \end{cases}, \begin{cases} f(I_i^-) - f(I_i^+) \\ f(I_i^+) - f(I_i^-) \end{cases} \right) \quad (2)$$

The fully connected layer is utilized to map them into 512-dimensional features respectively, the results can be described as $F(Fc2)$:

$$F(Fc2) = \left(\begin{cases} g(f(I_i^o) - f(I_i^+)) \\ g(f(I_i^+) - f(I_i^o)) \end{cases}, \begin{cases} g(f(I_i^o) - f(I_i^-)) \\ g(f(I_i^-) - f(I_i^o)) \end{cases}, \begin{cases} g(f(I_i^-) - f(I_i^+)) \\ g(f(I_i^+) - f(I_i^-)) \end{cases} \right) \quad (3)$$

The normalized layer is applied to normalize the three 512-dimensional feature-vectors to [0,1] respectively for FAS used for triplet-loss.

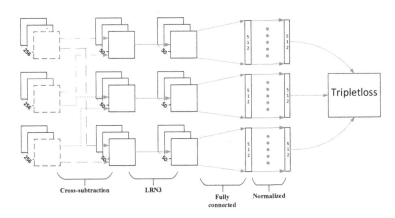

Fig. 3. Network of the second FAS.

Joint Features Classification. $f(I_i^o), f(I_i^+)$ and $\begin{cases} f(I_i^o) - f(I_i^+) \\ f(I_i^+) - f(I_i^o) \end{cases}$ are selected as the positive sample features. $f(I_i^o), f(I_i^-)$ and $\begin{cases} f(I_i^o) - f(I_i^-) \\ f(I_i^-) - f(I_i^o) \end{cases}$ are selected as negative sample features. The reconstructed sample features are extracted again using CNN. The network structure is shown in Fig. 4, the parameters of convolutional layer and pooling layer are shown in Table 3.

Fully-connected layer is used for mapping results of pool3 into 4096-dimensional feature vectors for Softmax classification.

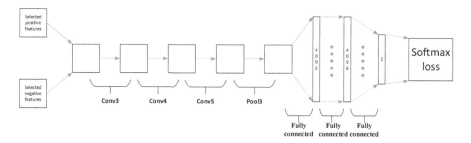

Fig. 4. Network of joint features classification.

Table 3. Parameters of convolution layer and pooling layer.

Name	Type	Output Dim	Kernel size	Stride
Conv3	Convolution	9*9*384	3*3	1
Conv4	Convolution	9*9*384	3*3	1
Conv5	Convolution	9*9*256	3*3	1
Pool3	Max-pooling	4*4*256	3*3	2

2.3   Training Algorithm

Triplet-loss can perform well for the classification of simple samples by the first FAS. The formula is as follows:

$$d^n\left(I_i^o, I_i^+, I_i^-\right) = d\left(g\left(f\left(I_i^o\right)\right), g\left(f\left(I_i^+\right)\right)\right) - d\left(g\left(f\left(I_i^o\right)\right), g\left(f\left(I_i^-\right)\right)\right) \leq \tau_1 \qquad (4)$$

It can be seen from Eq. (4), when I_i^+ and I_i^- are very similar, the angle between directional vectors, in the direction of distance increased between $g\left(f\left(I_i^o\right)\right)$ and $g\left(f\left(I_i^-\right)\right)$ and that of distance reduced between $g\left(f\left(I_i^o\right)\right)$ and $g\left(f\left(I_i^+\right)\right)$ is small. The distance between them is not optimized in this case.

An improved method is proposed as follows:

$$
\begin{aligned}
&d^n\left(I_i^o, I_i^+, I_i^-\right) \\
&= d\left(g\left(\left\{\begin{matrix} f\left(I_i^- - I_i^o\right) \\ f\left(I_i^o - I_i^-\right) \end{matrix}\right\}\right) - g\left(\left\{\begin{matrix} f\left(I_i^- - I_i^+\right) \\ f\left(I_i^+ - I_i^-\right) \end{matrix}\right\}\right)\right) - d\left(g\left(\left\{\begin{matrix} f\left(I_i^- - I_i^o\right) \\ f\left(I_i^o - I_i^-\right) \end{matrix}\right\}\right) - g\left(\left\{\begin{matrix} f\left(I_i^+ - I_i^o\right) \\ f\left(I_i^o - I_i^+\right) \end{matrix}\right\}\right)\right) \leq \tau_2
\end{aligned}
$$

$$(5)$$

The formula (5) is applied for the second FAS to ensure that the distance between $g(f(I_i^+))$ and $g\left(f\left(I_i^-\right)\right)$ in the first FAS is not reduced.

3   Experimental Results and Analysis

The testbed system for training consist of: Intel E5-2620 v3 (2.4 GHz processor), 128 GB memory, M40 graphics card, centos operating system, Caffe deep learning open source framework. The testing platform includes: Inter i5-4460 (3.2 GHz processor), GTX750Ti graphics, ubuntu14.04 operating system, Caffe deep learning open source framework.

3.1 Training and Validation Data Sets

CUHK01 data set [9]: The data set includes 971 individuals, there are four images for each from two cameras with different angles as shown in Fig. 5.

Fig. 5. CUHK01 data set sample.

VIPeR data set [10]: The data set includes 632 individuals, there are two images for each from two cameras with different angles as shown in Fig. 6. The data set shows a huge change in poses and lighting. This is challenging for pedestrian identification.

Fig. 6. VIPeR data set sample.

3.2 Training and Validation Data Sets

Firstly, each sample of the data set is resized to 227×227. Secondly, the data set is expanded by mirroring. Finally, the expanded data set is randomly generated into triples. In this work, the network structure is a three-input parallel network, so there are three LMDB files, two of them represent positive sample data and one of them represents negative sample data.

3.3   Experimental Results and Analysis

The proposed method is compared with AIDLA [4] and MCPB-TC [8]. CMC is adopted as evaluation method.

Experiment on CUHK01. 871 individuals of the CUHK01 data set are randomly selected as the training set. Leaving 100 of them are used as the testing set. The results are shown in Table 4.

For the proposed method, the accuracy of top1 is 73.5% in the CUHK01 data set. It is 8.5% and 24.2% higher than that of AIDLA [4] and MCPB-TC [8], respectively. The accuracy of top10 is 96.5% in the CUHK01 data set. It is 2.4% and 9.9% higher than that of AIDLA [4] and MCPB-TC [8], respectively. This suggests that the proposed method is more accurate than AIDLA [4] and MCPB-TC [8].

Compared with AIDLA [4], the method proposed improves the accuracy of top1 by 3.54 times of top10 and 2.44 times higher than that of top10 of MCPB-TC [8]. This proves the effectiveness of the proposed method for inter-class targets with high visual similarity.

Table 4. The CMC evaluation results on the CUHK01 data set.

Method	Top1	Top5	Top10	Top15	Top20
FPNN [3]	27.9	59.9	74.0	82.6	
Sakrapee [7]	53.4	76.4	84		90.5
AIDLA [4]	65.0	88.5	94.1		
MCPB-TC [8]	49.3	76.5	86.6	93.7	94.7
MCPB-TCP [8]	53.7	84.3	91.0	93.3	98.3
Proposed	73.5	93.5	96.5	98.5	99.0

Experiment on VIPeR. 316 individuals of the VIPeR data set are randomly selected as the training set for fine tuning the model generated by VIPeR. Leaving 100 of them are used as a testing set. The test results are shown in Table 5.

For the proposed method, the accuracy of top1 is 40.1% in the VIPeR data set. It is 5.3% and 2.9% higher than that of AIDLA [4] and MCPB-TC [8], respectively. The accuracy of top10 is 78% in the VIPeR data set. It is 2% and 10.9% higher than that of AIDLA [4] and MCPB-TC [8], respectively. This indicates that the proposed method in the VIPeR data set outperforms AIDLA [4] and MCPB-TC [8].

Table 5. The CMC evaluation results on the VIPeR data set.

Method	Top1	Top5	Top10	Top15	Top20
DML [1]	28.2	59.2	73.4	81.2	86.3
Improved DML [2]	34.4	62.1	75.8	82.5	87.2
LMLF [6]	29.1	52.3	66.0	73.9	79.9
Sakrapee [7]	45.9				
AIDLA [4]	34.8	63.0	76.0		
MCPB-TC [8]	37.2	55.6	67.1	76.5	75.3
MCPB-TCP [8]	47.8	74.7	84.8	91.1	94.3
Proposed	40.1	69.3	78.0	85.1	89.2

Compared with AIDLA [4], the method proposed improves the accuracy of top1 by 2.65 times of top10. This proves the effectiveness of the proposed method for inter-class targets with high visual similarity.

Compared with MCPB-TC [8], the method proposed improves the accuracy of top1, it is 3.75 times higher than that of top10. This suggests that the proposed method in the VIPeR data set provides better generalization capability.

4 Conclusions

In conclusion, a pedestrian identification method based on features twice FAS is proposed. In the case of samples of less variation, the first FAS performs well for inter-class separation and intra-class aggregation. For large variational samples, the second FAS achieves well inter-class separation and intra-class aggregation. It is more flexible to find the optimized inter-class distance and better intra-class aggregation by FAS. The proposed method achieves promising results in the test of CUHK01 and VIPeR data sets.

Acknowledgment. This work was supported by the 2016 Guangxi Science and Technology support program under Grant No. AB16380264 and 2016 Key Laboratory of Cognitive Radio and Information Processing (Guilin University of Electronic Technology), Ministry of Education Fund Project, Project No. CRKL160102.

References

1. Yi, D., Lei, Z., Liao, S., Li, S.Z.: Deep metric learning for person re-identification. In: ICPR, pp. 2666–2672, Stockholm (2014)
2. Yi, D., Lei, Z., Li, S.Z.: Deep metric learning for practical person re-identification. In: ICPR, pp. 3908–3916, Stockholm (2014)
3. Li, W., Zhao, R., Xiao, T., Wang, X.: Deepreid: deep filter pairing neural network for person re-identification. In: CVPR, pp. 152–159, Columbus (2014)
4. Ahmed, E., Jones, M., Marks, T.K.: An improved deep learning architecture for person re-Identification. In: CVPR, pp. 3908–3916, Boston (2015)
5. Schroff, F., Kalenichenko, D., Philbin, J.: FaceNet: a unified embedding for face recognition and clustering. arXiv preprint arXiv:1503.03832 (2015)
6. Zhao, R., Ouyang, W., Wang, X.: Learning mid-level filters for person re-identification. In: CVPR, pp. 144–151, Columbus (2014)
7. Paisitkriangkrai, S., Shen, C., van den Hengel, A.: Learning to rank in person re-identification with metric ensembles. arXiv preprint arXiv:1503.01543 (2015)
8. Cheng, D., Gong, Y., Zhou, S., Wang, J., Zheng, N.: Person re-Identification by multi-channel parts-based CNN with improved triplet loss function. In: CVPR, pp. 1335–1344, Las Vegas (2016)
9. Li, W., Zhao, R., Wang, X.: Human reidentification with transferred metric learning. In: ACCV, pp. 31–44, Daejeon (2012)
10. Gray, D., Brennan, S., Tao, H.: Evaluating appearance models for recognition, reacquisition, and tracking. In: Proceedings of the IEEE International Workshop on Performance Evaluation for Tracking and Surveillance (PETS), vol. 3 (2007)

Vehicle Type Recognition Based on Deep Convolution Neural Network

Lei Shi[1], Yamin Wang[2(✉)], Yangjie Cao[1], and Lin Wei[2]

[1] School of Information Engineering, Zhengzhou University,
Zhengzhou 450001, China
[2] School of Software, Zhengzhou University, Zhengzhou 450002, China
1245593616@qq.com

Abstract. The systems based on image processing for vehicle type recognition is becoming more and fiercer. It plays an important role in traffic safety. In order to improve the problems that traditional Convolutional Neural Network has low accuracy of feature extraction from the low-resolution image, a novel model based on Deep Convolutional Neural Network (DCNN) was proposed. In this paper, our work mainly contains two aspects both extraction of feature dimension and recognition of vehicle image. Firstly, the learning way was introduced, and the raw image of vehicle subsampled with several different sizes was operated with the filter corresponding each channel in a way of convolution to extract the feature dimension of image. Secondly, the features dimension obtained from every channel were merged by a full connected layer. Eventually, features used to recognize the type of vehicle is got. The experiment shows that the architecture of DCNN model has a efficient performance on the recognition of vehicle image. Compared with the traditional algorithm of CNN, the results of experiment show that the mode of DCNN can achieve 97.6% accuracy and a higher precision is got.

Keywords: Vehicle · Deep convolution neural network

1 Introduction

Since the 21st century, with the continuous improvement of living standards, automobiles have became a very popular means of transport and its number shows the trend of rapid increase. It provides a very convenient condition for people's lives, while it also brings a huge challenge to the traffic management. In recent years, as the number of cars shows is rapidly growing in the state, the mortalities of traffic accident have been ranking the top 1 in the world.

The traditional way relying on human or transportation facilities has been unable to meet the needs of the current development, and thus a real-time, efficient and accurate traffic management system-intelligent transportation system was put forward, which improved the transport efficiency of vehicles, eased the traffic congestion pressure, ensured traffic safety and reduced the environmental pollution and energy consumption. And vehicle identification is an important part of intelligent traffic management system. At present, The remarkably mature modes of vehicle identification are

© Springer Nature Singapore Pte Ltd. 2017
B. Zou et al. (Eds.): ICPCSEE 2017, Part II, CCIS 728, pp. 492–502, 2017.
DOI: 10.1007/978-981-10-6388-6_42

ultrasonic detection [1] and laser infrared detection [2], But those facilities are easily be damaged in installation these methods bring inconvenience to the traffic. At the same time, although their recognition is much more efficient, they requires a high costs in maintenance.

At present, with the rapid development of deep learning, the deep learning algorithm has achieved good results in image classification [3] and recognition of target objects [4]. Compared with the traditional algorithm of extraction, the deep neural network has better applicability and agility can use training data to adapt to different network construction features. And an important structure in deep learning is the convolution neural network, which has the characteristics of automatic learning and extraction of image features in deep learning and has been developed greatly in the fields of image recognition and search engine. Convolution neural network utilizes the local feelings, weight sharing and other methods to increase the generalization of the network capacity and robustness in this way to reduce the complexity of the network structure and the number of weights.

For the vehicle recognition, various methods are tried by domestic and foreign researchers. Early methods rely on artificial design features, such as SIFT [5], HOG [6], etc. But these extraction methods by feature need massive calculation and preprocess the input original image, which the convolution neural network does not need. The convolution neural network can make the machine autonomously learn the image features, thus avoiding the complexity of manually extracting the features as well as improving the operational efficiency.

Owing to the complexity of vehicle appearance and effects of shooting angle, distance ambient light and other factors, the recognition of vehicle appearance become difficult with image parameters scaling, rotation and translation. By comparison, the convolution neural network is invariant to the image translation, scaling, tilt and other visual deformation that can effectively overcome the problems bringing by the change in the appearance of the vehicle.

Single-channel CNN can only use grayscale image and a certain channel of RGB channel. Therefore, in the learning process it can not effectively take advantage of the color information. Moreover, the traditional CNN only extract the local features of the picture which loses some important image information. Extracting the local feature of the image is as important as extracting the global feature in recognition. Based on CNN, convolution neural network structure are presented for this paper, which inputs the final feature into the SVM classifier for recognition. In this structure, through a combination of multiple input image, more information of pictures can be extracted, thereby improving the accuracy of image recognition.

2 Deep Convolution Neural Network

2.1 Deep Convolution Neural Network Structure

In this paper, a Deep Convolution Neural Network (DCNN) is proposed based on the AlexNet structure in Caffe, a deep learning framework, and the final feature is input into SVM [7] classifier to complete the recognition of the target. This article introduces the AlexNet [8] model in Caffe.

2.2 AlexNet Model

An important model of Deep Learning, The structure of AlexNet based on Caffe illustrates in Fig. 1. AlexNet total contains eight layers excepting the input layer. The former 5 layers are used as the convolutional layer, and 6 to 8 including the final layer are the full connection layers. The whole structure can be considered as a 1000-dimension vector inputted to a softmax classifier. The structure of layer 1 and 2 includes convolutional layer, and normalization layer showing blue and pink section in the Fig. 1. And there is a ReLU function behind each convolutional layer. The operation of pooling is used to the layer 1, 2 and 5, and the operation of Dropout is utilized on the last two full connection layers.

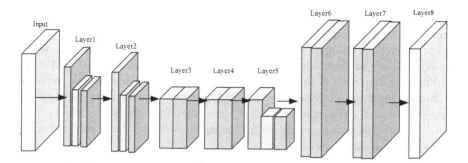

Fig. 1. AlexNet network architecture

2.3 Deep Convolution Neural Network Structure

The algorithm in this paper is based on the deep convolution neural network (DCNN) proposed by AlexNet algorithm. The structure of DCNN is shown in Fig. 2.

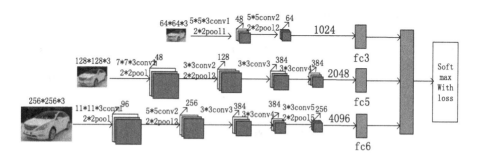

Fig. 2. Multi-scale convolutional neural network

Although the traditional CNN has a high degree of invariance in visual deformations such as translation, scaling, tilt, and so on. The final result of the network identification will still change [9] in the event of slight change in the size of the input image. In order to enhance the robustness of DCNN, the method of inputting multi-scale image is brought in the model of DCNN. During the experiment, the images with different sizes were inputted separately three channels, in which the operations of convolution and subsampled had been done. Finally, the features acquitting from the three channels were integrated at the full connection layer. The size of filters is different during the operation of multi-scale image so overall features of image can be extracted by the big size of filters and the features obtained by small size of filters greatly present the partial features.

The sizes of inputting images are 256 * 256 * 3, 128 * 128 * 3, 64 * 64 * 3 respectively. The convolutional kernel of convolution layer 1 also has three channels relative to three channels of inputting image, and they are done the operation of convolution with each other. When the size of inputting image is 256 * 256 * 3, the convolutional kernel with size of 11 * 11 * 3 was used at the first convolution layer, and its number is 96. Images with size of 256 * 256 * 3 were carried out the operation of convolution when the step was set at 4. In other words that the size of convolution kernel was 11 * 11, the images that were sampled at the 4 pixels interval were conducted the operation of convolution at three channels. After getting the basic features of image, does an operation of the ReLU rule and Norm shift. Then the subsampled layer, pool layer 1, outputs 96 feature images with size of 27 * 27, and these images were used as the input of the next layer.

The multi-channel convolution structure and single-channel convolution structure is different, as shown in Figs. 3 and 4, respectively, the single-channel convolution structure and the multi-channel convolution structure.

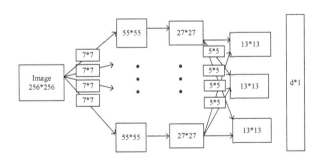

Fig. 3. Singel channel convolution layer structure

- Convolution Process of Multiple Convolution Kernels on Single Channel

One convolution kernel to extract features of image is used that can only get one feature graph is insufficient. Therefore, multiple convolution kernels should be used to extract features of image. For instance, using 96 convolution kernels can know 96 features of a image and get 96 feature graphs.

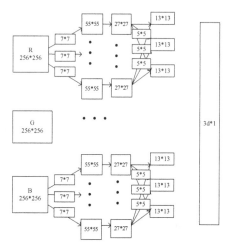

Fig. 4. Multi-channel convolution layer structure

- Multiple Convolution Kernels on Multi-channel

For example, Fig. 5 describes the convolution operation on three channels. There are two convolution kernels, and then generate two feature graphs. It is worth noting that each channe the three channels should check with a convolution. For example, the value at a certain position (i, j) at w1 is obtained by accumulating the convolution result at (i, j) on three channels and then taking the value of the activation function. If the number of input convolution kernels is 96, then the resulting feature graphs are also 96.

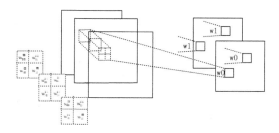

Fig. 5. Convolution in three channels

256 convolutional kernels with size of 5 * 5 were applied to the second convolution layer. After the pool layer 2, 256 features images with size of 13 * 13 were outputted.

384 convolutional kernels with size of 3 * 3 were applied to the third convolution layer. 384 features images with size of 13 * 13 were outputted.

After conv3 doing one ReLU, the forth convolution layer can be acquired by 384 convolutional kernels with size of 3 * 3. And 384 features images with size of 13 * 13 were got.

256 convolutional kernels with size of 3 * 3 were applied to the fifth convolution layer. After the pool layer 5, the size of map is int(13/2) = 6 inputting 356 features images with size of 6 * 6.

The sixth full connection layer was got by pooling previous fifth convolution layer, and its features dimensions were 4096.

The sixth full connection layer was got by pooling previous fifth convolution layer, and its features dimensions were 4096.

When the size of inputting image is 128 * 128 * 3, the structure with five layers except the inputting layer was adopted. There is a pooling layer behind the first and the second layers. The convolutional kernels have different size (including 7 * 7 * 3 and 3 * 3, and its number being 48, 128 respectively). There is only a convolution layer at the third and fourth layer. The sizes of convolutional kernels were 3 * 3, and its number is 384. The fifth full connection layer could extract image features with dimension 2048.

When the size of inputting image is 128 * 128 * 3, the structure with two layers except the inputting layer was adopted, and there is a pool layer behind each layer. The convolutional kernels have different size (including 5 * 5 * 3 and 3 * 3, and its number being 48, 64 respectively). The third full connection layer could extract image features with dimension 1024.

Ultimately, the feature dimensions learning from three channels were merged by the last full connection layer.

7168 dimensions of image feature were extracted and it could be used to recognize the goal image.

2.4 Optimize Network Parameters

When the training samples are few and the neural network model is relatively complex, there may will appear overfitting phenomenon in network training. As the number of iterations that increase the number of training samples, the error rate of the classification of the network is gradually reduced, but the error rate on the test set is gradually increased, this is the main performance of the over-fitting phenomenon. The main reason for the overfitting phenomenon is that the network has overfitted the training data set, but the data from the training set data are not good-fitting. As shown in the Fig. 6, the phenomenon of network over-fitting is intuitively described.

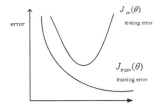

Fig. 6. Over fitting phenomenon

Over fitting is a common phenomenon existed in real deep network training. Therefore, the most important step in improving the performance of the network is to reduce the over fitting phenomenon in the network training. By optimizing the network parameters network degree of fitting is used in this paper.

- Setup of Batch Value

In network training, is generally based on the maximum iteration count or loss of thresholds to determine when to stop training. In the actual network training, each iteration there is a random batch of training samples in turn into the same size data set, and then each batch data set is input into the network at the same time for training and learning. Until all batch data sets are trained that the next iteration is carried out. In this paper, different batch data sets are chosen to carry out network training, and the range of batch value is summarized. When the network loss function value is lower than the predetermined threshold, the training is stopped. The test result is taken the average of five times to avoid accidental events.

- Local Response Normalization

Local Response Normalization means that the network layer in convolution neural normalizes the local area of the input, a lateral inhibition to neural cells. The calculation formula of the algorithm refer with Eq. 1.

$$b^i_{x,y} = a^i_{x,y} / (k + \alpha \sum_{j=\max(0,\,i-n/2)}^{\min(N-1,\,i+n/2)} (a^i_{x,y})^2)^\beta. \tag{1}$$

In the formula (1), $\alpha^i_{x,y}$ represents the output of the i-th convolution kernel in position (x, y), and the $b^i_{x,y}$ represents the output of local response normalization. N represents the number of convolution kernel in convolution layer. In this formula, k, α, n, β are constants in which k usually is set 2, n is set 5 and α is 10 to 4.

- Optimization of Dropout

Dropout is proposed to reduce the degree of over fitting in networks and effectively avoids the over fitting phenomenon in network due to the lack of training samples. Setting the output of each hidden layer to 0 with a probability of 50% makes these outputs invalid in forward and backward processes. Each input training samples corresponding to different network structure, but all of these different network structure weight is shared, and thus the parameters can be adapted to different circumstances of the network structure. Thus to a certain extent, error rate of classification in network is suppressed and the network performance is improved. The operating principle of Dropout is shown in the Fig. 7. The mathematical expression is: r = a ((M. * W) v).

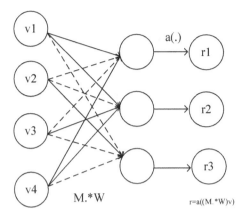

Fig. 7. The operating principle of dropout

3 Experiments and Results Analysis

3.1 Experiments Date

the vehicle images intercepted from the monitor of highway were used to validate the model, and all images would be changed to color images with 256 * 256 pixels. The experiment based on the environment of Ubuntu+Caffe+MATLAB. The data set includes five types of vehicles total having 8000 images. 80% images chosen randomly were used as the training samples, the rest images were used as the testing samples.

3.2 Experimental Setup

When the size of inputting image is 256 * 256 * 3, the network structure with eight layers was regarded the channel 1. Finally, the full connection layer could extract image features with dimension 4096.

When the size of inputting image is 128 * 128 * 3, the network structure with seven layers was regarded the channel 2. Finally, the full connection layer could extract image features with dimension 2048.

When the size of inputting image is 64 * 64 * 3, the network structure with five layers was regarded the channel 3. Finally, the full connection layer could extract image features with dimension 1024.

Ultimately, the feature images which were outputted from the full connection layer were sent to the softmax classifier to recognize the type of vehicles.

The images of vehicles with three different sizes were sent to three channels, and the operation of convolution and subsampled have been done in three channels, respectively. The feature dimensions got by the first full connection layer on three channels were merged by the last full connection layer. At last, the dimension of image is 7168.

3.3   Experimental Results

First, some common features on the data are classified by using the classification algorithm. The correct rate is shown in Table 1.

Table 1. Classification of common feature classification algorithm

	LDA [10]+NN	SVM [11]	Random forests [12]
Color	0.929	0.824	0.931
LBP [13]	0.781	0.773	0.817
SIFT [14]	-	-	0.802
HOG [15]	-	-	0.574

The experimental results show that the characteristics of color are more easily recognized than other features. But a grayscale image is input the traditional CNN, and DCNN structure of the input is a three-dimensional color image. Therefore, the DCNN structure recognition accuracy is higher.

The comparison between the method proposed in this paper and other experimental algorithms is shown in Table 2. CNN algorithm directly utilizes the image to operate the convolution in order to get the features of image, while the algorithm used in this paper is to input with different images and merge the features obtained. In this way the feature obtained is more accuracy with higher recognition rate.

Table 2. The comparison between our method and the others

Method	Recognition rates
CNN [16]	92.5%
HOG+SVM	93.4%
DCNN	97.6%

The recognition rate of different channel and united channels are showing in Table 3.

Table 3. Comparison of recognition rates in different paths

	Algorithm	Input dimension	Recognition rates
	Alexnet	4096	90.2%
DCNN	Channel 1	4096	90.5%
	Channel 2	2048	91.3%
	Channel 3	1024	89.4%
	Channel 1+2	6144	94.4%
	Channel 2+3	3072	92.6%
	Channel 1+3	5120	93.7%
	Channel 1+2+3	7168	97.6%

In terms of the data sets of vehicles, the channel 2 has the highest recognition rate, because the vehicles data set only contains the image of car. And the difference of overall feature of car with various types is not so obvious that the types of vehicles were largely recognized by the partial features. The recognition rate of channel 3 was slightly declined. The reason was the size of inputting image became smaller. Eventually, the recognition rate of DCNN combined the feature of channels has achieved 97.6%, so it meted the demand of recognizing vehicles.

Considered the Table 3, the algorithm of DCNN with three channels has a higher rate of recognition than that of one or two channels. Hence, the features extracted from three channels have stronger generalization ability and robustness.

4 Summary

In this paper, a variety of methods for vehicle identification are studied and explored. For the problem that the traditional deep convolution neural network can not effectively extract the obvious features of low-resolution images, a vehicle recognition model based on deep convolution neural network is proposed. In view of the size and complexity of the input image, the deep learning framework is improved. By inputting different types of image, more information about extracted feature increases and then the features obtained are merged and input into SVM classifier for recognition. The model proposed in this paper has higher excellent input image under different conditions, more effectively increasing the recognition rate and robustness of the image. It is proved that the structure used in the paper has strong learning ability and certain application value.

In order to solve the problems that the traditional CNN can't perfectly extract features from low-resolution image, a model of recognizing the type of vehicles based on Deep Convolutional Neural Network (DCNN) was proposed. The multi-scale inputting images enhance the invariance of partial information, and then combining the convolutional kernels which have several dimensions with the way of interval sub-sampled obtained the invariance of features. The detail information of the vehicles images was not losing, so the recognition rate of image and robustness were enhanced efficiently.

In the next step, we will explore and design other architecture of CNN, and further extend the self-learning ability. What's more, given the goal of improving accuracy of recognition, we will seek out new method to extract more plentiful information from the vehicles images.

References

1. Takamoto, M., Ishikawa, H., Shimizu, K., et al.: New measurement method for very low liquid flow rates using ultrasound. Flow Meas. Instrum. **12**(4), 267–273 (2001)
2. Warriach, E.U., Claudel, C.: Poster abstract: a machine learning approach for vehicle classification using passive infrared and ultrasonic sensors (2013)

3. Krizhevsky, A., Sutskever, I., Hinton, G.E.: ImageNet classification with deep convolutional neural networks. In: International Conference on Neural Information Processing Systems, pp. 1097–1105. Curran Associates Inc. (2012)

4. Erhan, D., Szegedy, C., Toshev, A., et al.: Scalable object detection using deep neural networks. 2155–2162 (2013)

5. Hua, L., Xu, W., Wang, T., et al.: Vehicle recognition using improved sift and multi-view model. J. Xi'an Jiaotong Univ. **47**(4), 92–99 (2013)

6. Taigman, Y., Yang, M., Ranzato, M., et al.: DeepFace: closing the gap to human-level performance in face verification. In: IEEE Conference on Computer Vision and Pattern Recognition, pp. 1701–1708. IEEE (2014)

7. Rahati, S., Moravejian, R., Mohamad, E., et al.: Vehicle recognition using contourlet transform and SVM. In: Proceedings of the 5th International Conference on Information Technology: New Generations [S. l.], pp. 894–898. IEEE Press (2008)

8. Jun, Z.: Research on image retrieval based on fusion feature of AlexNet. Chongqing University of Posts and Telecommunications (2016)

9. Goodfellow, I.J., Warde-Farley, D., Mirza, M., et al.: Maxout networks. Comput. Sci. **28**, 1319–1327 (2013)

10. Blei, D.M., Ng, A.Y., Jordan, M.I.: Latent dirichalet allocation. J. Mach. Learn. Res. **3**(1), 993–1022 (2003)

11. Andrew, A.M.: An introduction to support vector machines and other kernel-based learning methods. **32**(1), 1–28 (2001)

12. Fang, K., Wu, J., Zhu, J., et al.: A review of technologies on random forests. Stat. Inf. Forum **26**(3), 32–38 (2011)

13. Zhang, H., Xun, F., Chen, J.: Face recognition based on multi-scale LBP. Comput. Appl. Softw. **29**(1), 257–259 (2012)

14. Huang, F.C., Huang, S.Y., Ker, J.W., et al.: High-performance SIFT hardware accelerator for real-time image feature extraction. IEEE Trans. Circ. Syst. Video Technol. **22**(3), 340–351 (2012)

15. Dalal, N., Triggs, B.: Histograms of oriented gradients for human detection. In: IEEE Computer Society Conference on Computer Vision & Pattern Recognition, pp. 886–893. IEEE Computer Society (2005)

16. Liu, D.: Deep convolutional neural networks for vehicle classification. Southwest Jiaotong University (2015)

A Biomechanical Study of Young Women in High Heels with Fatigue and External Interference

Panchao Zhao and Zhongqiu Ji[✉]

Beijing Normal University, Beijing 100875, China
jizhongqiu61@bnu.edu.cn

Abstract. From the perspective of biomechanics, gait inquiry under the influence of fatigue on young women wearing high heels and wear flat shoes with a difference. Explore the intervention of external interference conditions mechanism young women wearing high heels to keep the body in balance, and when you wear flat shoes with a difference.

Keywords: High-heeled shoes · Fatigue · External interference · Gait · Biomechanics

1 Introduction

In everyday life, though not comfortable wearing high heels is very common occurrence, and the body has the potential pathogenic role, but based on the beauty of the heart and women live, work still needs to choose to wear high heels. Needless to say, women wearing high heels add to the beauty of the human body, muscles ankle, calf, thigh able tight, form a beautiful leg line, while the center of gravity forward, upright, tight hip muscles, the body naturally upright. However heels give women in beauty, but also to bring some influence women's health. Heel elevation, making the foot pressure redistribution lower extremity musculoskeletal system and the body's stress state will be changed accordingly.

Survey shows that for a long time people feel particularly tired wearing high heels, while human gait corresponding changes will occur in the state of fatigue, increase the probability of occurrence of sports injuries. Exercise-induced fatigue to temporarily decrease the ability of the human body functions as the main indicator, the action mainly as changes in kinematic structure, kinetic parameters [1–3]. Mizrahi et al. [3]. The study found that after running to fatigue, decreased average cadence, knee kicking degree decreases, Williams et al. [4] to run to fatigue kinematic data were compared before and after, found the biggest swing phase knee knee flexion angle increases, increase the maximum hip flexion angle. Christina [5] through the front of the ankle

---

The key technology of the old balance capability assessment and the development of training tools financial aid program (PXM2016-178215-000013).

B. Zou et al. (Eds.): ICPCSEE 2017, Part II, CCIS 728, pp. 503–515, 2017.
DOI: 10.1007/978-981-10-6388-6_43

joint biomechanics run local muscle fatigue analysis found that the degree of ankle flexion significantly changed.

These studies objectively reflects the impact of fatigue on the run action structure, but not from the perspective of the lower limb muscle function analysis of variation in the process of running to fatigue, nor consider the difference between left limb. Fatigue occurs movement will cause imbalance bilateral limb movements, or enlarged bilateral limb differences [6].

The human body movement is a holistic movement, uneven distribution of the results will inevitably lead to changes in movement efficiency and reduce movement structure, but also an important reason for damage arising sports.

In gait studies, research on methods of external interference effects on the human body in balance, there are two main, load and slip interference. Women wearing high heels will make the human lower limb joint mobility, prone to sprain while walking, falls, accidental falls while often leads to hip fractures [7], to reduce people's activity performance, and thus make the daily life and social activities restricted. 2020 by the United States to bring health care costs fall to more than 30 billion US dollars [8]. Our purpose was to determine the effect of heel height under fatigue induced and slip condition on net joint moments at the hip, knee, and ankle in the frontal plane, with particular attention placed on the knee moment. It was hypothesized that frontal plane net joint moments of the lower extremity increase systematically as heel height increases. We wished to clarify the changes in muscle activity from fatigue induced and slip condition by high-heeled shoes in controlling the foot and shank, and to determine the biomechanical effects on foot stability. In this study, surface EMG and FGP measurements were integrated to analyze the effects of muscular fatigue induced by high-heeled gait on the structural stability of the foot.

2 Subjects and Methods

2.1 Subjects

10 female students from Beijing Normal University were randomly recruited to participate in this study. All subjects were at least one year of experience in high heels, excluding organic diseases of major organs and systems of movement, normal vision, no foot deformities, abnormal gait and severe lower limb trauma, lower extremity joints during the test each normal; before the test has not been exposed to balance tester or special balance training. Before the test, the testers illustrative purposes, research methods and experimental procedures in this study, the test to fully understand the experiment of this study and post-adaptive test, test formally.

2.2 Methods

2.2.1 Test Program

In order to study the body in the real slipping process of balance adjustment mechanism, by simulating the real slipping environment to carry out experiments. In this study, the dynamic sliding track developed by the Department of Sports Biomechanics of Beijing Normal University. It allows subjects to produce a sudden forward movement when the swing legs touch the slide plane (trigger threshold adjustable, accurate to 0.1 N), resulting in balanced interference, so that the subjects have a tendency to fall backwards. The sliding table drives the force plate to do one-dimensional movement in the front-rear direction. Movement amplitude 1 cm–40 cm adjustable, movement speed 1 cm/s–60 cm/s adjustable.

All subjects were wearing flat shoes heel of 1 cm and 7 cm heels to complete the three gait tests. Test types include natural state walk NG (Normal Gait), the natural state it encountered while walking outside interference EDG (External disturbance Gait) and fatigue when walking under it encountered external interference (F-EDG). The tests were carried out separately, that is, all the same subjects after completion of a test sequence, and then unified under a test. Because the use of a rotating test method, so after the completion of the relevant sports fatigue intervention, subjects can get enough rest, avoid changes in the previous test organism occurred impact on later tests. Data collection is divided into three parts, kinematic data, kinetic data and EMG data.

2.2.2 Kinematics Test

TS using a three-dimensional infrared motion capture system (2000 Hz) and JVC camcorder to achieve the kinematic gait test data collection. The positioning and calculation of human gait in the course of experiment are carried out. The data acquisition frequency was 3000 Hz, and the infrared reflect identification mark was placed on the subject according to the David human standard model.

2.2.3 Kinetic Test

Using two dimensional Kistler force platform to achieve kinetic gait test data collection. The size of the measuring platform is 0.6 m × 0.4 m × 0.2 m, the sampling frequency is 300 Hz, the static detection error is less than 0.5%, and the real-time acquisition and display of the three-dimensional mechanical data can be realized. The three-dimensional dynamometer is placed on a dynamic rail with a size of 3.8 m × 0.8 m × 0.4 m.

2.2.4 Surface EMG Test

Use BTS Free EMG and Noranxon surface electromyography test system implementation muscle gait test signal acquisition. Noraxon's Telemyo is a surface EMG telemetry system with most technologically advanced. The process of transmission is wireless, so the experimental subjects can be measured in the real activities, which is particularly suitable for sports research, the system is 8-channel, wireless telemetry surface electromyography machine. Providing clear, consistent, and reliable data for different types of equal length or constant speed training.

2.2.5 Fatigue Test

The fatigue state in the test was achieved by the fatigue test in the Biodex multi-joint isokinetic force test system. Fatigue parts choose the knee and ankle, the degree of fatigue are quantitative for each participant's strength by 50%.

2.3   Statistical Analysis

To exclude height, weight and other factors for the results of the impact, so that the results of different test persons have certain data comparability, the present study the data were normalized data, and the resulting data in the input SPSS17.0 statistical analysis.

3   Result

3.1   Kinematic Data Results

3.1.1   Gait Cycle Time of the Three States of Normal, Sliding and Fatigue After Sliding

Gait Cycle refers to the time from any instantaneous heel contact with the ground to the side of the heel again contact the ground experienced by the time and space changes constitute a complete gait cycle.

By analyzing the data obtained by the three-dimensional capture system, it can be seen that the gait cycle of the high heels in each state is larger than the flat shoes, and the gait cycle time for wearing high heels in the case of sliding intervention is greater than the flat shoes and there was significant difference (p < 0.05). Figure 1 is the normal gait under the flat shoes and high heels gait cycle time. The gait time of the high heels group was slightly larger than that of the flat shoes group, but there was no significant difference. Figure 2 shows the gait time of the flat shoes and high heels in the sudden sliding state. The gait time of the high heels group is larger than that of the flat shoes group and has significant difference (p < 0.05). Figure 3 is under the state of lower limb fatigue and sudden sliding, the gait cycle time of flat shoes and high heels, high heels group gait time slightly larger than the flat shoes group, but no significant difference.

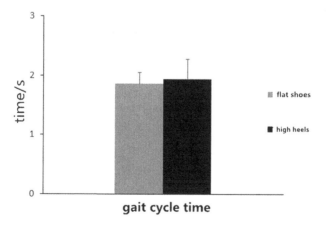

Fig. 1. No external interference state of flat shoes and high heels gait cycle time (s) *indicates P < 0.05 compared to the flat shoe group; **indicates P < 0.01 compared to the flat shoe group

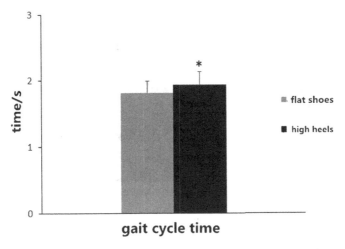

Fig. 2. Experimental platform sliding state of flat shoes and high heels gait cycle time (s) *indicates P < 0.05 compared to the flat shoe group; **indicates P < 0.01 compared to the flat shoe group

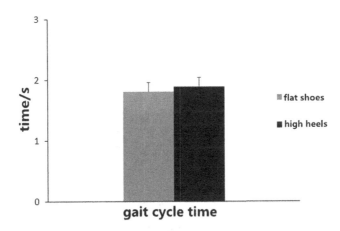

Fig. 3. Lower limb fatigue on the sliding platform of flat shoes and high heels gait cycle time (s) *indicates P < 0.05 compared to the flat shoe group; **indicates P < 0.01 compared to the flat shoe group

3.1.2 Knee Angle Data Results

The change of knee angle can be used as a measure of the kinematic index of human knee joint adjustment in gait. The study found that wearing high heels in the normal state and sliding state of the maximum range of activity of the knee were significantly less than wearing flat shoes. Figure 4 is knee maximum activity range under normal circumstances when wearing flat shoes and high heels. The maximum range of knee movement when wearing high heels is significantly less than the maximum range of motion when wearing flat shoes, and both have very significant differences (p < 0.01).

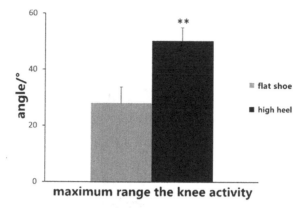

Fig. 4. The maximum range of motion of the knee joint in the gait in normal state (°) [*]indicates $P < 0.05$ compared to the flat shoe group; [**]indicates $P < 0.01$ compared to the flat shoe group

Figure 5 shows knee maximum activity range under the experimental platform sliding state of wearing flat shoes and high heels, the maximum range of knee movement when wearing high heels is significantly less than the maximum range of motion when wearing flat shoes, and the two sets of data have very significant differences ($p < 0.01$).

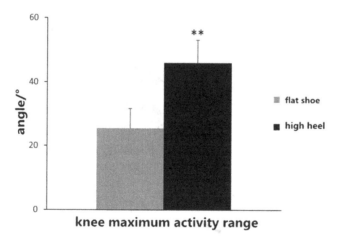

Fig. 5. Knee maximum activity range under the experimental platform sliding state (°) [*]indicates $P < 0.05$ compared to the flat shoe group; [**]indicates $P < 0.01$ compared to the flat shoe group

3.2   Kinetic Data Results

A gait cycle can be divided into six turning points. Take the left lower limb as an example: left foot with the ground, left foot completely flat, left toe leave the ground, right foot followed, right foot completely flat, right toe leave the ground were 6 turning points [9].

In a complete gait cycle, the component in the Z direction has an extreme value at each turning point. As shown in Fig. 6, when one side of the foot touch the ground, it appears a maximum value F1, with the side of the foot flat, the force gradually reduced until the foot completely flat, the force reached a minimum value of F2, and then the side of the foot leave from the ground, to the side of the toe leave from the ground and reached a maximum value of F3, it can be seen that the force curve in the Z direction has a typical bimodal characteristic. The maximum peak of the left foot force curve is about 1.04 times the gravity and the minimum peak is about 0.75 times the gravity.

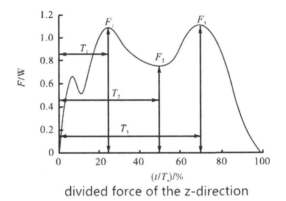

divided force of the z-direction

Fig. 6. Divided force of the z-direction

The force-time curves in the Y-direction are essentially symmetrically distributed. As shown in Fig. 7, it has a positive maximum and two negative minimum values, the positive maximum is about 7% of gravity; two negative minimums are about 3% and 15% of gravity. Since the positive direction of the Y direction is opposite to the walking direction, in the early stage of the support phase, the Y direction force is negative due to the backward horizontal friction force between the support foot and the ground when the test object travels forward [10].

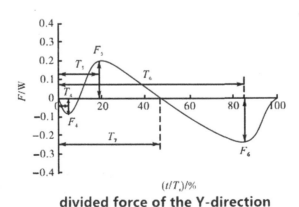

divided force of the Y-direction

Fig. 7. Divided force of the Y-direction

X direction of the force reflects the walking gait stability of the situation, the normal walking X direction of the smaller force, which shows the stability of normal walking [11]. In Fig. 8, the X direction is positive in the right direction of the walking direction, the force-time curve in the X direction is basically symmetrical and the value is small, and the maximum value is about 2.7% of the gravity.

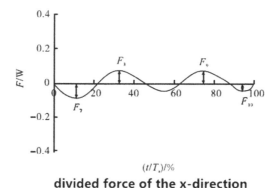

divided force of the x-direction

Fig. 8. Divided force of the x-direction

The maximum ground reaction force of the human body pressure center (COP) in the left and right direction (X), the front-rear direction (Y) and the vertical direction (Z), and the maximum displacement in the left and right direction were collected by the force plate. The data collected by the force plate are the changes of the human body pressure center during the single support phase in the gait. The force in the vertical direction Fz mainly comes from the influence of human body gravity, and the force Fx in the left and right direction and the force in the front and rear directions are mainly based on the force produced by the body during the movement to regulate the body posture. The force in the left and right direction reflects the stability of the gait when walking, if the force in the left and right direction is small, it indicating that the balance of the body is good [12]. Table 1 is the three forces in three directions of the force and the displacement in the left and right direction. The results shows that, in the front-rear and left-right direction, the force of high heels group is greater than the flat shoes group. And under the sliding state, fatigue sliding state, high heels group differences were significant, the sliding state of the high heels group was significantly different from the normal state (p < 0.01). There was also a significant difference in the fatigue sliding state compared with the normal state (p < 0.01). The normal state of the flat shoes group was significantly different from the high heel group (p < 0.01). There was also a significant difference between the sliding state and the high heel group (p < 0.01). And the same group of normal state also has a very significant difference (p < 0.01). Fatigue sliding state was also significantly different from that of high heels (p < 0.01) and had a significant difference from the same group (p < 0.01).

Table 1. Under normal, sliding and fatigue sliding state GRF-BW and pressure center displacement data

Group	Movement state	FzMax (%)	FyMax (%)	FxMax (%)	COPx (cm)
Flat shoes	NG	138.20 ± 6.17	31.50 ± 9.89	8.68 ± 1.99	5.02 ± 0.89
	EDG	123.44 ± 4.13	$49.66 \pm 11.28^{a+}$	$32.13 \pm 8.14^{a+}$	$13.47 \pm 3.16^{a+}$
	F-EDG	136.61 ± 11.33	$45.60 \pm 7.03^{a+}$	$28.47 \pm 1.56^{a+b+}$	$11.92 \pm 1.71^{a+}$
High heels	NG	131.21 ± 9.72	30.56 ± 14.96	6.10 ± 0.71	$3.50 \pm 0.44^{**}$
	EDG	148.02 ± 4.10	$41.13 \pm 8.33^{*}$	$31.97 \pm 9.05^{a+}$	$8.79 \pm 1.64^{a+**}$
	F-EDG	149.34 ± 5.38	42.19 ± 13.5^{a}	$27.21 \pm 1.01^{a+}$	$7.15 \pm 2.11^{a+**}$

*Indicates that compared with the high pressure heel group P < 0.05; **means compared with the high heels group P < 0.01; ^{a}means compared with the same group of normal state P < 0.05; ^{a} $^{+}$means the same group compared with the normal state P < 0.01; ^{b}means that compared with the same group of sliding state compared to P < 0.05; $^{b+}$means compared with the same group of sliding state P < 0.01.

3.3 Surface EMG Data Results

Integral electromyography (IEMG) is the electromyographic signal after rectifying filtering unit time total area under the curve, it can reflect the electromyographic signal varies with time change, the size of the unit for μV·s. It can reflect the electromyographic signal change with time for the size of the [13–15]. From Table 2 shows that starting from the heel strike to full foot touchdown, gastrocnemius signals in four muscle IEMG is strongest, is mainly due to the support process is supported by work of the gastrocnemius muscle weight, high with strongest electromyographic signal, wear high-heeled shoes to walk, integral overall electrical values higher than the other two shoes heel height, especially one of the gastrocnemius muscle IEMG. According to the analysis of variance shows that in the whole foot touchdown moment, two shoes exist significant differences in methods of electricity. Right foot high shares of rectus with group there was significant difference in the bottom of the peace group (P < 0.01). Left foot high pretibial muscle with significant differences between group and group (P = 0.006). This shows that the increase of the heel height for tibialis anterior muscle and gastrocnemius medial and shares, heel height increase, increase muscle discharge.

Table 2. Mean (SD) IEMG of flat shoes and high heels in the four test(unit: μV·s)

Group	Movement state	Tibialis anterior	Gastrocnemius medial	Rectus femoris	Biceps femoris
Flat shoes	NG	54.56 ± 10.63	54.65 ± 6.09	10.93 ± 1.86	42.76 ± 5.66
	EDG	$40.21 \pm 3.98^{a+}$	50.37 ± 10.83	11.67 ± 1.63	$35.63 \pm 5.52^{a+}$
	F-EDG	$31.79 \pm 2.97^{a+b+}$	60.14 ± 11.07^{b}	$17.78 \pm 4.24^{a+b+}$	$36.97 \pm 3.57^{a+}$
High heels	NG	45.92 ± 3.13	51.95 ± 5.33	$18.03 \pm 2.66^{*}$	35.03 ± 4.34
	EDG	$48.92 \pm 5.47^{*}$	59.07 ± 9.38^{a}	$22.83 \pm 3.36^{**a+}$	$43.65 \pm 6.96^{a+}$
	F-EDG	$44.46 \pm 3.58^{*}$	58.23 ± 9.03^{a}	$32.03 \pm 7.36^{*a+b+}$	$53.03 \pm 3.58^{*a+b+}$

*Indicates that compared with the state of the flat shoes with the same state P < 0.05; **said with the flat shoes with the same state P < 0.01; ^{a}means compared with the normal state P < 0.05; $^{a+}$indicates compared with the normal state of the same group P < 0.01; ^{b}means under the sliding state compared with the same group P < 0.05; $^{b+}$means under the sliding state compared with the same group P < 0.01.

4 Discussion

Wearing high heels, the equilibrium stability of the drop, the front control ability significantly reduced, fall index rose, suggests that people with high wear 7 cm high heels to walk, stand, the possibility of a fall, wear high-heeled shoes to walk the center of gravity is not stable also easy a sprained ankle, the ankle ligament, and even lead to fracture. Metatarsal area due to the stiletto heel high stress concentration, walking, standing there for a long time forefoot discomfort, it is easy to cause the plantar toe joint capsule relaxation or arch damage. Wear flat shoes, the foot plantar pressure distribution more uniform, can reduce the harm of the foot. Walking and wear high heels, the body center of gravity in the vertical direction of ups and downs, thus affecting the walking stability of the ups and downs, the greater the stability of the worse.

The study found that with the emergence and development of fatigue, subjects walking with the support of the left and right in a certain swing shift, showing from right to left along the ground to a single step length becomes smaller, step frequency gradually faster. In the initial stages of the movement did not produce fatigue, dominant side can take advantage of its strengths, the dominant role, but with the deepening of fatigue, actively participate in the advantages of the original side of the movement changes the structure of body movements obvious, non-dominant limbs gradually play a positive effect, the performance of the left foot to the right foot with a single step to increase the land of the long, slow down the cadence. Exercise-induced fatigue appeared bilateral limb movements cause imbalance, fatigue and tips will enlarge bilateral limb differences [16]. The human body movement is a holistic movement, uneven distribution of the results will inevitably lead to changes in movement efficiency and reduce movement structure, but also an important reason for damage arising sports.

With the increase in exercise time, the integrated EMG value overall upward trend, increasing the number of integrated EMG responses and participate in the work of each motor units of motor unit discharge size [17, 18] to some extent. As the movement, part of the motor unit fatigue, in order to maintain a predetermined strength, the body must be mobilized to participate in the new muscle fiber contraction, therefore, participate in the activities of the growing number of motor units, the discharge increases, the performance of integrated EMG gradually increase. Increased motor unit recruitment means an increase in the number of integrated EMG motor unit has been activated or accelerate discharge frequency, are nervous fatigue compensatory strategies to deal with change [12, 19, 20].

Factors that cause the body to slip backwards, many of them were located in the center of mass of the vertical projection of the body after the support area, and can not be returned within the support area of the timely adoption of effective regulation is an important reason causing slipping backwards. Therefore, the body to maintain balance on the premise that the human physique projection control center within the area of support, or when projected beyond the support area of the center of mass, angular momentum generated by the swing enough to offset part of the body to engender fallen trends angular momentum, and the people back into the center of mass within the support area of the [21]. This also resulted in a corresponding adjustment of the balance of the two methods, one is to fall rapidly in the direction towards recovery step, thereby expanding the support area, center of mass of people still fall within the projected area of support. Another step is not to rely on the recovery, but by swinging torso and arms back to the heart of the human physique within the support area.

As an important part of the hip joint control of human posture, hip joint plays a vital role in regulating the balance to avoid falling. Lower extremity kinematic chain as a whole, when the state of motion of the hip joint of the knee changes will make the appropriate changes.

The study found that, during the backward slipping balance adjustment, the body will produce a top general hip action, so that the maximum angular velocity of the hip than normal state is very significant increase. By this action, the hip extensors acting, so that the trunk can produce backward rotation of the hip forward and translational energy, this will allow the entire body system angular momentum transfer back to the trunk, reducing the whole human body down the trend, while also increasing the body's system forward momentum, thereby increasing the level of physical fitness heart rate of people forward, people end up back in the center of mass of the support area, to avoid the occurrence of falls.

Some studies show that the human body when subjected to an appropriate balance interference can make quick, strong, substantial support, is to decide whether there will be a key factor for falls [22]. Recovery step adjustment requires a lot of time slip joint torque [23], while the elderly because of the knee extensor muscle strength loss is large, resulting in the joints can not produce enough torque at slipping balance adjustment (either the magnitude or speed) to control the level of body and momentum in the vertical direction, resulting in the occurrence of fall [24–27].

When the body produces a tendency to slip, in order to avoid a further fall of the body, essential to recruit more motor units for physical posture adjusted to maintain the balance of the body [28] in continuous movement. Studies have shown that surface electromyography (sEMG) can be an important indicator of the degree of muscle activation, while the correlation between surface electromyography and muscle strength is very high [29]. The study found that, at the time of slipping backward balance adjustment, compared with the normal state compared to the measured 4 points EMG muscle have a very significant or significantly increased. This shows that the balance adjustment when slipping is a systemic activity, its force is not limited to a few parts of the muscles, but to achieve a good balance recovery effect by the coordinated movement of the body can be. Complete balance adjustment action active muscle, is mainly responsible for the foot brake, to enhance the role of the hip extensor position. Complete balance adjustment action antagonistic muscle contraction is mainly responsible for the initiative to increase the accuracy by adjusting the position of the [30].

Heels forefoot pressure is too concentrated can cause discomfort, there are a lot of people there forefoot pain, with long hallux valgus occurs, we recommend that if you want to wear high heels to stand or walk for a long time, can cushion insole forefoot position mitigation of damage to the metatarsal area. In terms of the impulse we found during exercise, thick soles can slow to face the impact of the foot, reducing the movement of the pressure, slowing fatigue can make the wearer line further. So heels soles plus a waterproof design reduces the height of the heel also greatly reduces the impact of the foot of the face.

Acknowledgements. This study is supported by the key technology of the old balance capability assessment and the development of training tools financial aid program (PXM2016-178215-000013). We would like to thank the patients for their participation in the study.

References

1. Lu, K.: The research of kinetic characteristics of once exhaustion sports before and after vertical jump. Sports Sci. **19**(1), 90–92 (1999)
2. Dierks, T.A., Davis, I.S., Hamill, J.: The effects of running in a exerted state on lower extremity kinematics and joint timing. J. Biomech. **43**, 2993–2998 (2010)
3. Mizrahi, J., Verbitsky, O., Isakov, E.: Effect of fatigue on leg kinematics and impace acceleration in long distancerunning. Hum. Mov. Sci. **19**, 139–151 (2000)
4. Williams, K.R., Snow, R., Agruss, C.: Changes in distancerunning kinematics with fatigue. Int. J. Sport Biomech. **7**, 138–162 (1991)
5. Christina, K.A., White, S.C., Gilchrist, L.: A effect of localized muscle fatigue on vertical ground reaction forces and ankle joint motion during running. Hum. Mov. Sci. **20**, 257–276 (2001)
6. Niu, W., Wang, Y., He, Y.: Kinematics, kinetics and electromyogram of ankle during drop landing: a comparison between dominant and non-dominant limb. Hum. Mov. Sci. **30**(3), 614–623 (2011)
7. Smeesters, C., Hayes, W.C., McMahon, T.A.: Disturbance type and gait speed affect fall direction and impact location. J. Biomech. **34**(3), 309–317 (2001)
8. Englander, F., Hodson, T., Terregrossa, R.: Economic dimensions of slip and fall injuries. J. Forensic Sci. **41**(5), 733–746 (1996)
9. Cernekova, M., Hlavacek, P.: The influence of heel height on plantar pressure. Clin. Biomech. **23**, 667–668 (2008)
10. Li, J., Wang, L.: Application and advances of sole pressure measurement in biomechanical research. J. Beijing Sport Univ. **28**(2), 255–259 (2005)
11. De Lateur, B.J., Giaconi, R.M., Questad, K.: Footwear and posture: compensatory strategies for heel height. Am. J. Phys.ˈMed. Rehabil. **70**, 246–254 (2009)
12. Yang, Y.: The EMG determination and analysis of quadriceps isokinetic concentric contraction. J. Beijing Sport Univ. **18**(4), 28–34 (1995)
13. Lockhart, T.E., Spaulding, J.M., Park, S.H.: Age-related slip avoidance strategy while walking over a known slippery floor surface. Gait Posture **26**(1), 142–149 (2007)
14. Siegmund, G.P., Heiden, T.L., Sanderson, D.J.: The effect of subject awareness and prior slip experience on tribometer-based predictions of slip probability. Gait Posture **24**(1), 110–119 (2006)
15. Kagawa, T., Uno, Y.: Necessary condition for forward progression in ballistic walking. Hum. Mov. Sci. **29**(6), 964–976 (2010)
16. Kagawa, T., Ohta, Y., Uno, Y.: State-dependent corrective reactions for backward balance losses during human walking. Hum. Mov. Sci. **30**(6), 1210–1224 (2011)
17. Bolton, D., Misiaszek, J.: Compensatory balance reactions during forward and backward walking on a treadmill. Gait Posture **35**(4), 681–684 (2012)
18. Qian, J., Song, Y.: The biomechanics principle of walking and analysis on gaits. J. Nanjing Inst. Phys. Educ. (Nat. Sci.) **5**(4), 35–40 (2006)

19. Song, Y., Kou, H., Zhang, X.: The study on changes of myo-electricity of planta pedis during walking with different hardness soles of shoes. Chin. J. Rehabil. Med. **25**(12), 1157–1165 (2010)
20. Dai, H., Cai, Y.: Research on the comfortability of different jogging shoes. J. Hunan Agric. Univ. (Nat. Sci.) **10**(1), 119–121 (2009)
21. Duchene, J., Hogrel, J.Y.: A model of EMG generation. IEEE Trans. Biomed. Eng. **47**(2), 192–201 (2000)
22. Jin, J.: The Advanced Tutorial of Sports Biomechanics, 2nd edn. Beijing Sports University Press, Beijing (2007)
23. Lockhart, T.E., Spaulding, J.M., Park, S.H.: Age-related slip avoidance strategy while walking over a known slippery floor surface. Gait Posture **26**(1), 142–149 (2007)
24. Siegmund, G.P., Heiden, T.L., Sanderson, D.J.: The effect of subject awareness and prior slip experience on tribometer-based predictions of slip probability. Gait Posture **24**(1), 110–119 (2006)
25. Kagawa, T., Uno, Y.: Necessary condition for forward progression in ballistic walking. Hum. Mov. Sci. **29**(6), 964–976 (2010)
26. Pratt, J., Carff, J., Drakunov, S.: Capture point: a step toward humanoid push recovery. In: 6th IEEE-RAS International Conference on Humanoid Robots, Italy, pp. 200–207 (2006)
27. Marigold, D.S., Patla, A.E.: Strategies for dynamic stability during locomotion on a slippery surface: effects of prior experience and knowledge. J. Neurophysiol. **88**(1), 339–353 (2002)
28. Chambers, A.J., Cham, R.: Slip-related muscle activation patterns in the stance leg during walking. Gait Posture **25**(4), 565–572 (2007)
29. Tang, P.-F., Woollacott, M.H., Chong, R.K.: Control of reactive balance adjustments in perturbed human walking: roles of proximal and distal postural muscle activity. Exp. Brain Res. **119**(2), 141–152 (1998)
30. Heiden, T.L., Sanderson, D.J., Siegmund, G.P.: Adaptations to normal human gait on potentially slippery surfaces: the effects of awareness and prior slip experience. Gait Posture **24**(2), 237–246 (2006)

Data Clustering Algorithm Based on Artificial Immune Network

Zongkun Li and Dechang Pi[✉]

College of Computer Science and Technology,
Nanjing University of Aeronautics and Astronautics,
29 Jiangjun Street, Nanjing 211106, Jiangsu, People's Republic of China
20121896@cqu.edu.cn, dc.pi@nuaa.edu.cn

Abstract. For the problem that the termination condition of artificial immune network algorithm aiNet is difficult to determine, an intelligent artificial immune network algorithm S-aiNet is proposed. The S-aiNet determines whether the network is saturated by monitoring the change trend of new generation population in the iterative process according to the affinity of the new generation of network cells and existing cells. The algorithm improves the adaptability of aiNet and reduces the number of parameters. For the problem that the network of aiNet updates slowly, a regional search optimization algorithm AS-aiNet is proposed. The AS-aiNet equally divides the antibody space where the network cells and antigen located, and only searches the antibody cells located in the same region as antigens in the immune response. The AS-aiNet reduces the workload of search in the process of immune response and effectively enhances the time efficiency of algorithm operation. Adopting public data set, experiments show that the time efficiency of AS-aiNet is 10% better than that of aiNet.

Keywords: Artificial immune · Data clustering · Stopping rule · Region searching

1 Introduction

K-means algorithm has been widely used in data clustering since it was proposed by McQueen in 1967 [1]. In the late 1980s, Famer et al. proposed the mathematical model of immune system, and thoroughly studied relationship between immune systems and other artificial intelligence methods [2]. Until now, the research in artificial immune network is aimed at two issues: One is how to determine the termination condition of the algorithm and how to improve running efficiency of the algorithm on the premise of ensuring the accuracy of clustering. On the issue about the running efficiency problem, Li propose an algorithm with immune secondary responses named IRA which improves the speed of adapting new patterns [3]. Liu et al. propose a dynamic local search based immune automatic clustering algorithm (DLSIAC) that automatically find the optimal number of clusters and a proper partition of datasets [4]. But the study on termination condition of artificial immune network algorithm still reaches a standstill here that most researchers determine it by experiential judge. The traditional approach presets a threshold as the termination point and adjusts the threshold by repeated

© Springer Nature Singapore Pte Ltd. 2017
B. Zou et al. (Eds.): ICPCSEE 2017, Part II, CCIS 728, pp. 516–527, 2017.
DOI: 10.1007/978-981-10-6388-6_44

experiments. But during the practical application, there is a great limitation in traditional method that the algorithm is not fit for different data set.

For the problem that the termination condition of artificial immune network algorithm aiNet is difficult to determine, an intelligent artificial immune network algorithm S-aiNet is proposed. On the basis of S-aiNet, for the problem that the network of aiNet updates slowly, a regional search optimization algorithm AS-aiNet is proposed.

2 Clustering Algorithm Based on Artificial Immune Network (aiNet)

The classical artificial immune network algorithm aiNet was proposed by de Castro and Von Zuben in 2000 [5]. The algorithm adopts an undirected graph with weights on vertices, ignoring the difference between B cells and antibodies, and mainly focusing on simulating the stimulation process between the immune network and antigens. As a typical immune network algorithm, aiNet conforms to population change rate formula, as follows:

$$RPV = New\ cells + Dead\ cells\ without\ antigen\ stimulation \\ + copied\ cells\ without\ antigen\ stimulation \tag{2.1}$$

The RPV represents change rate of population.

The affinity between cells and given antigens can be promoted by the formula as follows:

$$C = C - \alpha(C - X) \tag{2.2}$$

C is the matrix of network cells, X is the matrix of antigens, α is the mutation rate that decided by affinity between antibodies and antigens, the higher affinity becomes, the lower α is [6].

Ultimate aim of aiNet algorithm is to set up a memory set for recognizing and expressing data structure and organization. The more specific cells are, the less simplified (the less compressed) network will be; whereas, the more generalized cells are, the more simplified network of cell relations (promoted compression) will be [7]. By containment of threshold value of (σ_S), grade of specificity, cluster accuracy and network adaptability of cells are under control. Users should, at first, set a lower value ($\sigma_S \leq 10^{-3}$ for instance) before constantly making fine adjustment of network properties. Euclidean distance is applied in this algorithm to measure similarity and dissimilarity; shorter distance implies higher similarity and resultant network is drawn by clonal selection, affinity maturation and simulation of activities of immune network. To supplement general network structure presented in Expression (2.1), two steps (clone containment and network containment) exist in this algorithm to inhibit self-recognized cells.

Network output may also be memory cell coordinate matrix M and inter-cell affinity matrix S. Matrix M, on behalf of mapping in the network of antigen group, S is in charge of recognizing which cells are linked with other cells and describing overall network structure.

In order to extract final network structure from output, it is advised to compute the minimum spanning tree according to matrix M and set some threshold value in trimming to draw a forest. Each tree in the forest is exactly an aggregate of data recognized.

3 Artificial Immune Network Clustering Algorithm Based on Intelligent Termination Decision and Regional Search

The aiNet algorithm shows that many problems need to be improved in practical application. One issue is how to set the termination condition. Theoretically, every iteration step can enhance clustering effect by using the iterative method, however, in practical situations, the algorithm is not allowed to execute endlessly. Another issue is that the clustering speed of aiNet algorithm is seriously affected when the data set is multiplied. The aiNet algorithm needs to be further improved on time efficiency. In view of the above two questions, this paper separately proposes intelligent termination decision artificial immune network algorithm and regional search artificial immune network algorithm.

3.1 Relevant Definition

Definition 1. X: Data set consists of N_p vectors ($X \in \Re^p$).

Definition 2. C: Matrix includes all N_t network cells ($C \in \Re^{N_t \times p}$).

Definition 3. M: Matrix includes N memory cells ($M \subseteq C$).

Definition 4. N_c: The amount of clone that every stimulated cells produce.

Definition 5. D: Dissimilarity matrix includes element d_{ij} (antigens to antibodies).

Definition 6. S: Similarity matrix includes element s_{ij} (antibodies to antibodies).

Definition 7. n: n cells which have highest affinity.

Definition 8. ζ: The percentage of mature cells which is waiting to be chosen.

Definition 9. $\sigma_{d,s}$: Represent the Natural death threshold and network inhibition threshold.

3.2 Artificial Immune Network Algorithm Based on Intelligent Termination Decision (S-aiNet)

Main Idea of the Algorithm. The traditional method is to preset the upper limit of the number of iterations or upper limit of the number of network cells proactively, then determine a good value by trial and error. But this algorithm with the conditions of a great deal of adjusting work of the parameters is clearly inefficient. So this paper proposes an artificial immune network algorithm S-aiNet which can distinguish the termination condition intelligently.

The algorithm is based on the fact that: For the static input data, the new generation of antibodies and existed antibodies will become more and more similar based on iterations. In other words, the affinity between proliferating antibody cells and other network antibody cells will continually grow. So a threshold σ_T is preset, the algorithm will be terminated when the antibody affinity between new generation antibody cells and the existed cells which own the lowest affinity is still greater than σ_T.

But this algorithm still not reduce the number of parameters, needs further optimize. The termination threshold σ_T has a strong relationship with the existed network suppression threshold σ_S. So we consider using σ_S or the multiple of σ_S replace σ_T. Experiments prove that when $\sigma_T = 2\sigma_S$ the algorithm achieve good results and the correlation with data is not significant. From this, taking $Min(S_{ij}) < \sigma_T = 2\sigma_S$ as the termination condition of the algorithm, and the improved algorithm is named S-aiNet.

Algorithm description of S-aiNet

Algorithm1: S-aiNet
Input: Data set X which need clustering.
Output: Result network C.
1: for i=1 to upper limit of the number of iteration
2:    for each antigen cell ag
3:       calculate relevant factor d_{ij} between ag and every network cell
4:       select n network cells which have highest correlation
5:       reproduce (clone) the n selected cells, get N_c new cells
6:       apply formula (2.2) mutate N_c cells
7:    calculate the dissimilarity D between antibody and antigen of improved network
8:          chose ζ percent cells which have the highest affinity, establish local memory cell matrix M_p
9:    eliminate the cells whose $d_{ij} < \sigma_S$ (natural death)
10:       calculate the affinity S among antibodies
11:          eliminate the cells whose $s_{ij} < \sigma_S$ (clone suppression)
12:       connect C and M_p ($C \leftarrow [C; M_p]$)
13:    end for each
14:    calculate S, eliminate the cells whose $s_{ij} < \sigma_S$ (network suppression)
15:    replace r percent of worst cells to new cells which are randomly generated
16:    if $Min(S_{ij}) < \sigma_T = 2\sigma_S$ break
17: end for
18: output result network C
19: end

3.3 Artificial Immune Network Algorithm Based on Regional Search (aS-aiNet)

Main Idea of the Algorithm. The aiNet algorithm has a quicker speed in clustering, compared with some traditional clustering algorithm like k-means, especially in distributed environment. However, with the increasing demands of big data processing, the data clustering problems are facing a multiplied scale of data set. So the time efficiency of algorithm needs to be further improved.

Policy Description. In the aiNet (S-aiNet) algorithm, the most important time bottleneck appears in the step 11. This step will calculate the affinities between every antibody and one antigen. The global searching mode like this costs vast running time, so we import the concept of regional search to reduce the amount of searching in the period of antigen-antibody recognition. The specific method is dividing R-dimensional vector space into K^R sub-blocks. In other words, it divides every dimension into K blocks. The algorithm only searches the antibodies which locate in the same sub-block with antigens. With this method, we can reduce the search volume effectively.

In fact, this optimization method is an approximation algorithm; it saves time at the expense of clustering accuracy in some way. However, owing to the diversity of artificial immune network, there are many corresponding antibodies for the same data clustering, accordingly, by choosing appropriate K we can get the results which meet the precision requirements. From this, AS-aiNet algorithm is proposed.

AS-aiNet algorithm description

Algorithm2: AS-aiNet
Input: Data set X which need clustering.
Output: Result network C
1: divide vector space into K^R sub-blocks, create mappings
2: for i=1 to upper limit of the number of iteration
3:     for each antigen cell ag
4:         confirm the block ag located, calculate relevant factor d_{ij} between ag and every network cell in the same block
5:         select n network cells which have highest correlation
6:         reproduce (clone) the n selected cells, get N_c new cells
7:     apply formula (2.2) to mutate N_c cells
8:         calculate the dissimilarity D between antibody and antigen of improved network
9:         chose ζ percent cells which have the highest affinity, establish local memory cell matrix M_p
10:     eliminate the cells whose $d_{ij} < \sigma_S$ (natural death)
11:         calculate the affinity S among antibodies
12:             eliminate the cells whose $s_{ij} < \sigma_S$ (clone suppression)
13:         connect C and M_p ($C \leftarrow [C; M_p]$)
14:             map the new cells to the blocks they belong to
15:     end for each
16:     calculate S, eliminate the cells whose $s_{ij} < \sigma_S$ (network suppression)
17:     replace r percent of worst cells to new cells which are randomly generated
18:     if $Min(S_{ij}) < \sigma_T = 2\sigma_S$ break
19: end for
20: output result network C
21: end

4 Experiment

4.1 Data Set

The experiment adopts three public data sets, they are R15, Spiral and Wisconsin Breast Cancer from http://cs.joensuu.fi/sipu/datasets/ The data set R15 and Spiral have a clear clustering architecture in two-dimensional space (as illustrated in Figs. 1 and 2) which is intuitive to judge the result of clustering. The Wisconsin Breast Cancer data set is widely used in the tests of clustering algorithm and mainly used to evaluate the time efficiency of the algorithm.

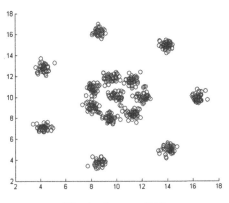

Fig. 1. Data set R15

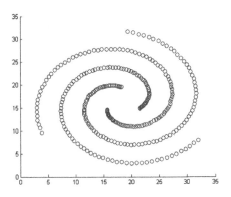

Fig. 2. Data set Spiral

4.2 Experimental Results and Analysis

The experiment consists of two parts: Part 1: Mainly study the impact of different parameters in AS-aiNet on the running result. Part 2: Compare the running result of AS-aiNet with the classical algorithm aiNet and evaluate the improvement.

Influence of Thresholds on AS-aiNet. Similar with aiNet, the clustering effectiveness and the running time of AS-aiNet depend on natural death threshold and network inhibition threshold. The experiment adopts data set R15 and Spiral; test the influence of thresholds on running effect by using different threshold. The threshold that each test used is shown in Table 1:

Table 1. Thresholds of the experiment

Serial number	T1	T2	T3	T4
Natural death threshold σ_d	0.03	0.05	0.1	0.15
Network inhibition threshold σ_s	0.05	0.1	0.25	0.5

Experimental results as follows:
The result adopting data set R15 is shown in Figs. 3, 4, 5 and 6.

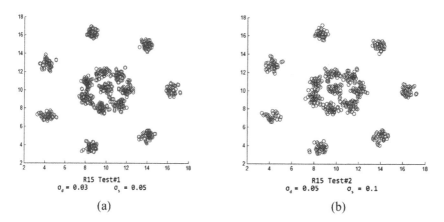

(a)   (b)

Fig. 3. Test results within R15 data set

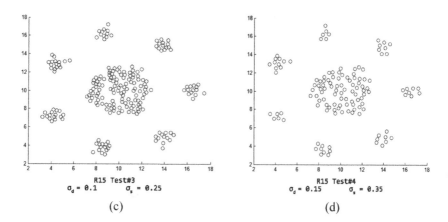

(c)   (d)

Fig. 4. Test results within R15 data set by changing thresholds.

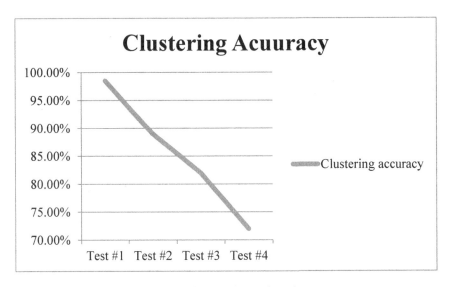

Fig. 5. Influence of threshold on clustering accuracy

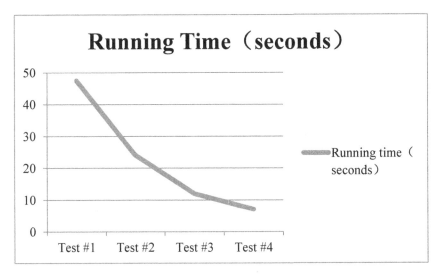

Fig. 6. Influence of threshold on time cost

From Figs. 3, 4, 5 and 6, it will achieve a good clustering accuracy and generate a larger network but cost more time by using a lower threshold. Instead, it will reduce the scale of the new network and time cost, but sacrifice the accuracy of clustering to some degree by using a higher threshold.

The result adopting data set Spiral is shown in Figs. 7, 8 and 9.

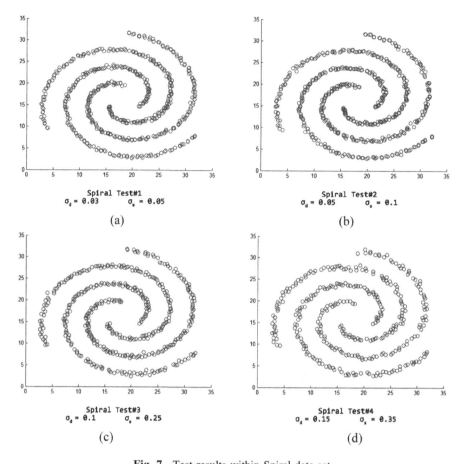

Fig. 7. Test results within Spiral data set

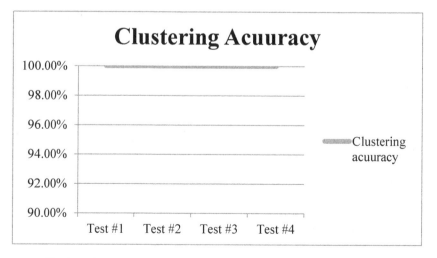

Fig. 8. Influence of threshold on clustering accuracy within R15 data set

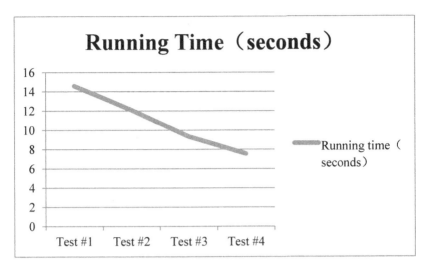

Fig. 9. Influence of threshold on time cost within R15 data set

The results from Spiral further confirm the results of R15, the difference is that the same threshold has a discrepancy of influence on different data set. For data set Spiral, the influence of threshold on clustering accuracy and time cost is not very significant.

From the above 2 sets of experiments, we can see that natural death threshold and network inhibition threshold have a noticeable effect on clustering accuracy and time cost. It will achieve a good clustering accuracy and generate a larger network but will spend more time by using a lower threshold. Instead, it will reduce the scale of the new network and time cost, but sacrifice the accuracy of clustering to some degree by using a higher threshold.

In the specific application condition, we shall first presuppose a lower threshold, and get a balance between time cost and clustering accuracy by gradually enlarging the threshold.

Evaluation Experiment of AS-aiNet. This experiment adopts Wisconsin Breast Cancer as the testing data set, and compares the running efficiency and clustering accuracy of AS-aiNet with aiNet to evaluate the improvement of the AS-aiNet.

In comparison test, we adopt same parameters to two algorithms. aiNet algorithm will be preset a upper limit of the number of iteration of 400 times. The K of AS-aiNet will be preset for 4.

The threshold that each test used is shown in Table 2:

Table 2. Thresholds of the experiment

Serial number	T1	T2	T3	T4
Natural death threshold σ_d	0.1	0.2	0.3	0.4
Network inhibition threshold σ_s	0.2	0.25	0.4	0.5

Experimental results is shown in Figs. 10 and 11.

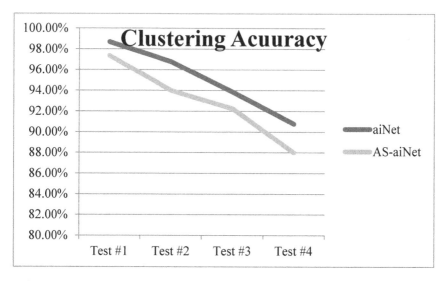

Fig. 10. Comparison between aiNet and AS-aiNet on clustering accuracy

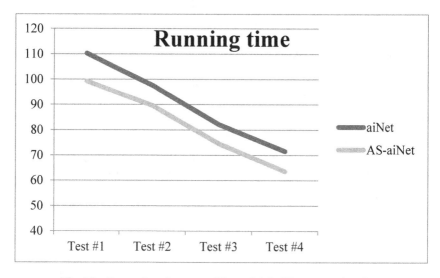

Fig. 11. Comparison between aiNet and AS-aiNet on running time

Through the Experimental results we can see that compared with aiNet the improved algorithm AS-aiNet can save about 10% running time in condition of losing 3% to 4% clustering accuracy.

5 Conclusion

This paper proposes S-aiNet algorithm based on aiNet, by presetting termination threshold which associated with network inhibition threshold, establishes intelligent termination determination mechanism, which solved the problem that termination condition of artificial immune network algorithm aiNet is difficult to determine. Based on S-aiNet, a regional search algorithm AS-aiNet is proposed which can reduce the time cost effectively. By divding the data set into blocks, the AS-aiNet algorithm reduces the amount of searching during the immune response. Comparing with classic algorithm aiNet, AS-aiNet proves its obvious improvement and optimization.

Acknowledgements. This work was supported by National Natural Science Foundation of China (U1433116), Foundation of Graduate Innovation Center in NUAA (kfjj20171603).

References

1. Kuo, R.J., Chen, S.S., Cheng, W.C., Tsai, C.Y.: Integration of artificial immune network and k-means for cluster analysis. Knowl. Inf. Syst. **40**(3), 541–557 (2014)
2. Farmer, J.D., Packard, N.H., Perelson, A.S.: The immune system, adaptation, and machine learning. Physica D-nonlinear Phenomena **2**(1–3), 187–204 (1986)
3. Li, X.H.: Research on incremental clustering algorithm based on artificial immune system and its optimization and application. (Doctoral dissertation, Jilin University) (2009)
4. Liu, R., Zhu, B., Bian, R., Ma, Y., Jiao, L.: Dynamic local search based immune automatic clustering algorithm and its applications. Appl. Soft Comput. **27**(C), 250–268 (2015)
5. De Casto, L.N., Von Zuben, F.J.: An evolutionary immune network for data clustering, pp. 84–89 (2000)
6. Rassam, M.A., Maarof, M.A.: Artificial immune network clustering approach for anomaly intrusion detection. J. Adv. Inf. Technol. **3**(3), 147–154 (2012)
7. Pan, Z.-M.: Artificial immune network clustering based on affinity accumulation. J. Comput. Appl. **31**(06), 1660–1663 (2011)

Multi-step Reinforcement Learning Algorithm of Mobile Robot Path Planning Based on Virtual Potential Field

Jun Liu$^{(\boxtimes)}$, Wei Qi, and Xu Lu

School of Automation, Guangdong Polytechnic Normal University,
Guangzhou 510665, China
liujun7700@163.com

Abstract. A algorithm of dynamic multi-step reinforcement learning based on virtual potential field path planning is proposed in this paper. Firstly, it is constructed the virtual potential field according to the known information. And then in view of Q learning algorithm of the $Q(\lambda)$ algorithm, a multi-step reinforcement learning algorithm is proposed in this paper. It can update current Q value used of future dynamic k steps according to the current environment status. At the same time, the convergence is analyzed. Finally the simulation experiments are done. It shows that the proposed algorithm and convergence and so on are more efficiency than similar algorithms.

Keywords: Robot · Path planning · Machine learning · Learning · Virtual potential field

1 Introduction

As robots are widely applied, robot path planning in complex environment is facing greater challenges [1, 2]. Especially in the complex dynamic environment, It is difficultly to meet the needs of the actual environment to set path in advance. When faced with complex tasks, it needs to robot has a reasonable feedback action by interaction online calculation with the environment [3]. Reinforcement learning is a kind of effective method for solving sequential optimization decision-making problems which can be used in the no environment model and the condition of no teachers' samples. It can obtain the optimal or sub-optimal behaviour strategy by maximized the cumulative returns through trial and error in interaction environment. It is suitable well for robot path planning in complex environment [4].

Q learning algorithm is a typical reinforcement learning method for solving the incomplete information Markov decision problem. In the standard Q learning algorithm, next a step information is used to update the value Q. The update speed is slow and the foresee ability is not strong. The efficiency is not high in robot path planning. And the $Q(\lambda)$ learning algorithm is used all the future information to update the current value. It has considered the influence of future decisions for the current value. The foresee ability is strong. However, when the space of action-state is with larger scale, which with large amount of calculation. And it generally assumed that robots did not know all the future global information when path planning in the working area. It difficult to met the algorithm virtual requirements. In view of above, this paper has put forward a mobile robot path planning algorithm based on virtual potential field. Combined with the virtual potential field, it built the mobile space in dynamic environment. And then combined with the surrounding known environmental information, it used dynamic $Q(k)$ learning algorithm to robot path planning. At the same time, prior knowledge is used to initialize the value Q. It can improve the learning efficiency in the initial stage as accelerate the algorithm convergence speed.

2 Related Works

At present, There are many scholars for mobile robot path planning research [5], and achieved fruitful results. A search algorithm based on computing complex figure proposed for mobile robot path planning problem in [6]. A beacon RSSI is used to navigate to realize path planning in [7] which needs to set up a variety of beacon nodes in the environment. A kind of dynamic path planning algorithm based on obstacle forecast was put forward in [8], which determined by the main obstacle motion and position accuracy. A heuristic path planning algorithm was proposed in [9], which compared with simulated annealing is better than other algorithm. And, it searched the shortest path in terms of computation time.

Mostly above are limited to global environmental information known or can more accurately predict obstacle scenarios. But, in most cases, mobile robots are mastery of environmental information incomplete, and even completely unknown. Meanwhile, mobile robot can only plan for step-by-step path planning through the local information of sensors. Machine learning for mobile robot path adaptive planning is a method of the numerous scholars generally accepted. Robot with perceiving environment autonomous can obtain new behavior by interactions with the environment of online learning. It enables the robot to choose according to specific tasks reached the goal of the optimal action [10, 11]. Reinforcement learning [12] is an important method of machine learning. Its used a method similar to the human mind of the trial and error (trial-error) to find the optimal behavioral strategy. It has demonstrated the good study performance [13] in the robot behavior study. In unknown environment, it showed stronger adaptive ability.

During traditional reinforcement learning, Agents adopted blind search at the beginning of learning, which will cause the target reward spread slowly. It is scarcely possible to solve more complex mass. In order to improve the algorithm convergence speed, researchers tried to use various methods to accelerate learning process. A priori knowledge used to the value Q initialized, improve the learning efficiency of initial stage and accelerate the algorithm convergence speed. At present, initialize methods for Q value including neural network [14], hierarchical reinforcement learning and self-organizing clustering methods. Neural networks use neural network to approximate the optimal value function. Prior knowledge mapped Q function table, which make the robot, accelerate the algorithm convergence speed by studied in the whole state space on a subset. Hierarchical reinforcement learning adopt the tactics of divide and rule by artificial decomposition or abstract to large scale of the reinforcement learning problem in state space or time domain. Self-organizing clustering method [15] also narrowed the state space, so as to accelerate the learning speed. However, these methods also complicate the problem, increase the difficulty of the algorithm.

The priori background knowledge induced is a effective way to guide learning from blind search and improve efficiency. Dyna-H algorithm is proposed by Santos et al. in [16]. The method that A* algorithm is introduced into the Dyna reinforcement learning framework is a kind of heuristic planning method. It is not reasonable in the different stages of computing resources allocation and efficiency is not ideal. Chen et al. in [17] divided reinforcement learning process into quantitative and qualitative. Quantitative layer accorded to the MDP model and qualitative layer is accordance with the SMDP model. Hierarchical control structure constituted a hybrid model. Because with a qualitative layer with the learn ability, learning efficiency and stability are larger.

Grzes and Kudenko in [18] put forward a kind of online reinforcement learning algorithm Sarsa-Rs, which can generate online Shaping return function. Konidaris et al. [19] proposed a algorithm that agent with reinforcement learning method get knowledge, and apply knowledge to more complex learning environment to improve the learning efficiency in the problem space. A multi-agent distributed hybrid response reinforcement learning technique is proposed in [20] which study exploratory steps to implement shortest path planning. They need multi-agent coordination.

Most of above methods aim to specific assumptions. Application scope is difficult to promotion. Some methods assumes that the robot know map global barrier model which are some limitations in the practical promotion. Aim to above insufficiency, a mobile robot path planning based on virtual potential field of initialization of reinforcement learning method was puts forward in this paper. According to known environment information formed a virtual space of potential energy field, it made obstacles area potential energy value as zero and the target as the global biggest potential value. The potential energy field form monotone increasing surface. Then the potential value of each state in the virtual potential field is the biggest cumulative returns on behalf of the state can be obtained. Then the initial value of all state-action of $Q(s; a)$ is defined as immediately returns the current state to perform the selected action and subsequent condition added to the biggest discount cumulative returns which optimal strategy obtain (maximum cumulative returns multiplied by a factor). Q initialization optimized the robot the initial stage of learning, which to provide a better learning foundation for robot. With Q initialization, improved algorithm makes the algorithm convergence speed faster and more stable convergence process.

As robots are widely applied, robot path planning in complex environment is facing greater challenges [1, 2]. Especially in the complex dynamic environment, It is difficultly to meet the needs of the actual environment to set path in advance. When faced

3   Modeling and Analysis

As assume that the target location unknown, the robot can only detect the surrounding local environment information according to own sensors to and other global information is unknown. How to efficient planning an optimal path from the starting position to the target position, and make sure to avoid all obstacles around is the issue of this paper to study. The main idea based on the reinforcement learning to solve this problem is through the current state of perception of the environment information and reasonable decision next move. Then, environment returns rewards and robots get a new status. The optimization goal is to obtain an optimal strategy which gained from the current state's largest equivalent cumulative returns by robots. Then, the following analysis problem work space model and decision-making model of path planning.

3.1   Virtual Potential Field Model

Target location information is known for mobile robot. To make sure the robots moved to the target direction during making decisions. A virtual potential field method is built a robot mobile space in this paper. Basic idea is to put the virtual robot moving space interval into corresponding potential space. The target produces a potential field in the global environment, and obstacles in the local produce repulsive force field. The virtual stack will constitute the whole area of potential energy field. Robots in virtual potential field receive virtual force by virtual potential field, which made robots moved from the starting position to reach the target point avoiding obstacles. At present, most methods of virtual potential field are based on calculation model of Coulomb forces theorem, which is proportional to the Euclidean distance between the target, and the obstacles between into reverse. The calculation formula is as follows:

$$U(s) = U_a(s) + U_r(s) \tag{1}$$

Among them: U(s) is the potential energy of the state point s. $U_a(s)$ is potential energy of the state point s of gravitational field. $U_r(s)$ is the potential energy in the state point s of the repulsive force field. $U_a(s)$ and $U_r(s)$ shown as follow:

$$U_a(s) = \frac{1}{2} k_a \rho_g^2(s) \tag{2}$$

Where, k_a is the scale factor. $\rho_g(s)$ is the shortest distance between the state point s and the target point.

$$U_r(s) = \begin{cases} \frac{1}{2} k_r \left(\dfrac{1}{\rho_{ob}(s)} - \dfrac{1}{\rho_o} \right)^2, & \rho(s) < \rho_0 \\ 0, & \rho(s) \geq \rho_0 \end{cases} \tag{3}$$

between the state point s and obstacles. ρ_o is obstacles influence coefficient. It can convert mobile robot space to a vector field by above methods. Then, it normalized vector field to the convenience of treatment:

$$U'(s) = \frac{U_{\max}(s) - U(s)}{|U_{\max}(s)|} \tag{4}$$

Where, $U_0(s)$ is the potential energy of state s in the potential energy field U_0. $U_{\max}(s)$ is the highest potential energy point in the area. Through conversion, it made the obstacle area potential energy value as zero and the target point potential energy value as 1. And the entire potential energy field formed a monotone increasing surface. Virtual potential field force $F(s)$ of robot in the potential field expressed as a negative gradient of potential function, namely

$$F(s) = -\nabla U(s) = F_a(s) + F_r(s) \tag{5}$$

Where, $F_a(s)$ and $F_r(s)$ respectively corresponding to the gravitational force of target and the repulsive force of obstacle. It can be deduced:

$$F_a(s) = -\nabla U_a(s) = -k_a \rho_g(s) n_t \tag{6}$$

$$F_r(s) = -\nabla U_r(s) = \begin{cases} k_r \left[\dfrac{1}{\rho_{ob}(s)} - \dfrac{1}{\rho_0} \right] \dfrac{1}{\rho_{ob}^2(s)} n_0 & \rho(s) \le \rho_0 \\ 0 & \rho(s) > \rho_0 \end{cases} \tag{7}$$

3.2   Markov Decision Process Model

The next path decision of robots for each state can be converted into a Markov Decision process (MDP, Markov Decision the Processes), the model may be expressed through the following:

$$MDP = \langle S, A(x), P_{xy}(a) \bullet r_{xy}(a), V \rangle \tag{8}$$

S is all of un-empty set of all possible states of A robot, also known as the system state space. x and y are as to states. x_t is as the state in time t. To $x \in S$, $A(x)$ is in the state of the available actions (decision) of the state x. it is un-empty. The available actions set of all the status expressed as $A = \cup_{x \in S} A(x)$. $P_{xy}(a)$ is the probability that the state x transfer to y by action a. $r_{x_t, x_{t+1}}(a)$ also simplified as r_t.

V is as the criterion function which a variety of learning. The expected total discount pay for unlimited period is toke here:

$$V(x) = E \left[\sum_{n=0}^{\infty} \gamma^n \bullet r_{t+n} \right] \tag{9}$$

Where, γ is the discount factor.

4   Multi-step Reinforcement Learning Algorithm

4.1   Algorithm Analysis

Because of incomplete state information of the robot working space, Q learning is a kind of effective reinforcement learning method that solve such incomplete information Markov decision problem. Q learning obtain the optimal strategy by maintaining a state-action space on the value function to. Based on Behrman formula is:

$$Q^{\pi}(x, a) = R(x, a) + \gamma \bullet \sum_{y \in S} P_{xy}(a) \bullet V^{\pi}(y) \tag{10}$$

$R_x(a) = \sum_{y \in S} P_{xy}(a) \bullet r_{xy}(a)$ is expectation of a transfer payment from state x took action a. It also known as reward of instant state expected.

Reinforcement learning goals is the optimal strategy π^* of maximize long term accumulated discount returns under the MDP model and the expected return R position:

$$\pi^*(s) = \arg\max_{a \in A} Q^*(s, a) \tag{11}$$

Where, $Q^*(s, a)$ is the optimal value function.

For work scene that lack of global information, it is difficult to get the optimal strategy. And Q learning algorithm used the current step information that is slow convergence speed. The improved incremental multi-step algorithm $Q(\lambda)$ used several steps to update the current information in the future. In actual scene, robots are difficult to predict future state. And the algorithm need to upgrade state-action of space matrix $|S| * |A|$ when is implemented. When the space is larger, the efficiency is low. Based on this, a compromise strategy is used by combining with the status of the robot itself information, In each state, it is used of step update current value Q in the future k dynamic according to their own environment information. The update rule for:

$$Q_t(x, a) = \begin{cases} Q_{t-1}(x, a) + a_t \bullet \left[e'_t + \sum_{t=1}^{k} (\lambda \gamma)^t \bullet e_{t+i} \right], & \text{if } x = x_t, a = a_t \\ Q_{t-1}(x, a), & \text{others} \end{cases} \tag{12}$$

Where, $e_t = r_t + \gamma \bullet V_{t-1}(x_{t+1}) - V_{t-1}(x_t)$, $e'_t = r_t + \gamma \bullet V_{t-1}(x_{t+1}) - Q_{t-1}(x_t)$.

Based on the status of the robot in environment, it adaptively adjusted value k. When there are big obstacles in the surrounding environment, it adopt long value k. It used more future information to optimize learning effect. When obstacles targets few, value k adopt small, and made robot moving fast.

In order to improve the convergence speed, Q value is initialized through a priori knowledge of robots to environment. It can improve the learning efficiency in initial stage and accelerate convergence speed. Initialized Q can used as the state space of the corresponding state value of the potential energy field. The relation is as follows:

$$Q_0(x_i, a) = r + \gamma |U'(x_i)| \tag{13}$$

Where, $|U'(x_i)|$ is potential energy value x_i of virtual potential energy field according to the known environment information formed.Steps of improved Q algorithm based on reinforcement learning to robot path planning are as follows:

(a) Construct the virtual potential field according to the target position information and current status information.

(b) Define the initial value according to the previously defined virtual potential field. And initialized the Q value according to the formula (7):

$$Q_0(x_i, a) = r + \gamma V_0(x_j) \tag{14}$$

(c) Adopt a value k according to current state x_i. Then select an action a and executed. Environmental status is updated for the new state x_{i+1}. And receive immediate return r.

(d) Observed the new state.

(e) Update table item value $Q(x, a)$ as follow:

$$Q_t(x, a) = \begin{cases} Q_{t-1}(x, a) + a_t \bullet \left[e'_t + \sum_{t=1}^{k} (\lambda\gamma)^t \bullet e_{t+i} \right], & \text{if } x = x_t, a = a_t \\ Q_{t-1}(x, a), & \text{other} \end{cases} \tag{15}$$

(f) Judged a robot reached the goal or reached the biggest number of learning set. If it's one or the other, meet the end of the study, or to return to the step (c) continue to learn.

It can explain reinforcement learning process by 25 states work space. Firstly, a virtual potential field was built as shown in Fig. 1. Each node corresponds to the discrete state of the current position. Red node represents the target state. Light gray nodes represent obstacles. Arrows represent their movements. Robots choose appropriate paths in each state between states transition by proposed algorithm until reach the end point 1.

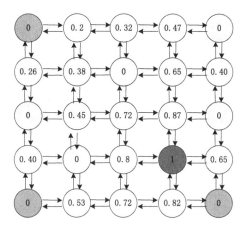

Fig. 1. It shown the state potential energy value and the initial Q value $Q_0(x, a)$ within robot target neighborhood (Color figure online)

4.2 Convergence Analysis

According to convergence theorem of the standard Q learning algorithm, there is a conclusion as:

Convergence Theorem: Given bounded reinforcement signal $|r_t| \leq R$, $0 \leq a_t \leq 1$, $\sum_{t=1}^{\infty} a_t(x, a) = \infty$, $\sum_{t=1}^{\infty} [a_t(x, a)]^2 \leq \infty$, $\forall x, a$. When $t \to \infty$, $Q_t(x, a)$ convergence in probability 1 and can obtained $Q_t^*(x, a)$.

When the target state and the current state is determined, the initial value is the only. The learning process is a deterministic Markov decision process. The pay is bounded in learning process. There is $(\forall x, a)|r(x, a)| \leq 1$, $\alpha(0 \leq \alpha < 1)$, $\gamma(0 \leq \gamma < 1)$, $\sum_{t=1}^{\infty} a_t(x, a) = \infty$. The convergence theorem conditions met, the algorithm certainly is convergent.

5   Simulation Analysis

5.1   Environment Modeling and Task Description

In order to verify the validity and advantage of the proposed algorithm, it took some experiment that robots moved in a two-dimensional grid map. The virtual space made by $M \times N = 20 \times 30$ grids. Each grid cell is square. Robots in any states can select a direction of four to movements. G is the target. It supposed that a number of obstacles with different sizes random deployed in maps. Pieces of a hollow square are free area for robots. Solid square area is obstacles. Walls are around the space in Fig. 2. Robots can't through the wall and obstacles. Assume that the robot can detect obstacles in the surrounding environment.

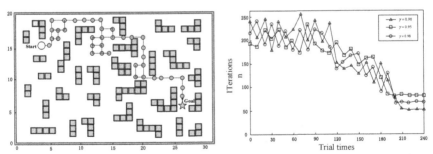

(a) Robot path planning based on standard Q-learning algorithm   (b) Standard Q-learning algorithm convergence process

Fig. 2. Simulation results of robot path planning of standards Q-learning algorithm

In experiments, learning process parameters settings are as follows. The biggest transfer steps for 150 times. The maximum number of iterations is for 2000 times. If the standard error of a robot tried 10 consecutive iterations is less than 0.25, the algorithm convergence. Parameters settings of virtual potential field and reinforcement learning are as follows (Table 1):

Table 1. Related parameters list of simulation

Parameter	k_a	k_r	ρ_0	α	γ	ε
Value	1.5	1.2	2.0	0.3	0.95	0.5

It took some experiments for every algorithm within the same environment condition to verify the proposed improved reinforcement learning algorithm. Experiment in each group randomly to run 50 times, and then shown by averaging the results.

5.2 Simulation and Experiments

Simulation results of standard Q-learning algorithm are shown in Fig. 2. Q value is initialized to 0. Robots start from the initial position Start. Then, mobile robot reached to target Goal by the shortest path. A small circle of continuous are represented of robots planning path as shown in Fig. 2. Figure 2(b) is convergence process based on the traditional Q-learning algorithm. Reinforcement learning algorithm is convergence after 183 attempts. At the beginning of the study, robot almost cannot reach the target in the maximum number of iterations. The reason is that the Q value is initialized to 0. Without any prior knowledge, the robot can only choose at random. It lead to the low efficiency of the initial study phase. The algorithm convergence speed is slow.

It also took some simulation on improved Q-learning algorithm. Using the proposed virtual potential field on environment map is described as shown in Fig. 3. According to the type (7)–(9), the Q value is initialized. Then it made robots learning in a new environment. Assume that the map size information and initial target location information is known. Obstacles random distributed indoor. Related information of obstacles are unknown. When robots moved a place, they can detect the surrounding environment information. Each potential energy value of the environment is defined as the biggest cumulative returns the state can get. Then, according to the current state of

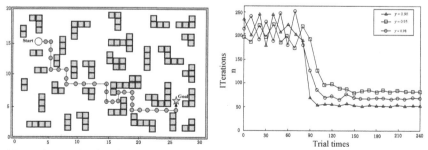

(a) Robot path planning based on improved Q-learning (b) Improved Q-learning algorithm convergence process

Fig. 3. Simulation results of robot path planning of improved Q-learning algorithm

returns immediately and the subsequent state of maximum equivalent cumulative returns, Q value is initialized. The priori knowledge about the environment is passed to the robot by this way. The robot has better learning foundation.

Convergence process of improved Q-learning algorithm for robot reinforcement learning is as shown in Fig. 3(b). The convergence rate of the improved algorithm obviously improved the learning process. The algorithm is convergence after 91 attempts. Basically it can reach the target within the largest number of iterations. It can be found that Q initialization method effectively improve the learning efficiency of the algorithm and the initial stage as shown in Fig. 3. It significantly improved the performance of the reinforcement learning algorithm for robot path planning (Fig. 4).

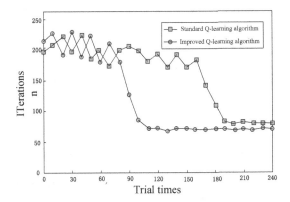

Fig. 4. Iteration times standard deviation of two Q-learning algorithms.

6 Summary and Outlook

The standard reinforcement learning algorithm for robot path planning needs to robots perfect forecasting results of any state transition immediately. It is very different in practice. Combined with the known part of environmental information to improve the Q algorithm, it proposed a robot path planning based on virtual potential field reinforcement learning algorithm. Combined with virtual potential field model and improved Q-learning algorithm, it realized mobile path planning with efficiently. Experiments show that Q initialization effectively improved the learning efficiency in the initial stage of the algorithm.

Acknowledgements. This work was supported in part by national natural science foundation of china (61602187), Science and Guangdong science and technology project (2016A040403122), Guangzhou science and technology project-science research project, (201707010482).

References

1. Scaramuzza, D., Achtelik, M.C., Doitsidis, L., et al.: Vision-controlled micro flying robots: from system design to autonomous navigation and mapping in GPS-denied environments. IEEE Robot. Autom. Mag. **21**(3), 26–40 (2014)

2. Chinnaiah, M.C., Savitri, T.S., Kumar, P.R.: A novel approach in navigation of FPGA robots in robust indoor environment. In: 2015 International Conference on Advanced Robotics and Intelligent Systems (ARIS), vol. 21, pp. 1–6, May 2015

3. Song, X., Fang, H., Jiao, X., Wang, Y.: Autonomous mobile robot navigation using machine learning. In: 2012 IEEE 6th International Conference on Information and Automation for Sustainability (ICIAfS), vol. 21, pp. 135–140, September 2012

4. Zuo, B., Chen, J., Wang, L., Wang, Y.: A reinforcement learning based robotic navigation system. In: 2014 IEEE International Conference on Systems, Man and Cybernetics (SMC), pp. 3452–3457, 5–8 October 2014

5. Raja, P., Pugazhenthi, S.: Optimal path planning of mobile robots: a review. Int. J. Phys. Sci. **7**(9), 1314–1320 (2012)

6. Niewola, A., Podsedkowski, L.: Nonholonomic mobile robot path planning with linear computational complexity graph searching algorithm. In: 2015 10th International Workshop on Robot Motion and Control (RoMoCo), pp. 217–222. IEEE (2015)

7. Kong, J., Ding, M., Li, X., et al.: An algorithm for mobile robot path planning using wireless sensor networks. In: 2015 IEEE International Conference on Mechatronics and Automation (ICMA), pp. 2238–2242. IEEE (2015)

8. Wu, Z., Feng, L.: Obstacle prediction-based dynamic path planning for a mobile robot. Int. J. Adv. Comput. Technol. **4**(3), 23–31 (2012)

9. Hussein, A., Mostafa, H., Badrel-din, M., et al.: Metaheuristic optimization approach to mobile robot path planning. In: 2012 International Conference on Engineering and Technology (ICET), pp. 1–6. IEEE (2012)

10. Lee, D.W., Seo, S.W., Sim, K.B.: Online evolution for cooperative behavior in group robot systems. Int. J. Control Autom. Syst. **6**(2), 282–287 (2008)

11. Schaal, S., Atkeson, C.: Learning control in robotics. IEEE Robot. Autom. Mag. **17**(3), 20–29 (2010)

12. Liu, C., Xu, X., Hu, D.: Multiobjective reinforcement learning: a comprehensive overview. IEEE Trans. Syst. Man Cybern. Part C Appl. Rev. **99**(4), 1–13 (2013)

13. Andersen, K.T., Zeng, Y., Christensen, D.D., et al.: Experiments with online reinforcement learning in real-time strategy games. Appl. Artifi. Intell. **23**(9), 855–871 (2009)

14. Lin, L., Xie, H., Zhang, D.: Supervised neural Q-learning based motion control for bionic underwater robots. J. Bionic Eng. **7**(Sup), 177–184 (2010)

15. Hwang, K.S., Lin, H.Y., Hsu, Y.P., et al.: Self-organizing state aggregation for architecture design of Q-learning. Inf. Sci. **181**(13), 2813–2822 (2011)

16. Santos, M., Martin, H.J.A., Lopez, V., et al.: Dyna-H: a heuristic planning reinforcement learning algorithm applied to role-playing game strategy decision systems. Knowl.-Based Syst. **32**(1), 28–36 (2012)

17. Chen, C., Dong, D., Li, H., et al.: Hybrid MDP based integrated hierarchical Q-learning. Sci. China Inf. Sci. **54**(11), 2279–2294 (2011)

18. Grezs, M., Kudenko, D.: Online learning of shaping rewards in reinforcement learning. Neural Netw. **23**(4), 541–550 (2010)

19. Konidaris, G., Scheidwasser, I., Barto, A.G.: Transfer in reinforcement learning via shared features. J. Mach. Learn. Res. **13**(1), 1333–1371 (2012)

20. Megherbi, D.B., Kim, M.A.: Collaborative distributed multi-agent reinforcement learning technique for dynamic agent shortest path planning via selected sub-goals in complex cluttered environments. In: 2015 IEEE International Inter-Disciplinary Conference on Cognitive Methods in Situation Awareness and Decision Support (CogSIMA), pp. 118–124. IEEE (2015)

A Novel Progressive Secret Image Sharing Method with Better Robustness

Lintao Liu$^{(\boxtimes)}$, Yuliang Lu, Xuehu Yan$^{(\boxtimes)}$, and Wanmeng Ding

Hefei Electronic Engineering Institute, Hefei 230037, China
`liuta1989@163.com`, `publictiger@126.com`

Abstract. Secret image sharing (SIS) can be utilized to protect a secret image during transmit in the public channels. However, classic SIS schemes, e.g., visual secret sharing (VSS) and polynomial-based scheme, are not suitable for progressive encryption of greyscale images in noisy environment, since they will result in different problems, such as lossy recovery, pixel expansion, complex computation, "All-or-Nothing" and robustness. In this paper, a novel progressive secret sharing (PSS) method based on the linear congruence equation, namely LCPSS, is proposed to solve these problems. LCPSS is simple designed and easy to realize, but naturally has many great properties, e.g., (k, n) threshold, progressive recovery, lossless recovery, lack of robustness and simple computation. Experimental results are given to demonstrate the validity of LCPSS.

Keywords: Secret sharing · Progressive secret sharing · Greyscale image · Linear congruence · Robustness

1 Introduction

Secret sharing (SS) is a significant branch in cryptography. By means of secret sharing techniques, the secret could be divided into several shares without any knowledge about secret information, and further the secret could be reconstructed if and only if sufficient shares participate in recovery. In comparison with the traditional cryptographic approaches, SS schemes have several advantages: (a) no need for key management, (b) loss-tolerant characteristic, (c) access control.

Along with the improvement of awareness of privacy and security, multimedia security (including images, audios and videos) has become an important research area combining methods and techniques coming from cryptography. Currently, there are many theoretical researches on how to utilize SS techniques to achieve security protection for images, and most of them are based on two basic SS methods: polynomial-based secret sharing and visual cryptography (VC).

Polynomial-based secret sharing was first proposed by Shamir [15] in 1979. In the beginning, this method was utilized to encrypt secrets in the integer field, and then Thien and Lin [17] proposed a specific method to achieve (k, n) threshold secret image sharing (SIS) scheme with smaller shares, the size of which is $\frac{1}{n}$

© Springer Nature Singapore Pte Ltd. 2017
B. Zou et al. (Eds.): ICPCSEE 2017, Part II, CCIS 728, pp. 539–550, 2017.
DOI: 10.1007/978-981-10-6388-6_46

times the secret image. Moreover, polynomial-based SIS schemes could be suitable for almost all types of images including binary, grayscale and color images. Hence, polynomial-based SIS schemes began developing rapidly, and they were designed to achieve several great features, such as smaller shadow images [19], lossless recovery [4,25] and meaningful shares combined with steganography [20]. However, these schemes have three vital drawbacks as follows.

- Complex computation. The algorithm complexity for decryption is $O(nlog^2n)$.
- Extra information for decryption, such as the serial number of the corresponding image.
- "All-or-Nothing". Either the original secret image or the noise-like image will be reconstructed for decryption.

Visual cryptography (VC) or visual secret sharing (VSS) was first proposed by Naor and Shamir [14] in 1994. Initial VC schemes were achieved based on OR operation, and their significant advantage is stacking-to-see property, that is, the secret image could be decrypted by stacking sufficient shares together and human visual system (HVS) with neither knowledge of cryptography nor complex computation. In order to reduce the pixel expansion of shadow images and improve the visual quality of reconstructed images, several basic VSS methods have been developed, including probalistic VSS (ProbVSS) [24], XOR-based VSS (XVSS) [13,18] and VSS based on random grids (RGVSS) [8,21]. At the same time, some researchers are devoted to the development of more interesting properties for VSS, e.g. general threshold [21], general access structure [1,16], meaningful shares [12,23], multiple secrets sharing [11], greyscale image sharing [2] and so on. In spite of rich theory achievements of VSS, they are not satisfactory from a practical point of view, and their drawbacks are listed as follows.

- Design for binary images, and be not suitable for digital images.
- Bad visual quality of reconstructed images and lossy recovery.
- Limited application. A pile of transparencies must be carried with you.

Along with the development of basic SS schemes, many exciting and useful properties are developed, such as progressive recovery. Progressiveness property, or so-called perceptual property [22], requires that the visual quality of media data could be controled by a factor which can be determined by the senders. Currently, there are many progressive secret sharing (PSS) schemes based on different basic methods, such as VCS [9], RGVSS [5], Shamir's polynomial-based SS [6] or transformation [10]. However, they overall suffer from the same disadvantages as corresponding basic methods.

Robustness is another useful properties of SS. Assumpt that the grayscale shadow images were printed on papers, smart phones were utilized for information acquisition and processing instead of eyes and brains. In comparison with digital images in storage device, papers are undoubtedly more safe and convenient during public transit. However, photos captured by cameras will be affected by the noise inevitably, hence it is important to restrust the recognizable secret images with noise shares. VSS schemes naturally have this kind of robustness property, and there are many researches about robustness based on polynomial-based SS [3,7]. However, none of them can satisfy the requirements in the given scenario above, where each of pixels in every shadow image will change because of the noise during transmission.

In this paper, a novel (k, n) threshold PSS method based on the linear congruence equation, namely LCPSS, is introduced. LCPSS makes up the shortage of classic SIS schemes which are applied in handling grayscale images, but reserves many good properties. LCPSS has large number of valuable features, including that: (a) design for the grayscale image, (b) no leakage of secret information with less than k shares, (c) progressive recovery with k or more shares, (d) lossless recovery with all shares, (e) robustness, (f) simple computation (needs only addition and module operations). Besides, LCPSS also has other advantages, such as no pixel expansion, alternative order of shadow images in recovery, and no need of extra information for recovery. Simulations are given to verify the security and effectiveness of these schemes.

2   Preliminaries

In the view of secret sharing, threshold means whether or not secret information could be recognized, while progressive property represents the improvement of visual quality of the reconstructed images. Progressiveness has the potential application in the multimedia area, where the visual quality needs to be controled by a factor. For example, in many business applications like pay-per-view videos, Pay-TV/Music, art-work image vending, and video on demand (VOD), progressiveness makes it possible for potential users to view low-quality copies of the media data products before buying them. By controlling the number of shares sent to users, current SS schemes with special construction may effectively achieve the goal of visual quality control. The definition of progressive secret sharing (PSS) is given as Definition 1.

Definition 1 (PSS). *A secret sharing scheme is progressive on t in interval $[T_1, T_2]$, when $T_1 \leq t_1 < t_2 \leq T_2$, $f(x, t_1) < f(x, t_2)$, where $f(x, t)$ denotes the visual quality of the recovered secret image with t shadow images and x indicates other parameters, such as k or n.*

Hou et al. [9] concluded that there are two basic PSS paradigms, which could be briefly summarized as follows: one is based on the whole secret image, while the other is based on image blocks which make up the whole secret image. Suppose each pixel in the secret image is encrypted independently. The former means that every shared pixel value approaches the corresponding secret value by a certain ratio with more shares participating in recovery. The other one implies that more corresponding pixels in certain image blocks are recovered precisely with one more shares, while pixels in other regions still keep noise-like. The latter is also named as block-based approach or region incrementing visual cryptography (RIVC).

LCPSS belongs to RIVC, but every image block is divided by the granularity of single pixel. That is, when one more share joins for recovery, more pixels are recovered precisely, meanwhile other pixels, whose reconstruction condition can't be satisfied, are still random value. Therefore, we introduce a simple evaluation criterion, namely Correct Recovery Probability (CRP), which can be used for theoretical proofs of the security and progressive property.

Definition 2 (Correct Recovery Probability). *Correct Recovery Probability (CRP), refers to the probability of precise recovery of a pixel, that is* $P_{y'(i,j)=y(i,j)}$. *CRP of a recovered image is the ratio of precise recovered pixels to all pixels.*

3   General Threshold LCPSS for the Grayscale Image

3.1   The Basic Idea

The proposed basic PSS method is based on the linear congruence equation as Eq. 1, so we call it LCPSS for short. Here, y means the secret value for input in encrypt process, while $x_1, x_2, \cdots, x_n$ represent a set of shared values for output. p should be a random positive integer which is to form a field and make related values in the same value range.

$$(x_1 + x_2 + \cdots + x_i + \cdots + x_k) \bmod p = y \tag{1}$$

(k, k) threshold LCPSS derives from Eq. 1. In the encryption process, random-number generation algorithm is used to obtain a set of shared values, which satisfies Eq. 1 with a known secret value y. According to Eq. 2, we just need to add the collected shared values $x_{i_1}, x_{i_2}, \cdots, x_{i_t}$ and then modulo the same p for decryption. In (k, k) threshold LCPSS, any secret will not be revealed with less than k shared values, while the recovered value y' equals to the original secret value y with all k shared values.

$$y' = (x_{i_1} + x_{i_2} + \cdots + x_{i_t}) \bmod p \tag{2}$$

The other problem is how to extend (k, k) threshold LCPSS to general (k, n) threshold LCPSS. In RGVSS, $(n - k)$ extra shared value are set to zero element "0" to achieve threshold property. On consideration, this zero-padding method can be also utilized in LCPSS. Although zero elements don't carry any useful information for our scheme, it can control the possibility for collecting the valid set, which consists of k valid shared values generated in (k, k) threshold LCPSS: the valid set will not be obtained until collecting k share values; with more shared values collected, the chance for obtaining the valid set increases. So (k, n) threshold LCPSS can be achieved, and the secret value y will be recovered precisely with all shared values.

3.2 Algorithm

In fact, LCPSS is a novel secret sharing scheme for number domain, and it can be directly utilized to encrypt the greyscale image. Considering that the grayscale image allows 256 different intensities (i.e., shades of gray) to be recorded, "p" in Eqs. 1 and 2 should be assigned to 256. The basic equation for sharing the grayscale image is shown as Eqs. 3 and 4, and the encryption and decryption algorithms are listed as Algorithms 1 and 2, respectively.

$$(x_1 + x_2 + \cdots + x_i + \cdots + x_k) \bmod 256 = y \qquad (3)$$

$$y' = (x_{i_1} + x_{i_2} + \cdots + x_{i_t}) \bmod 256 \qquad (4)$$

Algorithm 1. The encryption algorithm of general threshold LCPSS for the greyscale image

---

Input: (1) a $M \times N$ original grayscale secret image: S, (2) the required number of shares: n, (3) threshold: k.
Output: n shadow images $SC_1, SC_2, \cdots SC_n$.
Step1: For each position $(i, j) \in \{(i, j) | 1 \leq i \leq M, \ 1 \leq j \leq N\}$, repeat Steps 2–5;
Step 2: Let $y = S(i, j)$, and randomly generate the shared values $x_1, x_2, \cdots, x_n$ which satisfy Eq. 3.
Step 3: Determined whether $k == n$: if $k \neq n$, then go to Step 4; if $k = n$, then go to Step 5.
Step 4: Set the last $n - k$ shared values to zero element, e.g., $x_{k+1} = x_{k+i} = \cdots = x_n = 0$.
Step 5: Randomly distribute n shared values to n corresponding pixels of shadow images, e.g., $SC_p(i, j) = x_q, (1 \leq p \leq n; \ 1 \leq q \leq n; \ p, q \in Z^+.)$
Step 6: Output n shadow images $SC_1, SC_2, \cdots SC_n$.

Algorithm 2. The decryption algorithm of general threshold LCPSS for the greyscale image

---

Input: t shadow images SC_i of size $M \times N$ ($1 \leq i \leq t$).
Output: a $M \times N$ recovered secret image: S'.
Step1: For each position $(i, j) \in \{(i, j) | 1 \leq i \leq M, 1 \leq j \leq N\}$, repeat Steps 2–3;
Step 2: let $x_1 = SC_1(i, j), \ldots, x_t = SC_t(i, j)$, and compute the recovered values y' as Eq. 4.
Step 3: Assign the recovered value y' to the corresponding pixel of the recovered image, e.g., $S'(i, j) = y'$.
Step 4: Output the recovered image S'.

---

4   Experiments and Comparisons

In this section, experiments and analysis are conducted to evaluate the effectiveness of LCPSS. Firstly, the results of experiments are provided in order to illustrate the effectiveness of LCPSS intuitively. Secondly, security and progressiveness are analyzed with statistical results in detail. Finally, comparisons with polynomial-based SIS scheme are given to demonstrate robustness of LCPSS.

4.1   Image Illustration for General Threshold LCPSS

Considering that there are differences in algorithm design between (k, k) and (k, n) threshold LCPSS, simulation results of two schemes are provided, respectively. Experimental images could illustrate the effectiveness intuitively, while corresponding histograms draw the probability distribution of pixel values from the view of statistic analysis. For simplicity, only one image, such as only one share like Fig. 1(b) SC_1, is chosen out of the corresponding set of images.

Experimental results of $(3, 3)$ threshold LCPSS with the secret image "Lena" are shown as Fig. 1. In Fig. 1, both the shadow image as Fig. 1(b) and the recovered image with insufficient shares as Fig. 1(c) look like noise images and cannot reveal any secret information. When all shares participate in recovery as Fig. 1(c), the secret image can be reconstructed precisely.

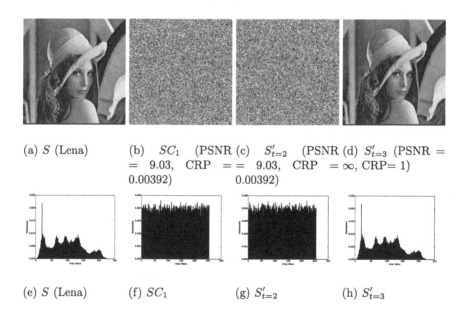

(a) S (Lena)   (b) SC_1 (PSNR (c) $S'_{t=2}$ (PSNR (d) $S'_{t=3}$ (PSNR = = 9.03, CRP = = 9.03, CRP = ∞, CRP= 1) 0.00392)   0.00392)

(e) S (Lena)   (f) SC_1 (g) $S'_{t=2}$ (h) $S'_{t=3}$

Fig. 1. The simulation results of the $(3, 3)$ threshold LCPSS using Lena. (a) The secret image S; (b) the shadow image SC_1; (c) the recovered image with 2 shares; (d) the recovered image with 3 shares; (e)–(h) are the corresponding histograms of above images.

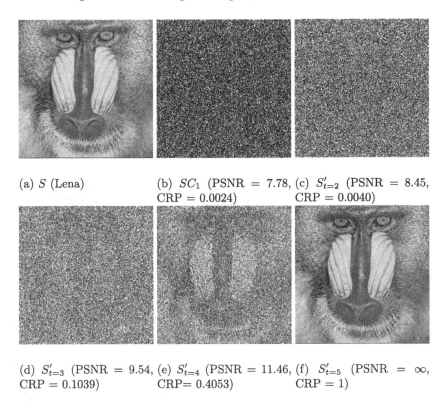

(a) S (Lena)　　(b) SC_1 (PSNR = 7.78, (c) $S'_{t=2}$ (PSNR = 8.45, CRP = 0.0024)　　CRP = 0.0040)

(d) $S'_{t=3}$ (PSNR = 9.54, (e) $S'_{t=4}$ (PSNR = 11.46, (f) $S'_{t=5}$ (PSNR = ∞, CRP = 0.1039)　　CRP= 0.4053)　　CRP = 1)

Fig. 2. The simulation result of the $(3, 5)$ threshold LCPSS using Baboon. (a) The secret image S; (b) one shadow image SC_1; (c)–(f) the recovered images with different number of shares, $t = 2, 3, 4$ and 5.

From Fig. 1(e)–(h), the same result can be concluded: the corresponding probability distributions of "SC_1" and "$S'_{t=2}$" follow uniform distribution, which means the information entropy of these images is max. Basing on the statistical analysis, it is unconditional security with insufficient shares.

"Baboon" is chosen as the secret image to verify the effectiveness of $(3, 5)$ threshold LCPSS. With the number of shares less than threshold, that is $t < k$, Fig. 2(b) and (c) are noise-like but more black than normal noise-like images like Fig. 1(b) due to zero-padding operation. From Fig. 2(d) with $t = k = 3$, the secret image begins to apear, but we can only distinguish the rough outline of baboon's face. The secret image becomes more clear and more details on baboon's face can be recognized in Fig. 2(e). With all shares, lossless recovery is achieved as Fig. 2(f).

As shown in Fig. 3(b) and (c), the probability of "0" is much larger than other values due to zero padding, while the probability of other values from "1" to "255" is equal but smaller. Generally, this kind of probability distribution should be security, and more zero elements can only darken the share images.

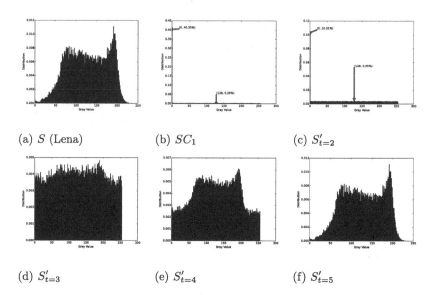

(a) S (Lena)   (b) SC_1 (c) $S'_{t=2}$

(d) $S'_{t=3}$ (e) $S'_{t=4}$ (f) $S'_{t=5}$

Fig. 3. The statistical histograms of experimental images for $(3, 5)$ threshold LCPSS. (a) The secret image S; (b) one shadow image SC_1; (c)–(f) the recovered images with different number of shares, $t = 2, 3, 4$ and 5.

When $t = 3$ and $CRP = 0.1039$, the outline of Fig. 3(d) becomes a little similar to that of Fig. 3(a): the peaks approximately apear in 85 and 190. However, the difference of probabilities of different pixel values is still small, Fig. 2(d) is still difficult to be recognized. With the number of shares increasing, as shown in Fig. 3(f) and (g), the outline of histogram become more and more similar to Fig. 3(a).

Based on experimental results above, the properties of threshold LCPSS can be concluded intuitively as follows.

1. **Security:** with less than k shares, no secert information can be revealed.
2. **Visual recognition:** with k or more shares, the reconstructed images can be visually recognized.
3. **Progressiveness:** with k or more shares, the visual quality of reconstructed images increases as the number of shares increases.
4. **Precise recovery:** with all shares, the reconstructed images is the same as the secret image.

4.2   Performance Analysis Based on Statistical Data

In this section, the significant features of LCPSS, such as security and progressiveness, will be demonstrated with sufficient simulations and statistic analysis in detail. Here, "Lena" is chosen as the test image, whose size is 256×256. Considering that there are totally 65536 pixels in the test image, it is large enough for the performance proof.

Table 1. The CRP and average PSNR between the secret image and recovered images as k, n and t change.

(k,n)	$t = 1$		$t = 2$		$t = 3$		$t = 4$		$t = 5$	
	CRP	PSNR	CRP	PSNR	CRP	PSNR	CRP	PSNR	CRP	PSNR
$(2,2)$	0.0040	8.45	1	∞						
$(2,3)$	0.0030	7.78	0.3364	10.00	1	∞				
$(3,3)$	0.0040	8.45	0.0040	8.45	1	∞				
$(2,4)$	0.0020	7.78	0.1693	8.64	0.5020	11.46	1	∞		
$(3,4)$	0.0028	7.78	0.0039	8.45	0.2531	9.54	1	∞		
$(4,4)$	0.0037	8.45	0.0040	8.45	0.0041	8.45	1	∞		
$(2,5)$	0.0015	6.99	0.1022	8.45	0.3022	9.54	0.6015	12.30	1	∞
$(3,5)$	0.0025	7.78	0.0036	7.78	0.1035	8.45	0.4022	10.41	1	∞
$(4,5)$	0.0032	7.78	0.0040	8.45	0.0040	8.45	0.2031	9.03	1	∞
$(5,5)$	0.0041	8.45	0.0040	8.45	0.0039	8.45	0.0039	8.45	1	∞

In Table 1, CRP and PSNR are used as the assessment criteria to prove the effectiveness of LCPSS, including security and progressiveness. The column named "CRP" represents the experimental value of CRP, while the column named "$PSNR$" means the experimental value of CRP. When $t < k$, CRP is less than or approximately equal to $\frac{1}{256}$, and $\frac{1}{256}$ is equal to CRP of two completely unrelated images, hence the security is proved. When $t = k$, a noticeable improvement of CRP could be seen, and CRP increases monotonously as t increases. In this case, the secret information could be recognized and become more and more clearly, hence visual recognization and progressiveness can be guaranteed. With all shares, the secret image will be recovered precisely. On the other hand, the same results can be obtained from the change rule of $PSNR$, and further verify the relationship between CRP and visual quality of recovered images.

4.3   Robustness Analysis

In order to test the robustness of LCPSS, we add Gaussian noise to the shadow images, obtain the reconstructed images recovered with these noise shares, and compare them with the normal reconstructed images.

In Fig. 4, Gaussian noise with $\mu = 0$ and $\sigma^2 = 10$ is added for test. Figure 4(a)–(d) are the normal reconstructed images recovered by threshold LCPSS, Fig. 4(e)–(h) are the reconstructed images with noise shares by LCPSS, and Fig. 4(i)–(l) are the recovered images with noise shares using Shamir's polynomial-based SS scheme.

For LCPSS, with insufficient shares, recovered images are still noise-like, hence this kind of simulation results is not listed here. With k or more shares, as shown in Fig. 4(e)–(h), although visual quality of recovered images reduces by the

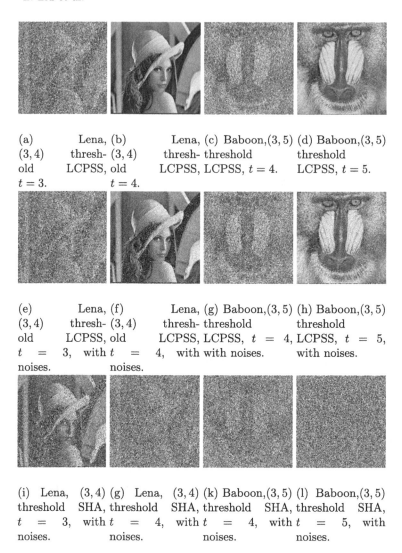

(a)   Lena, (b)   Lena, (c) Baboon,(3,5) (d) Baboon,(3,5)
(3,4)   thresh- (3,4) thresh- threshold   threshold
old   LCPSS, old   LCPSS, LCPSS, $t = 4$. LCPSS, $t = 5$.
$t = 3$. $t = 4$.

(e)   Lena, (f)   Lena, (g) Baboon,(3,5) (h) Baboon,(3,5)
(3,4)   thresh- (3,4) thresh- threshold   threshold
old   LCPSS, old   LCPSS, LCPSS, $t = 4$, LCPSS, $t = 5$,
$t = 3$, with $t = 4$, with with noises.   with noises.
noises. noises.

(i) Lena,   (3,4) (g) Lena,   (3,4) (k) Baboon,(3,5) (l) Baboon,(3,5)
threshold   SHA, threshold   SHA, threshold   SHA, threshold   SHA,
$t = 3$, with $t = 4$, with $t = 4$, with $t = 5$, with
noises.   noises.   noises.   noises.

Fig. 4. The simulation results for robustness comparison. (a)–(d) The normal reconstructed image using LCPSS; (e)–(h) the reconstructed images with noisy shares using LCPSS; (i)–(l) the reconstructed images with noisy shares using Shamir's polynomial-based SS scheme.

noise, the secret information could be still recognized and the basic properties, such as progressive, still remain. However, there exists a big error in certain pixels: some black pixels are transferred to white ones and vice versa. This is because Gaussian noise make part of recovered values "overflow" or "downflow", e.g. $y = 255, x_1 = 1, x_2 = 254, x_{1(noise)} = 3, x_{2(noise)} = 254, y' = 2$. So one of solutions to avoid "overflow" or "downflow" is to decrease the number of boundary values.

On the other hand, for polynomial-based SIS scheme, all the recovered images with k or more shares should have been the same as corresponding secret images, but in fact most of them could not even be recognized. Although the decryption of polynomial-based SIS scheme is based on solving the linear equations, small noises would be magnified several times in the decryption process.

In comparison with Shamir's polynomial-based SS scheme, LCPSS has better robustness and some trick can be utilized to improve the visual quality of recovered images in noisy environment. Therefore, it has a number of potential applicantions, where noises will be introduced during transmission.

5   Conclusion

This paper proposes a novel basic PSS method, namely LCPSS. The effectiveness of LCPSS has been proved via simulations. LCPSS possesses many great features similar to classic methods, such as: (a) no leakage of secret information with less than k shares, (b) lossless recovery with all shares, (c) simple computation (needs only addition and module). At the meanwhile, LCPSS also naturally has the interesting properties difficult to realize by traditional SIS methods, including: (a) progressive recovery, (b) robustness. Besides, other features are also realized, such as no pixel expansion, alternative order of shadow images and no need for extra information for decryption. Further theoretical proofs for security and progressiveness of LCPSS will be the future work.

Acknowledgement. The authors would like to thank the anonymous reviewers for their valuable comments. This work is supported by the National Natural Science Foundation of China (Grant Number: 61602491).

References

1. Ateniese, G., Blundo, C., De Santis, A., Stinson, D.R.: Visual cryptography for general access structures. Inf. Comput. **129**(2), 86–106 (1996)
2. Blundo, C., De Santis, A., Naor, M.: Visual cryptography for grey level images. Inf. Process. Lett. **75**(6), 255–259 (2000)
3. Cevallos, A., Fehr, S., Ostrovsky, R., Rabani, Y.: Unconditionally-secure robust secret sharing with compact shares. In: Pointcheval, D., Johansson, T. (eds.) EUROCRYPT 2012. LNCS, vol. 7237, pp. 195–208. Springer, Heidelberg (2012). doi:10.1007/978-3-642-29011-4_13
4. Chang, C.C., Hsieh, Y.P., Lin, C.H.: Sharing secrets in stego images with authentication. Pattern Recogn. **41**(10), 3130–3137 (2008)
5. Chen, S.K.: Friendly progressive visual secret sharing using generalized random grids. Opt. Eng. **48**(11), 117001 (2009)
6. Chen, S.K., Lin, J.C.: Fault-tolerant and progressive transmission of images. Pattern Recogn. **38**(12), 2466–2471 (2005)
7. Cramer, R., Damgård, I.B., Döttling, N., Fehr, S., Spini, G.: Linear secret sharing schemes from error correcting codes and universal hash functions. In: Oswald, E., Fischlin, M. (eds.) EUROCRYPT 2015. LNCS, vol. 9057, pp. 313–336. Springer, Heidelberg (2015). doi:10.1007/978-3-662-46803-6_11

8. Guo, T., Liu, F., Wu, C.: Threshold visual secret sharing by random grids with improved contrast. J. Syst. Softw. **86**(8), 2094–2109 (2013)

9. Hou, Y.C., Quan, Z.Y., Tsai, C.F., Tseng, A.Y.: Block-based progressive visual secret sharing. Inf. Sci. **233**, 290–304 (2013)

10. Huang, C.P., Hsieh, C.H., Huang, P.S.: Progressive sharing for a secret image. J. Syst. Softw. **83**(3), 517–527 (2010)

11. Li, P., Ma, P.J., Su, X.H., Yang, C.N.: Improvements of a two-in-one image secret sharing scheme based on gray mixing model. J. Vis. Commun. Image Represent. **23**(3), 441–453 (2012)

12. Liu, F., Wu, C.: Embedded extended visual cryptography schemes. IEEE Trans. Inf. Forensics Secur. **6**(2), 307–322 (2011)

13. Liu, F., Wu, C.K., Lin, X.J.: Colour visual cryptography schemes. IET Inf. Secur. **2**(4), 151–165 (2008)

14. Naor, M., Shamir, A.: Visual cryptography. In: De Santis, A. (ed.) EURO-CRYPT 1994. LNCS, vol. 950, pp. 1–12. Springer, Heidelberg (1995). doi:10.1007/BFb0053419

15. Shamir, A.: How to share a secret. Commun. ACM **22**(11), 612–613 (1979)

16. Shen, G., Liu, F., Fu, Z., Yu, B.: Perfect contrast XOR-based visual cryptography schemes via linear algebra. Des. Codes Crypt. **85**(1), 15–37 (2017)

17. Thien, C.C., Lin, J.C.: Secret image sharing. Comput. Graph. **26**(5), 765–770 (2002)

18. Tuyls, P., Hollmann, H.D., Van Lint, J.H., Tolhuizen, L.: XOR-based visual cryptography schemes. Des. Codes Crypt. **37**(1), 169–186 (2005)

19. Wang, R.Z., Su, C.H.: Secret image sharing with smaller shadow images. Pattern Recogn. Lett. **27**(6), 551–555 (2006)

20. Wu, X., Ou, D., Liang, Q., Sun, W.: A user-friendly secret image sharing scheme with reversible steganography based on cellular automata. J. Syst. Softw. **85**(8), 1852–1863 (2012)

21. Yan, X., Liu, X., Yang, C.N.: An enhanced threshold visual secret sharing based on random grids. J. Real-Time Image Process. 1–13 (2015)

22. Yan, X., Wang, S., El-Latif, A.A.A., Sang, J., Niu, X.: A novel perceptual secret sharing scheme. In: Shi, Y.Q., Liu, F., Yan, W. (eds.) Transactions on Data Hiding and Multimedia Security IX. LNCS, vol. 8363, pp. 68–90. Springer, Heidelberg (2014). doi:10.1007/978-3-642-55046-1_5

23. Yan, X., Wang, S., Niu, X., Yang, C.N.: Generalized random grids-based threshold visual cryptography with meaningful shares. Sig. Process. **109**, 317–333 (2015)

24. Yang, C.N.: New visual secret sharing schemes using probabilistic method. Pattern Recogn. Lett. **25**(4), 481–494 (2004)

25. Yang, C.N., Chen, T.S., Yu, K.H., Wang, C.C.: Improvements of image sharing with steganography and authentication. J. Syst. Softw. **80**(7), 1070–1076 (2007)

The NCC: An Improved Anonymous Method for Location-Based Services Based on Casper

Wenqi Liu[1(✉)], Mingyu Fan[1], Jie Feng[2], and Guangwei Wang[1]

[1] University of Electronic Science and Technology of China,
Chendu 610000, China
chinowen@foxmail.com
[2] Huawei Technologies Co., Ltd., Chendu 610000, China

Abstract. *Casper* Cloak is a privacy protection method based on *K*-anonymity algorithm. To be anonymous, *Casper* Cloak needs to search regional sibling and parent node, which requires a complex process and huge expenditure. In addition, the anonymous area has space redundancy and it is not accurate enough to achieve high Location-Based Services (LBS) quality. To address these problems, this paper proposes an improved privacy protection method—*NCC*, based on the *Casper* Cloak. To reduce the unnecessary search, *NCC* introduced the concept of the first sibling node. *NCC* also improves the LBS quality by considering the characteristics of user mobility. Moreover, the improved method, *NCC*, which is incorporated with a redundancy optimization processing strategy, realizing more precise in the anonymous area and accurately guaranteeing the related degree of privacy. Adopting *NCC* verification experiments reflects varied advantages as bellow: (1) By reducing 80% searching time, *NCC* highly improved searching process. (2) The anonymous area produced in *NCC* not only meet users' anonymous demands, but the direction of the mobility which improves 4 times accuracy of services in comparison with *Casper* mode. (3) According to optimization strategy, *NCC* can reach minimum anonymous area index, increasing the rates of anonymous optimization in original algorithm.

Keywords: Privacy protection · *Casper* cloak · Location-based services · Redundancy optimization

1 Introduction

Location-based applications use the positioning functions of a mobile device to determine the current location of a user and provide the user with services pertinent in the user's position [1–3]. Given the penetration of web-enabled mobile devices in the market, it is estimated to produce over 45 billion location-based service (LBS) queries daily in year of 2016. Note that the Swedish market research firm Berg Insight released the latest report predicts that the global LBS market size will be 22.5% compound annual growth rate (CAGR) from 30.3 billion euros in 2014 to 3.48 billion euros in 2020 [4, 5], the LBS-related revenues will be going forward. However, the actual query made to the LBS provider is tagged with the knowledge of location and other additional

© Springer Nature Singapore Pte Ltd. 2017
B. Zou et al. (Eds.): ICPCSEE 2017, Part II, CCIS 728, pp. 551–567, 2017.
DOI: 10.1007/978-981-10-6388-6_47

attributes about the query [6]. Examples of these case include traffic reports (*"is there any congestion on my way home within a quarter of an hour?"*), hospital finders (*"I got injured at home, where is the nearest hospital?"*) and so on. These information is often regarded as personal privacy [7]. Once the information leakage occurs, it will be a serious threat to the user.

The privacy breach in these cases has aroused public concerns, and a variety of methods have been proposed to protect the sensitive information. The *Casper* Cloak [12] which based on *K*-anonymity [8–11] algorithm proposed by Mokble is the most popular one. With the ability to deal with a mount of anonymous requests, *Casper* Cloak can quickly expand the anonymous area to meet the anonymous demands of different users simultaneously [13]. Besides, *Casper* is resistant to attacks and can protect users' privacy information in an effective way. Nevertheless, *Casper* Cloak has several disadvantages [14, 15]: (1) big spending on look-up table, which relies on the number of real-time users. When there are not enough other users in current grid in which user initiated the query, she must search for her sibling region or parent region to satisfy her anonymous needs. And it takes four look-table actions during a single checking operation, hence the question of how to simplify the checking operation may be open. (2) User access to the LBS by the geographic tag only identifying the real-time location coordinates of user, *Casper* doesn't take the characteristics of mobile user mobility into consideration while generates the anonymous area. Therefore, the anonymous area generated by such way cannot conform to the user's approximate direction of movement. (3) The size of anonymous area shows exponential growth with the times of checking operations. Although anonymity is well protected, it produces relatively high burden both in computing and storage capacity. Thus, putting forward a redundancy optimization processing strategy which can realize smaller and more precise in the anonymous area on the premise of guarantee related degree of privacy, remains an attractive problem.

To solve these deficiencies, several researches has been conducted to improve the *Casper* model. Based on the *Casper* model, Si, proposed a new privacy protection approach *V-grid* that considers the effects of users' mobility, making the anonymous area more accurate and LBS results more efficient [15]. Still, spatial redundancy is serious. And Jin, put forward a method, which can improve the accuracy of anonymous area [16]. The method which based on the quad-tree, uses the bottom-up approach in checking process, traversing the three subnodes of the same layer first, and then traversing the parent node upwards. Although the accuracy of the anonymous space is improved, the checking process is cumbersome and has the disadvantages of high cost and tardiness.

Based on the discussions above, we present an improved privacy protection method, named *NCC*, based on *Casper* Cloak, which can be more efficient in checking process, is both precise and simplified in the anonymous area on the premise of guaranteeing the related degree of privacy. Inspired by the [15], we modify the structure of the *GridTable* by introducing the concept of the first sibling node. Instead of inferring the adjacent sibling by retracing its parent node, we can immediately use first sibling to find the adjacent siblings by a simple reasoning. Thus, it can reduce the times of look-table and improve its efficiency. Since the *Casper* do not take into account the user mobility, it cannot achieve a high quality to meet user's actual situation in real-life. To address this problem, we also customize the expression of LBS query and modify the structure of *UserTable*, to consider the property of mobility when generate the anonymous area. Furthermore, the *NCC* which is incorporated with a redundancy optimization processing strategy, can realize more precise in the anonymous area on the premise of guaranteeing the related degree of privacy.

In addition, we have performed several experiments from the aspect of look-table time, cloaking cost, service accuracy and precision of anonymize requirement, which help to evaluate the efficiency and scalability of *NCC*. The experimental results prove that *NCC* reduces the time consumption of checking by almost 80%, increases the accuracy of services which is 4 times of the *Casper* and provides the effective quality assurance for LBS. The main contributions of this paper can be summarized as follows:

1. We introduce the concept of first sibling node, which can be used to infer its adjacent sibling nodes through a simple reasoning, to reduce the times of look-table and improve the efficiency of anonymous processing.
2. We take the user mobility into consideration while generate the anonymous area, which can help to conform to the user's approximate direction of movement, to meet users' actual situation in real-life.
3. Combining the redundancy optimization processing strategy proposed in this paper, we can guarantee users' anonymous expectations, reducing area consumption to make sure the efficiency and accuracy of location-based services.
4. The results in *NCC* experiment reflects its anonymous function; besides, on the base of *Casper* Cloak, it saves anonymous and provides users a more convenience service through improving efficiency in anonymous area searching and redundancy optimization.

The remainder of this paper is organized as follows. Section 2 summarizes the related works in the fields of privacy protection. Section 3 presents several basic expressions we'll use later in this paper and outlines the shortcomings of *Casper* in Sect. 4. Then we elaborately detail our improvements in Sect. 5. We also present experimental results demonstrating the effectiveness of these improvements in Sect. 6. Finally, we conclude this paper in Sect. 7.

2 Related Works

Motivated by the privacy breach in these cases, a variety of approaches have been studied to maintain the data privacy. The *K*-anonymity algorithm [8–10] has been widely utilized to protect the data privacy. The main idea of *K*-anonymous is to have each record in *K*-anonymous based architecture, i.e., not distinguishable among other *K*-1 records. The approaches based on *K*-anonymity algorithm are *spatial-temporal* cloaking [18], *Clique* Cloak [19] and *Casper* Cloak [12, 13]. The spatial-temporal cloaking assumes that all users have the same *K*-anonymity requirements. For each user's location update, the spatial space is recursively divided in a KD-tree format till a suitable subspace found [13]. This technique lacks scalability as it deals with each single movement of each user individually. The *Clique* Cloak algorithm grants a different *K*-anonymity requirement for each user, which combines a set of users together and constructs a clique graph to decide that some users can share the cloaked spatial area [12]. Yet, the approach suffers from a privacy threat when it reveals some information about the user locations. In addition, due to the computation overhead of computing the clique graph, this approach is limited to a small number of users with small K-anonymity requirements. The *Casper* Cloak [12] which was proposed by Mokble, is the most popular one. With the ability to deal with a mount of anonymous requests, *Casper* Cloak can quickly expand the anonymous area to meet the anonymous demands of different users simultaneously [13]. And *Casper* is resistant to attacks and can protect users' private informations in an effective way. Besides, *Casper* adopts the trusted third party based approaches as it requires less computation overheads and is more suitable for real-time query processing.

Trusted third party [20, 21] based approaches which utilize an anonymizer to create spatial cloaked regions, hiding the actual location of users have already been widely studied. Gruteser and Grunwald, 2003, proposed a method using spatial cloaking to obfuscate user locations [22]. Later, a location privacy architecture where the user specifies maximum spatial tolerances for cloaking area, was proposed by Gedik and Liu [23]. Inspired by the K-anonymity in database privacy, Gedik and Liu, took the requirements of K-anonymity into account when generating the spatial cloaked region. This enforces that the user will not be individually located within the area for a given period [24]. In [25], Ghinita et al., proposed a decentralized architecture to create the anonymous spatial areas which can help to reduce the need for centralized anonymizer. According to the architecture, mobile users can self-organize into a fault-tolerant overlay network with the use of a distributed protocol. Thus, from the fault-tolerant overlay network, a K-anonymous spatial cloaking set of users can be determined. Still, parameter specification is a big obstacle to the implement these techniques. Even if a user with advanced knowledge to understand the impact of parameter setting on location privacy, the implications of service is still unknown.

As *Casper* model is widely used in practical application, several researches have been conducted rely on *Casper*. In [15], based on the *Casper* model, Si proposed a new privacy protection approach *V-grid* which considers the effects of users' mobility, making the anonymous area more accurate and LBS results more efficient. Yet, spatial redundancy is serious. In [16], Jin put forward a method that can improve the accuracy of anonymous area. The method which based on the quad-tree, uses the bottom-up approach in the checking process, traversing the three subnodes of the same layer first, and then traversing the parent node upwards. Although, the accuracy of the anonymous space is improved, the checking process is still cumbersome and has the disadvantages of high cost and tardiness. Compared with previous approaches, we insist that the main contributions of our job are to reduce the overheads on checking process, increase the accuracy of LBS and provide the effective assurance for anonymous requirements.

3 Preliminaries

Casper is a framework in which mobile users can accept location-based services without revealing their private location information. As shown in Fig. 1, *Casper* employs a grid-based pyramid data structure to hierarchically decompose the spatial space into *n* levels where a level of n has 4^n grid-cells. The top level of the pyramid is the root node and has one grid cell which denotes the whole space [13]. *Casper* mainly consists of two components, the *privacy-aware query processor* and the *location anonymizer*. *Privacy-aware query processor*, which embedded inside the location-based sever, major deals with anonymous queries and cloaked spatial areas rather than the exact location information and outputs a *candidate-list* instead of a single exact answer. The *location anonymizer* is a trusted third party that acts as a middle layer between users and location-based server. Upon registration with *Casper*, mobile users can issue queries along with a user-specified privacy *profile* to location-based server. And job in this paper mainly in the *location anonymizer*.

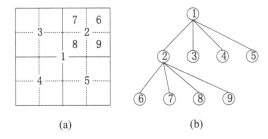

(a)   (b)

Fig. 1. (a) The method to number the *cell cid* and (b) a grid-based pyramid data structure

Each query *Q* is represented as three components (*loc, con, profile*); *loc* denotes the spatial coordinates of user's location, *con* is the type of service which users ask for, and *profile* denotes users' privacy *profile* which contains two parameters defined as a *tuple* (*k, A*), where *k* denotes that the user wants to be *k*-anonymous (i.e. not distinguishable among the *k* users), while *A* indicates that the user wants to hide her privacy location within a cloaked spatial region at least *A*. *Profile* is always used to specify their requirements of privacy [14]. When a mobile user issues a query *Q*, containing location information along with her privacy *profile*, the *location anonymizer* blurs the user's exact location information into cloaked spatial regions based on the users' specified privacy *profile*. Anonymization above can be described as: *f*(*Q*) = *CR*, where *f* is anonymizer's functionality, *CR* denotes the returned spatial cloaked region with the form (*id, uset, range*) where *id* is *CR* identifier, *uset* is the set of users in this region and *range* is the set of cells which constitutes this region together.

Moreover, in the use of *Casper*, *location anonymizer* maintains two structures, namely *UserTable* and *GridTable*, to keep track of each users' information and the information of grid cell, respectively. In *UserTable*, each registered mobile user has on entry with the form (*uid, cid, loc, profile*), *uid* is the mobile user identifier, *cid* is the cell identifier in which the user is located, *loc* and *profile* are extracted from query. Furthermore, in *GridTable*, each *cell* is described as four parts (*cid, parent, girdsite, total*); *parent* is the *cid* of the current cell's parent cell, *total* is the total numbers of users within this cell, *girdsite* is information of vertex coordinates of this cell, *cid* is this cell's identifier as it is in *UserTable*. If a *user* in *cell cid* issues a query along with her privacy *profile* (k_{min}, A_{min}), and it already meets the user privacy requirements (i.e., $k_{cid} \geq k_{min}$ and $A_{cid} \geq A_{min}$), then the *cell cid* can be treated as the spatial cloaked area. Otherwise, we should check whether the combination of the *cell cid* with its row (horizontal neighbor) or with column (vertical neighbor) contributes a region that satisfy. If it isn't the case, then the process above is recursively executed with the parent cell of *cid* until a valid cell is found.

4 Disadvantages

4.1 Overheads on Look-up Table

Figure 2(a) depicts a scenario that the *user* issued a query Q along with a privacy *profile* in which k equals 5. For clarity, the actual *user* location is marked as a black circle in area C_8 where only 3 users, which cannot meet K equals 5 users' demands. In this case, we should check for the vertical and horizontal neighbor cells, namely C_7 and C_9, to cell C_8.

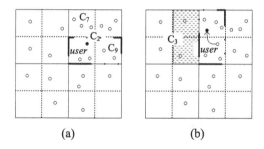

(a)   (b)

Fig. 2. (a) At time $t1$, the *user* issued a query Q along with a privacy *profile*. Furthermore, the user is moving forward to the north direction. By the time t_2, the *user*'s position can be shown as (b).

The detailed process is as follows: (1) after receiving the query from user, anonymizer utilizes the user identifier, *uid*, to find the entry for *user* in *UserTable*. (2) Later, according to the user-located cell identifier, *cid*, which lies in *user* entry, the entry of C_8 in *GridTable* is associated. (3) Then, anonymization extract the parent cell identifier, *parent*, from the entry of C_8 to keep track of the entry for *parent* cell in *GridTable*. Thus, it's easy to access the information in parent cell C_2, i.e., coordinates of the cell's each vertex and the number of the currently users. (4) Upon the vertex coordinates of both C_8 and C_2, we can infer the neighbor cells of C_2, namely C_7 and C_9, through a kind of complex process. (5) Finally, likewise, the entry of C_7 and C_9 will be accessed, and the *total* would be provided to check whether a combination of C_8 with C_7 or with C_9 can contribute a region that satisfy *user*'s anonymous requirements. Otherwise, the process above is recursively executed with the parent cell C_2 until a valid cell is found.

$$Cost_f = (4 \times Cs + Cb) \times x \qquad (1)$$

Assume that it takes x times of recursive check to finally get a valid cell, the cost of the anonymous operation can be computed as Eq. 1, where f denotes anonymizer's functionality, $Cost_f$ denotes the cost of function f, Cs is the cost of look-up table, and Cb is the cost of deducing the sibling node. Based on the discussions above, it takes at least four look-table actions during a single checking operation which turns out to be cumbersome and inefficiency.

4.2 Affection of Mobility

Again, as shown in Fig. 2(a), with the privacy requirement of $k = 5$, a spatial cloaked region, namely $C_{8,9}$, that composed by area C_8 and C_9 is returned to *user* as the valid area, instead of $C_{8,7}$ where k sequals 6. According to *Casper*, if both row-combined and column-combined areas satisfy the anonymous requirements of *user*, the combination that provides closer value of k to demands will be treated as the returned region. $C_{8,9}$ is plotted as bold dotted line in Fig. 2(a).

Furthermore, if Fig. 2(a) depicts a scenario at time t_1 that user is moving forward to the north direction, assuming it is possible to determine *user*'s position at time t_2, which is shown as Fig. 2(b). As it says that *user* may already entered the area C_7 with high probability. Thus, at time t_1, it is better to return area $C_{8,7}$ as spatial cloaked region instead of area $C_{8,9}$. Area $C_{8,7}$ is plotted as bold dotted line in Fig. 2(b).

4.3 Spatial Redundancy

In Fig. 2(a), assuming *user* in area C_8 issues a query Q with the privacy requirement of $k = 13$ and $A = 5$. Noted that area C_8 where k is less than 13, while A is less than 5, cannot meet the *user*'s privacy requirements, we will check for the neighbor cells, C_7 and C_9, to *cell* C_8. However, the neighbor-combined areas (i.e. $C_{8,9}$ and $C_{8,7}$), where $C_{8,9}.k$ equals 5 while $C_{8,7}.k$ equals 6, still cannot satisfy *user*'s demands. Therefore, the searching process will be recursively executed with the parent cell C_2. As shown in Fig. 2(b), when cell C_2 and C_3 are merged, cell $C_{2,3}$ where k equals 15 while A equals 8, can meet the anonymous demands of *user* and would be returned to *user* as the spatial cloaked region. Yet, cell $C_{2,3}$ is so big that exists spatial redundancy. For the combination of C_2 with the part of cell C_3 that is covered with shadow, where k is 13 and A is 6, can provide an acceptable resolution of the cloaked spatial region which achieves a better and a more accurate anonymity effect than that of $C_{2,3}$.

$$Sa = 2^y \times Sc \qquad (2)$$

Assume that it takes y times of recursive check to get a valid cell, and the size of this cell can be computed as Eq. 2, where Sa denote the area of the returned spatial cloaked region, and Sc is the area of the minimal cell at the lowest level of pyramid. Based on the discussions above, with the increase of y it is obviously that the Sa shows exponential growth. And there will be more and more spatial redundancy in spatial cloaked region.

5 Solutions

5.1 Simplify Look-Up Table

As discussed, it takes at least four look-table actions during a single checking operation which turns out to be cumbersome and inefficiency. To deal with such problem in practical implementation, instead of the hash function mentioned in [13], a new method is applied to number the *cell cid*, which is depicted as Fig. 1 (a). And we also introduce the concept of first sibling cell, which can be used to infer its adjacent sibling nodes through a simple reasoning, to reduce the times of look-table and improve the efficiency of anonymous processing.

Firstly, the first sibling cell is considered as the cell with a minimal *cid* among a group of cells which are in the same level. For example, in Fig. 1 (b), *cid* 3's first sibling is *cid* 2 which has adjacent neighbors *cid* 2 and *cid* 4 within [2, 5]; And *cid* 8's first sibling is *cid* 6 which has adjacent neighbors *cid* 7 and *cid* 9 within [6, 9]; However, *cid* 9 cannot follow the rule anymore, for its adjacent neighbor cells are *cid* 8 and *cid* 6, instead of *cid* 8 and *cid* 10. So, we need further consideration. Let us denote a cell in the grid by $N_{subnode}$, its first sibling cell is $N_{firstbro}$. The relationship between two of them is followed as Eq. 3.

$$N_{firstbro} = \begin{cases} 1 & N_{subnode} = 1 \\ 2 \times [\lfloor (N_{subnode} - 2)/4 \rfloor + 1] & N_{subnode} \succ 1 \end{cases} \tag{3}$$

Inferring the $N_{firstbro}$ from $N_{subnode}$, and the method which can be used to deduce its adjacent neighbor cells is computed as Eq. 4.

$$\langle N_{nodebro1}, N_{nodebro2} \rangle = \begin{cases} \langle \emptyset, \emptyset \rangle & N_{subnode} = 1 \\ \left\langle \begin{array}{l} N_{firstbro} + (N_{subnode} - 1 - N_{firstbro}) mod 4, \\ N_{firstbro} + (N_{subnode} + 1 - N_{firstbro}) mod 4 \end{array} \right\rangle & N_{subnode} \succ 1 \end{cases}$$
$$\tag{4}$$

Thus, given any cell in the gird, we can immediately deduce its neighbor cell at once, according to the method which gets rid of the traditional cumbersome process. And we customize the cell's entry in *GridTable*, i.e., *GridTable_new*, by introducing the concept of the first sibling node. The *GridTable_new* is represented as (*cid, parent, firstsibling, girdsite, total*). After optimization of *GridTable_new*, user only needs one time searching to find its sibling node among adjacent nodes, effectively reduce their searching consumption.

5.2 Introduce Direction of Movement

Noticing that the anonymizer fails to take into account the users' mobility when generates the spatial cloaked region, it cannot satisfy mobile users' exact demands in real-life. Aiming at this, we do the job as above, customizing both users' query and its entry in *UserTable*, i.e. Q_{user}' and *UserTable_new*, where Q_{user}' is represented as *(loc, con, profile, $\vec{v}$)* and *UserTable$_{new}$* is with the form *(uid, cid, loc, profile, direction)*. *loc* denote the users' current coordinates, $\vec{v}$ is the coordinates of users last update, *direcion* denotes the direction of *user*'s movement and is represented as *(way, θ)*, where *way* that initiated as zero has four bits which are on the behalf of *N* (North), *S* (South), *W* (West), *E* (East), respectively, while θ is the angle between the direction of movement and the east-west axis.

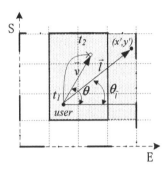

Fig. 3. The whole spatial cloaked region is denoted by *Range$_{user}$* where plotted as bold dotted square. Then, the parts that meets the direction of user is covered with shadow and is denoted by *Range$_{Direct}$*.

Assuming that *user* positioned in northern hemisphere range (0°N–90°N) issues queries that contain user coordinates (x_{t1}, y_{t1}) and (x_{t2}, y_{t2}) to server, respectively, at time *t1* and time *t2*, where *t1* < *t2*. In this case, *loc* is set to (x_{t2}, y_{t2}) and $\vec{v}$ is (x_{t1}, y_{t1}), *way* is computed as Eq. 5.

$$\begin{cases} |E| = 1 & x_{t_1} \prec x_{t_2} \\ \emptyset & x_{t_1} = x_{t_2} \quad AND \\ |W| = 1 & x_{t_1} \succ x_{t_2} \end{cases} \begin{cases} |N| = 1 & y_{t_1} \prec y_{t_2} \\ \emptyset & y_{t_1} = y_{t_2} \\ |S| = 1 & y_{t_1} \succ y_{t_2} \end{cases} \quad (5)$$

Setting any bit of way to 1, denotes that the user is moving forward to the direction it stands for. And θ is calculated as Eq. 6.

$$\theta = arctan \left| \frac{y_{t2} - y_{t1}}{x_{t2} - x_{t1}} \right|, \quad \theta \in \left[0, \frac{\pi}{2}\right] \quad (6)$$

As Fig. 3 depicts, let us denote the whole spatial cloaked region by *Range$_{user}$* where plotted as bold dotted square. Then, the parts that meets the direction of user is covered with shadow and is denoted by *Range$_{Direct}$*. The set of users in *Range$_{Direct}$*, namely *U*, is represented as $(u_0, u_1, u_2, \ldots, u_m)$, and $(x_{u_1}, x_{u_2}, x_{u_3}, \ldots, x_{u_m})$ is set of longitudes of users in *U*, while $(y_{u_1}, y_{u_2}, y_{u_3}, \ldots, y_{u_m})$ is the set of latitudes of users in *U*. Therefore, *set U* should meet the Eq. 7.

$$\begin{cases} Low\_xG_{user} \prec x_{u_i} & E = 1 \\ Up\_xG_{user} \succ x_{u_i} & W = 1 \end{cases} \quad AND \quad \begin{cases} Low\_yG_{user} \prec y_{u_i} & N = 1 \\ Up\_yG_{user} \succ y_{u_i} & S = 1 \end{cases} \quad (7)$$

Where u_i denotes user in U, G_{user} denotes the grid in which user current located, the (Low_xG, Low_yG) are coordinates of the vertex in the left-lower corner of G_{user}, and (Up_xG, Up_yG) are coordinates of the vertex in the upper-right corner of G_{user}.

5.3   Redundancy Optimization

As discussed above, although the parts that covered by shadow can satisfy users' requirements, still, it has spatial redundancy. In this section, a redundancy optimization processing strategy that realizes more precise in the anonymization, are proposed to achieve high location-based services quality and less redundancy on the premise of guaranteeing the related degree of privacy.

In Fig. 3, when *user* issues a location-based query Q along with the privacy *profile* (k, A), the $Range_{user}$ plotted as bold dotted square that denotes spatial cloaked region would be returned. Let us superimpose a matrix of $n \times n$, namely Ma, on the area $Range_{user}$, and define $m_{ij} = Ma[i, j]$. In addition, $|m_{ij}|$ is the number of users in *cell* m_{ij}, and Sm denotes the size of the *cell*. Let $Range_{user}$ be the input to the optimization processing, and the output region is represented as $Range_{user}'$. Moreover, the set of users in $Range_{user}'$ is represented as $U' = (u_0, u_1, u_2, \ldots, u_k)$, and k_{user}' denotes the anonymity degree that $Range_{user}'$ provide, which can be computed as Eq. 8.

$$k'_{user} = \sum_{\substack{i,j \in (0,n) \\ m_{ij} \in Range'_{user}}} |m_{ij}| \qquad (8)$$

A_{user}' denotes the size of $Range_{user}'$ within the user can hide her actual location, is computed as Eq. 9.

$$A'_{user} = \sum_{\substack{i,j \in (0,n) \\ m_{ij} \in Range'_{user}}} Sm \qquad (9)$$

Then, let $\vec{l}$ be the extended vector of $Range_{user}'$, which shows the direction of the expansion of area $Range_{user}'$. And the angle between the extended vector $\vec{l}$ and the east-west axis is denoted by $\theta_{\vec{l}}$. In addition, (x', y') contribute to $|x_{user} - x'|_{max}$ and $|y_{user} - y'|_{max}$, respectively, while x' belongs to $(x_{u_1}, x_{u_2}, x_{u_3}, \ldots, x_{u_k})$, y' belongs to $(y_{u_1}, y_{u_2}, y_{u_3}, \ldots, y_{u_k})$, and (x_{user}, y_{user}) is the coordinates of *user*.

$$\vec{l} = (x' - x_{user}, y' - y_{user}) \qquad (10)$$

$$\theta_{\vec{l}} = arctan \frac{|y_{user} - y'|}{|x_{user} - x'|}, \quad \theta_{\vec{l}} \in \left[0, \frac{\pi}{2}\right] \qquad (11)$$

Ideally, an optimal cloaked region should be both minimal and effective. For instance, as shown in Fig. 3, area $Range_{Direct}$ can meet the direction of user, but still fails to be minimal and effective. We see that $\theta_{\vec{i}}$ differs from θ, and the difference between $\theta_{\vec{i}}$ and θ are getting less and less, the $Range_{user}$' will get closer to idealization. On the other hand, the area that minimal enough maybe not effective to provide services accurately. Learning from the [11], the service similarity is tied to distance between users' location; i.e., the farther the distance, the bigger the differences among services. Hence, let *Simliarity* be a similarity function, defined as Eq. 12.

$$Similarity(con, user, u_i) = 1 - \sqrt{\frac{\left(y_{u_i} - y_{user}\right)^2 + \left(x_{u_i} - x_{user}\right)^2}{hRange_{user} + wRange_{user}}}, \quad u_i \in U \qquad (12)$$

And *Simliarity* (con,user,u_i) is the measure of the similarity of the result retrieved for the *user*'s position relative to that retrieved for u_i's position in the area $Range_{user}$. ($hRange_{user}$, $wRange_{user}$) are height and width of $Range_{user}$, respectively. When *user* and u_i's position are the same, and result of *Simliarity* is the biggest which equals 1. Hence the problem of being both minimal and effective can be model Eq. 13.

$$\begin{array}{c} Minimal\ A'_{user}\ and\ k'_{user}\ toMaxmize \\ Similarity\left(con,\ user,\ Range'_{user}\right) \end{array} \qquad (13)$$

To ensure *A* and *k* can guarantee the requirements of anonymous, and meet the direction of users' movement, we constrain the model as Eqs. 14–17.

$$A'_{user} \geq A \qquad (14)$$

$$k'_{user} \geq k \qquad (15)$$

$$\left|\theta - \theta_{\vec{i}}\right| \rightarrow 0 \qquad (16)$$

$$\begin{cases} Low\_x_{G_{user}} \prec x_{u_i} & E = 1 \\ Up\_x_{G_{user}} \succ x_{u_i} & W = 1 \end{cases} AND \begin{cases} Low\_y_{G_{user}} \prec y_{u_i} & N = 1 \\ Up\_y_{G_{user}} \succ y_{u_i} & S = 1 \end{cases} \qquad (17)$$

According to optimization strategy, *NCC* is able to reach minimum anonymous area index, increasing the rates of anonymous optimization in original algorithm. And we can guarantee users' anonymous expectations, reducing area consumption to make sure the efficiency and accuracy of location-based services.

6 Comparative Analysis

In this section, we compare the efficiency and scalability of the *Casper* and the *NCC* with respect to the look-table time, cloaking cost, service accuracy and precision of anonymize requirement. In all the experiments, we use the Network-Based Generator of Moving Objects [17] to generate a set of moving objects. We used the map of a certain area in Sichuan, China, as the input of generator. And a set of moving objects would be returned as output. Unless mentioned otherwise, the experiments use 50 K mobile users in a pyramid structure with 9 levels which is the same with [12]. And a random privacy *profile* (k, A) where k is set within (1–50) while A is set within (.005, .01)% of the space, respectively. Target objects are chosen as uniformly distributed in the spatial space.

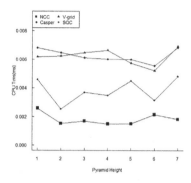

Fig. 4. A figure shows the effect of pyramid height on the average look-table time per user quest of *NCC*, *Casper*, *V-grid* and *SQC*.

As depicted in Fig. 4, the performance of the *NCC* is surely better than others. The experimental results show that, compared with *Casper* and *V-grid* [15], *NCC* reduces the time consumption of checking process by almost 80%. And by introducing the conception of first sibling cell, *NCC* can immediately deduce its neighbor cell and reduce the times of look-table. In addition, the checking process of *Casper and V-grid* are cumbersome and inefficiency, they take four look-table actions during a single checking operation. Moreover, the *SQC* [16] uses the bottom-up checking approach based on the quad-tree, which consumes about 60% of *Casper*.

Location-based services are included in a *candidate list* which is returned to *user*. In this section, we set the size of *candidate list* to 20, i.e., there are 20 service objects in *candidate list*. As shown in Fig. 5, the results statistics the proportion of the services in *candidate list* which meet the actual needs of users. It is clearly that, by taking into the consideration of users' mobility and using the redundancy optimization strategy, *NCC* increases the accuracy of location-based services which is 4 times of the *Casper* and ensures the efficiency and accuracy of services. In *Casper*, the average services accuracy is only about 20%, which is too lousy and users cannot be comfortable with. The services accuracy of *SQC* is about 45%, which is better than *Casper* but also unstable. Moreover, by considering the mobility of user, the services accuracy of *V-grid* is at least 57%, which makes the LBS results more efficient than *Casper* and *SQC*. However, the performance of *V-grid* is still lower than *NCC*.

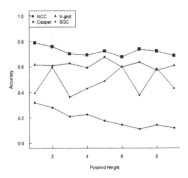

Fig. 5. A figure shows the effect of pyramid height on the services accuracy of *NCC*, *Casper*, *V-grid* and *SQC*.

Due to the resolution of pyramid structure, it is very difficult for a user to have a cloaked region which can exactly match her privacy requirements. Figures 6 and 7 show the effect of pyramid height on the precision of anonymize requirements. In Fig. 6, the *k* precision is measured as *k'/ k*, where *k'* is the number of users included in the spatial cloaked region while *k* is the exact requirement of *user*. After conducting

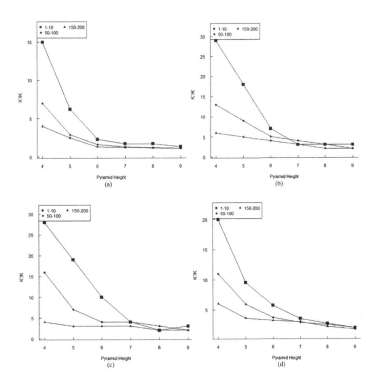

Fig. 6. A figure shows the effect of pyramid height on the precision of *k*. (a) is the *k* precision in *NCC*, (b) is the *k* precision in *Casper*, (c) is the *k* precision in *V-grid*, (d) is the *k* precision in *SQC*.

several experiments with relaxed privacy requirements (k within range from 1 to 10) to restrictive requirements (150–200), while setting A to 0, the results show that in these four methods, lower pyramid layer gives more inaccurate result for relaxed user than a higher level. Moreover, for both relaxed and restrictive users, results in *NCC* is better than others. In *NCC*, results given by higher pyramid levels is very close to optimal case even for relaxed users. Similar results are shown in Fig. 7 where the A accuracy is measured as A'/A, where A' is the computed cloaked spatial region while A is the requirement of users. Also, in experiment of A-precision, the K is setting to 1.

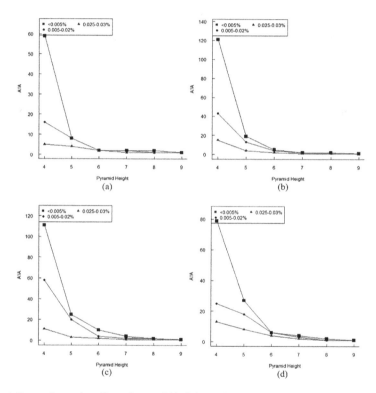

Fig. 7. A figure shows the effect of pyramid height on the precision of A. (a) is the A precision in *NCC*, (b) is the A precision in *Casper*, (c) is the A precision in *V-grid*, (d) is the A precision in *SQC*.

In Fig. 8, the results show the scalability of *Casper* and *NCC* with respect to number of users within (1–3 k) when (K, A) are setting to (10, 1). As for cloaking time, the performance of the *NCC* is enhanced with the increase of the user amounts. The reason is that the users' privacy requirements is likely to be meet at a lower level with the increase of number of registered users. And the trend of both *NCC*, *Casper*, *V-grid* and *SQC* are same.

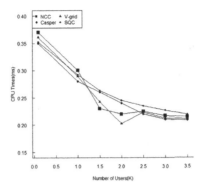

Fig. 8. A figure shows the scalability of *NCC*, *Casper*, *V-grid* and *SQC* with respect to number of users.

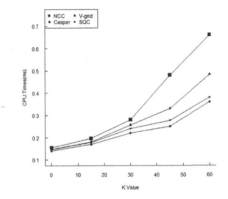

Fig. 9. A figure shows the effect of privacy *profile* on the performance of *NCC*, *Casper*, *V-grid* and *SQC*.

In Fig. 9, we conduct experiments with relaxed privacy requirements (k is within range from 1 to 10) to less restrictive requirements (40–60) when A is setting to 0. As shown in Fig. 9, when faced with relaxed requirements, the performance of *NCC* is similar with that of others, they have same cloaking cost. When faced with less restrictive requirements, the cost of *NCC* and *V-grid* is higher than others, but the difference in cost between *NCC* and *Casper* is still kind of acceptable. The *NCC* gets a poor performance for restrictive requirements, because the cost of *NCC*'s optimal process is growing with the increase of K. However, when it comes to relaxed and less restrictive requirements, the *NCC* can provide users with location-based services which improve 4 times accuracy of services in comparison with *Casper* mode under an acceptable cost. Also, we perform several experiments with various A requirement while k is setting to 1. For the cost trend in the case of changing A is like the results in Fig. 9, we decide not to show it.

7 Conclusion

This paper proposed an improved privacy protection method—*NCC*, based on the *Casper* Cloak. To reduce the times of look-table and improve the efficiency of anonymous processing, *NCC* introduces the concept of the first sibling node, and puts forward a new mothed to find its adjacent siblings. Besides, *NCC* takes the user mobility into consideration while generating the anonymous area, which can help to conform to the user's approximate direction of movement and meet users' actual situation in real-life. Moreover, combining the redundancy optimization processing strategy proposed in this paper, *NCC* can guarantee users' anonymous expectations, reducing area consumption to make sure the efficiency and accuracy of location-based services. We have performed several experiments from the aspect of look-table time, cloaking cost, service accuracy and precision of anonymize requirement, which help to evaluate the efficiency and scalability of *NCC*. The results in these experiments reflected *NCC*'s anonymous function; besides, on the base of *Casper* Cloak, *NCC* saved anonymous and provided users a more convenience service through improving efficiency in anonymous area searching and redundancy optimization. Yet, when faced with restrictive requirements, the *NCC* got a poor performance and the cost of the optimal process increased with privacy requirements (K, A). We believe our solution is just the first step, several areas remain to be addressed in our future work, such as reducing the cost of the optimal process.

References

1. Zhao, D.P., Liang, L., Tian, X.X., et al.: Privacy protection in location based services: model and development (2015)
2. Zhang, X.J., Gui, X.L., Wu, Z.D.: Privacy preservation for location-based services: a survey. J. Softw. **26**(9), 2373–2395 (2015)
3. Chen, X., Pang, J.: Protecting query privacy in location-based services. GeoInformatica **18**(1), 95–133 (2014)
4. 199IT: 08/09/2015. http://www.199it.com/archives/381843.html
5. Sythoff, J., Morrison, J.: Location-based services: market forecast, 2011–2015. Pyramid Research (2011)
6. Dewri, R., Thurimella, R.: Exploiting service similarity for privacy in location-based search queries. IEEE Trans. Parallel Distrib. Syst. **25**(2), 374–383 (2014)
7. Chow, C.Y., Mokbel, M.F., He, T.: A privacy-preserving location monitoring system for wireless sensor networks. IEEE Trans. Mob. Comput. **10**(1), 94–107 (2011)
8. Kaliyamurthie, K.P., Parameswari, D., Udayakumar, R.: k-anonymity based privacy preserving for data collection in wireless sensor networks. Indian J. Sci. Technol. **6**, 4604–4614 (2013)
9. Zuberi, R.S., Ahmad, S.N., Lall, B.: Privacy protection through k-anonymity in location-based services. Iete Tech. Rev. **29**(3), 196 (2012)
10. Xue, M., Liu, Y., Ross, K.W., et al.: Thwarting location privacy protection in location-based social discovery services. Secur. Commun. Netw. **9**(11), 1496–1508 (2016)
11. Dewri, R., Thurimella, R.: Exploiting service similarity for privacy in location-based search queries. IEEE Trans. Parallel Distrib. Syst. **25**(2), 374–383 (2014)

12. Chow, C.Y., Mokbel, M.F., Aref, W.G.: Casper: query processing for location services without compromising privacy. ACM Trans. Data base Syst. (TODS) **34**(4), 24 (2009)
13. Mokbel, M.F., Chow, C.Y., Aref, W.G: The new casper: query processing for location services without compromising privacy. In: International Conference on Very Large Data Bases, DBLP, Seoul, Korea, pp. 763–774, September 2006
14. Niu, B., Zu, X.Y., Chi, T.H.: Pseudo-location updating system for privacy-preserving location-based services. China Commun. **10**(9), 1–12 (2013)
15. Si, C., Xu, H.Y.: Location privacy protection method based on V-grid model. Comput. Eng. **38**(12), 276–278 (2012)
16. Jin, S.F., Ye, Z.S., Song, H.: A similar quadtree based on location K-Anonymity algorithm. Trans. Beijing Inst. Technol. **1**, 68–71 (2014)
17. Brink, H.T.: A framework for generating network based moving objects. GeoInformatica **6** (2), 153–180 (2002)
18. Gruteser, M., Grunwald, D.: Anonymous usage of location-based services through spatial and temporal cloaking. In: MobiSys (2003)
19. Gedik, B., Liu, L.: A customizable k-anonymity model for protecting location privacy. In: ICDCS (2005)
20. Jefferies, N., Mitchell, C., Walker, M.: A proposed architecture for trusted third party services. In: Dawson, E., Golić, J. (eds.) CPA 1995. LNCS, vol. 1029, pp. 98–104. Springer, Heidelberg (1996). doi:10.1007/BFb0032349
21. Aggarwal, G., et al.: Vision paper: enabling privacy for the paranoids. In: VLDB (2004)
22. Gruteser, M., Grunwald, D.: Anonymous usage of location-based services through spatial and temporal cloaking. In: Proceedings of the 1st International Conference on Mobile Systems, Applications, and Services, pp. 31–42 (2003)
23. Gedik, B., Liu, L.: Protecting location privacy with personalized k-anonymity: architecture and algorithms. IEEE Trans. Mob. Comput. **7**(1), 1–18 (2008)
24. Samarati, P.: Protecting respondents' identities in microdata release. IEEE Trans. Knowl. Data Eng. **13**(6), 1010–1027 (2001)
25. Ghinita, G., Kalnis, P., Skiadopoulos, S.: PRIVE: anonymous location-based queries in distributed mobile systems. In: Proceedings of the 16th International Conference on World Wide Web, pp. 371–380 (2007)

Baymax: A Mental-Analyzing Mobile App Based on Big Data

Fangyi Yuan[✉], Hongzhi Wang, Shucun Tian, and Xin Tong

Harbin Institute of Technology, Xidazhi Str. 92, Harbin 150001, China
yfy0829yfy@163.com

Abstract. Nowadays people are facing various psychological problems. Existing solution of evaluation and treatment of mental illness is only to see a psychiatrist, but most of the users has sense of resistance on psychiatrist. Meanwhile most of the existing systems are psychological tests for entertainment, whose results cannot be accurately analyzed. On basis of this phenomenon, we develop the Baymax system to construct a relation between the potential patients' motion and a specific models in order to analyze the user's mental conditions and predict his or her potentialities for suffering certain mental diseases, by exerting and exploring the user's short messages, online diaries, and communicative information on social media, as well as the questionnaires accomplished by the user. Our system uses big data analysis method, and allows users to use the system to determine whether there is a possibility of mental illness in the absence of a psychiatrist.

Keywords: Mental health · Data mining

1 Introduction

Nowadays, the universal fast-paced life-style drenches people with heavy stress in everyday life. According to survey, 17.5% of adults in China are likely to be attacked by mental disease; more than 50% of the citizens have symptoms of depression or distress in different degrees; and there is still another 11.5% should be attributed to those who are suffering severe depression in their workplace [1]. Adopting measures to protect and cure mental illnesses are in demand.

But in our country at present, the number of professional psychotherapist is still unable to meet the demand. At the same time, the cost of psychotherapy is relatively expensive, which lead to the situation that many people can hardly find a therapist and afford the cost of treatment. As a result, they delayed the psychological treatment. It causes great harm to themselves and their family. At the same time, due to the level of psychological doctors limit, some mental illness cannot be accurately identified, thereby delaying the illness.

Motivated by this, we developed "Baymax" system as a user's personal psychological doctor, whose name comes from Baymax in the film, is.

The information used in the system is mainly from text message and corresponding psychological questionnaires user fill in. Through deep delves of these data, the

© Springer Nature Singapore Pte Ltd. 2017
B. Zou et al. (Eds.): ICPCSEE 2017, Part II, CCIS 728, pp. 568–571, 2017.
DOI: 10.1007/978-981-10-6388-6_48

condition of psychological health of users is evaluated. We can also predict possible diseases, and provides suggestions to improve the health status in our system.

For users who want to know their own health condition, this system, under the permission of the users' authorized access, can make a deep data exploration of information from the users' texts, diaries, as well as their mental health trend which reflected by the psychological questionnaires they filled out recently. On the basis of the matching degree between their own features with disease feature library, the system can determine the users' health status and give relevant advices.

This system has the following characteristics:

(1) Accuracy: The system uses large amounts of data generated by social networking, data mining in order to achieve accurate analysis and forecasting.
(2) Efficiency: Efficient algorithms are designed for the system in order to achieve efficient mining and predictive.
(3) User-friendly: The system provides users with a simple user interface, which is easy to use.

The remaining parts of this paper are organized as followings. Section 2 proposes the architecture of our system. Section 3 describes the key technologies in the system. Section 4 introduces demo and the method of presentation of the system.

2   System Architecture

In this section, we discuss the architecture of our system.

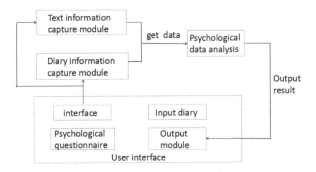

Fig. 1. System architecture

As shown in Fig. 1, our system has 4 key modules:

(1) Text information capture module: This module captures text information as required. We apply for permission to extract the sender and the information in texts, and then obtain keywords related to mental illness from the text with information extraction algorithms. These keywords are stored in our database, preparing for further analysis.

(2) Dairy information capture module: This module captures diary information as required. We also extract keywords from the dairy and store them in the database.
(3) Psychological data analysis module:This module is used to analyze users' psychological information. The implementation of the module will be discussed in Sect. 3.
(4) User interface: This component provides friendly interface to users including user interface, diary input module, psychological questionnaire module and output module.

The system firstly captures information through text information capture module and dairy information capture module. Then psychological data analysis module will analyze the data captured and display results in input module. According to how well the output results are, we decide whether the test of psychological questionnaire module is recommended for the user.

After authorization, we capture information of the user's texts and diaries. Meantime, psychological data analysis module is invoked in accordance with the analysis results of the psychological questionnaire, to make analyses and matches. Then we obtain results and show them in output module.

3 Key Technologies

This section discusses key techniques in our system.

Logistic Regression: To predict the probability of disease by risk factors and make our system sensitive with data changes, we use logical regression model. Similarly, we use logical regression model in the prediction of mental disease. We can predict the probability of depression by user's SMS and diary. We choose some keywords as the training data. With logistic regression, the determination function is obtained. With such function, the possible illness with the probability could be judged according to the extracted keywords. At the same time, the result of logistic regression is S-type. As a result, the probability in the middle could be in big changes and be sensitive [2].

Bayes Inference: To make the emotion analysis, we choose Naive Bayes. The result of Naive Bayes are P(yes) and P(no), which means the possibility of illness and health. P(yes) and P(no) are too small to calculate on computer. Meanwhile, P(yes) and P(no) can differ by orders of magnitude. $P(yes\_adaboost) = \frac{P(yes)}{P(yes) + P(no)}$ Therefore, P(yes_adaboost) is close to 1 or 0. And this probability will take a huge mistake to adaboost model.

Therefore, in each calculation and the process, we let P(yes) and P(no) divide by a number. Therefore, one of them is close to 1, and the other one is within [0, 1]. Therefore, we obtain the reasonable result of Naive Bayes on P(yes) and P(no).

Adaboost: In order to improve the accuracy of mental analysis, we use both of the logical regression model and Bayesian model. In order to combine them effectively, we adopt adaboost. In adaboost model, we give different weight to the sample according to the prediction accuracy. We improve the prediction accuracy by this way.

4   Demonstration

We plan to demonstration Baymax in understanding the principles of data mining as well as the impact of its accuracy (Fig. 4).

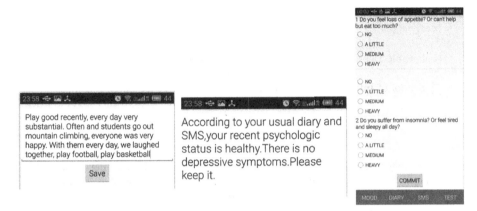

Fig. 2. Diary interface   **Fig. 3.** Feedback interface   **Fig. 4.**

Since the Baymax needs to obtain users' emotions, psychological activities, and oral expressions. Users can click diary button to record their moods and the things they want to remember which can reflect their mental condition and click save button to save diary records, as shown in Fig. 2.

When the recent mood button is clicked, the results of psychological state according to the analysis of the user's diary records, text messages are showed on the interface. If users' analysis results shows that she or he tends to have mental disease, the system shows the advices to the users. Meanwhile, our system will remind the user to do some psychological tests to ensure the results. An example for recent mood is shown in Fig. 3.

User can do some psychological tests by clicking test button shown in Fig. 4. When the results are submitted, the user's mental condition is saved and shown in our interface similar like Fig. 3.

References

1. West, R., White, R.W., Horvitz, E.: From cookies to cooks: insights on dietary patterns via analysis of web usage logs. In: WWW, pp. 1399–1410 (2013)
2. Cover, T.M., Thomas, J.A.: Elements of Information Theory. CMP, New York (2005)
3. Mitchell, T.M.: Machine Learning. CMP, New York (2003)

Ensemble Learning-Based Wind Turbine Fault Prediction Method with Adaptive Feature Selection

Shiyao Qin[1], Kaixuan Wang[2(⊠)], Xiaojing Ma[1], Wenzhuo Wang[1], and Mei Li[2]

[1] Renewable Energy Research Center, China Electric Power Research Institute, Beijing, China
[2] School of Information Engineering, University of Geosciences (Beijing), Beijing 100083, China
wangkaixuan@outlook.com

Abstract. In this paper we present a wind turbine (WT) fault detection method based on ensemble learning, WT supervisory control and data acquisition (SCADA) is used for model building. In feature selection process, random forest algorithm is applied to get the feature importances, this is much convenient compared with general feature selection by experience, also more accurate result is obtain. In model building, SVM based bagging algorithm is used, compared to individual SVM, out method is much faster and again with a better result.

Keywords: Wind turbine · Ensemble learning · Feature selection · Support vector machine · SCADA data

1 Introduction

For WTs the maintenance costs are high because of their remote location, and this can amount to as much as 25 to 30% of the total energy production [1]. The most effective way to reduce maintenance costs is to continuously monitor generators' status and predict the malfunction of WT. Then the system degradation problems can be found and responded in time. Maintenance can be carried out ahead of time before the system crash and excess maintenance is avoided. Therefore, fault prediction could maximize the normal production of wind power plant and greatly reduce maintenance costs.

The main methods of WT fault diagnosis include these types such as the fault mode analysis based on statistical data, the fault diagnosis based on time series prediction, model-based fault diagnosis of control system, the fault diagnosis based on vibration analysis, and other auxiliary diagnosis methods like

K. Wang—This paper is supported by Renewable Energy Research Center of China Electric Power Research Institute of STATE GRID,'s science and technology project: *Research on Key Technologies of condition monitoring and intelligent early detection of wind turbine based on big data.*

B. Zou et al. (Eds.): ICPCSEE 2017, Part II, CCIS 728, pp. 572–582, 2017.
DOI: 10.1007/978-981-10-6388-6_49

acoustic emission technique and ultrasonic capacitance liquid level detection [2]. Schlechtingen has proposed an approach for WT supervisory control and data acquisition (SCADA) data for condition monitoring purposes [3]. They apply adaptive neuro-fuzzy interference system (ANFIS) models to a wide range of different SCADA signals.

Since SCADA signals are recorded with a long interval, which was initially not for the purpose of condition monitering and fault detection (CMFD) [4], most dynamical features of WT faults that are useful for CMFD are lost. Therefore, the detailed information f most WT faults cannot be diagnosed by using SCADA signals via frequency or time-frequency SCADA signals have been mainly used by model-based methods and prediction methods for WT CMFD and prognosis [3,4]. Schlechtingen's successive work give detailed examples of the application of their method [5]. Many other data mining methods are utilized in fault prediction, for instance, SVM-based solution by Santos et al. [6], GA optimization method by Odofin Sarah et al. [7], probabilistic neural network by Malik and Mishra [8], k-means and neural net by Liu et al. [9].

Our proposed method can greatly reduce the work in feature selection stage, and with lower time spend getting a more precise result.

2   Principles

2.1   Ensemble Learning

The below picture give the idea of ensemble learning. For the training set of data, we train a number of individual learners, through a certain combination of strategies, you can eventually form a strong learner, in order to achieve a better learning result. Type 1. All individual learners are of one kind, or homogeneous (Fig. 1). For example, they are decision trees, or individual learners are all neural networks.

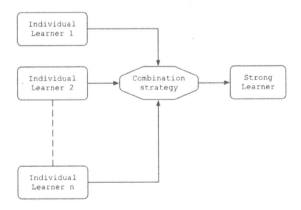

Fig. 1. Ensemble learning principle

Tpye 2. All individual learning devices are not of one kind. For example, we have a classification problem, the training set by using support vector machine learner, logistic regression and individual learners Naive Bayesian individual learners to learn, and then through a combination of strategies to determine the final classification of strong learner.

We using bagging learner in this paper, it belongs to tpye 1, the priciple of bagging is given in detail below.

Bagging. Bagging, or called bootstrap aggregating, is characterized by the lack of dependencies between the weak learners and can be fitted in parallel. This paper makes a summary of Bagging and random forest algorithm in ensemble learning. There is no dependency between the weak learners and can be generated in parallel. We can use a graph as a generalization: As you can see from the picture above, the training set of bagging's individual weak learner is obtained by random sampling. Through random sampling N times, we can get a set of N sampling, N sampling for this set, we can separate the trained n a weak learner, then this N a weak learner through the collection of strategies to get the final strong learner (Fig. 2).

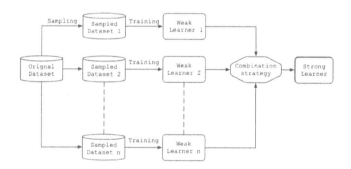

Fig. 2. Bagging principle

It is necessary to make further introduction for the random sampling, i.e. bootstrap. From our training set to collect a fixed number of samples, but after each sample, the sample will be returned. That is to say, the samples that were collected before are likely to continue to be collected after being returned. For our Bagging algorithm, a sample of the same number of samples of N will be collected and trained at random. The number of sample sets and training set samples is the same, but the sample contents are different. If we have random sampling of m times for N sample sets, the m sampling sets are different due to randomness.

For a training set containing m samples, the probability that a sample will be collected is $\frac{1}{m}$, and probability not be collected is $1 - \frac{1}{m}$. If the m times are not sampled, the probability in the collection is $(1 - \frac{1}{m})^m$. As $m \to \infty$, $(1 - \frac{1}{m})^m \to \frac{1}{e} \simeq 0.368$. That is to say in the random sampling of bagging about

36.8% of data will not be sampled, it is often called OOB (Out Of Bag). These data are not involved in the fitting of the training set model, so they can be used to detect the generalization ability of the model. The combine strategy of bagging is also simple. For classification problems, the simple voting method is usually used to get the most votes of the categories or categories, which is the final model output. For regression problems, the average result of regression results obtained by N weak learning is usually used.

Since the Bagging algorithm takes samples every time to train the model, the generalization ability is very strong, and it is very useful to reduce the variance of the model. Of course, the fit of the training set will be a little worse, that is, the bias of the model will be larger [10].

Random Forest. Random Forest (RF) is a special kind of bagging learner. RF uses the CART decision tree as a weak learner. Based on the decision tree, RF make improvement on the establishment of decision tree. For the ordinary decision tree we will choose a best feature from N node sample, but RF randomly select part of sample features on the node, the number is less than N, is assumed to be n_{rf}, and then select the best feature from the randomly selected n_{rf} sample features to partition subtrees. This step further enhances the generalization ability of the model.

If $n_{rf} = n$, the CART decision tree of RF is no different from the ordinary CART decision tree. The smaller the n_{rf}, the more robust the model, of course, and the degree to which the training sets fit declines. That is to say, the smaller the n_{rf}, the variance of the model will decrease, but the bias will increase. Thus a suitable n_{rf} value is generally obtained by cross validation.

2.2 Necessary Background of SVM

In the field of machine learning, SVM is a supervised learning model, which is usually used for pattern recognition, classification and regression analysis. In our work, we utilize SVM to train a normal behavior model and then give the prediction. A simple introduction to understand the principle of SVM here.

Logistic Regression. Popular speaking, SVM is a two class classification model, the basic model is a maximum linear classifier on feature space, the learning strategy is the maximum distance, and can be transformed into solving a convex quadratic problem [11]. To understand SVM, we must first understand a concept, linear classifier. Given a number of data points X, they belong to two different classes Y(Y can take 1 or -1, representing the different classes), and now to find a linear classifier to divide the data into two categories. The goal of a linear classifier is to find a hyperplane in the n-dimensional data space The equation of this hyperplane can be expressed as the following:

$$w^T x + b = 0 \tag{1}$$

The purpose of logistic regression is to learn a 0/1 classification model from the features, the model is a linear combination of features as independent variables.

Because the range of the independent variable is negative infinity. Therefore, the logistic function (or sigmoid function) is used to map the independent variables to (0, 1), and the value of the mapping is considered to be the probability of belonging to y = 1. The following function g is the logistic function, where x is an n-dimensional feature vector:

$$h_\theta(x) = g(\theta^T x) = \frac{1}{1 + e^{-\theta^T x}} \tag{2}$$

And its figure is (Fig. 3)

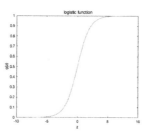

Fig. 3. Plot of logistic regression

As you can see, the function maps infinity to (0, 1). The function is the probability when y = 1

$$\begin{cases} P(y = 1|x; \theta) = h_\theta(x) \\ P(y = 0|x; \theta) = 1 - h_\theta(x) \end{cases} \tag{3}$$

Thus, when we want to identify a new feature that belongs to which class, only needs solve $h_\theta(x)$, if $h_\theta(x)$ greater than 0.5 is the 'y = 1' class, otherwise the 'y = 0' class. Next, try to make a change in logistic regression. First change the result label y = 0 and y = 1 to y = −1 and y = 1. And then set θ_0 in $\theta^T x = \theta_0 + \theta_1 x_1 + \theta_2 x_2 + \cdots + \theta_n x_n (x_0 = 1)$ to b, and lastly $\theta_0 + \theta_1 x_1 + \theta_2 x_2 + \cdots + \theta_n x_n (x_0 = 1)$.

That is to say, in addition to the Y from y=0 to y = −1, the linear classification function have no difference with the formal representation of logistic regression. Furthermore, the G (z) in the hypothesis function can be simplified to the y = −1 and y = 1. The mapping relationships are as follows:

$$g(z) = \begin{cases} 1, z \geqslant 0 \\ -1, z < 0 \end{cases} \tag{4}$$

The next question is, how to determine the hyperplane? Intuitively, this hyperplane should be the best fit for separating two types of data. The criterion for the 'best fit' is the largest interval between the straight line and the straight line. So, we have to look for the super plane with the largest distance.

Functional Margin and Geometrical Margin. When we have a certain hyperplane $w^T x + b$, $|w^T x + b|$ can represent the distance from the point x to the hyperplane. And through observation the sign of $w^T x + b$ whether or not is consistent with the sign of y to determine the correctness the classification. In this way, we derive the concept of functional margin $\widehat{\gamma}$:

$$\widehat{\gamma} = y(w^T x + b) = yf(x) \tag{5}$$

And the minimal functional margin of all the sample point (x_i, y_i) of training data set T in hyperplane (w, b) is the functional margin of training data set T:

$$\widehat{\gamma} = min\widehat{\gamma}_i \quad (i = 1, \cdots, n) \tag{6}$$

But there is a problem with the defined functional margin, that is, if proportional change the value w and b, the functional margin become twice of the Original, so only the functional margin is not enough. In fact, we can add some constraints to the normal vector w, which leads to the definition of the distance between the real definition of the point to the hyperplane - geometrical margin. Suppose that for a point x, the corresponding point of the vertical projection to the hyperplane is x_0, w is a vector perpendicular to the hyperplane, the distance between the sample x and the hyperplane, as shown in the following figure (Fig. 4):

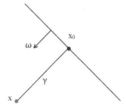

Fig. 4. Distance

According to the knowledge of plane geometry,

$$x = x_0 + \widehat{\gamma}\frac{w}{||w||} \tag{7}$$

And since x_0 is the point on the hyperplane, which satisfies $f(x_0) = 0$, substitute in the equation of hyper plane $w^T x + b = 0$, get $w^T x_0 + b = 0$ i.e. $w^T x_0 = -b$. Multiply both sides simultaneously $x = x_0 + \widehat{\gamma}\frac{w}{||w||}$ with w^T, according to $w^T x_0 = -b$ and $w^T w = ||w||$ we have:

$$\gamma = \frac{w^T x_0 + b}{||w||} = \frac{f(x)}{||w||} \tag{8}$$

It can be seen from the definition that the geometric margin is the functional margin divided by $||w||$ and the functional margin $y*(wx + b) = y*f(x)$ is in fact $|f(x)|$, it is only a measure of the margin, and the geometric margin $|f(x)|/||w||$ is the distance from the point to the hyperplane.

Maximum Margin Classifier. The classification of a data point, when the greater the distance from the data points to hyperplane, the greater confidence. So, in order to make sure that the classification confidence is as high as possible, it is necessary to maximize the value of this margin by selecting the hyperplane, which is half of the Gap in the figure below. By the previous analysis: Function margin $\widehat{\gamma}$ is not suitable to maximize the interval value, so the "gap" in the maximum interval hyperplane to be found here refers to the geometric margin $\widetilde{\gamma}$. The function of maximum margin classifier can be defined as: $max\widetilde{\gamma}$ Also, some conditions must be met (Fig. 5):

$$y_i(w^T x_i + b) = \widehat{\gamma}_i \geqslant \widehat{\gamma} \quad i = 1, \cdots, n \tag{9}$$

textwidthtextwidth

Fig. 5. Maximum margin classifier

As the following figure shows, the line in the middle is to find the optimal hyperplane, its distance to the two dotted lines are equal. The distance is the geometric margin $\widetilde{\gamma}$, the distance between the two dotted lines is equal to $2\widetilde{\gamma}$. The point on the boundary of the dotted line is the support vector. Because these support vectors are just on the dotted line boundary, so they satisfy $y(w^T + b) = 1$, for all points that are not support vectors, $y(w^T + b) > 1$. The actual SVM algorithm used is more complicated, we give only an introduction to SVM to help readers to understand the basic principle. In our work we use LibSVM toolbox to preform training and prediction (Fig. 6).

textwidthtextwidth

Fig. 6. Margin and support vectors

3   The Overall Process

Data Description. The SCADA data of MingYang wind farm is used, we used one year's 10 min average data for training, and according to fault event log, the genertor bearing fault append in 2015/7/17.

Data Preprocessing Null Value and Outliners: Null data in the SCADA system can last for a period of time, for such data can be directly removed or set to 0. In the comparison experiment of the training model, the results of the direct removal of the null value points are better than the 0 processing. An other case is that the existence of a null value is occasionally appears, set these data to zero.

Any of a large data set will have outliers, wind data set of the abnormal value is mainly manifested in some uncontrollable parameters such as wind speed, air temperature is higher than the abnormal range, the wind speed is less than 0, the temperature is less than 0, the wind speed is greater than the upper limit of history, air temperature is greater than the upper limit for the history of abnormal value, this kind of data is recorded should be deleted.

Normalization: Because of the different units, the degree of variation is different, in order to eliminate the influence of the dimension and the size of the variables, for different units or orders of magnitude can be compared and weighted, normalization must be applied. The normalization of the data is to scale the data so that it falls into a small specific interval. The most typical ways are 0–1 normalization and Z-score.

0–1 normalization applies linear transformation to the original data, so that the results fall to the $[0, 1]$ as the follows equation shows, where max is the maximum of the sample data, min is the minimum.

$$X^* = frac{x - min}{max - min} \tag{10}$$

Z-score normalization process data to conform to the standard normal distribution, the mean value is 0 and the standard deviation is 1:

$$X^* = frac{x - \mu}{\delta} \tag{11}$$

where μ is the mean for all sample data, δ is standard deviation for all sample data. This experiment uses the Z-score normalization method.

Model Building Feature Selection: A large number of features are obtained from the WT SCADA system, one should choose the suitable features for building models. We first chooses features by experience and then compare it with the features chosen by random forest method.

Error: The SVM output is the predict value of the label feature. Suppose A to be actual value of the label, P to be the prediction value of the label data, thus define error:

$$e = A - P \tag{12}$$

The training data is when WT running in good state, so that the prediction value will have the similar form. When the related components of WT have fault, actual value as well as error will rise, thus error can be used to detect fault.

4   Experiments and Result

4.1   Features Selected Comparation

The following two tables shows features selected by experience and by random forest. We can find that there are many same features selected (Tables 1 and 2).

With above selected features, using bagging method to build normal behavior model, we obtain the prediction value of generator bearing temperature (Fig. 7).

Table 1. Features selected by experience

No.	Feature name
0	Generator stator U temperature
1	Pitch angle
2	Rotor bear B temperature
3	Generator power
4	WindSpeed
5	Nacelle temperature
6	Generator stator U temperature
7	Outdoor temperature

Table 2. Feature importance by random forest

No.	Feature importance	Feature name
0	0.5699	Generator stator U temperature
1	0.3048	Generator cooling air temperature
2	0.0552	Rotor bear B temperature
3	0.0164	Outdoor temperature
4	0.0123	Nacelle temperature
5	0.0109	Gearbox oil temperature
6	0.0084	Rotor bear A temperature
7	0.0064	Generator stator W temperature
8	0.0045	Gearbox cooling water temperature
9	0.0018	Nacelle total position
10	0.0012	Blade1 motor temperature
11	0.0011	Generator stator V temperature
12	0.0008	Reactive power

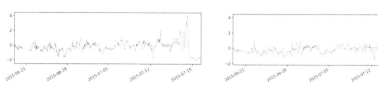

(a) Error of Feature Selection by Experience and Bagging Method

(b) Error of Random Forest Feature Selection and Bagging Method

Fig. 7. Error of different feature seletion method

4.2   General SVM Method

We use features selected by random forest in the following experiments to compare general SVM and bagging based SVM, the error of both method is shown below (Fig. 8).

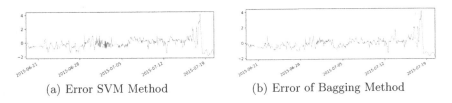

(a) Error SVM Method   (b) Error of Bagging Method

Fig. 8. Error of bagging and SVM

4.3   Discussion

According to steady state working error's mean and variance of proposed bagging method, we set warning lines as in Because we mainly concerns the case when temperature is too high, only two warning lines were set. The blue line and purple line stands for upperbound warning and upperbound alarming respectively, if the error keeps exceeding for a set time, the machine will halt. We can also find the error goes beyond the up alarm line for several times, this are some impulse which disappear quickly so that the alarm won't be triggered (Fig. 9).

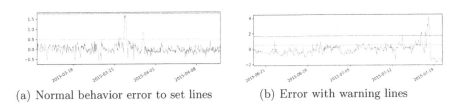

(a) Normal behavior error to set lines   (b) Error with warning lines

Fig. 9. Set the warning line (Color figure online)

Bagging method uses much less time than that of SVM. One reason is that bagging method only use part of the training data, another is that bagging separate the training data set to smaller ones. So compare to general SVM method, bagging based SVM take more advantages (Table 3).

Table 3. Run time

Method	Training time (s)	Applying time (s)
Bagging	1.301202	1.620391
SVM	22.618941	1.420256

5 Conclusions

SCADA signals have been used by model-based methods and prediction methods for WT CMFD and prognosis U. SCADA data contains hundreds of features with random forest method one no long need to select feature by experience. Bagging method can save a lot time and even get more precise model. This two method are both ensemble learning method which combines the advantages of multiple algorithms. Further research can combines more precise algorithms with different kinds of ensemble methods.

References

1. Aziz, M.A., Noura, H., Fardoun, A.: General review of fault diagnostic in wind turbines **20**(1) 1302-1307 (2010)
2. Amirat, Y., Benbouzid, M.E.H., Al-Ahmar, E., Bensaker, B., Turri, S.: A brief status on condition monitoring and fault diagnosis in wind energy conversion systems. RSER **13**, 2629–2636 (2009)
3. Schlechtingen, M., Santos, I.F., Achiche, S.: Wind turbine condition monitoring based on SCADA data using normal behavior models. Part 1 system description. App. Soft Comput. **14**, 447–460 (2013)
4. Yang, W., Court, R., Jiang, J.: Wind turbine condition monitoring by the approach of SCADA data analysis. Renew. Energy **53**(9), 365–376 (2013)
5. Schlechtingen, M., Santos, I.F.: Wind turbine condition monitoring based on SCADA data using normal behavior models. Part 2: application examples. Elsevier Science Publishers B.V. (2014)
6. Santos, P., Villa, L.F., Reñones, A., Bustillo, A., Maudes, J.: An SVM-based solution for fault detection in wind turbines. Sensors **15**, 5627–5648 (2015)
7. Odofin, S., Gao, Z., Sun, K.: Robust fault estimation in wind turbine systems using GA optimisation. Industrial Informatics (INDIN). In: 2015 IEEE 13th International Conference on IEEE, pp. 580–585 (2015)
8. Malik, H., Mishra, S.: Application of probabilistic neural network in fault diagnosis of wind turbine using FAST. TurbSim Simulink. Procedia Comput. Sci. **58**, 186–193 (2015)
9. Liu, X., Li, M., et al.: A predictive fault diagnose method of wind turbine based on K-means clustering and neural networks. JIT, **17** (2016). doi:10.6138/JIT.2016.17.7.20151027i
10. Géron, A.: Hands-On Machine Learning with Scikit-Learn and TensorFlow: Concepts, Tools, and Techniques to Build Intelligent Systems. O'Reilly Media, Sebastopol (2017)
11. Statnikov, A., et al.: A Gentle Introduction to Support Vector Machines in Biomedicine. World Scientific, Singapore (2014)

Author Index

Ai, Haojun I-534
Alagar, Vangalur I-276

Bao, Zhenshan I-594
Bi, Ran II-148
Bin, Xiao II-401

Cai, Xiaodong II-483
Cai, Zhao-hui I-52
Cao, Dejuan II-138
Cao, Han I-573, I-583
Cao, Nuan I-472
Cao, Yangjie II-492
Chen, Chunrong I-88
Chen, Lei I-729
Chen, Lin I-88
Chen, Meng I-242
Chen, Ming-rui I-729
Chen, Shanxiong I-88
Chen, Weiwei I-689
Chen, Xiao I-534, II-198
Chen, Xiaoli I-633
Chen, Yi II-230
Chen, Ying II-424
Chen, Yingwen II-256
Chen, Yongqiang I-741
Chen, Yujiao II-377
Chen, Yun II-483
Cheng, Yi I-558
Chuang, Yuelong II-424

Dang, Weichao II-350
Dang, Xiaochao II-212
Deng, Xiaoheng II-138
Deng, Zongji I-462
Ding, Jiaman I-39, II-28
Ding, Wanmeng I-305, I-331, II-539
Ding, Zhao I-549
Dong, Jinjin I-242
Dong, Xiaoju I-660
Du, Pengyu II-325

Fan, Hong I-718
Fan, Mingyu II-551

Fang, Liyuan I-317
Fang, Xiaolin II-148
Feng, Huamin I-549
Feng, Jie II-551
Feng, Qian I-573
Feng, Xiaoning I-433, I-447
Feng, Yi II-42, II-94
Fu, Guohong II-1
Fu, Li II-453

Gai, Zhen I-64
Gan, Liping II-424
Gan, Zhenye II-126
Gao, Hong I-110
Gao, Hui II-459
Gao, Kai II-115
Gao, Shen II-115
Ge, Jidong II-42, II-94
Gong, Xiaoli I-352
Guo, Chonghui II-78
Guo, Jianli II-271
Guo, Liting II-442
Guo, Wenping II-424
Guo, William I-153
Guo, Yan-Hui I-573, I-583
Guo, Yihan I-425

Hamid, Isma II-401
Han, Qilong I-207, I-412, II-459
Han, Siming I-573
Han, Yiliang I-220
Hao, Kaixue I-134
Hao, Yongtao I-73
Hao, Zhanjun II-212
He, Fazhi I-251, I-534
He, Kaiyuan II-365
He, Zhengyou II-388
Hu, Kuangsheng I-262
Hu, Zhigang I-179, II-335
Huang, Fei I-751
Huang, Guimin II-184
Huang, Heyan II-55
Huang, Jing I-462
Huang, Shuguang I-486

Huang, Weile   II-256
Huang, Xia   I-729

Ji, Zhongqiu   II-503
Jia, Dongli   II-365
Jia, Kuikui   I-166
Jia, Lianyin   I-39
Jia, Mengyu   I-751
Jiang, Bingting   I-179
Jiang, Taijiao   II-286, II-302
Jiang, Wanchun   I-317
Jiang, Ying   II-28
Jin, Ou   I-341
Jin, Yiqiao   I-341

Kang, Hui   II-335
Khatoun, Rida   II-198
Kong, Siyuan   II-94
Kui, Xiaoyan   II-256

Lei, Qi   I-674
Li, Chenghao   II-171
Li, Chuanyi   II-42, II-94
Li, Cuncai   I-425
Li, Cunfang   I-472
Li, Jia   II-335
Li, Jiali   I-549
Li, Jian   I-674
Li, Jianzhong   I-110
Li, Kenli   II-286, II-302
Li, Lijie   I-207
Li, Ling   I-64, I-558, I-621
Li, Mei   I-27, I-134, II-572
Li, Mengjuan   I-39
Li, Mingzhao   I-605
Li, Ning   II-311
Li, Qi   II-241
Li, Renfa   I-386
Li, Wei   I-153
Li, Xiaowei   I-605
Li, Xiao-yun   I-100
Li, Yating   II-256
Li, Yifei   I-573
Li, Yong-kai   I-52
Li, Zhijuan   II-271
Li, Zhitao   I-633
Li, Zhongjin   II-42, II-94
Li, Zongkun   II-516
Liang, Jing   I-605
Liang, Qidi   I-341

Lin, Junyu   II-459
Lin, Xi   I-220
Liu, Biao   I-549
Liu, Caihong   II-78
Liu, Guangqi   I-373
Liu, Haibin   I-166
Liu, Hanlin   I-231, I-305
Liu, Haodong   I-12
Liu, Jia   I-364
Liu, Jianqiao   I-674
Liu, Jun   II-528
Liu, Kesheng   I-331
Liu, Keyan   II-365
Liu, Lintao   I-305, I-331, II-539
Liu, Liu   I-500
Liu, Pingshan   II-184
Liu, Qi   II-230
Liu, Ruyue   I-352
Liu, Sanyang   II-442
Liu, Shenling   II-377
Liu, Shu-bo   I-52
Liu, Wenqi   II-551
Liu, Xia   I-729
Liu, Yuan   I-751
Long, Fang   II-138
Lou, Yuansheng   I-64, I-558, I-621
Lu, Dan   I-412
Lu, Mingming   I-425
Lu, Xu   II-528
Lu, Yuliang   I-231, I-305, II-539
Luo, Bin   II-42, II-94
Luo, Ling   II-414
Luo, Qi   I-594
Luo, Qiang   I-508
Luo, Shoushan   I-292
Lv, Tao   I-73

Ma, Fuxiang   I-400
Ma, Hua   I-179
Ma, Meixiu   II-212
Ma, Nan   I-472
Ma, Rucang   II-212
Ma, Rui   I-472
Ma, Xiaojing   I-27, II-572
Ma, Xiujuan   I-400
Ma, Yukun   II-241
Mao, Yingchi   I-633
Marir, Naila   I-192
Meng, Dan   I-425
Meng, Lingxi   I-660

Meng, Xiangxu   I-373
Meng, Xiao   I-242
Ming, Qian   II-13

Nawaz, Qamar   II-401

Oyikanmi, Peter   I-276

Pan, Da   II-1
Pan, Haiwei   I-433, I-447
Pan, Yan   II-138
Pan, Yang   I-549
Pan, Yiteng   I-534
Pan, Zhengrong   II-325
Pang, Liang   II-198
Peng, Lihua   I-741
Peng, Xiaoning   I-364, I-386
Peng, Yousong   II-286, II-302
Pi, Dechang   II-516

Qi, Wei   II-528
Qian, Long   II-171
Qian, Qingquan   II-388
Qin, Shiyao   I-27, II-572
Qin, Zheng   I-262
Qiu, Jiahui   II-230
Qiu, Zhao   I-729
Qiu, Zhaowen   I-462

Ran, Meng   II-28
Rao, Wenbi   I-646
Ren, Shengbing   I-751

San, Ying   II-459
Shen, Hailan   II-138
Shen, Jun   I-462
Shen, Li   II-311
Shi, Lei   II-492
Shi, Lianshuan   II-433
Shi, Shumin   II-55
Shi, Yuanquan   I-386
Song, Yang   I-110
Su, Rihai   II-55
Sun, Chengjun   I-352
Sun, Jianwei   I-508
Sun, Juanjuan   I-508
Sun, Ming-hui   II-108
Sun, Xiaoying   II-108
Sun, Xin-yue   II-108
Sun, Yuqing   I-207

Tan, Guozhen   II-148
Tan, Longdan   I-231
Tan, Yao   I-646
Tan, Zhiying   II-302
Tang, Linlin   I-242
Tao, Yong   II-335
Tian, Shucun   II-568
Tong, Xin   II-568

Wan, Kaiyu   I-276
Wan, Song   I-231, I-305
Wang, Baosheng   I-500
Wang, Chundong   II-161
Wang, Guangwei   II-551
Wang, Guansheng   I-718
Wang, Hongzhi   I-110, II-568
Wang, Huahui   II-433
Wang, Huiqiang   I-192
Wang, Jianhua   I-605
Wang, Jie   I-674
Wang, Jing-song   I-52
Wang, Kaixuan   I-27, II-572
Wang, Lianhai   I-373
Wang, Longbao   I-633
Wang, Meng   II-483
Wang, Mengqi   II-1
Wang, Qian   I-583
Wang, Qiong   II-311
Wang, Tiaodi   I-447
Wang, Wenzhuo   I-27, II-572
Wang, Yake   II-13
Wang, Yamin   II-492
Wang, Yan   II-388
Wang, Yanqing   I-689, I-708
Wang, Yonggang   II-325
Wang, Yue   II-42
Wang, Zhifang   I-242
Wang, Zhiying   II-311
Wei, Lin   II-492
Wei, Yan   II-414
Weisheng, Li   II-401
Wu, Junfeng   I-508
Wu, Peiwen   II-126
Wu, Qixin   I-262
Wu, Wei   II-271

Xiang, Qixin   II-28
Xie, Haiming   I-317
Xie, Wanyun   I-520
Xie, Xiaoqin   I-433, I-447

Xie, Zhuyang   I-251
Xing, Kaiyan   I-134
Xiong, Hao   I-660
Xiong, Xiaoyi   II-184
Xu, Beibei   II-302
Xu, Bin   II-241
Xu, Xiao-li   II-108
Xue, Di   I-486
Xue, Zhi   II-198

Yan, Xiaorong   I-621
Yan, Xuehu   I-231, I-305, I-331, II-539
Yang, Bin   II-293
Yang, Guangyi   I-262
Yang, Hongwu   II-126
Yang, Jiao   II-335
Yang, Rui   I-123
Yang, Xiaoyu   I-352
Yang, Yixian   I-292
Yang, You   II-414
Yao, Honglei   II-161
Yao, Jing   I-179
Ye, Feng   I-64, I-558
Ye, Gang   I-1
Yin, Ao   I-12
Yin, Hui   I-262
Yin, Jingwei   II-325
You, Jinguo   I-39
You, Silan   I-605
Yu, Bin   I-696
Yu, Hong   I-508
Yu, Ying   I-100
Yuan, Fangyi   II-568
Yuan, Ping   I-696
Yue, Tianbai   I-110

Zang, Tianlei   II-388
Zeng, Jianchao   II-350
Zeng, Yan   II-483
Zhang, Chunkai   I-12
Zhang, Chunyuan   II-377
Zhang, Dejun   I-251
Zhang, Haocheng   I-660
Zhang, Huaping   II-115
Zhang, Hu-yin   I-1
Zhang, Huyin   II-171
Zhang, Jian   I-341, I-674
Zhang, Jin   I-352
Zhang, Jing   I-292

Zhang, Jinxin   I-262
Zhang, Kejia   I-412
Zhang, Meiling   I-508
Zhang, Meishan   II-1
Zhang, ShaoHua   I-472
Zhang, Shaoqun   I-520
Zhang, Shiqing   II-424
Zhang, Shuhui   I-373
Zhang, Sining   I-144
Zhang, Suzhi   I-123
Zhang, Taohong   I-472
Zhang, Wei   II-293
Zhang, Wenbo   I-594
Zhang, Wenhao   II-230
Zhang, Xian   I-364
Zhang, Xinxin   II-335
Zhang, Yan   I-696
Zhang, Yilan   I-605
Zhang, Yiming   II-256
Zhang, Yongbin   I-39
Zhang, Yu   I-386
Zhang, Yunan   II-271
Zhang, Zhiqiang   I-433, I-447
Zhao, Bing   I-144, II-453
Zhao, Fengzhan   II-365
Zhao, Guiling   I-462
Zhao, Hongbin   I-207, I-412
Zhao, Jing   II-271
Zhao, Meng   II-55
Zhao, Nan   I-486
Zhao, Panchao   II-503
Zhao, Shengnan   I-433
Zhao, Tingting   II-365
Zhao, Wei-yu   II-108
Zhao, Wenyang   I-660
Zhao, Wenyu   II-66
Zhao, Xiaoming   II-424
Zhao, Yanan   I-123
Zhao, Yin   I-425
Zhao, Zhentang   II-161
Zhao, Ziping   I-352
Zheng, Hao   I-689, I-708
Zheng, Jianghua   I-718
Zheng, Jinghua   I-486
Zheng, Jun   I-144
Zheng, Meiguang   II-335
Zhong, Ping   II-256
Zhou, Dong   II-66
Zhou, Jingcai   II-171
Zhou, Rui   I-1

Zhou, Tianying   II-171
Zhou, Xiangqian   I-317
Zhou, Xiaoyu   II-42, II-94
Zhou, Yemao   II-42, II-94
Zhu, Jinghua   II-13
Zhu, Likun   II-161

Zhu, Min   I-605
Zhu, Qiuhui   I-605
Zhu, Tiejun   II-467
Zhu, Yuchen   I-88
Zou, Lu   I-251
Zou, Yuanqiang   II-286

Printed in the United States
By Bookmasters